Managing Biological and Ecological Systems

Environmental Management Handbook, Second Edition

Edited by
Brian D. Fath and Sven E. Jørgensen

Volume 1
Managing Global Resources and Universal Processes

Volume 2
Managing Biological and Ecological Systems

Volume 3
Managing Soils and Terrestrial Systems

Volume 4
Managing Water Resources and Hydrological Systems

Volume 5
Managing Air Quality and Energy Systems

Volume 6
Managing Human and Social Systems

Managing Biological and Ecological Systems

Second Edition

Edited by
Brian D. Fath and Sven E. Jørgensen

Assistant to Editor
Megan Cole

CRC Press
Taylor & Francis Group
Boca Raton London New York

CRC Press is an imprint of the
Taylor & Francis Group, an **informa** business

Cover photo: Lake Elsinore, California, USA, B. Fath

Second edition published 2021
by CRC Press
6000 Broken Sound Parkway NW, Suite 300, Boca Raton, FL 33487-2742

and by CRC Press
2 Park Square, Milton Park, Abingdon, Oxon, OX14 4RN

© 2021 Taylor & Francis Group, LLC

First edition published by CRC Press 2013
CRC Press is an imprint of Taylor & Francis Group, LLC

ISBN: 978-1-138-34264-4 (hbk)
ISBN: 978-0-367-51542-3 (pbk)
ISBN: 978-0-429-34617-0 (ebk)

Typeset in Minion
by codeMantra

Contents

SECTION I   APC: Anthropogenic Chemicals and Activities

SECTION II   COV: Comparative Overviews of Important Topics for Environmental Management

SECTION III   CSS: Case Studies of Environmental Management

SECTION IV   DIA: Diagnostic Tools: Monitoring, Ecological Modeling, Ecological Indicators, and Ecological Services

SECTION V   ENT: Environmental Management Using Environmental Technologies

SECTION VI NEC: Natural Elements and Chemicals Found in Nature

SECTION VII PRO: Basic Environmental Processes

Preface

Given the current state of the world as compiled in the massive Millennium Ecosystem Assessment Report, humans have changed ecosystems more rapidly and extensively during the past 50 years than in any other time in human history. These are unprecedented changes that need certain action. As a result, it is imperative that we have a good scientific understanding of how these systems function and good strategies on how to manage them.

In a very practical way, this multivolume *Environmental Management Handbook* provides a comprehensive reference to demonstrate the key processes and provisions for enhancing environmental management. The experience, evidence, methods, and models relevant for studying environmental management are presented here in six stand-alone thematic volumes as follows:

VOLUME 1 – Managing Global Resources and Universal Processes
VOLUME 2 – Managing Biological and Ecological Systems
VOLUME 3 – Managing Soils and Terrestrial Systems
VOLUME 4 – Managing Water Resources and Hydrological Systems
VOLUME 5 – Managing Air Quality and Energy Systems
VOLUME 6 – Managing Human and Social Systems

In this manner, the handbook introduces in the first volume the general concepts and processes used in environmental management. The next four volumes deal with each of the four spheres of nature (biosphere, geosphere, hydrosphere, and atmosphere). The last volume ties the material together in its application to human and social systems. These are very important chapters for a wide spectrum of students and professionals to understand and implement environmental management. In particular, the features include the following:

- The first handbook that demonstrates the key processes and provisions for enhancing environmental management.
- Addresses new and cutting-edge topics on ecosystem services, resilience, sustainability, food–energy–water nexus, socio-ecological systems, etc.
- Provides an excellent basic knowledge on environmental systems, explains how these systems function, and gives strategies on how to manage them.
- Written by an outstanding group of environmental experts.

Since the handbook covers such a wide range of materials from basic processes, to tools, technologies, case studies, and legislative actions, each handbook entry is further classified into the following categories:

APC: Anthropogenic chemicals: The chapters cover human-manufactured chemicals and activities
COV: Indicates that the chapters give comparative overviews of important topics for environmental management

CSS: The chapters give a case study of a particular environmental management example

DIA: Means that the chapters are about diagnostic tools: monitoring, ecological modeling, ecological indicators, and ecological services

ELE: Focuses on the use of legislation or policy to address environmental problems

ENT: Addresses environmental management using environmental technologies

NEC: Natural elements and chemicals: The chapters cover basic elements and chemicals found in nature

PRO: The chapters cover basic environmental processes.

Overall, these volumes will be a valuable resource for all libraries supporting programs in environmental science and studies, earth science, geography, and policy.

In this second volume, over 40 entries focus on managing biological and ecological systems. The coverage ranges from species to ecosystems, with chapters on biodiversity, sustainability, and conservation, to name a few. Specific case studies using biological controls are included, with new entries on sustainable fisheries and recovery of fish populations.

Brian D. Fath
Brno, Czech Republic
December 2019

Editors

Brian D. Fath is Professor in the Department of Biological Sciences at Towson University (Maryland, USA) and Senior Research Scholar at the International Institute for Applied Systems Analysis (Laxenburg, Austria). He has published over 180 research papers, reports, and book chapters on environmental systems modeling, specifically in the areas of network analysis, urban metabolism, and sustainability. He has co-authored the books *A New Ecology: Systems Perspective* (2020), *Foundations for Sustainability: A Coherent Framework of Life–Environment Relations* (2019), and *Flourishing within Limits to Growth: Following Nature's Way* (2015). He is also Editor-in-Chief for the journal *Ecological Modelling* and Co-Editor-in-Chief for *Current Research in Environmental Sustainability*. Dr. Fath was the 2016 recipient of the Prigogine Medal for outstanding work in systems ecology and twice a Fulbright Distinguished Chair (Parthenope University, Naples, Italy in 2012 and Masaryk University, Czech Republic in 2019). In addition, he has served as Secretary General of the International Society for Ecological Modelling, Co-Chair of the Ecosystem Dynamics Focus Research Group in the Community Surface Modeling Dynamics System, and member and past Chair of Baltimore County Commission on Environmental Quality.

Sven E. Jørgensen (1934–2016) was Professor of environmental chemistry at Copenhagen University. He received a doctorate of engineering in environmental technology and a doctorate of science in ecological modeling. He was an honorable doctor of science at Coimbra University (Portugal) and at Dar es Salaam (Tanzania). He was Editor-in-Chief of *Ecological Modelling* from the journal inception in 1975 until 2009. He was Editor-in-Chief for the *Encyclopedia of Environmental Management* (2013) and *Encyclopedia of Ecology* (2008). In 2004, Dr. Jørgensen was awarded the Stockholm Water Prize and the Prigogine Medal. He was awarded the Einstein Professorship by the Chinese Academy of Sciences in 2005. In 2007, he received the Pascal Medal and was elected a member of the European Academy of Sciences. He had published over 350 papers, and has edited or written over 70 books. Dr. Jørgensen gave popular and well-received lectures and courses in ecological modeling, ecosystem theory, and ecological engineering worldwide.

Contributors

G. J. Ash
E.H. Graham Center for Agricultural Innovation,
    Industry and Investment, NSW
Charles Sturt University
Wagga Wagga, New South Wales, Australia

William Au
Department of Preventive Medicine and
    Community Health
University of Texas Medical Branch
Galveston, Texas

Simone Bastianoni
Ecodynamics Group
Department of Chemistry
University of Siena
Siena, Italy

James R. Brandle
School of Natural Resource Sciences
University of Nebraska—Lincoln
Lincoln, Nebraska

Benjamin Burkhard
Institute for the Conservation of Natural
    Resources
University of Kiel
Kiel, Germany

Nídia Sá Caetano
Chemical Engineering Department
School of Engineering (ISEP)
Polytechnic Institute of Porto (IPP)
and
Laboratory for Process, Environmental and
    Energy Engineering
Porto, Portugal

Vera Lucia S.S. de Castro
Ecotoxicology and Biosafety Laboratory
Brazilian Agricultural Research Corporation
    (Embrapa Environment)
São Paulo, Brazil

Ioan M. Ciumasu
CEARC Center, Observatory of Versailles
    Saint-Quentin-en-Yvelines
University of Versailles
    Saint-Quentin-en-Yvelines
Guyancourt, France

Patrick De Clerq
Department of Crop Protection
Ghent University
Ghent, Belgium

Luca Coscieme
Ecodynamics Group
Department of Chemistry
University of Siena
Siena, Italy

Mihai Costica
Faculty of Biology
Alexandru Ioan Cuza University
Iasi, Romania

Marc A. Cubeta
Center for Integrated Fungal Research,
    Plant Pathology
North Carolina State University
Raleigh, North Carolina

Keith Culver
Faculty of Management and STAR Research
    Initiative
University of British Columbia (Okanagan)
Kelowna, British Columbia, Canada

Katja Enberg
Department of Biological Sciences
University of Bergen
Bergen, Norway

Shannon Estenoz
Office of Everglades Restoration Initiatives
U.S. Department of the Interior
Davie, Florida

David N. Ferro
Department of Entomology
University of Massachusetts
Amherst, Massachusetts

J. R. de Freitas
Department of Soil Science
University of Saskatchewan
Saskatoon, Saskatchewan, Canada

Stephen R. Gliessman
Department of Environmental Studies
University of California—Santa Cruz
Santa Cruz, California

Ragini Gothalwal
Institute of Microbiology and Biotechnology
Barkatullah University
Bhopal, India

Simon Gowen
Department of Agriculture
University of Reading
Reading, United Kingdom

Simon Grenier
Functional Biology, Insects and Interactions
National Institute for Agricultural Research (INRA)
Villeurbanne, France

Ann E. Hajek
Department of Entomology
Cornell University
Ithaca, New York

Lise Stengård Hansen
Danish Pest Infestation Laboratory
Danish Institute of Agricultural Sciences
Kongens Lyngby, Denmark

Peter Harris
Agriculture and Agri-Food Canada
Lethbridge, Alberta, Canada

James D. Harwood
Department of Entomology
University of Kentucky
Lexington, Kentucky

Laurie Hodges
Department of Agronomy and Horticulture
University of Nebraska—Lincoln
Lincoln, Nebraska

Heikki Hokkanen
Department of Applied Biology
University of Helsinki
Helsinki, Finland

Hei-Ti Hsu
Agricultural Research Service (USDA-ARS)
Floral and Nursery Plants Research
U.S. Department of Agriculture
Beltsville, Maryland

Sven Erik Jørgensen
Institute A, Section of Environmental Chemistry
Copenhagen University
Copenhagen, Denmark

Marion Kandziora
Institute for the Conservation of Natural
    Resources
University of Kiel
Kiel, Germany

Anthony P. Keinath
Coastal Research and Education Center
Clemson University
Clemson, South Carolina

Peter Kerr
CSIRO Health and Biosecurity
Black Mountain Laboratories
Canberra ACT, Australia

Tomaz Langenbach
Federal University of Rio de Janeiro
Rio de Janeiro, Brazil

David B. Langston, Jr.
Tidewater Agricultural Research and
    Extension Center
Virginia Tech
Suffolk, Virginia

Pierre Mineau
Science and Technology Branch
Environment Canada
Ottawa, Ontario, Canada

Felix Müller
Ecology Center
University of Kiel
Kiel, Germany

Joji Muramoto
University of California Cooperative Extension
Center for Agroecology and Sustainable Food
    Systems
Department of Environmental Studies
University of California—Santa Cruz
Santa Cruz, California

John J. Obrycki
Department of Entomology
University of Kentucky
Lexington, Kentucky

Philip Oduor-Owino
Department of Botany
University of Kenyatta
Nairobi, Kenya

Maurizio G. Paoletti
Department of Biology
University of Padova
Padova, Italy

Timothy Paulitz
Department of Plant Science
MacDonald Campus of McGill University
Ste-Anne-de-Bellevue, Quebec, Canada

Catalina Chaparro Pedraza
Swiss Federal Institute of Aquatic Science and
    Technology EAWAG
Dübendorf, Switzerland

Meir Paul Pener
Department of Cell and Developmental Biology
The Hebrew University of Jerusalem
Jerusalem, Israel

Julie A. Peterson
Department of Entomology
University of Kentucky
Lexington, Kentucky

Odo Primavesi
Brazilian Agricultural Research Corporation
    (Embrapa)
São Paulo, Brazil

Federico M. Pulselli
Ecodynamics Group
Department of Chemistry
University of Siena
Siena, Italy

Jennifer Ruesink
Department of Biology
University of Washington
Seattle, Washington

Alka Sapat
School of Public Administration
Florida Atlantic University
Boca Raton, Florida

Claus Schimming
Institute for the Conservation of Natural Resources
University of Kiel
Kiel, Germany

Cetin Sengonca
Department of Entomology and Plant Protection
Institute of Plant Pathology
University of Bonn
Bonn, Germany

Eric B. Spurr
Department of Wildlife Ecology
Landcare Research New Zealand, Ltd.
Lincoln, New Zealand

F. Craig Stevenson
Department of Soil Science
University of Saskatchewan
Saskatoon, Saskatchewan, Canada

Tanja Strive
CSIRO Health and Biosecurity
Black Mountain Laboratories
Canberra ACT, Australia

Alberto Traverso
Thermochemical Power Group
Department of Mechanical, Energy, Management
    and Transportation Engineering (DIME)
University of Genoa
Genoa, Italy

David Tucker
National Energy Technology Laboratory
Department of Energy
Morgantown, West Virginia

Jean-Paul Vanderlinden
CEARC Center
Observatory of Versailles
    Saint-Quentin-en-Yvelines
University of Versailles
    Saint-Quentin-en-Yvelines
Guyancourt, France

Denise Vienne
School of Public Administration
Florida Atlantic University
Boca Raton, Florida

A. Wang
E.H. Graham Center for Agricultural Innovation,
    Industry and Investment, NSW
Charles Sturt University
Wagga Wagga, New South Wales, Australia

Gerald E. Wilde
Department of Entomology
Kansas State University
Manhattan, Kansas

Wilhelm Windhorst
Institute for the Conservation of Natural
    Resources
University of Kiel
Kiel, Germany

Xinhua Zhou
School of Natural Resources
University of Nebraska—Lincoln
Lincoln, Nebraska

Fabian Zimmermann
Institute of Marine Research
Bergen, Norway

and

Institute of Marine Sciences
and
Department of Applied Mathematics
University of California—Santa Cruz
Santa Cruz, California

I

APC:
Anthropogenic
Chemicals and
Activities

1

APC:
Anthropogenic
Chemicals and

1

Animals: Sterility from Pesticides

William Au

Mechanisms of Action of Pesticides

Pesticides can be subdivided into several major categories: organophosphates, carbamates, organochlorines, synthetic pyrethroids, and others. Their principal mechanisms of action include inhibition of cholinesterase, perturbation of microsomal enzyme production, and damage to nervous systems.[1] Therefore, it is possible that excessive exposure to pesticides can interfere with gametogenesis, sexual activities, and reproduction leading to the expression of sterility.

Observed Effects in Native Animals and in Experimental Systems

Among the pesticides, organochlorines are characterized by their persistence in the environment, and the potential for both bioaccumulation and transfer of the pesticides up the food chain. Therefore, the widespread use of organochlorines in the past has been documented to cause contamination of wildlife and reduction of their populations. For example, the exposure is associated with a significant reduction of fish populations such as trout and salmon.[2] Subsequently, the populations of predatory birds were significantly reduced.[2] The devastating effects in migratory birds and in other wildlife were also demonstrated.[3,4] In these cases, failure to reproduce appropriately has been shown to be a major cause for the decline of the populations.

In studies using experimental animals under controlled exposure conditions to pesticides, organochlorine pesticides such as methoxychlor have been reported to cause reduction of fertility and litter size,[5,6] and kepone to cause anovulation.[7,8] Organophosphate pesticides have been shown to induce premature ovulation and to perturb oocyte development.[9]

Observed Effects in Humans

Very few pesticides have been shown systematically and consistently to cause sterility in humans. An exception is the exposure to a nematocide, 1,2-dibromo-3-chloropropane (DBCP). In the 1970s, workers in several pesticide manufacturing plants were reported to have fertility problems. From a systematic investigation, the infertility based on reduced sperm counts was shown to be associated with testicular function alteration and with exposure to DBCP rather than to other pesticides.[10] Subsequently, the

TABLE 1   Reproductive Problems from Exposure to Dibromochloropropane (DBCP)

A. Oligospermia

Months of exposure to DBCP[a]	% Workers with oligospermia
0	2.9
1–6	8.3
7–24	28.6
25–42	66.7
>42	76.5

B. Offspring

Years after recovery from oligospermia[b]	% Females in offspring
5	84.6
8	78.9
17	58.6

[a] Data derived from 10.
[b] Data derived from 13–15.

same group of scientists found that the reduction of sperm count was associated with an occupational exposure as short as 3 months[11] and with an employment duration-dependent effect (Table 1A). As shown in the table, as many as 76.5% of the workers who had been exposed to DBCP for more than 42 months were oligospermic. Among all the affected workers, many were azospermic or sterile.

Long-term follow-up studies of DBCP production workers showed that some of the affected workers did regain fertility and testicular function, and many of them were able to have children. Their offspring appeared normal and healthy. However, these workers predominantly had female offspring (Table 1B), ranging from 58.6 to 84.6% for the recovery duration from 5 to 17 years.[12] As shown in the table, the highest female to male offspring ratio was found among workers within 5 years of recovery.[13,14] It appears that the recovery is a slow process and complete recovery with respect to the sex ratio was achieved only after 17 years.[15] The observation confirms the previous recommendation of using altered sex ratios as an indication of reproductive hazards associated with pesticides.[16] In another study, males infertile due to poor sperm quality were more likely than expected to be in the agricultural occupations with exposure to pesticides.[17] Papaya fumigant workers with exposure to ethylene dibromide were reported to have significantly reduced sperm count per ejaculate (the percentage of viable and motile sperm) and increases in the proportion of sperm with abnormalities.[18] Abnormal pregnancy outcomes (miscarriages and pre-term deliveries) were associated with exposure to a variety of chemicals in combination with pesticides (atrazine, glyphosate, organophosphates, 4-[2,4-dichlorophenoxy]butyric acid) in males.[19] On the other hand, fertility in traditional male farmers, compared with organic farmers (who do not use pesticides), was not influenced by exposure to pesticides, based on the time taken to have the youngest child.[20]

Among females, exposure to DBCP in pesticide manufacturing plants appears to have no effects on their fertility based on a limited survey.[21] A study was conducted to investigate the relationship between the plasma level of organochlorine pesticides and the diagnosis of endometriosis, and no association was found.[22] On the other hand, among women with medically confirmed infertility, exposure to pesticides was shown to be a significant contributing factor.[23] Furthermore, the mechanism appears to be due to abnormal ovulation.

Future Considerations

Based on the mechanisms of action of pesticides[1,24] and on observations in animals, it is highly likely that overexposure and/or prolonged exposure to pesticides can cause reproductive problems in human. However, adverse reproductive effects have not been demonstrated unequivocally with modern pesticides.

One reason is that the human population is usually exposed to much lower doses of pesticides, except in accidental exposure conditions, than those used in animals that have been shown to cause sterility. Under this condition, any adverse effects in human would be very small. Therefore, investigations using inappropriate study protocols may have generated inconsistent observations. Future studies should be conducted by using large enough populations and by minimizing multiple confounding factors. At this stage of our knowledge, it is fair to state that the potential impact of modern pesticides on sterility in humans has not been clearly demonstrated yet. However, based on the known biological activities of pesticides, they should be considered hazardous chemicals and should be handled with extreme caution.

References

1. Kaloyanova, F.P.; el Batawi, M.A. *Human Toxicity to Pesticides;* CRC Press: Boca Raton, FL, 1991.
2. Pimentel, D. *Ecological Effects of Pesticides on Non-target Species;* U.S. Government Printing Office: Washington, DC, 1971.
3. Gard, N.; Hooper, M. An Assessment of Potential Hazards of Pesticides and Environmental Contaminants. In *Ecology and Management of Neotropical Migratory Birds*; Martin, T., Finch, D., Eds.; Oxford University Press: Oxford, 1995; 294–310.
4. Stinson, E.; Bromely, P. *Pesticides and Wildlife: A Guide to Reducing Impacts on Animals and Their Habitat*; Publication No. 420–004, Virginia Department of Game and Inland Fisheries: Virginia, 1991.
5. Gray, L.E.; Ostby, J.S.; Ferrell, J.M.; Sigman, E.R.; Goldman, J.M. Methoxychlor induces estrogen-like alterations of behavior and the reproductive tract in the female rat and hamster: effects on sex behavior, running wheel activity, and uterine morphology. Toxicol. Appl. Pharmacol. **1988**, *96*, 525–540.
6. Gray, L.E.; Ostby, J.S.; Ferrel, J.M.; Rehnberg, G.; Lindler, R.; Cooper, R.; et al. A dose-response analysis of methoxychlor-induced alterations of reproductive development and function in the rat. Fundam. Appl. Toxicol. **1989**, *12*, 92–108.
7. Eroschenko, V.P. Estrogenic activity of the insecticide chlordecone in the reproductive tract of birds and mammals. J. Toxicol. Environ. Health **1981**, *8*, 731–742.
8. Guzelian, P.S. Comparative toxicology of chlordecone (kepone) in humans and experimental animals. Annu. Rev. Pharmacol. Toxicol. **1982**, *22*, 89–113.
9. Rattner, B.A.; Michael, S.D. Organophosphorous insecticide induced decrease in plasma luteinizing hormone concentration in white-footed mice. Toxicol. Lett. **1985**, *24*, 65–69.
10. Whorton, D.; Milby, T.H.; Krauss, R.M. Testicular function in DBCP exposed pesticide workers. J. Occup. Med. **1979**, *21*, 161–66.
11. Whorton, D.; Krauss, R.M.; Marshall, S. Infertility in male pesticide workers. Lancet **1977**, *ii*, 1259–1261.
12. Goldsmith, J.R. Dibromocholorpropane: epidemiological findings and current questions. Ann. of the New York Academy of Sci. **1997**, *831*, 300–306.
13. Potashnik, G.; Goldsmith, J.; Insler, V. Dibromochloropro-pane-induced reduction of the sex-ratio in man. Andrologia **1984**, *16*, 213–218.
14. Potashnik, G.; Yanai-Inbar, H. Dibromochloropropane: an eight-year re-evaluation of testicular function and reproductive performance. Fertil. Steril. **1987**, *47*, 317–322.
15. Potashnik, G.; Porath, A. Dibromochloropropane: a 17-year reassessment of testicular function and reproductive performance. J. Occup. Environ. Med. **1995**, *37*, 1287–1292.
16. James, W.H. Offspring sex ratio as an indicator of reproductive hazards associated with pesticides. Occup. Environ. Med. **1996**, *52*, 429–430.
17. Strohmer, H.; Boldizsar, A.; Plockinger, B.; Feldner-Busztin, M.; Feichtinger, W. Agricultural work and male infertility. Am. J. Ind. Med. **1993**, *24*, 587–592.
18. Ratcliff, J.M.; Schrader, S.M.; Steenland, K. Semen quality in papaya workers with long term exposure to ethylene dibromide. Br. J. Ind. Med. **1987**, *44*, 317–326.

19. Savitz, D.A.; Arbuckle, T.; Kaczor, D.; Curtis, K.M. Male pesticide exposure and pregnancy outcome. Am. J. Epidemiol. **1997**, *146,* 1025–1036.

20. Larsen, S.B.; Joffe, M.; Bonde, J.P. The asclepiod study group. Time to pregnancy and exposure to pesticides in Danish farmers. Occup. Environ. Med. **1998**, *55,* 278–283.

21. Marshall, S.; Whorton, D.; Krauss, R.M.; Palmer, W.S. Effect of pesticides on testicular function. Urology **1978**, *11,* 257–259.

22. Lebel, G.; Dodin, S.; Ayotte, P.; Marcoux, S.; Ferron, L.A.; Dewailly, E. Organochlorine exposure and the risk of endometriosis. Fertil. Steril. **1998**, *69,* 221–228.

23. Smith, E.M.; Hammonds-Ehlers, M.; Clark, M.K.; Kirchner, H.L.; Fuortes, L. Occupational exposures and risk of female infertility. J. Occup. Environ. Med. **1997**, *39,* 138–147.

24. Sharara, F.I.; Seifer, D.B.; Flaws, J.A. Environmental toxicants and female reproduction. Fertil. Steril. **1998**, *70,* 613–622.

2

Bacillus thuringiensis: Transgenic Crops

Julie A. Peterson,
John J. Obrycki, and
James D. Harwood

Introduction

Genetically modified organisms have been widely adopted in many parts of the world, prompting debate about the implications that this technology may have for environmental health. Transgenic crops have been genetically engineered to incorporate genes derived from another species that confer nutritional and agronomic benefits, such as resistance to insect pests, viruses, herbicides, or environmental conditions, such as low water availability. Among insect- resistant transgenic crops, the most widespread are those that express Bt toxins, coded for by genes from the naturally occurring soil bacterium *Bacillus thuringiensis.* Commercialized Bt crops include corn, cotton, and rice that are protected against Coleoptera and Lepidoptera pests. Bt toxins are recognized as having a narrower range of toxicity than many insecticides, including pyrethroids and neo- nicotinoids, and may therefore pose less risk to non-target organisms; however, potential environmental impacts of Bt toxins need to be examined and documented. This entry will therefore examine the environmental risk assessment of Bt crops, focusing on sources and fate of Bt toxins in exposure pathways for non-target organisms, impact of Bt crops on the environment, and approaches to environmental management of Bt crops.

What Are Bt Crops?

Transgenic Bt crops are genetically engineered to express insecticidal proteins that cause mortality of several common agricultural pests. The genes that code for these proteins, from a naturally occurring bacterium, *Bacillus thuringiensis* (Berliner) (Bacillaceae: Bacillales), are inserted into the genome of the desired crop plant. Genetic transformation is achieved by insertion of the target gene, its promoter and termination sequences, and a marker gene into the crop genome using the microprojectile bombardment method ("gene gun") or the *Agrobacterium tumefaciens* (Smith and Townsend) (Rhizobiales: Rhizobiaceae) bacterium (vector-mediated transformation).

Bt Toxins

Bacillus thuringiensis bacterial strains can produce a series of different toxins; however, only a few have been bioengineered into agricultural crops, including crystalline (Cry) and vegetative insecticidal (VIP) proteins.[1,2] These Bt toxins vary in their range of toxicity to invertebrates, with targeted pests dominated by larval insects in the orders Lepidoptera (moths) and Coleoptera (beetles). The insecticidal mode of action occurs when the Bt toxins bind to receptors on the midgut lining of susceptible insects, causing lysis of epithelial cells on the gut wall and perforations in the midgut lining. This damage to the insect's digestive tract induces cessation of feeding and death by septicemia. An important component of the insecticidal mechanism is its specificity, which is greater than that of many currently used insecticides. Additionally, Bt toxins degrade rapidly in the digestive tract of vertebrates,[3] contributing to their selective nature.

Bt Crops and Their Targeted Pests

Many crop plants have been genetically engineered to express Bt toxins, including field and sweet corn, cotton, potato, rice, eggplant, oilseed rape (canola), tomato, broccoli, collards, chickpea, spinach, soybean, tobacco, and cauliflower. However, only corn and cotton have seen widespread commercialization. Bt potatoes were grown commercially in the United States starting in 1995, but were withdrawn from the market in 2001 following pressure from anti-biotechnology groups and the decision of the global fast-food chain McDonalds to ban the use of genetically modified potatoes in their products.[4] This crop may see a resurgence in planting in Russia and eastern Europe in the near future,[5] as small-scale and subsistence farmers in these regions seek alternatives to expensive insecticide applications.[4] Bt rice has also been approved in certain regions of China,[5] thereby facilitating increased production worldwide.

Global Prevalence

The planting of Bt crops has increased dramatically since the mid-1990s, becoming a prevalent component of agroecosystems worldwide[5–10] (Table 1). For example, Bt cotton and corn in the United States comprised just 1% of total area planted in 1996, their first year of commercial release; however, planting rates have increased rapidly, with areas of Bt cotton and corn in 2010 comprising 73% and 63% of total U.S. production, respectively.[11] Genetically modified crops are grown on 134 million hectares of land in 25 countries by 14.0 million farmers[5]; approximately 40% of that area is planted to corn and cotton expressing Bt insecticidal toxins.[12]

Sources and Fate of Bt Toxins in the Environment

Toxin distribution and expression levels within a transgenic plant vary depending on the type of Bt protein, transformation event, gene promoter used, crop phenology, and environmental and geographical effects.[13–17] Most Bt crops employ a constitutive promoter, such as the cauliflower mosaic virus (CaMV 35S), that expresses insecticidal proteins throughout the life of the plant in nearly all tissues,

TABLE 1 Commercialized Bt Crops, Years Marketed, Bt Toxins Most Commonly Expressed in Commercial Lines, Their Targeted Pests, and Countries That Have Adopted This Technology

Crop	Marketed	Bt Toxins Expressed	Targeted Pest/s	Countries
Corn	1996– present	CrylAb, Cry1A.105, CrylF, Cry2Ab2, Cry9C (withdrawn in 2000), VIP3A	European corn borer *Ostrinia nubilalis* Hübner, southwestern corn borer *Diatraea grandiosella* Dyar (Lepidoptera: Pyralidae), corn earworm *Helicoverpa zea* (Boddie), fall armyworm *Spodoptera frugiperda* Smith (Lepidoptera: Noctuidae)	United States, Brazil, Argentina, Canada, South Africa, Uruguay, Philippines, Spain, Chile, Honduras, Czech Republic, Portugal, Romania, Poland, Egypt, Slovakia
	2003– present	Cry3Bb1, Cry34Ab1, Cry35Ab1, Cry3Aa	Corn rootworm *Diabrotica* spp. (Coleoptera: Chrysomelidae)	
Cotton	1996– present	CrylAc, CrylF, Cry2Ab, VIP3A	Bollworm complex: *Heliothis, Helicoverpa* (Lepidoptera: Noctuidae), and *Pectinophora* (Lepidoptera: Gelechiidae)	United States, Brazil, Argentina, India, China, South Africa, Australia, Burkina Faso, Mexico, Colombia, Costa Rica
Potato	1995–2000	Cry3Aa	Colorado potato beetle *Leptinotarsa decemlineata* Say (Coleoptera: Chrysomelidae)	United States, Canada, Romania

Source: Data from James[5] and Duan et al.[146]

which may include foliage, roots and root exudates, phloem, nectar, and pollen, creating the potential for a multitude of sources for environmental exposure. These pathways to exposure of non-target organisms include, but are not limited to, direct consumption of Bt toxins via ingestion of live or detrital plant material, as well as indirect consumption of Bt toxins via soil contamination from root exudates and persistence in the soil, or consumption of Bt-containing prey in tritrophic interactions (Figure 1). These

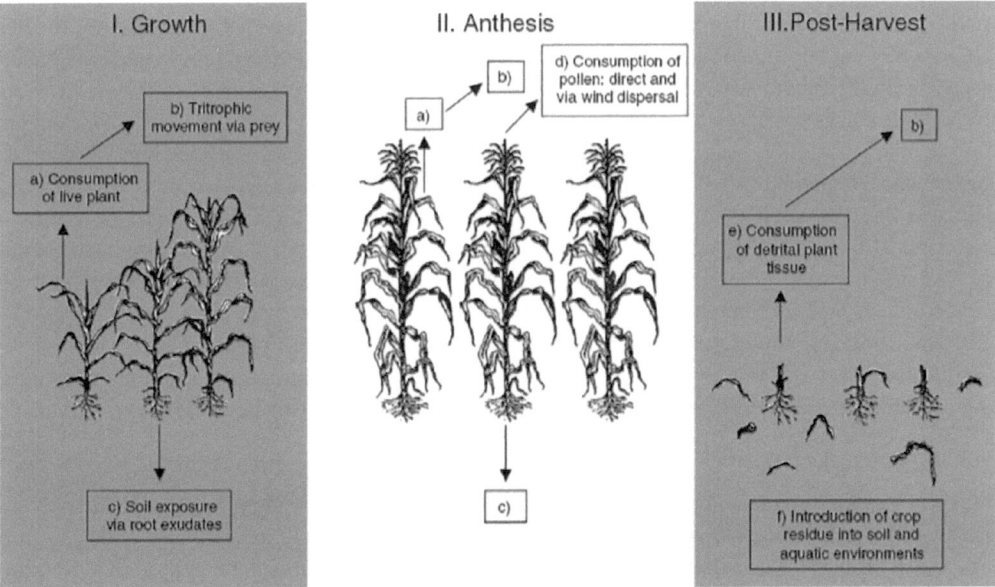

FIGURE 1 Sources for Bt toxin movement in a transgenic corn agroecosystem over the course of a growing season, including (I) growth, (II) anthesis, and (III) post-harvest time periods.

pathways allow for multiple routes to exposure, even potentially within a given taxonomic group, such as ground beetles (Coleoptera: Carabidae), which have been documented to take up Bt toxins in the field.[18] Certain agronomic practices may also create unexpected routes to exposure. For example, following harvest in China, cottonseed hulls may be used as substrate for growing edible oyster mushrooms before being incorporated into cattle feed.[19] Other cotton gin by-products from transgenic plants are used in a variety of ways, including as catfish feed,[20] mulch, and fuel for wood-burning stoves.[21] Although transfer of Bt toxins from cottonseed hulls into mushrooms or cattle feed was not detected,[22] investigation of these complex and non-conventional pathways for Bt toxin movement is critically important.

Direct Consumption of Bt Toxins

Consumption of Live Plant Tissue

Ingestion of plant material, including foliage, roots, phloem, nectar, or pollen may be the most obvious pathway to Bt toxin exposure for targeted pests species, as well as non-target herbivores and natural enemies. Uptake of Bt toxins by herbivores feeding on transgenic plants is well documented (e.g., Dutton et al.,[23] Harwood et al.[24] Meissle et al.,[25] Obrist et al.,[26] and Obrist[27]). However, ingestion of Bt crop tissue may not always result in exposure to toxins. For example, phloem-feeding insects and their honeydew have tested positive for Bt toxins in some transgenic agroecosystems, including certain rice, oilseed rape, and corn events,[28–30] while failing to take up toxins from selected Bt corn events.[31] Exposure pathways of Bt toxins to herbivorous arthropods in transgenic agroecosystems are variable and may therefore be difficult to predict.

Many natural enemies are facultatively phytophagous during some or all of their life stages, consuming plant material or feeding on plant liquids to meet their nutritional and moisture requirements (reviewed in Lundgren).[32] Despite an abundant supply of moisture and prey items, many predatory insects, including ground beetles (Coleoptera: Carabidae), damsel bugs (Hemiptera: Nabidae), stink bugs (Hemiptera: Pentatomidae), and ladybird beetles (Coleoptera: Coccinellidae) will also ingest plant leaf tissue, nectar, or phloem to supplement a prey-based diet.[33]

Pollen Feeding

Another potential route of Bt toxin flow in the environment is through direct pollen feeding or consumption of pollen-contaminated material. Pollen is a component of the diets of many organisms, including spring- tails (Collembola)[34,35] and Western corn rootworms *Diabrotica virgifera virgifera* LeConte (Coleoptera: Chrysomelidae),[36] as well as natural enemies, including ladybird beetles (Coleoptera: Coccinellidae),[37] ground beetles (Coleoptera: Carabidae),[38] green and brown lacewings (Neuroptera: Chrysopidae, Hemerobiidae),[39] hoverflies (Diptera: Syrphidae),[40] and spiders (Araneae).[41,42] In wind-pollinated Bt crops, such as corn, pollen is an abundant resource during anthesis and is deposited in large quantities (up to 1400 grains/cm^2 on plant surfaces[43] and more than 250 grains/cm^2 in ground-based spider webs[44]). Some pollen-feeding omnivores, such as *Orius insidiosus* (Say) (Hemiptera: Anthocoridae), may also maximize their exposure to Bt toxins by aggregating at corn silks and leaf axils, where pollen grains accumulate during anthesis.[45,46] Pollen consumption can therefore represent a significant direct and indirect (through consumption of pollen-feeding prey) route of exposure for both predators and prey in transgenic agroecosystems, particularly during periods of crop anthesis.

Consumption of Detritus

Bt toxins can persist in plant detritus beyond a single growing season[47,48] thereby exposing detritivores, such as earthworms, slugs, nematodes, protozoa, bacteria, and fungi, to Bt toxins through the consumption of such litter.[49–51] Crop detritus may also enter aquatic environments; for example, in agricultural systems where crop detritus is left in the field to prevent erosion, plant residues may account for up to 40% of non-woody vegetation entering streams.[52] Bt-containing crop tissue may then be consumed by aquatic detritivores, such as larval caddis- flies (Trichoptera), crane flies (Diptera: Tipulidae), midges

(Diptera: Chironomidae), and isopods. However, the bioactivity of Bt toxins in senesced plant material may be relatively short; lepidopteran-specific toxins were absent after 2 weeks in aquatic systems, while coleopteran-specific toxins decayed in as few as 6 days.[53] The harsh environmental conditions and constant physical abrasion experienced by plant tissue in flowing water were suggested as mechanisms stimulating such rapid breakdown.[54] Thus, while detritus provides a potential route of exposure, the functional consequence of Bt toxins in detritivore food webs remains unclear. However, what is evident is the persistence of Bt toxins in the environment following harvest and the possibility for long-term exposure of non-target organisms to this material.

Indirect Consumption of Bt Toxins

Soil Contamination via Root Exudates

One potential pathway of indirect exposure to Bt toxins is through contamination of the soil and therefore to soildwelling arthropods via root exudates. Bt corn, potato, and rice all release transgenic proteins from their roots during plant growth.[55,56] The soil-dwelling fauna, including beneficial non-target organisms, may therefore be exposed to Bt toxins via their secretion in plant root exudates. Bt toxin exposure to epigeal predators, ground beetle larvae and adults, and certain spiders [e.g., wolf spiders (Araneae: Lycosidae)] may also occur because of their feeding habits. Several studies have quantified the persistence of Bt toxins in the soil,[47,57–59] with results indicating persistence of these insecticidal proteins ranging from 2 to 32 weeks after introduction into the soil. This wide discrepancy may partially reflect differences in microbial activity of soils,[57,60,61] which is in turn affected by pH and mineral content.[59] Bt toxins may bind to humic acids, organic supplements, or soil particles, protecting the toxins from degradation by microbes and extending the persistence of insecticidal activity in the soil.[2] Thus, the persistence of Bt toxins may vary significantly due to their differential rate of decay based on microbial activity, soils, and environmental factors.

Consumption of Prey Containing Bt Toxins

The movement of Bt toxins from plant tissue into herbivores and subsequently into their natural enemies has been well documented. Concentrations of Bt toxins typically decrease as they move through a food chain, indicating little evidence for bioaccumulation effects as seen in other insecticidal compounds[62]; however, two-spotted spider mites, *Tetranychus urticae* Koch (Acari: Tetranychidae), show evidence for the bioaccumulation of Bt toxins.[63] Although in a more typical example, Cry1Ac proteins expressed in transgenic cotton are ingested by beet armyworm caterpillars *Spodoptera exigua* (Hübner) (Lepidoptera: Noctuidae) and are also detectible, but at lower concentrations, in predatory stink bugs *Podisus maculiventris* (Say) (Hemiptera: Pentatomidae) when these prey are consumed.[63] However, not all tritrophic pathways facilitate the uptake of Bt toxins; Cry1Ab toxins are present in the marsh slug *Deroceras laeve* (Müller) (Pulmonata: Agriolimacidae) following consumption of Bt corn tissue, but are not taken up by the predatory ground beetle *Scarites subterraneus* (F.) (Coleoptera: Carabidae) in laboratory studies[64]; accordingly, field- collected specimens of this species did not test positive for Bt toxins.[18] Additionally, the concentration of Bt toxins transferred via trophic connections may vary based on the identity of the prey. In a laboratory experiment, two prey species of the wolf spider *Pirata subpiraticus* (Bösenberg and Strand) (Araneae: Lycosidae), the striped stem borer *Chilo suppressalis* (Walker) (Lepidoptera: Crambidae), and the Chinese brushbrown caterpillar *Mycalesis gotama* Moore (Lepidoptera: Nymphalidae) were allowed to feed on transgenic rice expressing Cry1Ab Bt toxins. These prey were subsequently fed to the wolf spider, and assays of each trophic level indicated that Bt toxins were transferred up the food chain; Cry1Ab concentration diminished with each additional trophic step, and the two prey species transferred Cry1Ab with significantly different efficiencies, having approximately 60 times the Cry1Ab concentration in brushbrown caterpillar-fed spiders compared with striped stem borer-fed spiders.[65] Adult ladybird beetles (Coleoptera: Coccinellidae) showed greatest uptake of Bt toxins in a corn agroecosystem post-anthesis, indicating that tritrophic movement of toxins was a greater pathway for toxin uptake than direct pollen consumption.[66] It is therefore clear that consumption of Bt-containing prey could be a major source of Bt toxin flow in non-target food webs, although the extent of toxin uptake and its concentration will depend on the strength of specific trophic pathways that occur within a given food web in the field.

Impacts of Bt Crops on the Environment

Bt crops have become a dominant fixture in selected agroecosystems worldwide. Their planting on cultivated lands globally allows for large potential impacts of this technology on the environment. These impacts include both benefits and potential risks, the consequences to the environment of using Bt technology are intensely debated.

Benefits of Bt Technology

Reduced Risk Compared with Conventional Insecticides

The insecticidal toxins produced by transgenic Bt crops are considered to have fewer non-target effects than many insecticides due to their narrow range of toxicity and, therefore, to be advantageous to traditional methods of control. For example, populations of many natural enemies responded negatively to foliar applications of broad-spectrum pyrethroids compared with more selective insecticides, such as Bt toxins, indoxacarb, and spinosad, used to combat lepidopteran pests in sweet corn agroecosystems.[67] Field studies comparing Bt crops with their non-transgenic isolines that have been treated with broad-spectrum insecticides almost always reveal higher populations of beneficial arthropods in the Bt crops. A meta-analysis of these studies found that total non-target invertebrate abundance was higher in lepidopteran-targeting corn and cotton compared with non-transgenic crops managed with insecticides; however, no differences for coleopteran-targeting corn were reported.[68] Non-transgenic control plots treated with insecticides had lower predator and herbivore abundance compared with unsprayed Bt fields; this result was particularly strong for predator populations in non-transgenic plots treated with pyrethroids, such as lambda-cyhalothrin, cyfluthrin, and bifenthrin.[69] Similarly, spiders were more abundant in Bt corn, cotton, and potato when compared with conventionally managed crops employing a range of insecticides, including foliar pyrethroid sprays, systemic neonicotinoid seed treatments, and organophosphate soil applications at planting.[70] Due to their selectivity, Bt crops are therefore safer for non-target organisms when compared with many insecticides, particularly those with broad-spectrum action.

Economic Savings

A reduction in the quantity and frequency of insecticide applications are economically beneficial, in addition to reduced exposure to chemical insecticides for farm workers and the environment. Bt cotton has significantly reduced insecticide inputs in numerous regions of the world, including the United States,[71,72] China,[73] and South Africa.[74] The adoption of Bt corn in the midwestern United States has provided an estimated $6.9 billion in benefits to growers of both Bt and non-Bt corn in the past 14 years, due to area-wide suppression of European corn borer *Ostrinia nubilalis* (Hübner) (Lepidoptera: Pyralidae), a key pest of this crop.[75] With more than 53 million hectares of Bt crops now planted worldwide, there are significant economic considerations, and it is evident that Bt- based production systems are not only more sustainable in the context of pest management but also have the capacity to enhance agricultural diversity through reduced chemical inputs.

Global Food Security

The human population is projected to reach 10 billion by 2050, and concomitant to this is the need for augmented global food security and production.[76] The employment of Bt crop technology may aid in this goal by increasing quantity and consistency of crop yields; for example, corn yields are increased or protected because of season-long control of European corn borer.[71] Additionally, stored corn grain is protected against lepidopteran pests[77] and mycotoxin levels, which pose a threat to the health of humans and livestock if introduced into the food supply,[78] are lower because of reduced feeding activity of European corn borer, which are associated with the fungal causal agents.[71,79] Bt crops may therefore confer significant beneficial effects for the global drive to increase agricultural productivity and safety.

Potential Risks of Bt Crops

Impacts on Non-Target Organisms

Despite the specificity of Bt toxins toward target pests, questions have been raised concerning their effects on abundance, diversity, or fecundity of some components of the non-target food web, including beneficial species such as pollinators, natural enemies, and/or detritivores. Given the important ecosystem services provided by the abovementioned non-target organisms, the risk assessment of these groups is essential in the context of understanding environmental health. Lundgren et al.[17] identified four main approaches that risk assessment researchers have used to study the impact of Bt crops on non-target invertebrates: direct toxicity, tritrophic interactions, community level studies, and meta-analyses of data.

Direct toxicity. Feeding non-target organisms a diet that contains Bt toxins and measuring resulting parameters of development, fitness, and fecundity are done to examine the potential for directly toxic effects of Bt crops. The literature (reviewed in Lundgren et al.[17] and Lövei and Arpai[80]) provides contrasting evidence of non-target effects, ranging from no discernable effects of consumption of transgenic crops (e.g., Harwood et al.,[64] Pilcher et al.,[81] Al-Deeb et al.,[82] Lundgren and Wiedenmann,[83] and Anderson et al.[84]) to reports of a variety of negative effects (e.g., increased mortality, delay in development, reduction in weight gain, or changes in behavior) on beneficial organisms, such as pollinators,[85] predators,[86] parasitoids,[87] and other non-target arthropods.[50,88–91] Differing results of studies of direct toxicity of Bt toxins to non-target organisms exist for many groups; for example, in caddisflies (closely related to the target order Lepidoptera), studies have been published that report both sublethal negative effects[91] and the absence of negative impacts of Bt toxins.[54] Such laboratory toxicity studies may be extrapolated to the field, although toxicity studies should address all ecologically relevant routes to exposure for non-target organisms.[92]

Tritrophic interactions. These studies test for effects of Bt crops on natural enemies via consumption of Bt-containing prey; any observed effects may be due to ingestion of toxins or through prey-quality-mediated effects. Several studies have reported no tritrophic effects of Bt crops on natural enemies[63,93–95]; however, negative effects have been observed in other cases,[96,97] although these results are often attributed to prey-mediated effects whereas prey quality is lower when fed Bt crop tissue. Meta-analyses of tritrophic studies revealed that using prey items that were totally or partially susceptible to Bt toxins (and therefore were likely to be of lesser quality) had a negative effect on the performance of natural enemies, while using nonsusceptible prey (whose quality should be unaffected by consuming Bt toxins) had no effect on the performance of the natural enemies that consumed it.[98,99]

Community level. To study the effect of Bt crops on non-target organisms at the community level, arthropods are sampled from Bt and conventional crops to observe any differences in abundance, diversity, or community structure. Such studies have examined a variety of nontarget organism populations, including soil microarthropods, nematodes, decomposers, pollinators, and natural enemies.[81,100–110] Results of such studies often report no significant differences between populations of non-target organisms in Bt and non-Bt crops; however, a lack of taxonomic resolution in some studies can weaken these results.[70]

Meta-analysis data. This quantitative method addresses effects of Bt crops across multiple published studies and has been widely used to infer the consequence of Bt crops on a series of different parameters. For example, a meta-analysis of 42 field experiments revealed that the overall mean abundance of non-target invertebrates was significantly lower in lepidopteran-targeting Bt corn fields compared with nontransgenic fields when neither is treated with insecticides; no differences were found between coleopteran-targeting Bt and non-transgenic corn.[68] Unsurprisingly, the abundance of non-target arthropods was significantly higher in transgenic corn versus non-transgenic corn that had been treated with insecticides.[68] Additional meta-analyses have reported the effects of Bt crops on functional guilds of non-target organisms,[69] honey bees,[111] and spiders,[70] generally finding no differences in non-target arthropod populations between Bt and non-Bt crops. When examined at further taxonomic resolution, such analyses may reveal differential responses of functionally distinct taxa to Bt crops, as is the case with spider families. Meta-analysis revealed positive effects of Bt crops on the abundance of certain groups (Clubionidae, Linyphiidae, Thomisidae), no effect on others (Lycosidae, Oxyopidae, Araneidae), and negative effects on several families (Anyphaenidae, Philodromidae) relative to non-transgenic crops untreated with insecticides.[70]

Presence in Human Food Supply

Concerns about the presence of Bt toxins in the human food supply do not stem from any direct toxic effects, as vertebrates lack the midgut receptors for binding of Bt toxins, but from the possibility that a portion of the population will exhibit an allergic reaction to ingestion of Bt proteins.[112] Most Bt toxins will readily break down in the acidic environment of a vertebrate digestive tract.[3] Bt corn expressing Cry9C proteins, marketed under the commercial name StarLink™, was planted in the United States from 1998 to 2000, but approved only for animal feed and ethanol production due to the persistence of Cry9C in the vertebrate gut.[113] When traces of Cry9C proteins were found in cornmeal destined for human consumption, several food items were recalled, including Taco Bell® taco shells, and StarLink was voluntarily removed from the market.[114] However, no confirmed allergenic reactions due to Cry9C contamination were reported. Despite the lack of evidence for any true risk to humans based on consumption of Bt food products, sentiment against transgenic agricultural products destined for human consumption exists, especially in Europe, and has influenced the commercial acceptance of some products such as Bt potatoes.[4] Therefore, despite these limited effects on the human (and vertebrate) population, safeguards need to be in place to prevent the presence of unapproved genetically modified products entering the human food chain.

Pleiotropic Effects of Genetic Transformation

Insertion of a Bt gene complex into a crop plant may result in unpredicted and unintended pleiotropic effects that change the plant from its non-transgenic counterpart in ways beyond just the expression of Bt toxins.[49,115–117] For example, a reported pleiotropic effect in Bt corn is an increase in the lignin content in transgenic plant tissue,[49] a trait that could lead to reduced decomposition rates in the soil.[118] However, other studies have contested this conclusion and shown no differences in rate of decomposition.[119] An additional pleiotropic effect of transformation in Cry1F corn may be an increase in attractiveness as an oviposition site for corn leafhoppers *Dalbulus maidis* (DeLong and Wolcott) (Hemiptera: Cicadellidae), a pest that is not targeted by Bt toxins, possibly due to altered plant traits that influence oviposition, such as leaf vein characteristics, foliar pubescence, or plant chemistry.[120] There is a lack of understanding of how these pleiotropic effects will affect ecosystem processes, although the potential consequences merit further examination in the context of their environmental impacts.

Gene Escape

The transfer of genes from populations of domesticated crops into wild plants has been documented for many years.[121] The "escape" of Bt transgenes into wild plants could have undesirable effects by reducing genetic diversity and fecundity in wild populations or increasing fecundity and creating an invasive weed through reduction or elimination of herbivory. The presence of transgenic material from the CaMV 35S promoter used in Bt crops was reported in native maize landraces grown in remote areas of Oaxaca, Mexico, in 2001.[122] However, these results have been highly debated[123–125] and additional studies are conflicting, reporting both the presence[126] and absence[127] of transgenic DNA in traditional maize lines in Mexico. Additionally, transgene escape into weedy rice may increase the fecundity of this plant, as well as its ecological interactions with surrounding organisms.[128] The implications of transgene escape are yet to be fully understood, particularly in the context of ecological risk assessment.

Approaches to Environmental Management of Bt Crops

To safely incorporate Bt crop technology into agroecosystems, approaches to environmental management should address the issue at multiple scales. These include engineering at the level of the individual plant genome, field- and farm-level modifications to reduce exposure of Bt toxins and escape of transgenes, and large-scale incorporation of Bt technology into integrated pest and resistance management programs. Finally, continued research concentrating on the non-target impacts of Bt crops should be conducted at multiple tiers across crop and toxin types, geographic regions, non-target organism taxa, and temporal and spatial scales, studying non-target organisms at the greatest taxonomic resolution possible. Regulation of transgenic crops in the effort to mitigate risk is complex; further recommendations and discussion of this topic can be found elsewhere.[129]

Within-Plant Modifications

Selection of Low-Risk Promoters

As the gene promoter used in a transgenic event can have a strong impact on the eventual concentration and distribution of Bt toxins within the plant, the choice of promoter should be made within the context of environmental safety. Certain promoters have been identified as having greater non-target risks than others; for example, harmful effects of Bt corn event 176 on non-target Lepidoptera larvae [monarchs *Danaus plexippus* (L.) (Nymphalidae) and black swallowtails *Papilio polyxenes* F. (Papilionidae)] have been reported, while other events expressing the same Cry toxin (e.g., Bt11 and MON810) had no effect. [90] Event 176 has increased expression of Bt proteins in the pollen compared with the other events[130] and therefore poses a greater risk to non-target organisms.

Tissue- and Time-Specific Expression

The use of gene promoters that are tissue- or time-specific to express toxins only in plant tissues when they are susceptible to feeding has been introduced.[131] This technique has been employed in the transgenic expression of snowdrop lectin, a plant-derived protein with insecticidal properties, in rice. To target phloem-feeding pests such as brown planthoppers, lectins are selectively expressed in the vascular tissue.[132,133] Such selective expression of Bt toxins in tissue and time to target susceptible pests and reduce exposure to non-target beneficial arthropods could potentially increase environmental safety, thereby reducing the pathways for Bt toxin movement through non-target food webs.

Reduction in Bt Toxin and Transgene Escape

At the field or farm level, management practices may be implemented that reduce the movement of Bt toxins or transgenes from their source (Bt plants) into surrounding habitats. Current practices may depend on the crop and agronomic aims of the grower; for example, large quantities of crop residue may be incorporated into the soil during the harvesting process, although this is not the practice when crop material is removed for ethanol production or under reduced-tillage practices.[134] Although Bt toxins may degrade quickly following the incorporation of Bt crop plant detritus into aquatic systems, this potential pathway for transgenic protein movement may be avoided through the employment of practices that prevent movement of transgenic crop tissue beyond field borders. The establishment of riparian buffer zones and filter strips may reduce the quantity of crop detritus and other compounds originating in cropland (e.g., fertilizers, insecticides) that enter nearby streams and waterways.[135]

Similarly, reducing exposure pathways for gene flow into wild plant populations via physical methods, such as isolation of crops or plant destruction, may delay transgene escape. However, controlling gene flow via pollen and seeds in the environment can be very difficult; a physical separation of 200 m between transgenic corn still yields contamination levels of 0.1% between plant populations due to cross-pollination.[136] Seeds are additionally difficult to control owing to their persistence in the soil seed bank, as well as ability to sometimes germinate and persist outside of cultivated fields.[137] Management of the movement of Bt toxins and genetic material from cultivated fields into the surrounding environments warrants additional research. Interestingly, technology that could have reduced the spread of transgenes, called the "terminator gene," was abandoned in 1999 because of the criticism that the gene prevents farmers from harvesting viable seed and thereby exclusively benefits the seed companies.[138]

Large-Scale Integrated Pest and Resistance Management

Although Bt crops allow reductions in the application of certain insecticides compared with conventionally managed crops (while other insecticidal practices persist, such as neonicotinoid seed treatments on corn[139]), it should not be assumed that this technology will readily fit into integrated pest management practices.[17] Considerations of compatibility with biological control and delaying resistance in pest populations are also necessary.

Compatibility with Biological Control

Integrated pest management practices attempt to incorporate mechanical, physical, chemical, and cultural controls; host resistance (including transgenic crops); and auto-cidal, biochemical, and biological controls in a synergistic manner. Increased attention has focused on conservation biological control: the modification of the environment or existing practices to protect and enhance specific natural enemies or other organisms to reduce effects of pests (e.g., Landis et al.[140] and Eilenberg et al.[141]). Natural enemies can be abundant in agricultural systems and often play an essential role in pest suppression. Maintenance of relevant natural enemy populations via conservation biological control is a practical and sustainable option for high-acreage field crops, such as corn and cotton,[142] which are dominated by Bt varieties. Any negative effects of Bt toxins on natural enemies could reduce their effectiveness as biological control agents and therefore limit natural pest suppression in agroecosystems. Understanding the potential impacts of transgenic crops on non-target arthropods is essential in order to provide a framework for integrating natural enemies into sustainable methods of pest control in the agricultural environment.

Resistance Management Techniques

The development of resistance to Bt toxins by pest populations is a major concern. Integrated resistance management programs must continue to be developed and followed to promote the sustainable use of Bt crops. This is of critical importance given that resistance to Bt sprays has occurred in multiple populations of the pestiferous diamond back moth *Plutella xylostella* (L.) (Lepidoptera: Plutellidae)[143] and three instances of field-evolved resistance to transgenic Bt crops have been reported in moth larvae: *Spodoptera frugiperda* (Smith) (Lepidoptera: Noctuidae) to Cry1F corn in Puerto Rico, *Busseola fusca* (Fuller) (Lepidoptera: Noctuidae) to Cry1Ab corn in South Africa, and *Pectinophora gossypiella* (Saunders) (Lepidoptera: Gelechiidae) to Cry1Ac in the southwestern United States.[144] Current resistance management employs structured refuges and high-dose toxin crops, as well as monitoring for resistance development in the field and monitoring for compliance of growers to refuge protocol. Additional attempts to delay resistance include creating transgenic plants that express more than one type of Bt toxin that targets the same pest, called gene pyramiding.[131] Improved resistance management would include increased education for growers and the public about the importance of resistance management and refuge compliance, as well as continued monitoring of field populations for the development of resistance. Future strategies to passively achieve resistance compliance include mixed seed refuges, in which transgenic and non-transgenic seeds are sold in combination within the bag.[145]

Conclusions

The sources and fate of Bt toxins in the environment can be complex and variable depending on crop, transgenic event, geography, and other environmental variables. The effects of Bt crops and their toxins on the environment have been widely debated, particularly the potential implications associated with ecological impacts such as gene escape and non-target risks. Approaches to the environmental management of Bt crops and their integration into integrated pest and resistance management systems warrant further study. Despite the concerns associated with Bt crops, significant reductions in chemical input are evident and this technology is environmentally safer when compared with many approaches to pest suppression, particularly those using broad-spectrum insecticides.

Future of Bt Technology

The focus of current transgenic technology has been on stacking and pyramiding of events. Stacking incorporates multiple transgenic traits into the crop genome in order to express more than one type of insecticidal toxin, therefore targeting multiple pest species. Pyramiding of transgenes allows for the crop to express multiple types of Bt toxins that target the same pest. Additionally, several other Bt crops are expected to be approved for commercial availability by 2015, including potatoes for planting in eastern Europe and eggplant in India.[5] The global adoption of biotechnology in agriculture is projected to continue with estimates that genetically modified crops will reach 200 million hectares, grown by 20 million farmers in 40 countries by 2015.[5]

Acknowledgments

Funding for this project was provided by the U.S. Department of Agriculture Cooperative State Research Education and Extension Service Biotechnology Risk Assessment Grant 2006-39454-17446. JDH is supported by the University of Kentucky Agricultural Experiment Station State Project KY008043. This is publication number 10-08-132 of the University of Kentucky Agricultural Experiment Station.

References

1. Yu, C.G.; Mullins, M.A.; Warren, G.W.; Koziel, M.G.; Es-truch, J.J. The *Bacillus thuringiensis* vegetative insecticidal protein Vip3 lyses midgut epithelium cells of susceptible insects. Appl. Environ. Microbiol. **1997**, *63*, 532–536.
2. Glare, T.R.; O'Callaghan, M. *Bacillus thuringiensis: Biology, Ecology, and Safety*; John Wiley & Sons, Ltd.: West Sussex, U.K., 2000.
3. Mendehlson, M.; Kough, J.; Vaituzis, Z.; Matthews, K. Are *Bt* crops safe? Nat. Biotechnol. **2003**, *21* (9), 1003–1009.
4. Kaniewski, W.K.; Thomas, P.E. The potato story. AgBio- Forum **2004**, *7*, 41–46.
5. James, C. *Global Status of Commercialized Biotech/GM Crops: 2009;* ISAAA Brief No. 41. ISAAA: Ithaca, NY, 2009.
6. Cannon, R.J.C. Bt transgenic crops: Risks and benefits. In- tegr. Pest Manag. Rev. **2000**, *5*, 151–173.
7. Pray, C.E.; Huang, J.; Hu, R.; Rozelle, S. Five years of Bt cotton in China—The benefits continue. Plant J. **2002**, *31*, 423–430.
8. Shelton, A.M.; Zhao, J.Z.; Roush, R.T. Economic, ecological, food safety, and social consequences of the deployment of Bt transgenic plants. Annu. Rev. Entomol. **2002**, *47*, 845–881.
9. Lawrence, S. AgBio keeps on growing. Nat. Biotechnol. **2005**, *63*, 3561–3568.
10. James, C. *Executive Summary of Global Status of Commercialized Biotech/GM Crops: 2006;* ISAAA Brief No. 36. ISAAA: Ithaca, NY, 2006.

11. United States Department of Agriculture National Agricultural Statistics Service. Acreage report, 2010, available at http://usda.mannlib.cornell.edu/usda/current/Acre/Acre-06-30-2010.pdf. (Accessed, November 2010).

12. GMO Compass. *Rising Trend: Genetically Modified Crops Worldwide on 125 Million Hectares*, 2009, available at http://www.gmocompass.org/eng/agri_biotechnology/gmo_planting/257.global_gm_ planting_2008.html.

13. Fearing, P.L.; Brown, D.; Vlachos, D.; Meghji, M.; Privalle, L. Quantitative analysis of CryIA(b) expression in *Bt* maize plants, tissues, and silage and stability of expression over successive generations. Mol. Breed. **1997**, *3* (3), 169–176.

14. Duan, J.J.; Head, G.; McKee, M.J.; Nickson, T.E.; Martin, J.H.; Sayegh, F.S. Evaluation of dietary effects of transgenic corn pollen expressing Cry3Bb1 protein on a nontarget ladybird beetle, *Coleomegilla maculata*. Entomol. Exp. Appl. **2002**, *104*, 271–280.

15. Grossi-de-Sa, M.F.; Lucena, W.; Souza, M.L.; Nepomu- ceno, A.L.; Osir, E.O.; Amugune, N.; Hoa, T.T.C.; Hai, T.N.H.; Somers, D.A.; Romano, E. Transgene expression and locus structure of Bt cotton. In *Environmental Risk Assessment of Genetically Modified Organisms*. *Methodologies for Assessing Bt Cotton in Brazil*; Hilbeck, A., Andow, D.A., Fontes, E.M.G., Eds.; CAB International: Wallingford, U.K., 2006; 93–107.

16. Obrist, L.B.; Dutton, A.; Albajes, R.; Bigler, F. Exposure of arthropod predators to Cry1Ab toxin in Bt maize fields. Ecol. Entomol. **2006**, *31* (2), 143–154.

17. Lundgren, J.G.; Gassman, A.J.; Bernal, J.; Duan, J.J.; Ru- berson, J. Ecological compatibility of GM crops and biological control. Crop Prot. **2009**, *28*, 1017–1030.

18. Peterson, J.A.; Obrycki, J.J.; Harwood, J.D. Quantification of Bt-endotoxin exposure pathways in carabid food webs across multiple transgenic events. Biocontrol Sci. Technol. **2009**, *19* (6), 613–625.

19. Li, X.; Pang, Y.; Zhang, R. Compositional changes of cotton-seed hull substrate during *P. ostreatus* growth and the effects on the feeding value of spent substrate. Bioresour. Technol. **2001**, *80*, 157–161.

20. Li, M.H.; Hartnell, G.F.; Robinson, E.H.; Kronenberg, J.M.; Healy, C.E.; Oberle, D.F.; Hoberg, J.R. Evaluation of cottonseed meal derived from genetically modified cotton as feed ingredients for channel catfish, *Ictalurus punctatus*. Aquac. Nutr. **2008**, *14* (6).

21. Robertson, R. *Cotton Gin Trash Now Valuable By-Product*, Southeast Farm Press, 2009, available at http://southeastfarmpress.com/grains/cotton-gin-trash-now-valuable-product.

22. Jiang, L.; Tian, X.; Duan, L.; Li, Z. The fate of Cry1Ac Bt toxin during oyster mushroom (*Pleurotus ostreatus*) cultivation on transgenic Bt cottonseed hulls. J. Sci. Food Agr. **2008**, *88*, 214–217. Accessed November 1, 2010.

23. Dutton, A.; Klein, H.; Romeis, J.; Bigler, F. Uptake of Bt-toxin by herbivores feeding on transgenic maize and consequences for the predator *Chrysoperla carnea*. Ecol. Entomol. **2002**, *27* (4), 441–447.

24. Harwood, J.D.; Wallin, W.G.; Obrycki, J.J. Uptake of Bt endotoxins by nontarget herbivores and higher order arthropod predators: Molecular evidence from a transgenic corn agroecosystem. Mol. Ecol. **2005**, *14* (9), 2815–2823.

25. Meissle, M.; Vojtech, E.; Poppy, G.M. Effects of Bt maize- fed prey on the generalist predator *Poecilus cupreus* L. (Cole- optera: Carabidae). Transgenic Res. **2005**, *14* (2), 123–132.

26. Obrist, L.B.; Klein, H.; Dutton, A.; Bigler, F. Effects of Bt maize on *Frankliniella tenuicornis* and exposure of thrips predators to prey-mediated Bt toxin. Entomol. Exp. Appl. **2005**, *115* (3), 409–416.

27. Obrist, L.B.; Dutton, A.; Romeis, J.; Bigler, F. Biological activity of Cry1Ab toxin expressed by Bt maize following ingestion by herbivorous arthropods and exposure of the predator *Chrysoperla carnea*. BioControl **2006**, *51* (1), 31–48.

28. Raps, A.; Kehr, J.; Gugerli, P.; Moar, W.J.; Bigler, F.; Hilbeck, A. Immunological analysis of phloem sap of *Bacillus thuringiensis* corn and of the nontarget herbivore *Rhopalo- siphum padi* (Homoptera: Aphididae) for the presence of Cry1Ab. Mol. Ecol. **2001**, *10* (2), 525–533.

29. Bernal, C.C.; Aguda, R.M.; Cohen, M.B. Effect of rice lines transformed with *Bacillus thuringiensis* toxin genes on the brown planthopper and its predator *Cyrtorhinus lividipen-nis*. Entomol. Exp. Appl. **2002**, *102* (1), 21–28.

30. Burgio, G.; Lanzoni, A.; Accinelli, G.; Dinelli, G.; Bonetti, A.; Marotti, I.; Ramilli, F. Evaluation of Bt-toxin uptake by the non-target herbivore, *Myzus persicae* (Hemiptera: Aphi- didae), feeding on transgenic oilseed rape. Bull. Entomol. Res. **2007**, *97* (2), 211–215.

31. Head, G.; Brown, C.R.; Groth, M.E.; Duan, J.J. Cry1Ab protein levels in phytophagous insects feeding on transgenic corn: Implications for secondary exposure risk assessment. Entomol. Exp. Appl. **2001**, *99* (1), 37–45.

32. Lundgren, J.G. *Relationships of Natural Enemies and Non-prey Foods*; Springer International: Dordrecht, the Netherlands, 2009.

33. Hagen, K.S.; Mills, N.J.; Gordh, G.; McMurtry, J.A. Terrestrial arthropod predators of insect and mite pests. In *Handbook of Biological Control*; Bellows, T.S., Fisher, T.W., Eds.; Academic Press: San Diego, CA, 1999; 383–504.

34. Kevan, P.G.; Kevan, D.K.M. Collembola as pollen feeders and flower visitors with observations from the high Arctic. Quaest. Entomol. **1970**, *6*, 311–326.

35. Chen, B.; Snider, R.J.; Snider, R.M. Food consumption by Collembola from northern Michigan deciduous forest. Pedobiologia **1996**, *40*, 149–161.

36. Kim, J.H.; Mullin, C.A. Impact of cysteine proteinase inhibition in midgut fluid and oral secretion on fecundity and pollen consumption of western corn rootworm *(Diabrotica virgifera virgifera)*. Arch. Insect Biochem. **2003**, *52*, 139–154.

37. Lundgren, J.G.; Wiedenmann, R.N. Nutritional suitability of corn pollen for the predator *Coleomegilla maculata* (Coleoptera: Coccinellidae). J. Insect Physiol. **2004**, *50* (6), 567–575.

38. Larochelle, A.; Larivière, M.C. *A Natural History of the Ground-Beetles (Coleoptera: Carabidae) of America North of Mexico*; Pensoft: Sofia, Bulgaria, 2003.

39. Canard, M. Natural food and feeding habits of lacewings. In *Lacewings in the Crop Environment*; McEwen, P.K., New, T.R., Whittington, A.E., Eds.; Cambridge University Press: Cambridge, U.K. 2001; 116–129.

40. Olesen, J.M.; Warncke, E. Predation and potential transfer of pollen in a population of *Saxifraga hirculus*. Holarctic Ecol. **1989**, *12*, 87–95.

41. Smith, R.B.; Mommsen, T.P. Pollen feeding in an orbweaving spider. Science **1984**, *226* (4680), 1330–1332.

42. Ludy, C. Intentional pollen feeding in the garden spider *Araneus diadematus*. Newsl. Br. Arachnol. Soc. **2004**, *101*, 4–5.

43. Pleasants, J.M.; Hellmich, R.L.; Dively, G.P.; Sears, M.K.; Stanley-Horn, D.E.; Mattila, H.R.; Foster, J.E.; Clark, P.; Jones, G.D. Corn pollen deposition on milkweeds in and near cornfields. Proc. Natl. Acad. Sci. U. S. A. **2001**, *98*, 11919–11924.

44. Peterson, J.A.; Romero, S.A.; Harwood, J.D. Pollen interception by linyphiid spiders in a corn agroecosystem: Implications for dietary diversification and risk-assessment. Arthropod Plant Interact. **2010**, *4* (4), 207–217.

45. Isenhour, D.J.; Marston, N.L. Seasonal cycles of *Orius in- sidiosus* (Hemiptera: Anthocoridae) in Missouri soybeans. J. Kans. Entomol. Soc. **1981**, *54*, 129–142.

46. Coll, M.; Bottrell, D.G. Microhabitat and resource selection of the European corn-borer (Lepidoptera, Pyralidae) and its natural enemies in Maryland field corn. Environ. Entomol. **1991**, *20*, 526–533.

47. Zwahlen, C.; Hilbeck, A.; Gugerli, P.; Nentwig, W. Degradation of the Cry1Ab protein within transgenic *Bacillus thuringiensis* corn tissue in the field. Mol. Ecol. **2003**, *12* (3), 765–775.

48. Zwahlen, C.; Andow, D.A. Field evidence for the exposure of ground beetles to Cry1Ab from trans-genic corn. Environ. Biosafety Res. **2005**, *4* (2), 113–117.

49. Saxena, D.; Stotzky, G. *Bacillus thuringiensis* (Bt) toxin released from the root exudates and bio-mass of Bt corn has no apparent effect on earthworms, nematodes, protozoa, bacteria, and fungi in soil. Soil Biol. Biochem. **2001**, *33* (9), 1225–1230.

50. Zwahlen, C.; Hilbeck, A.; Howald, R.; Nentwig, W. Effects of transgenic Bt corn litter on the earthworm *Lumbricus ter- restris*. Mol. Ecol. **2003**, *12* (4), 1077–1086.

51. Harwood, J.D.; Obrycki, J.J. The detection and decay of Cry1Ab Bt-endotoxins within non-target slugs, *Deroceras reticulatum* (Mollusca: Pulmonata), following consumption of transgenic corn. Biocontrol Sci. Technol. **2006**, *16* (1), 77–88.

52. Stone, M.L.; Whiles, M.R.; Webber, J.A.; Williard, K.W.J.; Reeve, J.D. Macroinvertebrate communities in agriculturally impacted southern Illinois streams: Patterns with riparian vegetation, water quality, and in-stream habitat quality. J. Environ. Qual. **2005**, *34*, 907–917.

53. Prihoda, K.R.; Coats, J.R. Aquatic fate and effects of *Bacillus thuringiensis* Cry3Bb1 protein: Toward risk-assessment. Environ. Toxicol. Chem. **2008**, *27* (4), 793–798.

54. Jensen, P.D.; Dively, G.P.; Swan, C.M.; Lamp, W.O. Exposure and nontarget effects of transgenic Bt corn debris in streams. Environ. Entomol. **2010**, *39* (2), 707–714.

55. Saxena, D.; Stotzky, G. Insecticidal toxin from *Bacillus thuringiensis* is released from roots of transgenic Bt corn in vitro and in situ. FEMS Microbiol. Ecol. **2000**, *33* (1), 35–39.

56. Saxena, D.; Stewart, C.N.; Altosaar, I.; Shu, Q.Y.; Stotzky, G. Larvicidal Cry proteins from *Bacillus thuringiensis* are released in root exudates of transgenic *B-thuringiensis* corn, potato, and rice but not of *B-thuringiensis* canola, cotton, and tobacco. Plant Physiol. Biochem. **2004**, *42* (5), 383–387.

57. Koskella, J.; Stotzky, G. Microbial utilization of free and clay-bound insecticidal toxins from *Bacillus thuringiensis* and their retention of insecticidal activity after incubation with microbes. Appl. Environ. Microb. **1997**, *63* (9), 3561–3568.

58. Stotzky, G. Persistence and biological activity in soil of the insecticidal proteins from *Bacillus thuringiensis*, especially from transgenic plants. Plant Soil **2004**, *266* (1–2), 77–89.

59. Icoz, I.; Stotzky, G. Fate and effects of insect-resistant *Bt* crops in soil ecosystems. Soil Biol. Biochem. **2008**, *40*, 559–586.

60. Palm, C.J.; Schaller, D.L.; Donegan, K.K.; Seidler, R.J. Persistence in soil of transgenic plant-produced *Bacillus thuringiensis* var. *kurstaki* Δ-endotoxin. Can. J. Microbiol. **1996**, *42*, 1258–1262.

61. Crecchio, C.; Stotzky, G. Insecticidal activity and biodegradation of the toxin from *Bacillus thuringiensis* subsp. *kurstaki* bound to humic acids from soil. Soil Biol. Biochem. **1998**, *30*, 463–470.

62. Skarphedinsdottir, H.; Gunnarsson, K.; Gudmundsson, G.A.; Nfon, E. Bioaccumulation and bio-magnification of organochlorines in a marine food web at a pristine site in Iceland. Arch. Environ. Contam. Toxicol. **2009**, *58* (3), 800–809.

63. Torres, J.B.; Ruberson, J.R. Interactions of *Bacillus thuringiensis* Cry1Ac toxin in genetically engineered cotton with predatory heteropterans. Transgenic Res. **2008**, *17* (3), 345–354.

64. Harwood, J.D.; Samson, R.A.; Obrycki, J.J. No evidence for the uptake of Cry1Ab Bt-endotoxins by the generalist predator *Scarites subterraneus* (Coleoptera: Carabidae) in laboratory and field experiments. Biocontrol Sci. Technol. **2006**, *16* (4), 377–388.

65. Jiang, Y.-H.; Fu, Q.; Cheng, J.-A.; Zhu, Z.-R.; Jiang, M.-X.; Ye, G.-Y.; Zhang, Z.-T. Dynamics of Cry1Ab protein from transgenic Bt rice in herbivores and their predators. Acta Entomol. Sin. **2004**, *47*, 454–460 [in Chinese with English abstract].

66. Harwood, J.D.; Samson, R.A.; Obrycki, J.J. Temporal detection of Cry1Ab-endotoxins in coccinellid predators from fields of *Bacillus thuringiensis* corn. Bull. Entomol. Res. **2007**, *97*, 643–648.

67. Musser, F.R.; Shelton, A.M. Bt sweet corn and selective insecticides: Impacts on pests and predators. J. Econ. Entomol. **2003**, *96* (1), 71–80.

68. Marvier, M.; McCreedy, C.; Regetz, J.; Kareiva, P. A metaanalysis of effects of Bt cotton and maize on nontarget invertebrates. Science **2007**, *316* (5830), 1475–1477.

69. Wolfenbarger, L.L.; Naranjo, S.E.; Lundgren, J.G.; Bitzer, R.J.; Watrud, L.S. Bt crop effects on functional guilds of non-target arthropods: A meta-analysis. PLoS One **2008**, *3* (5), e2118.

70. Peterson, J.A.; Lundgren, J.G.; Harwood, J.D. Interactions of transgenic *Bacillus thuringiensis* insecticidal crops with spiders (Araneae). J Arachnol., **2011**, 39(1), 1–21.
71. Betz, F.S.; Hammond, B.G.; Fuchs, R.L. Safety and advantages of *Bacillus thuringiensis*-protected plants to control insect pests. Regul. Toxicol. Pharmacol. **2000**, *32* (2), 156–173.
72. Gianessi, L.P.; Carpenter, J.E. Agricultural biotechnology: Insect control benefits; National Center for Food and Agricultural Policy, 1999, available at https://research.cip.cgiar.org/confluence/download/attachments/3443/AG7.pdf.
73. Pray, C.; Ma, D.; Huang, J.; Qiao, F. Impact of Bt cotton in China. World Dev. **2001**, 29, 813–825.
74. Thirtle, C.; Beyers, L.; Ismael, Y.; Piesse, J. Can GM- technologies help the poor? The impact of Bt cotton in Makhathini Flats, KwaZulu-Natal. World Dev. **2003**, *31*, 717–732.
75. Hutchinson, W.D.; Burkness, E.C.; Mitchel, P.D.; Moon, R.D.; Leslie, T.W.; Fleischer, S.J.; Abrahamson, M.; Hamilton, K.L.; Steffey, K.L.; Gray, M.E.; Hellmich, R.L.; Kaster, L.V.; Hunt, T.E.; Wright, R.J.; Pecinovsky, K.; Ra-baey, T.L.; Flood, B.R.; Raun, E.S. Areawide suppression of European corn borer with Bt maize reaps savings to non- Bt maize growers. Science **2010**, *330*, 222–225.
76. Lutz, W.; Samir, K.C. Dimensions of global population projections: What do we know about future population trends and structures? Philos. Trans. R. Soc. Lond. B **2010**, *365* (1554), 2779–2791.
77. Giles, K.L.; Hellmich, R.L.; Iverson, C.T.; Lewis, L.C. Effects of transgenic *Bacillus thuringiensis* maize grain on *B-thuringiensis*-susceptible *Plodia interpunctella* (Lepi- doptera: Pyralidae). J. Econ. Entomol. **2000**, *93* (3), 1011–1016.
78. Hussein, H.S.; Brasel, J.M. Toxicity, metabolism, and impact of mycotoxins on humans and animals. Toxicology **2001**, *167* (2), 101–134.
79. Munkvold, G.P.; Hellmich, R.L.; Rice, L.G. Comparison of fumonisin concentrations in kernels of transgenic Bt maize hybrids and nontransgenic hybrids. Plant Dis. **1999**, *83* (2), 130–138.
80. Lövei, G.L.; Arpaia, S. The impact of transgenic plants on natural enemies: A critical review of laboratory studies. En- tomol. Exp. Appl. **2005**, *114*, 1–14.
81. Pilcher, C.D.; Obrycki, J.J.; Rice, M.E.; Lewis, L.C. Prei- maginal development, survival, and field abundance of insect predators on transgenic *Bacillus thuringiensis* corn. Environ. Entomol. **1997**, *26*, 446–454.
82. Al-Deeb, M.A.; Wilde, G.E.; Higgins, R.A. No effect of *Bacillus thuringiensis* corn and *Bacillus thuringiensis* on the predator *Orius insidiosus* (Hemiptera: Anthocoridae). Environ. Entomol. **2001**, *30* (3), 625–629.
83. Lundgren, J.G.; Wiedenmann, R.N. Coleopteran-specific Cry3Bb1 toxin from transgenic corn pollen does not affect the fitness of a nontarget species, *Coleomegilla maculata* DeGeer (Coleoptera: Coccinellidae). Environ. Entomol. **2002**, *31*, 1213–1218.
84. Anderson, P.L.; Hellmich, R.L.; Sears, M.K.; Sumerford, D.V.; Lewis, L.C. Effects of Cry1Ab-expressing corn anthers on monarch butterfly larvae. Environ. Entomol. **2004**, *33*, 1109–1115.
85. Ramirez-Romero, R.; Desneux, N.; Decourtye, A.; Chaf- fiol, A.; Pham-Delègue, M.H. Does Cry1Ab protein affect learning performances of the honey bee *Apis mellifera* L. (Hymenoptera, Apidae)? Ecotoxicol. Environ. Saf. **2008**, *70* (2), 327–333.
86. Hilbeck, A.; Moar, W.J.; Pusztai-Carey, M.; Filippini, A.; Bigler, F. Toxicity of *Bacillus thuringiensis* Cry1Ab toxin to the predator *Chrysoperla carnea* (Neuroptera: Chrysopi- dae). Environ. Entomol. **1998**, *27*, 1255–1263.
87. Ramirez-Romero, R.; Bernal, J.S.; Chaufaux, J.; Kaiser, L. Impact assessment of Bt-maize on a moth parasitoid, *Cote- sia marginiventris* (Hymenoptera: Bracondiae), via host exposure to purified Cry1Ab protein or Bt-plants. Crop Prot. **2007**, *26*, 953–962.
88. Losey, J.E.; Rayor, L.S.; Carter, M.E. Transgenic pollen harms monarch larvae. Nature **1999**, *399*, 214.
89. Jesse, L.C.H.; Obrycki, J.J. Field deposition of *Bt* transgenic corn pollen: Lethal effects on the monarch butterfly. Oecologia **2000**, *1125*, 241–248.

90. Zangerl, A.R.; McKenna, D.; Wraight, C.L.; Carroll, M.; Ficarello, P.; Warner, R.; Berenbaum, M.R. Effects of exposure to event 176 *Bacillus thuringiensis* corn pollen on monarch and black swallowtail caterpillars under field conditions. Proc. Natl. Acad. Sci. U. S. A. **2001**, *98* (21), 11908–11912.

91. Rosi-Marshall, E.J.; Tank, J.L.; Royer, T.V.; Whiles, M.R.; Evans-White, M.; Chambers, C.; Griffiths, N.A.; Pokelsek, J.; Stephen, M.L. Toxins in transgenic crop byproducts may affect headwater stream ecosystems. Proc. Natl. Acad. Sci. U. S. A. **2007**, *104* (41), 16204–16208.

92. Duan, J.J.; Lundgren, J.G.; Naranjo, S.E.; Marvier, M. Extrapolating non-target risk of *Bt* crops from laboratory to field. Biol. Lett. **2010**, 6, 74–77.

93. Zwahlen, C.; Nentwig, W.; Bigler, F.; Hilbeck, A. Tritrophic interactions of transgenic *Bacillus thuringiensis* corn, *Anaphothrips obscurus* (Thysanoptera: Thripidae), and the predator *Orius majusculus* (Heteroptera: Anthocori- dae). Environ. Entomol. **2000**, *29* (4), 846–850.

94. Lundgren, J.G.; Wiedenmann, R.N. Tritrophic interactions among Bt (CryMb1) corn, aphid prey, and the predator *Co- leomegilla maculata* (Coleoptera: Coccinellidae). Environ. Entomol. **2005**, *34* (6), 1621–1625.

95. Ferry, N.; Mulligan, E.A.; Majerus, M.E.N.; Gatehouse, A.M.R. Bitrophic and tritrophic effects of Bt Cry3A transgenic potato on beneficial, non-target, beetles. Transgenic Res. **2007**, *16,* 795–812.

96. Hilbeck, A.; Baumgartner, M.; Fried, P.M.; Bigler, F. Effects of transgenic *Bacillus thuringiensis* corn-fed prey on mortality and development time of immature *Chrysoperla carnea* (Neuroptera: Chrysopidae). Environ. Entomol. **1998**, *27*, 480–487.

97. Hilbeck, A.; Moar, W.J.; Pusztai-Carey, M.; Filippini, A.; Bigler, F. Prey-mediated effects of Cry1Ab toxin and protoxin and Cry2A protoxin on the predator *Chrysoperla carnea*. Entomol. Exp. Appl. **1999**, *91,* 305–316.

98. Naranjo, S.E. *Risk Assessment: Bt Crops and Invertebrate Non-Target Effects—Revisited;* ISB News Report: Agricultural and Environmental Biotechnology, December 2009, 1–4.

99. Naranjo, S.E. Impacts of *Bt* crops on non-target invertebrates and insecticide use patterns. CAB Rev. Perspect. Ag- ric. Vet. Sci. Nutr. Nat. Resour. **2009**, *4* (11), 1–23.

100. Orr, D.B.; Landis, D.A. Oviposition of European corn borer (Lepidoptera: Pyralidae) and impact of natural enemy populations in transgenic versus isogenic corn. J. Econ. Entomol. **1997**, *90,* 905–909.

101. Reed, G.L.; Jensen, A.S.; Riebe, J.; Head, G.; Duan, J.J. Transgenic Bt potato and conventional insecticides for Colorado potato beetle management: Comparative efficacy and non-target impacts. Entomol. Exp. Appl. **2001**, *100*, 89–100.

102. Al-Deeb, M.A.; Wilde, G.E.; Blair, J.M.; Todd, T.C. Effect of *Bt* corn for corn rootworm control on nontarget soil microarthropods and nematodes. Environ. Entomol. **2003**, *32,* 859–865.

103. Sisterson, M.S.; Biggs, R.W.; Olson, C.; Carriére, Y.; Den- nehy, T.J.; Tabashnik, B.E. Arthropod abundance and diversity in Bt and non-Bt cotton fields. Environ. Entomol. **2004**, *33,* 921–929.

104. Bhatti, M.A.; Duan, J.; Head, G.P.; Jiang, C.; McKee, M.J.; Nickson, T.E.; Pilcher, C.L.; Pilcher, C.D. Field evaluation of the impact of corn rootworm (Coleoptera: Chrysomeli- dae)-protected *Bt* corn on foliage-dwelling arthropods. Environ. Entomol. **2005**, *34,* 1336–1345.

105. de la Poza, M.; Pons, X.; Farinós, G.P.; López, C.; Ortego, F.; Eizaguirre, M.; Castañera, P.; Albajes, R. Impact of farm-scale Bt maize on abundance of predatory arthropods in Spain. Crop Prot. **2005**, *24,* 677–684.

106. Meissle, M.; Lang, A. Comparing methods to evaluate the effects of *Bt* maize and insecticide on spider assemblages. Agric. Ecosyst. Environ. **2005**, *107,* 359–370.

107. Naranjo, S.E. Long-term assessment of the effects of transgenic *Bt* cotton on the abundance of nontarget arthropod natural enemies. Environ. Entomol. **2005**, *34,* 1193–1210.

108. Naranjo, S.E. Long-term assessment of the effects of transgenic Bt cotton on the function of the natural enemy community. Environ. Entomol. **2005**, *34,* 1211–1223.

109. Ludy, C.; Lang, A. A 3-year field-scale monitoring of foliage-dwelling spiders (Araneae) in transgenic *Bt* maize fields and adjacent field margins. Biol. Control **2006**, *38,* 314–324.

110. Torres, J.B.; Ruberson, J.R. Abundance and diversity of ground-dwelling arthropods of pest management importance in commercial Bt and on-Bt cotton fields. Ann. Appl. Biol. **2007**, *150* (1), 27–39.

111. Duan, J.J.; Marvier, M.; Huesing, J.; Dively, G.; Huang, Z.Y. A meta-analysis of effects of Bt crops on honey bees (Hymenoptera: Apidae). PLoS One **2008**, *3* (1), e1415.

112. Bernstein, J.A.; Bernstein, I.L.; Bucchini, L.; Goldman, L.R.; Hamilton, R.G.; Lehrer, S.; Rubin, C.; Sampson, H.A. Clinical and laboratory investigation of allergy to genetically modified foods. Environ. Health Perspect. **2003**, *111* (8), 1114–1121.

113. Environmental Protection Agency. *Cry9C Food Allergenicity Assessment Background Document,* 1999, available at http://www.epa.gov/pesticides/biopesticides/cry9c/cry9c-peerreview.htm.

114. Carter, C.A.; Smith, A. Estimating the market effect of a food scare: The case of genetically modified StarLink corn. Rev. Econ. Stat. **2007**, *89* (3), 522–533.

115. Picard-Nizou, A.L.; Pham-Delegue, M.-H.; Kerguelen, V.; Douault, P.; Marillaeu, R.; Olsen, L.; Grison, R.; Toppan, A.; Masson, C. Foraging behaviour of honey bees (*Apis mellifera* L.) on transgenic oilseed rape *(Brassica napus* L. var. *oleifera).* Transgenic Res. **1995**, *4,* 270–276.

116. Birch, A.N.E.; Geoghegan, I.E.; Griffiths, D.W.; McNicol, J.W. The effect of genetic transformations for pest resistance on foliar solanidine-based glycoalkaloids of potato *(Solanum tuberosum).* Ann. Appl. Biol. **2002**, *140,* 143–149.

117. Faria, C.A.; Wäckers, F.L.; Pritchard, J.; Barrett, D.A.; Turlings, T.J.C. High susceptibility of Bt maize to aphids enhances the performance of parasitoids of lepidopteran pests. PLoS One **2007**, *2,* e600.

118. Flores, S.; Saxena, D.; Stotzky, G. Transgenic *Bt* plants decompose less in soil than non-Bt plants. Soil Biol. Biochem. **2005**, *37,* 1073–1082.

119. Zurbrugg, C.; Honemann, L.; Meissle, M.; Romeis, J.; Ne- ntwig, W. Decomposition dynamics and structural plant components of genetically modified Bt maize leaves do not differ from leaves of conventional hybrids. Transgenic Res. **2010**, *19,* 257–267.

120. Virla, E.G.; Casuso, M.; Frias, E.A. A preliminary study on the effects of a transgenic corn event on the non-target pest *Dalbulus maidis* (Hemiptera: Cicadellidae). Crop Prot. **2010**, *29* (6), 635–638.

121. Ellstrand, N.C.; Prentice, H.C.; Hancock, J.F. Gene flow and introgression from domesticated plants into their wild relatives. Annu. Rev. Ecol. Syst. **1999**, *30,* 539–563.

122. Quist, D.; Chapela, I.H. Transgenic DNA introgressed into traditional maize landraces in Oaxaca, Mexico. Nature **2001**, *414* (6863), 541–543.

123. Christou, P. No credible scientific evidence is presented to support claims that transgenic DNA was introgressed into traditional landraces in Oaxaca, Mexico. Transgenic Res. **2002**, *11* (1), III–V.

124. Kaplinsky, N.; Braun, D.; Lisch, D.; Hay, A.; Hake, S.; Freeling, M. Biodiversity (communications arising): Maize transgene results in Mexico are artefacts. Nature **2002**, *416* (6881), 601.

125. Quist, D.; Chapela, I.H. Maize transgene results in Mexico are artefacts—Reply. Nature **2002**, *416* (6881), 602.

126. Serratos-Hernández, J.-A.; Gómez-Olivares, J.-L.; Salinas- Arreortua, N.; Buendía-Rodríguez, E.; Islas-Gutiérrez, F.; de-Ita, A. Transgenic proteins in maize in the soil conservation area of Federal District, Mexico. Front. Ecol. Environ. **2007**, *5* (5), 247–252.

127. Ortiz-García, S.; Ezcurra, E.; Schoel, B.; Acevedo, F.; So- berón, J.; Snow, A.A. Absence of detectable transgenes in local landraces of maize in Oaxaca, Mexico (2003–2004). Proc. Natl. Acad. Sci. U. S. A. **2005**, *102* (35), 12338–12343.

128. Xia, H.; Lu, B.R.; Su, J.; Chen, R.; Rong, J.; Song, Z.P.; Wang, F. Normal expression of insect-resistant transgene in progeny of common wild rice crossed with genetically modified rice: Its implication in ecological biosafety assessment. Theor. Appl. Genet. **2009**, *119* (4), 635–644.

129. Committee on Genetically Modified Pest-Protected Plants, Board on Agriculture and Natural Resources, National Research Council. *Genetically Modified Pest-Protected Plants: Science and Regulation;* National Academy Press: Washington, D.C., 2000.

130. Wraight, C.L.; Zangerl, A.R.; Carroll, M.J.; Berenbaum, M.R. Absence of toxicity of *Bacillus thuringiensis* pollen to black swallowtails under field conditions. Proc. Natl. Acad. Sci. U. S. A. **2000**, *97* (14), 7700–7703.

131. Gould, F. Integrating pesticidal engineered crops into Mesoamerican agriculture. In *Transgenic Plants in Meso- american Agriculture: Bacillus thuringiensis*; Hruska, A.J., Pavón, M.L., Eds.; Zamorano Academic Press: Zamorano, Honduras, 1997; 6–36.

132. Shi, Y.; Wang, W.B.; Powell, K.S.; Van Damme, E.; Hilder, V.A.; Gatehouse, A.M.R.; Boulter, D.; Gatehouse, J.A. Use of the rice sucrose synthase-1 promoter to direct phloem- specific expression of β-glucoronidase and snowdrop lectin genes in transgenic tobacco plants. J. Exp. Bot. **1994**, *45*, 623–631.

133. Rao, K.V.; Rathore, K.S.; Hodges, T.K.; Fu, X.; Stoger, E.; Sudhaker, D.; Williams, D.; Christou, P.; Bharathi, M.; Brown, D.P.; Powell, K.S.; Spence, J.; Gatehouse, A.M.R.; Gatehouse, J.A. Expression of snowdrop lectin (GNA) in transgenic rice plants confers resistance to rice brown plan- thopper. Plant J. **1998**, *15*, 469–477.

134. Giampietro, M.; Ulgiati, S.; Pimentel, D. Feasibility of large-scale biofuel production—Does an enlargement of scale change the picture? Bioscience **1997**, *47*, 587–600.

135. Mayer, P.M.; Reynolds, S.K.; McCutchen, M.D.; Canfield, T.J. *Riparian Buffer Width, Vegetative Cover, and Nitrogen Removal Effectiveness: A Review of Current Science and Regulations*; EPA/600/R-05/118; U.S. Environmental Protection Agency: Cincinnati, OH, 2006.

136. National Academy of Sciences. *Genetically Modified Pest- Protected Plants: Science and Regulation*; National Academy Press: Washington D.C., 2000.

137. Pessel, F.D.; Lecomte, J.; Emeriau, V.; Krouti, M.; Mes- sean, A.; Gouyon, P.H. Persistence of oil-seed rape (*Brassica napus* L.) outside of cultivated fields. Theor. Appl. Genet. **2001**, *102*, 841–846.

138. Terminator gene halt a 'major U-turn.' *BBC News* 1999, available at http://news.bbc.co.uk/2/hi/science/nature/465222.stm.

139. Leslie, T.W.; Biddinger, D.J.; Mullin, C.A.; Fleischer, S.J. Carabidae population dynamics and temporal partitioning: Response to couples neonicotinoid-transgenic technologies in maize. Environ. Entomol. **2009**, *38* (3), 935–943.

140. Landis, D.A.; Wratten, S.D.; Gurr, G.M. Habitat management to conserve natural enemies of arthropod pests in agriculture. Annu. Rev. Entomol. **2000**, *45*, 175–201.

141. Eilenberg, J.; Hajek, A.; Lomer, C. Suggestions for unifying the terminology in biological control. Biocontrol **2001**, *46*, 387–400.

142. Thorbek, P.; Sunderland, K.D.; Topping, C.J. Reproductive biology of agrobiont linyphiid spiders in relation to habitat, season and biocontrol potential. Biol. Control **2004**, *30*, 193–202.

143. Tabashnik, B.E. Genetics of resistance to *Bacillus thuringiensis*. Annu. Rev. Entomol. **1994**, *34*, 47–79.

144. Tabashnik, B.E.; Patin, A.L.; Dennehy, T.J.; Liu, Y.-B.; Carrière, Y.; Sims, M.A.; Antilla, L. Frequency of resistance to *Bacillus thuringiensis* in field populations of pink bollworm. P. Natl. Acad. Sci. U. S. A. **2000**, *97* (24), 12980–12984.

145. Environmental Protection Agency, Office of Pesticide Programs, Biopesticides and Pollution Prevention Division. *Optimum AcreMax Bt Corn Seed Blends: Biopesticides Registration Action Document,* 2010, available at http://www.epa.gov/oppbppd1/biopesticides/ingredients/tech_docs/brad_006490_oam.pdf.

146. Duan, J.J.; Head, G.; Jensen, A.; Reed, G. Effects of *Bacillus thuringiensis* potato and conventional insecticides for Colorado potato beetle (Coleoptera: Chrysomelidae) management on the abundance of ground-dwelling arthropods in Oregon potato ecosystems. Environ. Entomol. **2004**, *33*, 275–281.

3

Biopesticides

G. J. Ash and
A. Wang

Introduction

Inappropriate use of synthetic pesticides in agriculture can lead to environmental degradation, health risks, and loss of biodiversity. Additionally, overuse of particular pesticides may also lead to a loss in their effectiveness through pesticide resistance. Several studies have highlighted the accumulation of pesticides in soils in agricultural systems and the resultant environmental damage.[1] The short-term effects of pesticides on mammalian health are relatively well known and documented through pesticide approval processes. However, the longer-term effects are less well understood but may include cancer, neurotoxic effects, and reproductive disorders.[2] In developing countries, it is estimated that there may be 25 million cases of occupational pesticide poisoning annually.[3] Biopesticides are environmentally sensitive alternatives to the use of synthetic pesticides for the management of a wide range of agricultural pests including weeds, insects, pathogens, and other invertebrate pests. Biopesticides are a class of pesticides that contain plant, animal, or microbial products or organisms.[4] The International Union of Pure and Applied Chemistry[5] further classifies biopesticides as plant-incorporated protectants (protectants produced from genetically modified plants), biochemical pesticides (natural products that affect pests), and microbial biopesticides. Microbial biopesticides are a type of biological control in which a living organism is included as the active ingredient in an inundative application of a formulated product. A range of organisms have been used in biopesticide applications including fungi, nematodes, bacteria, and viruses for the control of weeds, insects, acarines, plant diseases, and molluscs.[6–16]

Formulations of biopesticides contain the organism, a carrier, and adjuvants, which may contain compounds such as nutrients and/or chemicals that aid in the survival of the pathogen or help in protecting the active ingredient from adverse environmental conditions.[17] The formulation of the biocontrol agents is affected by the type of biocontrol organism but must ensure that the agent is delivered in a form that is viable, virulent, and with sufficient inoculum potential to be effective in the field. Furthermore, the formulation may also dictate the means of delivery of the final product, for example, as a seed dressing or a foliar-applied formulation. To be successful, a formulation must be effective, economical, and practical to use.[18] Formulation of biopesticides can be in the form of dry products (dusts, granules, pellets, wettable powders, and encapsulated products) and liquid formulations (suspensions, emulsions, and encapsulated products).[19] The biopesticide is usually packaged, handled, stored, and applied in a similar way to traditional, synthetic pesticides.[20]

Due to the enormous costs involved in the production of synthetic pesticides, global companies have tended to focus on the registration of pesticides in the major crop/cropping systems. This has led to a number of attempts to develop biopesticides in these non-core or niche markets.[21] These markets could be of considerable size and include those that have been created by synthetic pesticide withdrawal and the organic food movement. Philosophically, the use of biopesticides is compatible with organic food production, provided the agent had not been genetically modified, the carriers and adjuvants are natural products, and the host range of the biopesticide is not considered to be too wide.[22]

The use of biopesticides as a strategy in pest management can be applied to both native and introduced pests. However, the success of this type of biocontrol revolves around the costs of production, the quality of the inoculum, and, most importantly, the field efficacy of the product.[18] Biopesticides are usually developed by collaboration with commercial companies in an expectation that they will recoup their costs through the sale of the product.

The broad range of biopesticides may be further subdivided based on the type of target pest and/ or the active ingredient. For example, a bioherbicide is a biopesticide developed for weed management whereas a mycoherbicide is a bioherbicide that contains a fungus as an active ingredient. The goal of this entry is to introduce the broad range of biopesticides available and their targets.

Bioinsecticides and Bioacaricides

More than 1500 species of pathogens have been shown to attack arthropods and include representatives from bacteria, viruses, fungi, protozoa, and nematodes.[23] Diseases caused by insects have been known since the early 1800s with the first attempts at inundative applications of fungi to control insects being developed in 1884, when the Russian entomologist Elie Metchnikoff mass produced the spores of the fungus *Metarhizium anisopliae*. A mycoacaricide for the management of citrus rust mite was first registered in the United States in 1981.[24] This was based on the fungus *Hirsutella thompsonii*. Currently, the main fungal species formulated as mycoinsecticides and mycoacaricides include *M. anisopliae* (33.9%), *Beauveria bassiana* (33.9%), *Isaria fumosorosea* (5.8%), and *Beauveria brongniartii* (4.1%).[15] De Faria and Wraight[15] reported a total of 171 fungi-based products with the majority being for the control of insects (160) and mites.[28] Their review indicated that the main formulation types are fungus-colonized substrates, wettable powders, and oil dispersions containing conidia (asexual spores). In the United States, there are seven bioinsecticide products based on various strains of *B. bassiana* registered. These include products such as BotaniGard® and Mycotrol® (available as emulsifiable suspensions and wettable powders).

Bacteria that attack insects can be divided into nonspore-forming and spore-forming bacteria. The nonspore-forming bacteria include species in the Pseudomonadaceae and the Enterobacteriaceae. The spore-forming bacteria belong to the Bacillaceae and include species such as *Bacillus popilliae* and *Bacillus thuringiensis*. *B. thuringiensis* (Bt) have primarily been developed as biopesticides to control Lepidopteran larva. However, other serotypes of Bt produce toxins that kill insects in the Coleoptera and Diptera as well as nematodes. The bacterium produces δ-endotoxin, an insecticidal crystal protein, which is converted into proteolytic toxins on ingestion. Up to nine different toxins, which have different host ranges, have been described.[25] Commercial formulations of the bacteria contain living spores of the bacteria. Biopesticides based on Bt are the most widely available of the bacterially based products.[25] Up to 90% of the Microbial Pest Control Agent market is Bt or Bt-derived products. The most well known is Dipel.

Entomopathogenic nematodes of the families Steinernematidae and Heterorhabditidae, in conjunction with bacteria of the genus *Xenorhabdus*, have been successfully deployed as biopesticides, for example, BioVECTOR.[23] They are usually applied to control insects in cryptic and soil environments. The nematodes harbor the bacteria in their intestines. The infective third-stage larvae enter the host through natural openings and penetrate into the hemocoel. The bacteria are voided in the insect and cause septicemia, killing the insect in approximately 48 hr.

Entomopathogenic viruses have also shown promise as bioinsecticides. They were first used to control populations of *Lymantria monacha* in pine forests in Germany in 1892 (Huber, 1986, from Moscardi[26]), but the first commercial viral insecticide registered was called Viron/H for the control of *Helicoverpa (Heliothis) zea* in 1971.[27] Viruses from the family Baculoviridae have been isolated from more than 700 invertebrates, with the virus group not common outside of the Lepidoptera and Hymenoptera.[26,28] The nucleo-polyhedroviruses (NPVs) are rod-shaped, double-stranded DNA viruses that are produced in polyhedral proteinaceous occlusion bodies[28] that are ingested by the insect. Granulosis viruses (GVs) are also members of the Baculoviridae but are restricted to the Lepidoptera and have capsular protein-aceous occlusion bodies.[26] These authors[26] provide a table of products that have been developed for the control of insects using these two viral groups.

Biofungicides

Biological control of fungi that cause plant disease can be accomplished by a number of mechanisms including antibiosis, hyperparasitism, or competition. Additionally, weak pathogens may induce systemic acquired resistance in the host, giving a form of cross-protection. Biofungicides have been used in both the phylloplane and rhizosphere to suppress disease. A biological control agent for the control of foliar pathogens in the phylloplane must have a high reproductive capacity, the ability to survive unfavorable conditions, and the ability to be a strong antagonist or be very aggressive. A wide range of bacteria and fungi are known to produce antibiotics that affect other microorganisms in the infection court. Most often, these organisms are sought from a soil environment, as this environment is seen as the richest source of antibiotic-producing species. Species of *Bacillus* and *Pseudomonas* have been successfully used as seed dressings to control soil-borne plant diseases.[29] Serenade', marketed by BASF, is a formulation of *Bacillus subtilis* (strain QST713), which has claimed activity against a wide range of plant diseases.[30] It is applied as a foliar spray to crops such as cherries, cucurbits, grapes, leafy vegetables, peppers, potatoes, tomatoes, and walnuts.[31] Fluorescent pseudomonads are also often seen as a component of suppressive soils. These bacteria may prevent the germination of fungi by the induction of iron competition through the production of siderophores (ironchelating compounds). These are effective only in those soils where the availability of iron is low. Control of foliar and fruit pathogens such as *Botrytis cinerea,* a pathogen of strawberries, has been accomplished by the foliar application of the soil-inhabiting fungus *Trichoderma viride.*[32] This fungus inhibits *Botrytis* using a combination of antibiosis and competition. On grapevines, *Trichoderma harzianum* competes with *B. cinerea* on senescent floral parts, thus preventing the infection of the ovary. It has also been shown to coil around the hyphae of the pathogen during hyperparasitism.[33] *T. harzianum* has also been reported to induce systemic resistance in plants.[34] One of the earliest commercial successes using *T. harzianum* is the product Rootshield°. Rootshield contains the T-22 strain of *T. harzianum* and is produced and marketed by Bioworks Inc. This strain of the fungus was first registered by the U.S. Environmental Protection Agency in 1993. The product is available as a granular formulation and is usually applied to soil mixtures in glasshouse situations.[35–37]

Bioherbicides

Fungi are the most important group of pathogens causing plant disease. Therefore, fungi (or oomycetes) are most commonly used as the active ingredient in bioherbicides and as such the formulated organism is referred to as a mycoherbicide.[38] However, there are examples of bacteria[7–11] and viruses being used or proposed to be used as bioherbicides.[39,40] The aim of bioherbicide development is to overcome the natural constraints of a weed–pathogen interaction, thereby creating a disease epidemic on a target host.[41] For example, the application of fungal propagules to the entire weed population overcomes the constraint of poor dissemination. After removal of the host weed, the pathogen generally returns to background levels because of natural constraints on survival and spread.

The first commercially available biopesticide for the control of weeds was DeVine*, a bioherbicide for the control of strangler vine in citrus groves in the United States. It was released in 1981.[42] In 1982, a formulation of *Colletotrichum gloeosporioides* f. sp. *aeschynomene* was released to control northern jointvetch in soybean crops in the United States. Since then, there have been a number of products commercialized[12,21] as well as numerous examples of pathogen–weed combinations that had been reported as having potential as bioherbicides in countries including in Canada, United States, Europe, Japan, Australia, and South Africa.[18,21] Necrotrophic or hemibiotrophic fungi are usually used as the basis of mycoherbicides, as they can be readily cultured on artificial media and so lend themselves to mass production. Other desirable characteristics of fungi under consideration as mycoherbicides include the ability to sporulate freely in artificial culture, limited ability to spread from the site of application, and genetic stability. In most cases, these biopesticides are applied in a similar fashion to chemical herbicides using existing equipment, although the development of specialized application equipment and formulation may improve their efficacy and reliability. Since 2000, there have been two successful registrations for bioherbicides in Canada. In 2002, a product called Chontrol*, based on the fungus *Chondrostereum purpureum* for the control of trees and shrubs, was registered.[12] This was based on the research of Hintz and colleagues.[43–45] A more recent success in the area of bioherbicides includes the registration of Sarritor* for dandelion control by the company of the same name in Canada. This product is based on the phytopathogenic fungus *Sclerotinia minor*, which has been extensively researched by Professor Alan Watson at McGill University.[46–50]

Biomolluscicides

Biomolluscicides are a type of molluscicide derived from natural materials such as animals, plants, and microorganisms (e.g., bacterium, fungus, virus, protozoan, or nematode). They are usually used in the fields of agriculture and gardening to control pest slugs and snails. In some circumstances, biomolluscicides are also used in the health area to control molluscs acting as vectors of harmful parasites to human beings.

Currently, the most widely used biomolluscicide is Nemaslug*, a successful biomolluscicide developed by Becker Underwood (U.K.). The active ingredient of Nemaslug is *Phasmarhabditis hermaphrodita*, a nematode species from the family of Rhabditidae. The pathogenicity of *P. hermaphrodita* against slugs had not been recognized until 1994 when Wilson et al.[51] discovered that *P. hermaphrodita* could infect and kill a wide variety of pest slugs under laboratory conditions. Like entomopathogenic nematodes, *P. hermaphrodita* kills slugs by penetrating into the hemocoel of hosts through natural openings and releasing its associated bacteria, which kill the host eventually.

P. hermaphrodita was found to be associated with several different bacteria rather than one particular species, but its association with *Moraxella osloensis* proved to be highly pathogenic to gray garden slug *(Deroceras reticulatum)*.[51] This bacterium was used in the mass production of *P. hermaphrodita* via monoxenic culture.[51]

The host of *P. hermaphrodita* is not restricted to only one slug species (*D. reticulatum*). It can attack and kill several species of slugs including *Arion ater, Arion intermedius, Arion distinctus, Arion silvaticus, Deroceras caruanae, Tandonia budapestensis,* and *Tandonia sowerbyi*.[51] Moreover, *P. hermaphrodita* can also parasitize several species of snails including *Cernuella virgata, Cochlicella acuta, Helix aspersa, Monacha cantiana, Lymnaea stagnalis,* and *Theba pisana*.[52]

Nemaslug is now sold in many European countries, including U.K., Ireland, France, the Netherlands, Belgium, Germany, Denmark, Norway, Finland, Poland, Spain, the Czech Republic, Italy, and Switzerland. In 2005, the retail sale of this biomolluscicide was up to £1 million in Europe and approximately 500 ha horticultural crops (e.g., lettuce and strawberries) and field crops (e.g., wheat, potatoes, and oilseed) were treated with this biomolluscicide. At the dose rate of 3×10^9/ha, *P. hermaphrodita* provides protection against slug damage similar to, if not better than, methiocarb pellets.[16]

Bacteria-based biomolluscicides are now in the process of development. *Streptomyces violaceoruber* and *Xanthobacter autotrophicus* have been examined for their molluscicidal activity against *Oncomelania hupensis* (a unique host of schistosomiasis blood fluke parasite) under laboratory conditions.[53] The results revealed that both bacteria were effective in killing *O. hupensis*, with *S. violaceoruber* causing more snail mortality than *X. autotrophicus* (90% vs. 85%).

Biomolluscicides of plant origin have also been studied extensively in recent years when the environmental pollution caused by chemical molluscicides was realized increasingly. More than 1400 plant species have been screened for their molluscicidal properties against pest snail species.[54] Several groups of compounds present in various plants have been found to be poisonous to snails at acceptable doses, ranging from <1 to 100 ppm, including saponins, tannins, alkaloids, alkenyl phenols, glycoalkaloids, flavonoids, sesquiterpene lactones, and terpenoids.[54] The molluscicidal activity of the dried root latex powder of *Ferula asafoetida*, the flower-bud powder of *Syzygium aromaticum*, and the seed powder of *Carum carvi* against the snail *Lymnaea acuminata* was proved.[55] Similarly, acetogenin (extracted from the seed powder of custard apple) presented promising and stable molluscicidal activity against *L. acuminata*.[54] When sodium alginates was used as a binding matrix for the formulation of acetogenin, the release of this biomolluscicide extended over 25 days, which set up a good example for the development of biomolluscicide delivery system.[54]

The combination of bacteria-based biomolluscicides and plant-based biomolluscicides may lead to a synergistic effect between plant and microbe extracts as molluscicides. Zhang and coworkers[56] reported that higher snail mortality was produced when a mixture of *Arisaema erubescens* tuber extracts and *S. violaceoruber* dilution was applied against the snail *O. hupensis*. The mechanisms of snail toxicosis might be that the combination of *A. erubescens* tuber extract and *S. violaceoruber* dilution reduced the detoxification ability of liver and increased the oxidative damage in liver cells of snails.

Conclusion

Biopesticides are a viable alternative to synthetic pesticides in a number of crops. The development of microbial biopesticides relies on agent discovery and selection, development of methods to culture the pathogen, creation of formulations that protect the organism in storage as well as aid in its delivery, studies of field efficacy, and methods of storage. Each microbial biopesticide is unique, in that not only will the organism vary but so too will the host, the environment in which it is being applied, and economics of production and control.

There are a number of advantages of the use of biopesticides over the use of conventional pesticides, including the minimal residue levels, control of pests already showing resistance to conventional pesticides, host specificity, and the reduced chance of resistance to biopesticides. This indicates an emerging, strong role for biopesticides in any integrated pest management strategy and an important involvement in sustainable farming production systems in the future.

There have been some spectacular successes in the use of microbial biopesticides, despite the perceived constraints to their deployment.[57] In the past, biopesticides have been expected to behave in the same way as synthetic pesticides. For the ultimate success of biopesticides, microorganisms developed for biological control must be viewed by researchers, manufacturers, and end users in a biological paradigm rather than a chemical one. The business model for the commercialization of the products may also vary significantly from that used for traditional synthetic pesticides.[18]

The efficacy and reliability of many microbial biopesticides may be affected by environmental parameters as well as the aggressiveness of the pathogen. Furthermore, the narrow host range of many pathogens may restrict their commercial attractiveness. Both of these issues can be addressed by research into the use of genetic engineering and formulation.[18,58–60] As research into the molecular basis of host specificity and pathogenesis continues, it will become possible to produce more aggressive pathogens with the desired host range for biological control. The survival and efficacy of these pathogens will be enhanced through the use of novel formulations.

References

1. Alletto, L.; Coquet, Y.; Benoit, P.; Heddadj, D.; Barriuso, E. Tillage management effects on pesticide fate in soils. A review. Agron. Sustainable Dev. **2010**, *30*, 367–400.

2. Komarek, M.; Cadkova, E.; Chrastny, V.; Bordas, F.; Bollinger, J.C. Contamination of vineyard soils with fungicides: A review of environmental and toxicological aspects. Environ. Int. **2010**, *36*, 138–151.

3. Jeyaratnam, J. Acute pesticide poisoning: A major global health problem. World Health Stat. Q. **1990**, *43*, 139–144.

4. Available at http://www.epa.gov/opp00001/biopesticides/whatarebiopesticides.htm (accessed July 30, 2010).

5. Available at http://agrochemicals.iupac.org/index.php?p=biopesticides (accessed July 30, 2010).

6. Jaronski, S.T. Ecological factors in the inundative use of fungal entomopathogens. Biocontrol **2010**, *55*, 159–185.

7. Imaizumi, S.; Nishino, T.; Miyabe, K.; Fujimori, T.; Yamada, M. Biological control of annual bluegrass (*Poa annua* L.) with a Japanese isolate of *Xanthomonas campestris* pv *poae* (JT-P482). Biol. Control **1997**, *8*, 7–14.

8. Daigle, D.J.; Connick, W.J.; Boyetchko, S.M. Formulating a weed-suppressive bacterium in "Pesta". Weed Technol. **2002**, *16*, 407–413.

9. Weissmann, R.; Uggla, C.; Gerhardson, B. Field performance of a weed-suppressing *Serratia plymuthica* strain applied with conventional spraying equipment. Biocontrol **2003**, *48*, 725–742.

10. Anderson, R.C.; Gardner, D.E. An evaluation of the wilt-causing bacterium *Ralstonia solanacearum* as a potential biological control agent for the alien kahili ginger (*Hedy-chium gardnerianum*) in Hawaiian forests. Biol. Control **1999**, *15*, 89–96.

11. DeValerio, J.T.; Charudattan, R. Field testing of *Ralstonia solanacearum* [Smith] Yabuuchi et al. as a biocontrol agent for tropical soda apple *(Solanum viarum* Dunal). Weed Sci. Soc. Am. Abstr. **1999**, *39*, 70.

12. Bailey, K.L.; Boyetchko, S.M.; Langle, T. Social and economic drivers shaping the future of biological control: A Canadian perspective on the factors affecting the development and use of microbial biopesticides. Biol. Control **2010**, *52*, 221–229.

13. Bailey, K.L. Canadian innovations in microbial biopesticides. Can. J. Plant Pathol. **2010**, *32*, 113–121.

14. Hartman, C.L.; Markle, G.M. IR-4 biopesticide program for minor crops. In *Biopesticides;* Humana Press Inc.: Totowa, 1999; Vol. 5, 443–452.

15. de Faria, M.R.; Wraight, S.P. Mycoinsecticides and mycoa-caricides: A comprehensive list with worldwide coverage and international classification of formulation types. Biol. Control **2007**, *43*, 237–256.

16. Rae, R.; Verdun, C.; Grewal, P.; Robertson, J.F.; Wilson, M.J. Biological control of terrestrial molluscs using *Phasmarhabditis hermaphrodita*—Progress and prospects. Pest Manage. Sci. **2007**, *63*, 1153–1164.

17. Hynes, R.K.; Boyetchko, S.M. Research initiatives in the art and science of biopesticide formulations. Soil Biol. Bio-chem. **2006**, *38*, 845–849.

18. Ash, G.J. The science, art and business of successful bioherbicides. Biol. Control **2010**, *52*, 230–240.

19. Auld, B.A.; Hertherington, S.D.; Smith, H.E. Advances in bioherbicide formulation. Weed Biol. Manage. **2003**, *3*, 61–67.

20. Van Driesche, R.G.; Bellows, T.S. *Biological Control;* Chapman and Hall: New York, 1996.

21. Charudattan, R. Biological control of weeds by means of plant pathogens: Significance for integrated weed management in modern agro-ecology. Biocontrol **2001**, *46*, 229–260.

22. Rosskopf, E.; Koenig, R. Are bioherbicides compatible with organic farming systems and will businesses invest in the further development of this technology? VI International Bioherbicide Group Workshop, Canberra, Australia, 2003; 2003.

23. Kaya, H.K.; Gaugler, R. Entomopathogenic nematodes. Annu. Rev. Entomol. **1993**, *38*, 181–206.

24. McCoy, C.W. Factors governing the efficacy of *Hirsutella thompsonii* in the field. In *Fundemental and Applied Aspects of Invertebrate Pathology;* Samson, R.A., Vlak, J.M., Peters, D., Eds.; Foundation of the Fourth International Colloquim of Inverstbrate Pathology: Wageningen, the Netherlands, 1986; 171–174.

25. Rosell, G.; Quero, C.; Coll, J.; Guerrero, A. Biorational insecticides in pest management. J. Pestic. Sci. **2008**, *33*, 103–121.

26. Moscardi, F. Assessment of the application of baculoviruses for control of lepidoptera. Annu. Rev. Entomol. **1999**, *44*, 257–289.

27. Ignoffo, C.M.; Rice, W.C.; Mcintosh, A.H. Inactivation of occluded baculoviruses and baculovirus-DNA exposed to simulated sunlight. Environ. Entomol. **1989**, *18*, 177–183.

28. Fuxa, J.R. Ecology of insect nucleopolyhedroviruses. Agric. Ecosyst. Environ. **2004**, *103*, 27–43.

29. Johnsson, L.; Hokeberg, M.; Gerhardson, B. Performance of the *Pseudomonas chlororaphis* biocontrol agent MA 342 against cereal seed-borne diseases in field experiments. Eur. J. Plant Pathol. **1998**, *104*, 701–711.

30. Available at http://www.agro.basf.com/agr/AP-Internet/en/content/solutions/solution_highlights/serenade/bacillussubtilis (accessed December 1, 2010).

31. Available at http://www.epa.gov/pesticides/biopesticides/ingredients/factsheets/factsheet_006479.htm (accessed December 1, 2010).

32. Sutton, J.C.; Peng, G. Biocontrol of *Botrytis cinerea* in strawberry leaves. Phytopathology **1993**, *83*, 615–621.

33. Oneill, T.M.; Elad, Y.; Shtienberg, D.; Cohen, A. Control of grapevine grey mould with *Trichoderma harzianum* T39. Biocontrol Sci. Technol. **1996**, *6*, 139–146.

34. Shoresh, M.; Harman, G.E. The relationship between increased growth and resistance induced in plants by root colonizing microbes. Plant Signaling Behav. **2008**, *3*, 737–9.

35. Larkin, R.P.; Fravel, D.R. Efficacy of various fungal and bacterial biocontrol organisms for control of fusarium wilt of tomato. Plant Dis. **1998**, *82*, 1022–1028.

36. Brewer, M.T.; Larkin, R.P. Efficacy of several potential biocontrol organisms against *Rhizoctonia solani* on potato. Crop Prot. **2005**, *24*, 939–950.

37. Gravel, V.; Menard, C.; Dorais, M. Pythium root rot and growth responses of organically grown geranium plants to beneficial microorganisms. HortScience **2009**, *44*, 1622–1627.

38. Crump, N.S.; Cother, E.J.; Ash, G.J. Clarifying the nomenclature in microbial weed control. Biocontrol Sci. Technol. **1999**, *9*, 89–97.

39. Ferrell, J.; Charudattan, R.; Elliott, M.; Hiebert, E. Effects of selected herbicides on the efficacy of tobacco mild green mosaic virus to control tropical soda apple (*Solanum viarum*). Weed Sci. **2008**, *56*, 128–132.

40. Charudattan, R.; Hiebert, E. A plant virus as a bioherbicide for tropical soda apple, *Solanum viarum*. Outlooks Pest Manage. **2007**, *18*, 167–171.

41. TeBeest, D.O., Ed. Biological control of weeds with plant pathogens and microbial pesticides. In *Advances in Agronomy*; Academic Press Inc.: San Diego, 1996; Vol. 56, 115–137.

42. Tebeest, D.O.; Yang, X.B.; Cisar, C.R. The status of biological control of weeds with fungal pathogens. Annu. Rev. Phytopathol. **1992**, *30*, 637–657.

43. Harper, G.J.; Comeau, P.G.; Hintz, W.; Wall, R.E.; Prasad, R.; Hocker, E.M. *Chondrostereum purpureum* as a biological control agent in forest vegetation management. II. Efficacy on Sitka alder and aspen in western Canada. Can. J. Forest Res. **1999**, *29*, 852–858.

44. de la Bastide, P.Y.; Zhu, H.; Shrimpton, G.; Shamoun, S.F.; Hintz, W.E. *Chondrostereum purpureum:* An alternative to chemical herbicide brush control. Seventh International Symposium on Environmental Concerns in Rights-of-Way-Management **2002**, 665–672.

45. Becker, E.M.; Ball, L.A.; Hintz, W.E. PCR-based genetic markers for detection and infection frequency analysis of the biocontrol fungus *Chondrostereum purpureum* on Sitka alder and Trembling aspen. Biol. Control **1999**, *15*, 71–80.

46. Shaheen, I.Y.; Abu-Dieyeh, M.H.; Ash, G.J.; Watson, A.K. Physiological characterization of the dandelion bioherbicide, *Sclerotinia minor* IMI 344141. Biocontrol Sci. Technol. **2010**, *20*, 57–76.

47. Li, P.; Ash, G.J.; Ahn, B.; Watson, A.K. Development of strain specific molecular markers for the *Sclerotinia minor* bioherbicide strain IMI 344141. Biocontrol Sci. Technol. **2010**, *20*, 939–959.

48. Abu-Dieyeh, M.H.; Watson, A.K. Efficacy of *Sclerotinia minor* for dandelion control: Effect of dandelion accession, age and grass competition. Weed Res. **2007**, *47*, 63–72.

49. Abu-Dieyeh, M.H.; Watson, A.K. Effect of turfgrass mowing height on biocontrol of dandelion with *Sclerotinia minor.* Biocontrol Sci. Technol. **2006**, *16*, 509–524.

50. Abu-Dieyeh, M.H.; Watson, A.K. The significance of competition: Suppression of *Taraxacum officinale* populations by *Sclerotinia minor* and grass overseeding. Can. J. Plant Sci. **2006**, *86*, 1416–1416.

51. Wilson, M.J.; Glen, D.M.; Hughes, L.A.; Pearce, J.D.; Rodgers, P.B. Laboratory tests of the potential of entomo-pathogenic nematodes for the control of field slugs (*Deroc-eras reticulatum*). J. Invertebr. Pathol. **1994**, *64,* 182–187.

52. Coupland, J.B. Susceptibility of helicid snails to isolates of the nematode *Phasmarhabditis hermaphrodita* from southern France. J. Invertebr. Pathol. **1995**, *66*, 207–208.

53. Li, Y.D.; Yang, J.M. The study on effect of microbe and microbial pesticides killing *Oncomelania hupensis.* Acta Hydro. Sin. **2005**, *29*, 203–205.

54. Singh, A.; Singh, D.K.; Kushwaha, V.B. Alginates as binding matrix for bio-molluscicides against harmful snails *Lymnaea acuminata.* J. Appl. Polym. Sci. **2007**, *105*, 1275–1279.

55. Kumar, P.; Singh, D.K. Molluscicidal activity of *Ferula asafoetida, Syzygium aromaticum* and *Carum carvi* and their active components against the snail *Lymnaea acuminata.* Chemosphere **2006**, *63*, 1568–1574.

56. Zhang, Y.; Ke, W.S.; Yang, J.L.; Ma, A.N.; Yu, Z.S. The toxic activities of *Arisaema erubescens* and *Nerium indicum* mixed with *Streptomycete* against snails. Environ. Toxicol. Pharmacol. **2009**, *27*, 283–286.

57. Auld, B.A.; Morin, L. Constraints in the development of bioherbicides. Weed Technol. **1995**, *9*, 638–652.

58. Wang, C.S.; St Leger, R.J. A scorpion neurotoxin increases the potency of a fungal insecticide. Nat. Biotechnol. **2007**, *25*, 1455–1456.

59. Ash, G.J. Biological control of weeds with mycoherbicides in the age of genomics. Pest Technol. Rev. **2010**, *in press.*

60. St Leger, R.J.; Joshi, L.; Bidochka, M.J.; Roberts, D.W. Construction of an improved mycoinsecticide overexpressing a toxic protease. Proc. Natl. Acad. Sci. U. S. A. **1996**, *93*, 6349–6354.

4

Birds: Chemical Control

Eric B. Spurr

Toxicants

The earliest toxicants were formulations containing arsenic, antimony, phosphorus, and various botanical extracts.[1] Other toxicants used previously include chlorinated hydrocarbons such as endrin, metallic salts such as thallium, organometallic salts such as sodium monofluoroacetate (1080), alkaloids such as nicotine, and anticoagulants such as coumatetralyl and brodifacoum. Most are highly toxic to both birds and mammals. More than 2000 chemicals were evaluated as avicides between the 1940s and the 1980s, and some were found that were selectively toxic to birds. Some were even selectively toxic to certain species of birds. Since the 1980s, however, little effort has been put into finding new toxicants. Instead, most effort has been spent gathering toxicological and environmental data to ensure continued registration of existing products. Recently, international attention has focussed on the animal welfare aspects of toxicants.[2]

Strychnine was once used widely as an oral toxicant for control of birds such as rock pigeons (*Columba livia*) and house sparrows (*Passer domesticus*), and is still used by certified operators in some countries today (Table 1). It is mainly applied in grain baits (e.g., Sanex Poison Corn in Canada). Strychnine is highly toxic to both birds and mammals, and poses a high risk of both primary and secondary poisoning to nontarget species. Time to death varies from 5 to 50 minutes. It causes extreme pain in poisoned animals and is considered inhumane.

Fenthion was previously used as an oral and dermal toxicant, but is currently used only as a dermal toxicant. Its use is restricted to certified operators. It is applied to wicks in artificial perches or other surfaces, for control of birds such as rock pigeons, house sparrows, and starlings (*Sturms vulgaris*). It was available previously as Rid-A-Bird˙ in the United States, and currently as Control-A-Bird˙ and Avigrease˙ in Australia. Fenthion (Queletox˙) is also aerially sprayed onto birds, especially red-billed quelea (*Quelea quelea*), in their nighttime roosts, to protect ripening grain crops in some African countries. It is highly toxic to birds and moderately toxic to mammals. Death occurs in 3 to 12 hours. The risk of nontarget bird mortality (from both primary and secondary poisoning) and environmental contamination is high, especially following aerial application. The symptoms of poisoning (e.g., convulsions) indicate that fenthion is likely to be inhumane.

4-Aminopyridine (Avitrol˙ in the U.S. and Canada, Avis Scare˙ and Scatterbird˙ in Australia) is often described as a frightening agent, but it is also an oral toxicant. Birds that ingest it die, but before dying they exhibit erratic behavior and alarm calling (often termed distress behavior) that supposedly frightens away other birds in the flock before they are able to ingest it. Time to death ranges from 15 min to 3 days.

TABLE 4.1   Chemicals Currently Used for Bird Control in United States of America (U.S.A.), Canada, United Kingdom (U.K.), France, Israel, Australia, and New Zealand (N.Z.)

Compound	Activity	Countries
Strychnine	Oral toxicant	Canada, Australia
Fenthion	Oral and dermal toxicant	Some African countries, Australia
4-Aminopyridine	Oral toxicant, frightening agent	U.S.A., Canada, Australia
DRC-1339	Oral toxicant	U.S.A., N.Z.
Alpha-chloralose	Oral toxicant, immobilizing agent	U.S.A., France, U.K., Israel, Australia, N.Z.
Seconal (+ alpha-chloralose)	Immobilizing agent	U.K.
Polybutene	Tactile repellent	U.S.A., Canada, U.K., Israel, Australia, N.Z.
Denatonium saccharide	Taste repellent	U.S.A., Canada
Aluminium ammonium sulfate	Taste repellent	U.K., Australia
Thiram	Taste repellent	France, Israel
Endosulfan	Taste repellent	France
Triacetate guazatine	Taste repellent	France
Methyl anthranilate	Irritant	U.S.A., Canada
Capsaicin	Irritant	U.S.A.
Naphthalene	Irritant	U.S.A.
Methiocarb	Secondary repellent	U.S.A., Canada, Israel, Australia, N.Z.
Ziram	Secondary repellent	U.K., France
Anthraquinone	Secondary repellent	U.S.A., France, N.Z.
Azacosterol	Reproductive inhibitor	Canada
Corn oil	Reproductive inhibitor	U.S.A.
Paraffin oil	Reproductive inhibitor	U.K.

Source: Adapted from Schafer,[1] Ministry of Agriculture[2] and Clark.[3]

It is used to control birds such as rock pigeons, house sparrows, starlings, and in the U.S., red-winged blackbirds *(Agelaius phoeniceus)*. It is available as a concentrate or as ready-to-use treated grain to certified operators. It is highly toxic to both birds and mammals, and may cause both primary and secondary poisoning of nontarget species. Despite appearances to the contrary, it has been claimed that death from the compound is relatively painless. However, this needs to be verified because severe symptoms of intoxication may last for up to 3 days.

DRC-1339 (3-chloro-p-toluidine hydrochloride) (Starlicide') is an oral toxicant used for the control of birds such as rock pigeons, starlings, and in the U.S., red-winged blackbirds. It is available as a concentrate or as a ready-to-use cereal-based bait to certified operators. Time to death varies from 3 to 50 hr, depending upon the amount of toxicant ingested. DRC-1339 is not suitable as a toxicant for all pest bird species because it is not highly toxic to all species. For example, it has only low toxicity to sparrows (Ploceidae) and finches (Fringillidae). It also has low toxicity to most mammals. This selective toxicity is unique. DRC-1339 is rapidly metabolized, so there is little risk of secondary poisoning. The death of birds from DRC-1339 has been described as painless, but symptoms such as difficult breathing indicate that this might not be so.

Alpha-chloralose is used in some countries (e.g., Australia and New Zealand) as an oral toxicant, but in other countries only as an immobilizing agent (see below). It is available to certified operators as a concentrate or as ready-to-use treated grain, for the control of birds such as rock pigeons and house sparrows. It is generally more toxic to birds than to mammals, and is relatively fast-acting. The first signs of narcosis may occur 10 min after ingestion, and immobilization may last for up to 27 hr, though it generally lasts less than 1 hr, after which birds may recover. However, death may result from hypothermia if sufficient active ingredient is ingested, and/or the weather is inclement. Alpha-chloralose is only slowly metabolized, and so may cause secondary poisoning of nontarget species. It is considered to be relatively humane on the basis of the generally short time to insensitivity.[2]

Lethal Stressing Agents

PA-14 (Tergitol') is a surfactant that was used as a lethal stressing agent in the U.S., but is no longer available for this purpose. It was sprayed onto birds, such as starlings and red-winged blackbirds, in their nighttime roosts, resulting in a break-down of the oil in the birds' feathers, destroying their natural waterproofing, and causing death from hypothermia.

Immobilizing Agents

Immobilizing agents, administered in baits, are used to make birds easier to capture for removal from areas where they cause problems, or for killing humanely by other methods (e.g., by breaking their necks, or gassing them with carbon dioxide). Nontarget birds that become immobilized can be revived and released. However, the effectiveness of immobilizing agents depends upon the amount ingested and environmental conditions. All known immobilizing agents are lethal to birds if they ingest a sufficient quantity. The most commonly used immobilizing agent worldwide is alpha-chloralose, which is also used in some countries as a lethal toxicant (see above). In the U.S., it is available as an immobilizing agent only to approved operators, mainly to capture rock pigeons and waterfowl in nuisance situations. In the U.K., seconal is also used as an immobilizing agent, in combination with alpha-chloralose, to enhance its speed of action.

Repellents

Chemical repellents can be primary or secondary in effect. Primary repellents are avoided reflexively because of an unpleasant sensation (e.g., touch, taste, smell, irritation). Tactile repellents include polybutene-based products (e.g., 4 The Birds', Hot Foot', and Tanglefoot' in the U.S., Bird-X, Buzz-Off', Shoo, Super Hunter, and Waco in Canada). They are applied to buildings and other structures, modifying the surface so that it becomes sticky or slippery and discouraging birds from landing or roosting. They are all available to the general public.

Taste repellents, which discourage birds from eating potential food sources to which they are applied, include denatonium saccharide (Ro-Pel' in the U.S. and Canada) and aluminium ammonium sulfate (Curb, Guardsman, and Rezist in the U.K., D-ter, Gaard, and Scat in Australia). Ro-Pel' also contains thymol, a fungicide that imparts a secondary repellent effect. Irritants include methyl anthranilate (ReJeX-iT' and Bird Shield' in the U.S., Avigon in Canada), capsaicin (Sevana), and naphthalene (Dr. T's), although there is no evidence that the latter two, by themselves, are effective.[3] Methyl anthranilate may be applied to grassy areas such as parks and golf courses to deter feeding by birds such as Canada geese, and also to ripening fruit to deter birds such as house sparrows and starlings.

Secondary repellents cause post-ingestional illness, resulting in conditioned aversion to the treated food source. Examples include methiocarb (Mesurol'), ziram (AAprotect), and anthraquinone (Flight Control™ in the U.S., Avex™ in New Zealand). Methiocarb and ziram are moderately toxic to birds and mammals. In some countries, methiocarb may be applied to seeds and seedlings, but in the U.S. it may be used only in dummy egg baits to condition crows (*Corvus* spp.) not to prey on the eggs of endangered birds. Ziram and anthraquinone may be sprayed onto grass, field crops, ornamentals, conifers, and dormant fruit trees, but not onto products for immediate human consumption.

4-Aminopyridine is sometimes described as a frightening agent, and classified as a repellent, because it induces behavioral changes in birds. However, it is highly toxic to birds, and should be considered as a toxicant (see above).

Reproductive Inhibitors

Reproductive inhibitors have the potential to reduce bird populations by preventing or reducing the production of young. Azocosterol (Ornitrol') is one of a number of chemicals that have been investigated for this purpose. It is applied to baits and fed to females daily for 10 to 15 days before egg-laying. It is no longer available in the U.S., but is still available for the control of rock pigeons in Canada. Corn oil (in the U.S.) and paraffin oil (in the U.K.) are two chemicals used to destroy the eggs of birds, such as gulls (*Larus* spp.) and Canada geese *(Branta canadensis)*, after they have been laid. The oil may be sprayed onto the eggs in the nest, or the eggs may be temporarily removed, immersed in oil, and then returned to the nest. The oil occludes the pores in the shell, asphyxiating the developing embryo. The technique is considered humane.[2]

Future Developments

No existing products are ideal for the control of pest birds. Toxicants are becoming increasingly publicly unacceptable worldwide from environmental and animal welfare perspectives. Currently, research is being done on the effectiveness of an oral toxicant/anaesthetic combination that reduces the time to unconsciousness, as a means of improving the animal welfare aspects of lethal bird control. Research is also being done on potential new repellents, including other derivatives of anthranilate, acetophenone, benzoate, cinnamamide, and d-pulegone. The use of nonlethal methods of bird control, especially repellents, may be a better option for the future than the use of toxicants.

References

1. Schafer, E.W., Jr. Bird Control Chemicals—Nature, Modes of Action, and Toxicity. In *CRC Handbook of Pest Management in Agriculture*, 2nd Ed.; Pimentel, D., Ed.; CRC Press: Boca Raton, Florida, USA, 1991; Vol. 2, 599–610.
2. Ministry of Agriculture, Fisheries and Food. *Assessment of Humaneness of Vertebrate Control Agents*; Ministry of Agriculture, Fisheries and Food, Pesticides Safety Directorate: York, United Kingdom, 1997.
3. Clark, L. Review of Bird Repellents. In *Proceedings of the 18th Vertebrate Pest Conference*; Baker, R.O., Crabb, A.C., Eds.; University of California: Davis, California, USA, 1998; 330–337.

5

Birds: Pesticide Use Impacts

Pierre Mineau

Birds inhabiting our farmland are in decline. We know this to be the result of agricultural intensification in which pesticide use plays a large direct and indirect role. Most pesticides used in developed countries no longer accumulate in birds, but they can poison birds and make them more susceptible to other causes of mortality.

More than 30 pesticides registered in North America or Europe have been known to result in kills of wild terrestrial vertebrates even when used according to the relatively stringent regulations in force in those countries. Among the species affected, birds figure prominently in the kill record and this, for several reasons. Birds are ubiquitous and visible. In North America, as in several other countries, most species are federally protected from unlicensed taking or kill. Birds are extremely mobile and cannot be excluded from areas that have been treated with pesticides. Some bird species are attracted to agricultural fields, and many are economically important to the control of agricultural pests, notably insects. Finally, birds, as a group, are particularly sensitive to some of the more toxic classes of pesticides such as the organophosphorus and carbamate insecticides (fortunately, the use of these compounds is in decline), and their reproduction has been found to be potentially affected by a wide range of pesticides. New pesticides developed in part for their relative safety to humans have been found to be especially toxic to birds.

In addition, pesticides are known to alter birds' basic requirements of food and shelter. Loss of food, especially, has been linked to population declines of several farmland bird species in Europe where this has been studied extensively.

Several different strategies are employed to study bird impacts, ranging from monitoring of pesticide applications to surveys of birds in farms subject to different pesticide regimes. Modeling has now given us an estimate of the yearly losses to acutely toxic pesticides; the full impact of pesticide use on birds is more difficult to assess and remains controversial.

How Serious Is the Impact?

Most of our farmland bird species appear to be declining globally and even common species are experiencing longterm declines, both in North America and in Europe.[1,2] For example, 76% of common grassland species in Canada are declining. The proportion of species declining or showing range contractions in the United Kingdom is higher still. Less is known about common farmland bird species outside of Europe or North America, but farming worldwide has been implicated in declines of specific groups of birds such as raptors. It is difficult to isolate the specific factors responsible for these declines: it is likely that a combination of factors is to blame and each species must be considered on a case-by-case basis. Agricultural landscapes have changed dramatically in the 20th century. There has been a shift from mixed agriculture including row crops, field crops, and livestock to more specialized farming where monocultures are regionally dominant. Field size has increased to accommodate larger machinery, and this increase has been often at the expense of marginal non-crop habitats such as fencelines, ditches, hedgerows, windbreaks, and remnant woodlots. In Europe especially, a shift to autumn sowing of grain crops has meant that much of the waste grain traditionally available to birds postharvest and throughout the lean winter months is no longer available.

Agricultural inputs in the form of synthetic fertilizers and pesticides have increased dramatically also and, increasingly, have been found to be contributing to bird declines. Two decades ago, a long-term study of declining grey partridge (*Perdix perdix*) populations in the United Kingdom identified insect prey reductions resulting from both insecticide and herbicide use as the main contributing factor. More recently, researchers in several regions of North America and Europe have shown that organic farms tend to support a higher diversity and abundance of birds even when matched for habitat characteristics. The reproductive success of some farmland species such as the Eurasian skylark (*Alauda arvensis*) is higher on organic or reduced input farms than it is on more "conventional" ones. The use of toxic granular insecticides for oilseed production has contributed to grassland bird population declines in the Canadian prairies.[3] It has been recently suggested[4] that broad aquatic contamination by the new neonicotinoid class of seed-treatment insecticides could be the main reason behind the decline of insectivorous bird species; this is expected to be a hot research area and point of debate in coming years. Taken as a whole, these results implicate current agricultural practices, and pesticide use in particular, in the decline of several farmland species. This is doubly unfortunate because, with a few exceptions, birds can play a useful role in integrated pest management systems.[5]

Types of Bird Impacts

There are several mechanisms through which pesticides can affect birds. The case of the grey partridge, Eurasian skylark, and other European species has shown that the effect can be an indirect one, mediated through "weed" removal and loss of insect biomass at critical times of the breeding season.[6] Herbicide use has increased dramatically in the past decades, and herbicide sales far surpass insecticide sales in North America and Europe at least. However, several direct mechanisms through which birds are impacted are also recognized.

Persistent Organochlorine Pesticides

Historically, several species of raptors such as the Eurasian sparrowhawk (*Accipiter nisus*) and peregrine falcon (*Falco peregrinus*) as well as fish-eating species such as the brown pelican (*Pelecanus occidentalis*) faced serious difficulties and regional extinction as a result of persistent organochlo\1–\2rine pesticides such as DDT (dichlorodiphenyltrichloroethane), aldrin/dieldrin, chlordane, and heptachlor. These were poorly metabolized and poorly excreted by birds and accumulated in fatty tissue. The impact of such substances was twofold. Some, such as aldrin and dieldrin, caused frequent poisonings, especially during lean times when birds metabolized their fat reserves and the pesticides reached extreme concentrations in the brain. Others, such as DDE (dichlorodiphenyldichloroethylene), a breakdown product of DDT, interfered with the bird's ability to lay eggs with normal shells. These substances were banned or severely restricted in most of the developed world in the early 70s. Yet, lower reproduction of birds breeding in areas with high historical usage is still being documented because of long persistence in soils. Several of these pesticides continue to be used massively in parts of the world such as the Indian subcontinent although current impacts on bird life are poorly documented. By and large, modern pesticides do not show such extreme persistence, at least in warmblooded organisms.

Lethal Effects

The acute oral toxicity (LD_{50}) of a pesticide and the extent of its use are good predictors of wildlife kills.[7] The dietary toxicity test currently carried out on young birds (dietary LC_{50}) can seriously mislead however. As a rule, insecticides and vertebrate control agents are much more likely than herbicides or fungicides to give rise to wildlife kills. Two groups of pesticides, the organophosphorus and carbamate insecticides, were initially introduced to replace persistent organochlorines. Unfortunately, they proved particularly toxic to birds. Their mode of action (inhibition of the enzyme acetylcholinesterase in the nervous system and at neuromuscular junctions) is not specific to the pests and affects a broad range of vertebrates and invertebrates alike. Birds are especially vulnerable because their ability to detoxify these pesticides is generally much lower than that of mammals. The more toxic products such as the carbamate insecticide carbofuran killed thousands of individuals in a single application. Reports of large numbers of North American birds being poisoned on their wintering grounds in Latin America by the insecticide monocrotophos have emphasized the need to consider bird impacts in a hemispheric, if not global, context.[8] The poisoning of birds is largely inevitable where acutely toxic pesticides are registered at high rates of application and used broadly. Of particular concern have been granular insecticides and seed treatments because birds are often attracted to them.[9] Pesticide poisonings can be a significant source of mortality relative to other factors, especially in the case of long-lived species such as birds of prey.[10]

There have been very few attempts to estimate the total incidental take resulting from direct intoxications following the use of toxic pesticides anywhere. Pimentel,[11] in an oft-cited study, estimated that pesticide-induced direct mortality totaled approximately 67 million per year in the United States. He based this estimate on the fact that 160 million ha of cropland received a very heavy dose of pesticides per year (3 kg a.i./ha on average—including a number of very toxic pesticides), a breeding density of 4.2 birds/ha (from census plot data), and a conservative kill estimate of 10% of exposed birds. This estimate ignores kills of wintering birds, which could be substantial. Also, some of the largest kills recorded in North America have been of migrants (e.g., Lapland long- spurs (*Calcarius lapponicus*)), which would not be captured in estimates based on breeding densities in farmland.

The carbamate insecticide carbofuran (Furadan™) was very broadly used in North America and has been studied more than any other insecticide.[12] The manufacturer's own studies on a granular formulation of carbofuran as well as search efficiency and scavenging studies were used to provide an estimate of bird mortality per treated surface.[13] Two major field studies, both from the United States, were retained for purposes of extrapolation. Estimated kill rates were 3.05 birds/ha for an Iowa site (once raw carcass counts were corrected for scavenging and for unsearched areas of the field) and 15.9 birds/ha for an Illinois site with better off-field habitat nearby. A third study gave estimates that were simply too high to lead to a kill rate that could safely be extrapolated; fully 799 carcasses of a single species (horned lark— *Eremophila alpestris*) were recovered from slightly more than 100 ha of crop. Based on the two lesser kill rates, it was estimated that, at the height of its popularity, in the late 70s to mid-80s, this single pesticide was killing approximately 17 to 91 million songbirds annually in the 32 million ha of U.S. corn (maize) fields alone.[13]

Fortunately, several of the more toxic organophosphorous and carbamate insecticides are being phased out and replaced in North America and Europe—although their use may still be increasing in the developing world. Their cancellation was not out of a concern for birds[14] but rather an attempt to reduce risks to consumers and applicators under new legislation that demanded the assessment of cumulative impacts from pesticides with the same mode of action. The result of these product cancellations, however, has been a definite reduction in the proportion of our crop area where birds are at risk of lethal poisoning.[15] The authors estimated that the cumulative number of cropped hectares over which avian mortality was probable decreased from about 17 million ha in 1997 to about 6 million ha 5 years later.

The measurement of brain cholinesterase levels was an extremely useful (although certainly not fool-proof) diagnostic tool for bird mortality from these classes of pesticides. The test has the advantage of being economical and relatively easy to carry out. Wildlife kills resulting from newly developed insecticides will be harder to elucidate in the absence of such a convenient biomarker. Diagnosis will hinge on sophisticated and costly residue analyses without the benefit of the "smoking gun," which cholinesterase titers represented.

Secondary Poisoning

Secondary poisoning occurs when predators, such as hawks or owls, consume prey contaminated by pesticides. Such predators are few because of their position at the top of the food chain. Therefore, the death of one predator may constitute a significant reduction in the local population of that species. Historically, researchers have associated secondary poisoning with persistent organochlorine insecticides and other substances that are not readily metabolized and that accumulate in tissues. However, other currently registered pesticides can cause secondary poisoning when the predator encounters the pesticide in a high concentration on the surface or in the gastrointestinal tract of its prey. Also, predators capture birds debilitated by insecticides much more easily.

Sublethal Effects and Delayed Mortality

Many pesticides can affect the normal functioning of exposed individuals at doses insufficient to kill them directly. At high doses, the organophosphorus and carbamate pesticides previously described cause respiratory failure and death. However, wild birds exposed to these agents in lesser amounts have experienced impaired coordination, weight loss, an inability to maintain body temperature, and loss of appetite. Also, exposed birds may spend less time at the nest, provide less food for their young, be less able to escape predation, and be more aggressive with their mates. Finally, exposure to some pesticides may reduce resistance to disease.

Effects on Reproduction

A high proportion of pesticides currently registered have the potential to affect reproduction by reducing egg production, hatching, or fledging success, although the extent to which this actually happens in the wild is not known.[16,17] A few products cause embryonic mortality when sprayed directly onto eggs.

Routes of Pesticide Exposure in Birds

Birds ingest pesticides through their food or through preening or grooming. Despite being feathered, they absorb pesticides through their skin, encountering droplets directly or by rubbing against foliage and other contaminated surfaces. Birds are also exposed through their feet. Finally, they have a very high ventilation rate and inhale vapor and fine droplets. The degree to which each of these routes of exposure contributes to the total dose depends on the crop being sprayed, the chemical, the species exposed, and environmental factors. Evidence to date suggests that the dietary route is not necessarily the dominant route of exposure in birds under most situations. Yet, this is the only route currently assessed by regulators worldwide—a mind-set that clearly needs to change if we want to be proactive in protecting birds from pesticides.

Although the relative importance of different exposure routes is difficult to ascertain on a case-by-case basis, it is possible to recognize different situations that arise where birds are massively exposed to pesticides and are often poisoned as a result.

Abuse and Misuse

Pesticide abuse is the deliberate use of a pesticide in a non-authorized fashion, usually to poison wildlife species considered to be pests. In the United Kingdom, as well as in several European countries, officials estimate that deliberate bird kills due to pesticide abuse outnumber cases where label instructions were strictly followed. Between 1978 and 1986, officials in the United Kingdom estimate that, on average, 71% of incidents were the result of abuse. For birds of prey alone, more than 90% of cases recorded between 1985 and 1994 in the United Kingdom were abuse cases. On the other hand, for raptors in the United States during the same period, kills involving labeled uses of pesticides were almost as frequent as abuse cases. This difference appears to be wholly attributable to the high toxicity of insecticides used in the United States. Abuse generally involves baits of some kind, the only limit being the imagination of the perpetrator. Typically, liquid insecticides are poured or injected and applied to seed, bread, meat, etc., and granules are sprinkled or mixed into a paste. Because of the high concentration of pesticide involved in abuse cases, carcasses are usually found in close proximity to the site of baiting, thus biasing the kill record through a higher recovery of carcasses. The choice of chemicals used in abuse cases reflects availability and toxicity. Pesticides typically used in deliberate poisoning attempts include carbofuran, aldicarb, monocrotophos, parathion, mevinphos, diazinon, and fenthion, chemicals that are all recognized as being inherently very toxic to vertebrates in general and birds in particular. The main problem of course is that the baits are often indiscriminate in the species that they kill. Secondary poisoning is also frequent when predators or scavengers take dead or debilitated prey with highly concentrated bait in their gut.

Pesticide misuse refers to a pesticide application that is not exactly as specified by the label. This may be an application at a rate that is higher than specified or an application to a crop or pest other than those listed. Alternatively, the user may not have the legal permission to use a certain product even if he followed label directions to the letter. Pesticide misuse is difficult to establish, especially after the fact. In many cases, it becomes very difficult to distinguish a misuse from a normal agronomic use when the label contains instructions that are vague, difficult, or impossible to follow. What constitutes a misuse in one jurisdiction may indeed be an approved use elsewhere.

Granular Formulations and Treated Seed

Granular insecticides and treated seeds are frequent routes of exposure and intoxication in birds. Granular insecticides were designed for convenience, safety to applicators, and time release of the chemical, yet for birds, granular formulations of the more toxic insecticides such as aldicarb, parathion, carbofuran, fensulfothion, phorate, terbufos, fonofos, disulfoton, diazinon, and bendiocarb are repeatedly associated with bird mortality. Several bird and small mammal species have a fatal attraction to granular formulations, mistaking them either as dietary grit or as a food source. The most attractive granules are those made of sand (silica) or an organic base such as dried corn (maize) cobs. Somewhat less attractive are clay, gypsum, and coal granules. Exposure can also occur via invertebrates, especially earthworms to which granules easily adhere. Secondary toxicity is likely in predators and scavengers that eat their prey whole or ingest their gastrointestinal tract contents. In Canada and the United States, there have been cases of poisoning of waterfowl foraging in puddles in fields more than 6 months after applications of granular insecticides because of specific soil conditions that may retard break- down.[18] Granules that are friable and disintegrate quickly when exposed to moisture are best for birds, but they are the products least convenient to farmers. Regardless of the type of carrier, a pesticide granule is likely to be a problem if a lethal dose can be obtained in a few granules only.

To date, *no* agricultural machinery or application technique can achieve complete incorporation of granular insecticides below the soil horizon. Birds have also been known to probe the soil for granules or to pull up germinating seeds with granules attached. The worst applications are those made above the soil surface and "banded" or "side dressed" over or to the side of the seed furrow. In carefully controlled engineering trials, between 6% and 40% of applied granules were left on the soil surface. The same equipment can achieve radically different soil incorporation when used by different individuals under different conditions.

Treated seeds present a similar engineering problem. As with granules, more seeds are left on the surface wherever the seeders have to turn or negotiate obstacles. Small spills are part of normal farm-ing practice and can occur anywhere depending on topography and soil conditions but more often at field edge. Historically, seed dressings were one of the main sources of bird exposure to organochlorine and mercurial compounds. Poisoning incidents with seed dressings are still relatively frequent because several bird species make heavy use of waste (or even planted) grain in fields. The size and type of seed dictate which bird species are at risk. Since use of organochlorines and organomercurials has declined, kills have been recorded with cholinesterase-inhibiting insecticides such as carbophenothion, chlorfen-vinphos, isofenphos, bendiocarb, disulfoton, furathiocarb, and fonofos. Some kills have been recorded with newer insecticides as well, e.g., the neonicotinoid imidacloprid, although it is not yet known how serious or frequent a problem this will become.

Liquid Formulations on Vegetation: The Grazing Problem

Grazing birds are particularly vulnerable to foliar applications of pesticides. Kills have been recorded with several cholinesterase-inhibiting pesticides such as parathion, diazinon, carbofuran, isofenphos, dimethoate, and triazophos. Grazers typically include geese, ducks, and coots (families Anatidae and Rallidae). These birds eat large quantities of foliage because they do not digest cellulose. Fertilized areas are particularly attractive to grazing species that can detect the high nitrogen levels. Golf courses attract grazers because the turf is cut frequently, watered, and fertilized, and courses often have other attractions such as ponds and drainage streams. More than 100 cases of waterfowl mortality were recorded due to the use of diazinon on turf [19] before the pesticide was withdrawn from golf courses and sod farms in the United States. Other well-documented problems are kills of ducks and geese in alfalfa fields treated with carbofuran and of sage grouse (*Centrocercus urophasianus*) feeding on alfalfa crops treated with dimethoate or on potato foliage and weeds in potato fields sprayed with methamidophos.

Liquid Formulations on Insect Prey: The Gorging Problem

Bird species that feed on agricultural pests such as grasshoppers, leatherjackets (larvae of the crane fly), grubs, and cutworms are at high risk of poisoning. Kills of these species are all the more tragic because they are beneficial to agriculture. Some species are particularly vulnerable because they specialize in insect outbreaks. These birds take advantage of pest control operations that result in insects becoming either debilitated or more visible following treatment. In a well-studied case in Argentina, approximately 20,000 Swainson's hawks were poisoned within the span of a few weeks after feeding on grasshoppers sprayed with monocrotophos. As with carbofuran, the extreme toxicity of this product means that it is difficult to find use patterns that do not result in bird kills.

Vertebrate Control Agents: Unintended Victims

Rodenticides as a rule are not specific to their intended targets and cause direct impacts to non-target species. Only a detailed knowledge of the habits of the target species and use of specific baiting locations or specialized bait holders can reduce kills of non-target species. More problematic is secondary poisoning. Historically, the use of thallium and endrin to control rodents has had disastrous consequences on raptors. Recently, the trend has been to use more efficacious "single feed" anticoagulants; these present a greater hazard to predators than the older products (e.g., warfarin, diphacinone, chlorophacinone). The new "super coumarin"-type products include compounds such as difenacoum, brodifacoum, bromadiolone, difethialone, and flocoumafen—all extremely toxic and very long lived in liver tissue, thereby increasing the likelihood of secondary poisoning. A recent analysis of Canadian data suggests that approximately 11% of the great horned owl (*Bubo virginianus*) population of southern inhabited Canada is at risk of fatal poisoning from anticoagulants.[20]

Fenthion, an organophosphorus "insecticide" used to control pest birds in Africa (e.g., *Quelea quelea*) and in North America (e.g., by means of the Rid-a-Bird™ perch system), has given rise to frequent secondary poisoning.[21] Secondary poisoning is also very likely following the use of toxic organophosphorous or carbamate products for the control of parakeets, doves, and other seed eaters. The use of organophosphorus pesticides such as famphur and fenthion for the treatment of parasites in livestock frequently leads to wildlife kills. Famphur, which was one of the leading causes of eagle poisonings in the American Southwest, persists on the hair of cattle up to 100 days after treatment. Magpies are poisoned when they eat the hair, and eagles are poisoned when they scavenge the magpies. Medicated feed at livestock feed yards is another high-exposure situation. Sparrows, starlings, and other birds pick up the feed and subsequently are scavenged by hawks and eagles.

Forestry Insecticides

Forestry uses of toxic insecticides deserve special consideration because the terrain and method of application result in kills being difficult or impossible to detect. In a forestry situation, critical wildlife habitat is sprayed directly, and a large number of individuals of many species are exposed to the chemical. In Canada, the forestry insecticides phos-phamidon and fenitrothion were canceled after impacts on birds were judged unacceptable. Although fenitrothion is not as acutely toxic as a number of other anti-cholinesterase insecticides used in agriculture, its use in forestry led to severe and widespread inhibition of brain acetylcholinesterase in a number of songbird species as well as some reports of kills.

Measuring Bird Impacts

Incident Monitoring

Incident monitoring refers to the capacity of competent authorities to investigate reported kills or conduct spot checks of use conditions. Even if a pesticide has been studied extensively under controlled conditions, unforeseen problems and situations often arise following commercialization of the product. An absence of incident reports does not necessarily mean there are no problems but, conversely, well-investigated incidents and kills can reveal unforeseen aspects of a pesticide or reinforce a suspicion that arose in the course of laboratory or field testing. An incident monitoring scheme will require a network of individuals trained in carrying out pesticide investigations and in proper handling of carcasses and tissue samples, as well as access to a laboratory equipped to perform the required chemical and biochemical analyses.[22]

Even where relatively efficient incident monitoring systems are in place, only a very small proportion of kills are ever uncovered. There are several reasons for this: affected wildlife are often dispersed and at relatively low density in farm fields, they often leave the treated area to die, they are likely to seek cover and hide when overcome by the pesticide, they are often cryptic and hard to see, and their carcasses are scavenged rapidly after death. Typical rates of carcass removal by scavengers are 40%-90% in the first 24 hr. Farm fields are large; the mechanization and sheer size of modern agricultural machinery often remove the farmer from any "close contact" with the land. The increasing size of farms also means that, when kills occur, often in the few days that follow a pesticide application, the farmer is busy elsewhere, treating another part of the farm. Pesticide intoxication may be a causal factor in a kill visibly caused by something else—e.g., intoxicated birds hitting fences, utility wires, cars, or buildings—and not be recognized as a pesticide kill. Also, there is a large difference between casual searching of fields and a well-organized intensive search effort. An intensive search effort consisting of several trained individuals, transects, and repeated, well-timed searches have produced between 10- and 500-fold improvements in carcass detection rates over field inspections carried out once or a few times only by single individuals. Equally important is the motivation and training of the search teams. Finally, a proper investigation of kills can be expensive and out of the reach of many jurisdictions despite the availability of inexpensive biomarkers—currently at least.[23]

Even when incidents are uncovered, they are often not reported. If the kills involve only one or a few individuals, not much importance is attached to the incident even though, for reasons just outlined, these few carcasses likely represent the "tip of the iceberg." Even if the kills are reported, the information is often not centralized and made available to national pesticide regulatory bodies. It is important to understand and recognize biases inherent in any incident reporting system. Some of those biases will depend on how the incident monitoring system is set up and which persons/organizations are responsible. Some biases can be reduced over time, but others are unavoidable. Common biases relate to body size and color of the casualties, numbers and density of the species in any given area, "status" of the species, and individual and institutional interests and sensitivities. We expect most kills to be of small-bodied birds widely dispersed in field margins. Yet, such kills are seldom reported.

Despite these limitations, it is important for countries to investigate wildlife kills and make the information available.[24] Registration decisions are made on the basis of very limited information. There are large differences in toxicological and ecological vulnerability among species. The ways in which wildlife species are exposed to pesticides are varied and sometimes difficult to predict or study. The behavior of pesticides depends on local conditions although pesticides are often tested under standardized conditions only. The outcome of exposure is also much more variable in the wild. Pesticide exposure can interact with weather, the condition or health of the animal, etc. Therefore, whether or not pesticides are routinely field tested to look for environmental impacts, it is essential to have a good incident monitoring system in place. An incident monitoring system can also be useful to warn manufacturers if their products are abused or used incorrectly.

Field Testing: Active Monitoring

Carrying out a field study to measure the impact (or lack of impact) of a specific pesticide usually consists of the surveillance of a group of birds prior to, during, and after the application of the pesticide according to label instructions. Researchers observe or count individuals of one or more species within and outside the treated area and record their behavior. Frequently, they search for carcasses in order to determine the extent of pesticide-induced mortality.[25] for an example of how these data can yield useful predictive models.) They may capture birds to ascertain the health of individuals or to collect samples, for example, blood or brain tissue for biochemical assays or feathers and foot rinses for residue determinations. Agricultural engineering studies (e.g., measurements of granular insecticides or treated seed remaining on the surface) or monitoring of pesticide residues remaining on avian food items over time provides valuable information on expected exposure levels. The most sophisticated field studies will involve monitoring nests, as well as banding, marking, or radio-tagging individuals in order to assess turnover rates and help locate sick and dead birds. Rare, vulnerable, or ecological keystone species can be used as indicator species where relevant and feasible.

It is not always feasible to investigate the effects of a single pesticide. In a number of cropping situations, several pesticides are used as a mix or in quick succession, making the identification of compound-specific impacts difficult. In agricultural systems, the mosaic of treated fields can be so complex as to make it difficult to assess exposure to any one pesticide. Two approaches then suggest themselves: 1) treated sites or landscapes are compared to non-treated areas, provided those can be found, and 2) the "severity" of treatment (the *a priori* expectation of toxicity) for any given site is used as a variable against which a number of different parameters (such as reproductive success) are regressed. Great care must be taken in comparing treated to non-treated areas because they are likely to differ in other ways as well.

Surveys

Data from regional or national surveys of bird population levels are rarely adequate to demonstrate specific pesticide impacts, although surveys can point to a general situation of bird declines in farmland. In order to carry out wildlife monitoring in treated areas, it is necessary to have a good knowledge of the normal complement of species for the area of concern and to be able to assess the vulnerability of each of these species during and after pesticide treatments. The diversity or abundance of species may already have been affected by previous pesticide use so that only a complement of the more insensitive species remains available for testing.

Regardless of the strategy employed, more attention needs to be paid to the impact of pest control practices on bird species if we are to reverse the current trends of population declines.

Modeling

The probability of finding a bird kill (of any size) following a pesticide application was derived from models based on a large sample of empirical field studies where known insecticides were applied, and searching was carried out to detect casualties.[25] Models were developed for field and orchard crops separately. Because few of the studies were quantitative in nature, logistic modeling was used and the output of the models is the likelihood that a kill of undefined size would occur and be found assuming an adequate search effort.

Species most frequently implicated in kills are those that are cosmopolitan, closely associated with agriculture, and reasonably common, e.g., mourning doves (*Zenaida macroura*); several sparrows, horned larks, and meadowlarks (*Sturnella* spp.), American robins (*Turdus migratorius*); house sparrows (*Passer domesticus*); and several blackbird species. However, the sheer diversity of birds potentially killed by pesticides is impressive and suggests that toxicological or ecological susceptibility is less important than being simply in the wrong place at the wrong time.

Conclusion: The Way Forward

The pesticide industry has clearly shown that it is incapable of policing itself when it comes to reducing or eliminating impacts on birds. Elsewhere,[14,26] I have reviewed some of the more egregious cases, showing how it often takes decades of legal wrangling to remove clear problem pesticides from the market in developed countries while the use of those same compounds continues to increase in the developing world, sometimes affecting the same bird populations (e.g., neotropical migrants). The extent to which birds are protected from pesticide impacts in various countries depends very much on public opinion and on the effectiveness of bird conservation groups. The pesticides responsible for most of the impacts on birds around the world (at least the direct impacts) tend to be the same group of depressingly familiar products. Fortunately, they usually can be replaced by better alternatives without risk to the livelihood of farmers or food security. Adopting better laws and regulations to protect the environment against pesticide use is part of the answer; enforcing those regulations in the face of a very strong pesticide lobby is undoubtedly the biggest hurdle.

References

1. Askins, R.A. Population trends in grassland, shrubland, and forest birds in eastern North America. Curr. Ornithol. **1993**, 11, 1–34.
2. Sirawardena, G.; Baillie, S.R.; Buckland, S.T.; Fewster, R.M.; Marchant, J.H.; Wilson, J.D. Trends in the abundance of farmland birds: A quantitative comparison of smoothed Common Birds Census indices. J. Appl. Ecol. **1998**, 35, 24–43.
3. Mineau, P.; Downes, C.M.; Kirk, D.A.; Bayne, E.; Csizy, M. Patterns of bird species abundance in relation to granular insecticide use in the Canadian prairies. Ecoscience **2005**, 12 (2), 267–278.
4. Tennekes, H. The Systemic Insecticides: A Disaster in the Making; Weevers Walburg Communicatie, Zutphen, Netherlands, 2010, 72 pp.
5. Kirk, D.A.; Evenden, M.D.; Mineau, P. Past and current attempts to evaluate the role of birds as predators of insect pests in temperate agriculture. Curr. Ornithol. **1997**, 13, 175–269.
6. Campbell, L.H.; Avery, M.I.; Donald, P.; Evans, A.D.; Green, R.E.; Wilson, J.D. A review of the indirect effect of pesticides on birds. Joint Nature Conservation Committee Report No. 227. JNCC: Peterborough, England, U.K., 1997.
7. Mineau, P.; Baril, A.; Collins, B.T.; Duffe, J.; Joerman, G.; Luttik, R. Reference values for comparing the acute toxicity of pesticides to birds. Rev. Environ. Contam. Toxicol. **2001**, 170, 13–74.
8. Hooper, M.J.; Mineau, P.; Zaccagnini, M.E.; Winegrad, G.W.; Woodbridge, B. Monocrotophos and the Swainson's hawk. Pestic. Outlook **1999**, 10 (3), 97–102.
9. Stafford, T.R.; Best, L.B. Bird response to grit and pesticide granule characteristics: Implications for risk assessment and risk reduction. Environ. Toxicol. Chem. **1999**, 18 (4), 722–733.
10. Mineau, P.; Fletcher, M.R.; Glazer, L.C.; Thomas, N.J.; Brassard, C.; Wilson, L.K.; Elliott, J.E.; Lyon, L.A.; Henny, C.J.; Bollinger, T.; Porter, S.L. Poisoning of raptors with organophosphorous pesticides with emphasis on Canada, U.S. and U.K. J. Raptor Res. **1999**, 33, (1), 1–37.
11. Pimentel, D.; Acquay, H.; Biltonen, M.; Rice, P.; Silva, M.; Nelson, J.; Lipner, V.; Giordano, S.; Horowitz, A.; D'Amore, M. Environmental and economic costs of pesticide use. BioScience **1992**, 42, 750–760.
12. Richards, N., Ed. *Carbofuran and Wildlife Poisoning: Global Perspectives and Forensic Approaches*; Wiley-Blackwell, Chichester, U.K. 2012, 277 pp.
13. Mineau, P. Direct losses of birds to pesticides—Beginnings of a quantification. In *Bird Conservation Implementation and Integration in the Americas*, Proceedings of the Third International Partners in Flight Conference 2002; Ralph, C.J., Rich, T.D., Eds.; U.S.D.A. Forest Service, GTR-PSW-191, Albany, CA, 2005; Vol. 2, 1065–1070.
14. Mineau, P. Birds and pesticides: Are pesticide regulatory decisions consistent with the protection afforded migratory bird species under the Migratory Bird Treaty Act? William Mary Environ. Law Policy Rev. **2004**, 28 (2), 313–338.
15. Mineau, P.; Whiteside, M. The lethal risk to birds from insecticide use in the U.S.—A spatial and temporal analysis. Environ. Toxicol. Chem. **2006**, 25 (5), 1214–1222.
16. Mineau, P.; Boersma, D.C.; Collins, B. An analysis of avian reproduction studies submitted for pesticide registration. Ecotoxicol. Environ. Saf. **1994**, 29, 304–329.
17. Mineau, P. A review and analysis of study endpoints relevant to the assessment of "long term" pesticide toxicity in avian and mammalian wildlife. Ecotoxicology **2005**, 14 (8), 775–799.

18. Elliott, J.E.; Wilson, L.K.; Langelier, K.M.; Mineau, P.; Sinclair, P. Secondary poisoning of birds of prey by the organophosphorus insecticide, phorate. Ecotoxicology **1996**, *5*, 1–13.

19. Frank, R.; Mineau, P.; Braun, H.E.; Barker, I.K.; Kennedy, S.W.; Trudeau, S. Deaths of Canada geese following spraying of turf with diazinon. Bull. Environ. Contam. Toxicol. **1991**, *46*, 852–858.

20. Thomas, P.J.; Mineau, P.; Shore, R.F.; Champoux, L.; Martin, P.; Wilson, L.; Fitzgerald, G.; Elliott, J.E. Second generation anticoagulant rodenticides in predatory birds: Probabilistic characterisation of toxic liver concentrations and implications for predatory bird populations in Canada. Environ. Int. **2011**, *37* (5), 914–920.

21. Hunt, K.A.; Bird, D.M.; Mineau, P.; Shutt, L. Secondary poisoning hazard of fenthion to American Kestrels. Arch. Environ. Contam. Toxicol. **1991**, *21*, 84–90.

22. ASTM. *Standard Guide for Fish and Wildlife Incident Monitoring and Reporting*; American Society for Testing and Materials: Philadelphia, PA, Standard E 1997; 1849–1896.

23. Mineau, P.; Tucker, K.R. Improving detection of pesticide poisoning in birds. J. Wildl. Rehab. **2002**, Part 1: *25* (2), 4–13; Part 2: *25* (3), 4–12.

24. Greig-Smith, P.W. Understanding the impact of pesticides on wild birds by monitoring incidents of poisoning. In *Wildlife Toxicology and Population Modeling*, SETAC Special Publication Series; Kendall, R.J., Lacher, T.E., Eds.; CRC Press, Inc.: Boca Raton, FL, 1994; 301–319.

25. Mineau, P. Estimating the probability of bird mortality from pesticide sprays on the basis of the field study record. Environ. Toxicol. Chem. **2002**, *24* (7), 1497–1506.

26. Mineau, P. Birds and pesticides: Is the threat of a silent spring really behind us? 2009 Rachel Carson Memorial Lecture, Pesticide News **2009**, *86*, 12–18, available at http://www.pan-uk.org/pestnews/Free%20Articles/PN86/Birds%20and%20pesticides.pdf.

6

Insect Growth Regulators

Meir Paul Pener

Introduction

Insect growth regulators, abbreviated as IGRs, are chemical compounds that interfere with insect-specific physiological systems that do not exist in vertebrates. The body of insects and related arthropods is covered by a more or less hard, often very hard, exoskeleton, named "cuticle". Once produced, the cuticle is fixed and it cannot grow larger. To allow growth of the insect, the old cuticle is shed and a new, larger, often different cuticle is formed. This change of the old cuticle to a new one is named "molt" or "molting" and a stage between two consecutive molts is named an "instar". To reach the adult stage, insects molt several times, i.e. they have several instars and they undergo "metamorphosis", from egg to several larval or nymphal stages, then to adult ("incomplete metamorphosis"), or from egg to several larval stages then to pupa and finally to adult ("complete metamorphosis"). Both molting and metamorphosis are controlled by hormones. Through molecular processes, the effect of these hormones is upregulation or downregulation of the expression of many enzymes, some of which are specific to insects or arthropods. These physiological systems constitute the targets of the IGRs as they do not exist in vertebrate animals, including humans. Most IGRs used today in insect control are either interfering with the formation of the new cuticle, disturbing the molting process, or interfering with metamorphosis. The name "insect growth regulators" is somewhat unfortunate; IGRs rather "deregulate" insect development. An attempt made by Pener and Dhadialla[1] to use the name "insect growth disruptors" instead of "insect growth regulators" turned out to be unsuccessful. It seems that the name "insect growth regulators" is deeply embedded in the literature and is difficult to uproot.

An Overview of IGRs

There are thousands of scientific publications on IGRs. ISI Web of Knowledge lists over 2,300 articles on the subject that appeared in peer-reviewed scientific journals. This list does not include books, book chapters, conference proceedings volumes, encyclopedia entries, and journals not covered by ISI. For comparison, in my former chapter of 2013,[2] this number was 1400.

Reviews on IGRs are numerous but usually deal with one or two kinds (categories) of IGRs and not with all categories. Also, in most instances, reviews are either applied oriented with effect(s) on whole organism and/or tissues, or related to basic science, focused on the mode of action of IGRs at cellular, biochemical, and more recently at molecular levels. Additionally, IGR research often constitutes an approach or method to study the mode of action of insect hormones and/or enzymes, again mostly in relation to basic research.

A review of 2012[1] presents a detailed account of the applied aspects of all categories of IGRs. A slightly earlier review[3] is comprehensive, and it is a remarkable example of discussion of all categories of IGRs, from both the applied and basic research aspects. It lists the most important literature on IGRs. Another review devoted to all categories of IGRs was published in 2016.[4] Some other reviews also deal with all categories of IGRs but aim to a specific insect group or biotope; for example, IGRs against stored product insects[5] or storage mites.[6] Also, tens of different studies were published on the mode of action of IGRs on a single harmful species, exemplified by *Rhodnius prolixus*.[7]

This chapter is an updated and enlarged revision of my former chapter on IGRs[2] and is application oriented, as this aspect is more important from the standpoint of environmental management. The presented list of references is based mostly on reviews, literature from the last two decades, and on publications with historical importance. Basic research on IGRs is not detailed; however, reviews relevant to basic research are mentioned. These reviews, as well as the comprehensive review mentioned,[3] cite a detailed literature on relevant basic research.

As targeted against insect-specific physiological processes, IGRs exhibit no or mild toxicity to vertebrates. In some cases, lack of susceptibility of vertebrates is so marked that IGRs are administered by the "feed-through" method, i.e. applied orally to a vertebrate animal against insects that are harmful as adults which in the larval stages feed on the feces of that animal. For example, cyromazin is fed to poultry for controlling housefly (*Musca domestica*) maggots in the manure.[8,9] Methoprene is fed to cattle to control horn fly (*Haematobia irritans*) larvae that develop in cattle dung.[10,11] In a more recent example,[12,13] novaluron and pyriproxyfen were fed to hamsters and gerbils in the laboratory, to control sand fly (*Phlebotomus papatasi* and *Phlebotomus doboscqi*) larvae; in nature, these larvae live in burrows of rodents. Adult sand flies are vectors of leishmaniasis, an infection caused by protozoan parasites.

Although the feed-through method is very convenient, in most instances, IGRs are administered for pest control by contact or oral application. For social insects (for example, termites), IGRs mixed with food serve as a bait, which is carried by the workers to the colony. These are the practical methods of application, though it was reported that injection of an IGR may be up to thousand times more efficacious than its topical application.[14]

The effect of IGRs on insects is often delayed, becoming overt at the next molt or at one of the next stages of metamorphosis. Up to the effect, the pest inflicts further damage. Therefore, the ideal cases for usage of IGRs are insects that are harmless in the larval stages and only the adults are harmful (mosquitoes, horn flies, sand flies, etc.).

IGRs are often very selective. A compound may be lethal to an insect species but may have no effect on another. For example, buprofezin well affects sweet potato whitefly (*Bemisia tabaci*), but it has no effect on cotton bollworm (*Helicoverpa armigera*), whereas lufenuron affects both.[15] The selectivity is extremely important in practical pest control. Insect pests usually have specific or nonspecific insect predators and/or parasitoids ("parasitoid" means that only the larvae are parasites, the adults are not).

If some predators or parasitoids are more susceptible to an IGR than some pests, instead of eliminating pests, the natural enemies of some pests may be eliminated by using that IGR.[16] Larvae of Chrysopidae are beneficial insects, predators that feed on small insects, mostly on harmful aphids. *Chrysoperla externa* is highly susceptible to chitin synthesis inhibitors (CSIs) though not affected by a juvenile hormone analog (JHA).[17] There are also other beneficial insects; especially pollinators and their possible susceptibility must be taken into account. Pests may develop resistance and even cross-resistance to IGRs, and further application may be ineffective. Integrated pest management (IPM) tries to find the best solutions to these problems.[18,19]

It is important to note that usage of any IGR (like any insecticide or acaricide) needs an official permit. Such permits may be specific to a compound used against a specific pest or may be more general. Different countries and often different states within the same country may issue different permits. Therefore, the legal situation should be made clear before usage of an IGR. Thoms et al.[20] report a good example of bureaucratic troubles concerning the usage of an IGR as termite bait in Florida.

IGR Categories

IGRs are usually classified according to their mode of action.[1–4] Sometimes, however, terminology related to their chemical structure is also practiced. For example, most CSIs are benzoylphenyl urea derivatives, and this term is often used in the literature.[1–4,21] However, some other CSIs are not related to benzoylphenyl ureas. Juvenile hormone analogs (JHAs), also termed juvenoids or juvenile hormone (JH) mimics, constitute another category of IGRs. JHAs disturb metamorphosis,[1–4,22] resulting in nymph-adult, or larva-pupa, or pupa-adult, intermediate creatures that are not viable or at least unable to reproduce. Anti-JH agents also exist,[23,24] but these have not been developed to commercial insecticides.

Ecdysone agonists (EAs) constitute the relatively most recent category of IGRs[1,25]; they induce an untimely molt, which cannot be completed by the insect, eventually resulting in death.

There are also insecticides of botanical origin that interfere with insect-specific physiological processes. Azadirachtin[26,27] extracted from the seed kernels of the neem tree, *Azadirachta indica*, is an antifeedant and/or molt inhibitor, and it is often toxic to insects. Azadirachtin is not discussed here because it is a botanical insecticide, like natural pyrethrins or nicotine.

Naming IGRs

IGR names may confuse nonprofessionals. In early stages of development, a commercial firm uses a code, usually constructed from the abbreviated company name and a product number. Later, a name (sometimes called "common name" and after commercial production termed as "active ingredient") is given to the compound and its "chemical name", reflecting its chemical structure is revealed. The commercial product is a formulation that has a "trade name". Different formulations, often directed against different pests, but based on the same active ingredient, may have different trade names. For example, the JHA, pyriproxyfen, was developed by Sumitomo Chemical Company, Japan, under the code "S-31183". It received the common name "pyriproxyfen", and its chemical name is "2-[1-methyl-2-(4-phenoxyphenoxy)ethoxy] pyridine", or "4-phenoxyphenyl (*RS*)-2-(2-pyridyloxy) propyl ether", according to two different naming systems of chemicals. Its formulations were registered under the trade names "Sumilarv", "Admiral", and "Knack". Sometimes, production of an IGR is transferred to another commercial body that may come up with other trade names. For example, the firm Valent gave the trade names "Distance", "Esteem", and "Seize" to pyriproxyfen formulations. The label on a formulation should include the name and amount of the active ingredient. Details are found in *The Pesticide Manual*[28] under the name of this active ingredient (always use the latest editions).

For each IGR listed in this chapter, some (but not necessarily all) trade names are given in brackets.

Chitin Synthesis Inhibitors (CSIs)

As the name of these IGRs indicates, they inhibit chitin synthesis. Chitin is a major component of the insect body wall (cuticle), and it is present in the peritrophic membrane.[29] Chitin is a linear homopolymer of the aminosugar, N-acetyl-D-glucosamine. The molecules of the aminosugar are linked together, like a chain, in microfibrils, and these are arranged in layers (named "lamella"), constituting the chitin component of the cuticle. The biosynthesis of chitin from the aminosugar is catalyzed by the enzyme "chitin synthase".[30–32] CSIs inhibit the action of chitin synthase or interfere with insect chitin production by other ways, resulting in abnormal, malformed, cuticle. Consequently, the insect dies in or after the molt.[21,30,31] CSIs also have ovicidal effects; see the article by Boiteau and Noronha[33] for example. Insect embryos undergo molt within the eggs and/or at hatching, and chitin is an important component of the embryonic cuticle.

Vertebrates do not have chitin. However, chitin is present in many other living organisms; in some microbes, fungi, and in several animal phyla. There are many CSIs, but some of them are not effective as insecticides. CSIs and the spectrum of their activity are summarized by Merzendorfer.[34]

From the chemical aspect, two groups of CSIs may be considered affecting arthropods, benzoylphenyl ureas and non-benzoylphenyl ureas.

Benzoylphenyl Urea CSIs

There are several reviews on benzoylphenyl urea CSIs,[21,31] and some of them are more recent.[1,3,35,36] The latest review by Sun et al.[37] is very comprehensive. It discusses the history of discovering commercial benzoylphenyl ureas, starting with the diflubenzuron (under the trade name, Dimilin in 1975), structure–activity relationship, environmental fate, ecotoxicology, and action mechanism of CSIs. The review concludes [37] that "…the exact action site of benzoylurea insecticides has not been identified". By the way these CSIs do not inhibit chitin production in fungi.

A short description of the CSIs is presented below under the common name of each compound and in brackets some of the trade names and commercial formulations. This description was compiled from data in the above reviews, research articles on the subject in the last decades, and reliable Internet sources. References are usually not presented, except in some specific instances. Figure 1 shows the chemical structure of most benzoylphenyl urea CSIs.

Diflubenzuron [Dimilin, Micromite, Adept, and other trade names]. It is the oldest CSI with insecticidal activity, discovered in the early 1970s. It was also the first registered commercial CSI under the trade name Dimilin. It affects insects by topical application and/or ingestion. Diflubenzuron is used against a wide range of insects, especially against leaf-chewing species. A review is devoted to its use against locusts and grasshoppers.[38]

Bistrifluron [DBI-3204]. It is a relatively new CSI, which is used against pests from Lepidoptera (moths) and Hemiptera (whiteflies). Recently, it was tried successfully as a bait component against termites.[39]

Chlorbenzuron [Chlorbenzuron]. It is a Chinese product. Its common name is approved in China as a trade name and sold by several Chinese firms. It is not listed in *The Pesticide Manual*. Chlorbenzuron is used in China against insect pests, especially against pine caterpillars with little effect on parasitoids. Chlorbenzuron is essentially harmless to humans, but its degradation product, 2-chlorobenzamide, is suspected of being carcinogenic.[40]

Chlorfluazuron [Atabron, Ishipron, Aim, Helix, Jupiter]. It is a CSI developed in Japan. It is active through ingestion and used against chewing pests on cotton and vegetables, as well as against pests on fruits, potatoes, and tea. Interestingly, bait formulation with chlorfluazuron increased the soldiers/workers ratio in the subterranean termite, *Coptotermes curvignathus*, in Indonesia; colony elimination took about 6–8 weeks.[41]

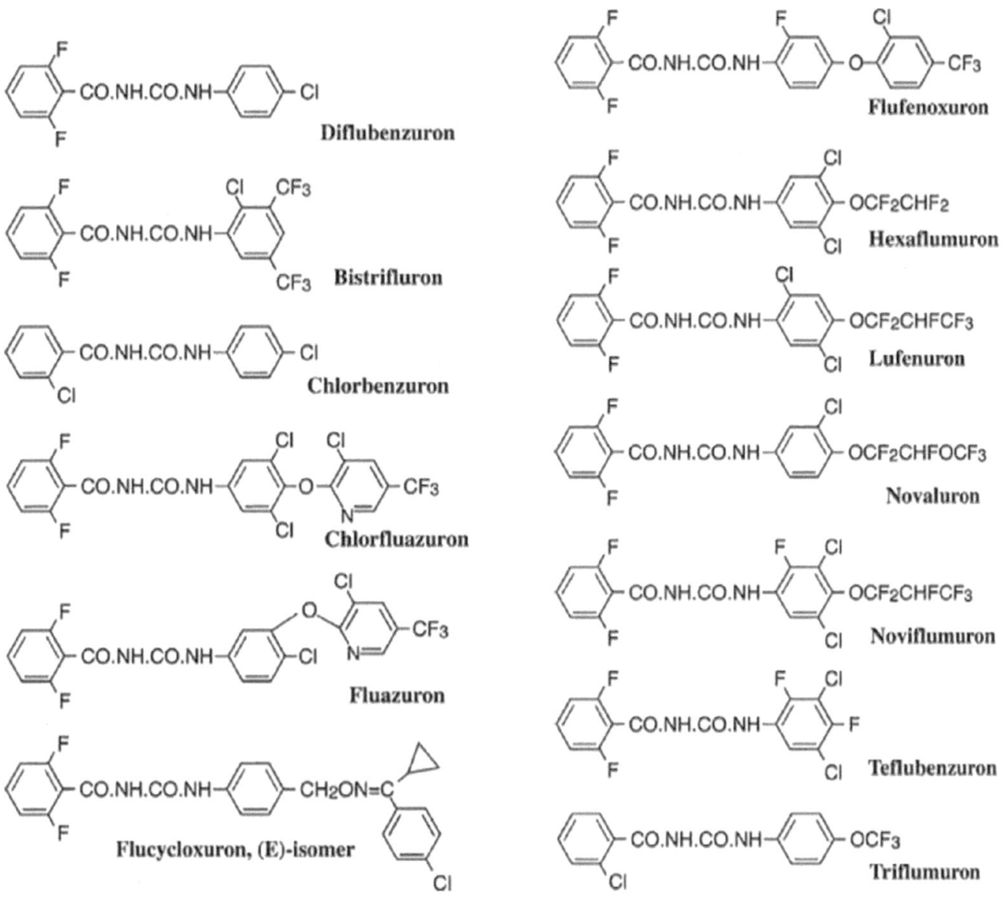

FIGURE 1 The chemical structure of benzoylphenyl urea CSIs.

Dichlorbenzuron [Dichlorbenzuron]. Although developed by Philips–Duphar, it is not listed in *The Pesticide Manual*. However, like chlorbenzuron (see above), it is approved and used as an insecticide in China.

Fluazuron [Acatak]. It is used mainly against ticks of the genus *Boophilus* and mites; therefore, it may be considered as an acaricide. Application of a pour-on formulation of fluazuron to cattle exerted no negative effects on the dung beetle, *Onthophagus gazella* in South Africa.[42] It is also used against ectoparasites, like fleas, on mammals.[43]

Flucycloxuron [Andalin]. This compound is both an acaricide and an insecticide. As an acaricide, it is effective against phytophagous mites. As an insecticide, it affects leaf-munching and leaf-rolling insects, mainly moth larvae on fruits, vegetables, cotton, and ornamentals.

Flufenoxuron [Cascade] (sometimes abbreviated as "flurox"). It is both an acaricide and an insecticide. The compound was found to be the most effective IGR against stored product mites; it also affects phytophagous mites. It controls many insect pests on fruit, tea, cotton, maize, and vegetables crops. In an artificial diet, it was lethal to the plant-sucking pea aphid, *Acyrthosiphon pisum*.[44]

Hexaflumuron [Consult, Recruit II, Trueno, Shatter, SentriTech as termite bait]. This compound is used mostly as an efficacious bait toxicant against subterranean termites. It is effective for control of insect larvae of Lepidoptera (including the Asiatic rice borer, also named striped stem borer, *Chilo suppressalis*), Coleoptera (including the Colorado potato beetle *Leptinotarsa decemlineata*), Homoptera, and Diptera.

Lufenuron [Match, Axor, Luroner, Luster, Manyi]. It is used mostly as a bait toxicant against subterranean termites, and in some instances, it was found more efficacious than hexaflumuron. Lufenuron is effective against larvae of moths (cotton bollworm, *H. armigera*, among others) and beetles on maize, vegetables, fruit, and cotton. It is effective against fleas and also acts as an acaricide.

Novaluron [Rimon, Diamond]. This compound acts by contact or ingestion. It is effective against whiteflies (*B. tabaci*), moth larvae (including species of the highly harmful genus, *Spodoptera*), and a wide range of insect pests on cotton, maize, certain fruits, citrus, potato, and vegetables. Its administration to adult female insects often reduces egg viability, up to complete ovicidal action. By feed-through method, it affects sand fly larvae in rodents' burrows.[12] Novaluron may also act as a mosquito larvicide.

Noviflumuron [Recruit III, Recruit IV, T-max as termite baits]. It is a relatively new (2004) CSI. It is active mostly as bait toxicant against several species of subterranean termites. Noviflumuron also affects cockroaches. In the German cockroach (*Blattella germanica*), it acts as a larvicide, and its ingestion as a bait by adults induces up to 100% ovicidal effect.

Teflubenzuron [Nomolt, Nemolt, Nobelroc, Dart, Diamond, Teflurate]. It is effective against whiteflies (*B. tabaci*), moth larvae (genera *Heliothis, Spodoptera*), and other pests on cotton. It also affects pests on fruits and potatoes (good effect on the Colorado potato beetle, *L. decemlineata*). An undesirable effect was observed on some sediment processing Polychaete worms in water.[45]

Triflumuron [Alsystin, Baycidal, Certero, Poseidon]. It affects pests, from various orders of insects, on fruits, soybean, cotton, and vegetables. It is used against ectoparasites on mammals, sheep lice, and others. Pour-on application was found highly effective against the biting louse, *Werneckiella equi*, of horses. It affected the blood-sucking bug, *R. prolixus* by oral, topical, or continuous contact (in coated Petri dishes) application. It affected larvae of mosquitoes.[46]

Search for novel benzoylphenyl urea derivatives. Research to discover novel benzoylphenyl urea-related CSIs has been maintained,[36,37] especially in China.[47–50] The usual procedure is to test the effect of the newly synthesized compound on certain insect(s) and compare the effect of one or more benzoylphenyl urea CSIs on the same insect(s).

Non-Benzoylphenyl Urea CSIs

These compounds chemically differ from benzoylphenyl urea CSIs. They inhibit chitin biosynthesis, not necessarily by inhibiting chitin synthase. Their exact mode of action is rather blurred, but the eventual effect is interference with chitin production and molting. This chapter shows the chemical structure of three compounds (Figure 2).

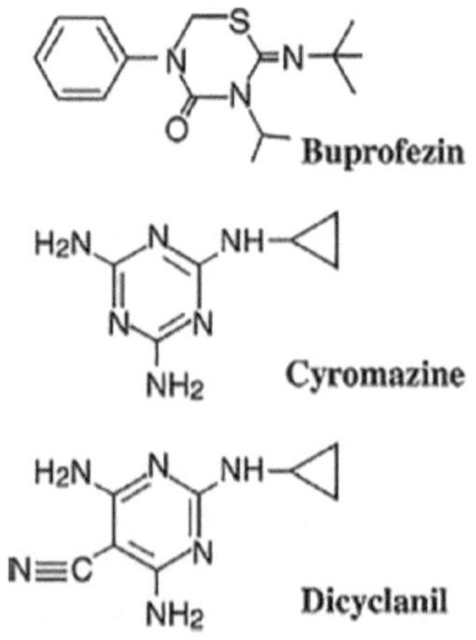

FIGURE 2 The chemical structure of non-benzoylphenyl urea CSIs.

Buprofezin [Applaud, Courier, Maestro, Mercao]. This compound is an insecticide that inhibits molt by contact or ingestion. It is used against homopteran insects, such as leafhoppers, plant hoppers, including the notorious rice plant hopper, *Nilaparvata lugens*, mealy bugs, jumping plant lice, and whiteflies, especially *B. tabaci*, which easily develop resistance.[51] Considerable resistance was found also in *N. lugens* in China.[52] It did not affect the ability of a predatory mite to suppress a pest mite in Californian vineyards.[53]

Cyromazine [Larvadex, Trigard, Neporex, Cliper, Ciromate, Vetrazin, Garland]. The most frequent use of cyromazine is against the housefly, *M. domestica*, in poultry farms. It is used by feed-through application or spray-on the manure.[8,9] It is also used to control leaf miners in vegetables, potatoes (including the Colorado potato beetle, *L. decemlineata*, and mushroom sciarid fly, *Lycoriella ingenua*).

Dicyclanil [Clik]. It is used mostly as a veterinary pharmaceutical agent against maggots (fly larvae) that infect traumatic myiasis in livestock. Dicyclanil is employed for prophylactic treatment against flesh flies, screwworm flies, Australian blowfly, and other flies, the maggot of which infect and/or live in wounds. Effective by pour-on-formulation induces hepatocarcinogenesis (liver cancer) in mice.

Etoxazole [Baroque, Sorado]. It was shown that etoxazole inhibits chitin biosynthesis.[54] Etoxazole is used as an acaricide; it affects eggs, larvae, and nymphs but not adults (which do not molt) of the spider mite species belonging to the genera *Tetranychus* and *Panonychus*. Unfortunately, resistance is reported in Japan, South Korea, and Australia.[55]

Based on genetic studies, Demaeght et al.[56] concluded that **hexythiazox** and **clofentezine**, well-known acaricides, like etoxazole, affect chitin synthase 1. If so, these may be considered as CSIs.

Juvenile Hormone and Its Analogs (JHAs)

The Roles of Juvenile Hormone in Insects

Insect metamorphosis is under endocrine control. The presence of JH in subadult stages (nymphs or larvae) prevents metamorphosis to adult. JH is secreted by a pair of endocrine glands, the corpora allata. In the last preadult instar, cessation of the activity of the corpora allata results in a temporary absence of JH, allowing completion of the metamorphosis. In the adult, the corpora allata resumes activity and the JH is involved in reproduction. There are several natural JHs in insects, the most common is JH III (Figure 3), but all have a similar sesquiterpenoid basic structure. No JH exists in vertebrates.

JH and its mode of action have been extensively reviewed[57–61] from the level of organism, including effects on cast differentiation in some social insects, to the level of molecular biology.

Juvenile Hormone Analogs

Juvenile hormone analogs (JHAs), also termed "JH mimics", or "juvenoids", are compounds that exert similar effects to that of the natural JH. Treatment of late nymphal or larval instars with JHAs prevents metamorphosis to adults. The chemical structure of JHAs may or may not resemble natural JH. There are thousands of chemically different compounds that exert JHA activity, many of them known for long time.[62] A JHA is often specific to a group of insects. Only some JHAs were developed to commercial products.

In the course of normal development, JH disappears only in the last larval instar and this instar constitutes the target of JHAs. However, treatment of earlier instars may also be effective as environmental degradation of JHAs takes some time and meanwhile the insect may reach the last larval instar. Also, metabolic inactivation of JHAs is usually slower than that of natural JH, and a JHA absorbed in an earlier instar is not necessarily inactivated before the insect reaches the last larval instar. Often sustained (slow) release formulations are used. JHAs are usually effective by contact; they penetrate the cuticle. JHAs prevent development of normal adults; larva- or pupa-adult intermediate creatures are obtained

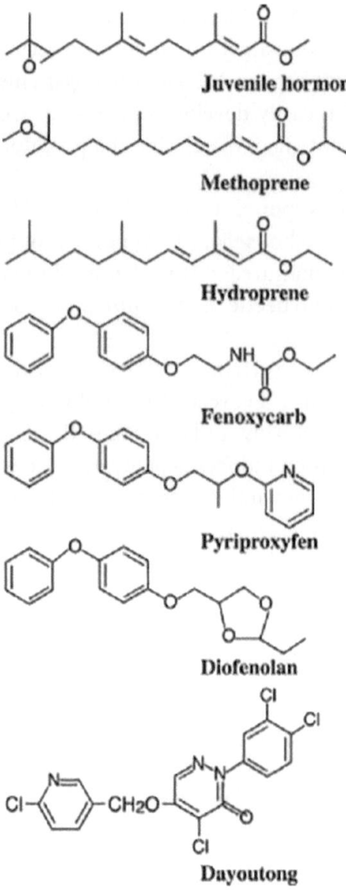

FIGURE 3 The chemical structure of the most common JH in insects and JHAs.

that die in the molt, or in the next (supernumerary) molt, or survive but are unable to reproduce. JHAs do not prevent damage caused by larvae and in some instances induce viable supernumerary larvae with prolonged life span and increased damage.[63]

JHAs, their effects, and their role in insect pest management were repeatedly reviewed.[1,3,22,25,62,64–70] These reviews also outline the roles of JH to explain the effects of JHAs. JHAs are also used for molecular basic research related to metamorphosis.[63,67–70]

Methoprene [Altosid, Precor, Apex, Dianex, Aquaprene, Kabat, Pharrorid, Fleatrol, Ovitrol, Extinguish, Diacon]. There are many formulations of methoprene, such as wettable powder, tossits, granules, and briquettes. Together with hydroprene, it was developed by Zoecon Company, constituting the first JHA approved for commercial usage. The first commercial products were racemic mixtures of the isomers of methoprene, but later only the 7S-isomer has been used. It is effective against mosquitoes, horn flies (feed-through method), sciarid flies (pests in mushroom cultures), fleas, some stored product insects, and fire ants, *Solenopsis invicta*. The compound has a low persistence in most field conditions; therefore, usually slow release technology is employed.[11] Methoprene was reviewed by Henrick.[11] Recently, control of stored product insects by methoprene was reviewed.[71]

Hydroprene [Altozar, Gencor, Gentrol, Raid]. It was approved at the same time together with methoprene for commercial usage. The 7S-isomer is used. It affects cockroaches, stored product insects, fruit flies, bed bugs, and some beetles. For stored product insects, pyriproxyfen (see below) seems to be more efficacious than hydroprene.[72]

Kinoprene [Enstar II]. The 7S-isomer is used. It is effective against homopterous pests. It was reported that kinoprene affects aphid polymorphism.[73]

Fenoxycarb [Insegar, Logic, Torus, Pictyl, Varikill, Comply]. It affects a very wide range of insects, disturbing metamorphosis and/or acting as an ovicide, directly applied to the eggs or parent females. Susceptible species were reviewed already in 1993.[74] It may be toxic to non-target species, possibly including honey bee brood.[75] Its chemistry and environmental fate were recently reviewed.[76]

Pyriproxyfen [Sumilarv, Admiral, Knack, Distance, Esteem, Seize, Tiger, Nylar, Intracure]. It is used against scale insects, whiteflies, aphids, pear psylla, thrips, cockroaches, codling moth, stored product insects, mosquitoes, and fire ants. The literature on its effect on the whitefly, *B. tabaci*, is especially rich, including resistance-related publications. The compound disturbs metamorphosis and/or interferes with reproduction. It may be toxic to some aquatic organisms. Its properties and environmental fate were reviewed.[77]

Diofenolan [Aware]. This JHA is less widely used than the above compounds. It was classified by some authors as a "molt inhibitor", but it is a JHA. Insects that are susceptible to diofenolan were listed by Streibert et al.[78]

Dayoutong. This compound has no trade name. It is a recently developed JHA approved under the name dayoutong in China.

Others. Triprene and Epofenonane are superseded JHAs.

Anti-Juvenile Hormone Agents

These are compounds that interfere with JH production,[23,24] but none of them have been developed to a commercial insecticide. Only the case of precocenes is mentioned here, because of their specific mode of action. The anti-JH effects of precocenes 1 and 2 (Figure 4) were discovered by Bowers and coworkers[79]; they extracted it from the plant *Ageratum houstonianum*. Many chemical derivatives, including the more active 7-ethoxy precocene (Figure 4), also called precocene 3, were synthesized and tested in several laboratories. Precocenes do not affect JH but selectively causes atrophy of the corpora allata,[80] the endocrine glands that secrete the JH. Susceptible species are limited, mostly hemipteran and orthopteran insects. Applied topically to early preadult nymphal instars, miniature nymph-adult intermediate creatures are obtained. Application of precocene to young adults also leads to the absence of JH that interferes with reproduction, especially with oocyte development. Precocenes are not mutagenic, but they are hepatotoxic and nephrotoxic to vertebrates and suspected to be carcinogenic.

Interference with JH biosynthesis, or enzymatic JH regulation, is considered as possible anti-JH procedures.[66]

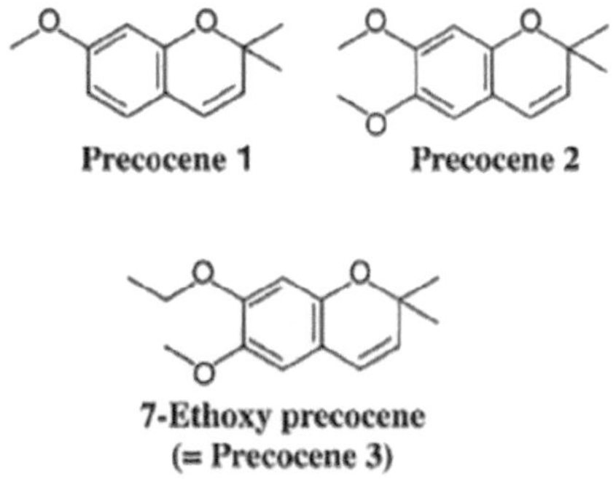

Precocene 1   **Precocene 2**

7-Ethoxy precocene
(= Precocene 3)

FIGURE 4   The chemical structure of precocenes that act as anti-juvenile hormone agents.

Ecdysteroids and Their Agonists

The Role of Ecdysteroids in Insects

Ecdysone, a steroid, serves as a prohormone of the molting hormone. It is secreted by a pair of endocrine glands, termed "prothoracic glands" (though in some insects they are not located in the prothorax but in the ventro-posterior part of the head). In certain tissues, ecdysone is converted to 20-hydroxyecdysone (Figure 5) which is the actual molting hormone. To initiate molting, a peak titer of 20-hydroxyecdysone is necessary, but for the full process of molting, a complete decline of this peak is needed. This decline leads to activation of certain genes that control production of necessary enzymes and secretion of additional hormones, which induce molting-related behavior. Ecdysone, 20-hydroxyecdysone, and chemically related substances, collectively termed as "ecdysteroids" are present in many plants and non-vertebrate animals.

The prothoracic glands usually degenerate in adult insects; if persist, they are inactive. In adult insects, the ovaries become the source of ecdysteroids and they transfer ecdysteroids, usually as conjugates, to the developing oocytes. An outline of the chemistry and physiological role of ecdysteroids, aimed primarily to under- or postgraduate students, is presented in Klowden's book.[81] A review dealing mostly with the mode of action of ecdysteroids at the molecular level was recently published.[82]

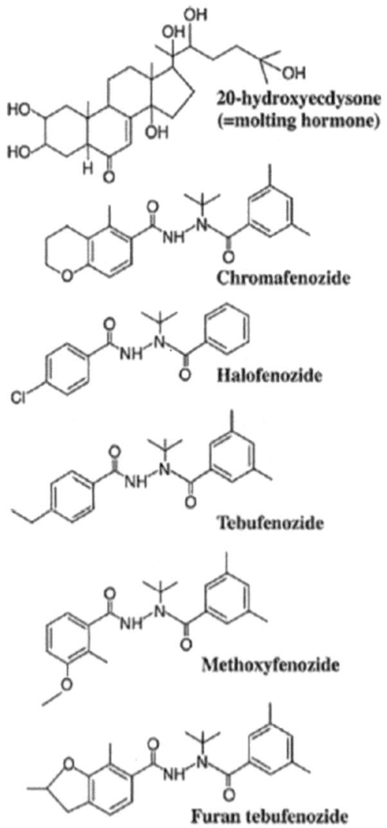

FIGURE 5 The chemical structures of 20-hydroxyecdysone (molting hormone) and nonsteroidal ecdysone (ecdysteroid) agonists.

Ecdysone (Ecdysteroid) Agonists (EAs)

Reviews devoted to EAs[1,3,25,83] usually describe and discuss the role of ecdysone in normal molting. As already outlined, a peak level and then a decline of the peak of ecdysone are needed for molting. The presently existing commercial EAs (Figure 5) bind strongly to ecdysteroid receptors, conveying a false message that no decline of 20-hydroxyecdysone occurred. Consequently, these EAs induce upregulation of all genes that occur in normal molting due to the 20-hydroxyecdysone peak; however, they do not activate those genes, which are upregulated in normal molting by the decline of the peak. The result is an incomplete, unfinished molt which is lethal. EAs are used against larvae, but they often also exert ovicidal effect.

From the chemical standpoint, commercial EAs are bisacylhydrazines, also termed diacylhydrazines. In contrast to ecdysteroids, bisacylhydrazines are not steroids. *In vivo* they are more stable than ecdysteroids and their effect may be regarded as "hyperecdysonism".[3,84] EAs are usually applied by ingestion. Their application is effective any time, except at natural molting. An advantage of the EAs is feeding inhibition of susceptible larvae, preventing further damage. Stopping of feeding is one of the first events in normal molting. The current commercial EAs are very selective and affect only some groups of insects. The reason for this selectivity seems to be differences in the ligand-binding domain of ecdysone receptors in different insects.

Chromafenozide [Matric, Killot, Pestanal]. This compound was developed in Japan, and it specifically affects highly harmful lepidopteran pests on rice, vegetables, fruit, vine, and other crops.

Halofenozide [Mach 2]. It was developed by Rohm and Haas Company. Halofenozide has a broader spectrum of susceptible insects than other commercial EAs. It affects both lepidopteran and coleopteran larvae. However, specific selectivity exists even within these two orders of insects. Despite its activity on some coleopteran larvae, halofenozide has a relatively low binding affinity to coleopteran ecdysteroid receptors.

Methoxyfenozide [Intrepid, Runner, Prodigy, Falcon]. Again it was developed by Rohm and Haas Company. Similar to chromafenozide and tebufenozide, it controls only lepidopteran larvae. Like in other EAs, efficacy depends on the species. Methoxyfenozide is effective against bollworms, leaf miners, and diamond moth. Recently, it was claimed that methoxyfenozide has some potential for mosquito control.[85]

Tebufenozide [Confirm, Mimic, Fimic, Romdan]. This EA was again developed by Rohm and Haas. It affects lepidopteran larvae, similar to methoxyfenozide, but species-dependent differences may exist. Tebufenozide has poorer ovicidal activity than methoxyfenozide.

Fufenozide [Fuxian]. This compound was developed in China under the code name JS-118; it is also named as furan tebufenozide. It is effective against the Asiatic rice borer, *C. suppressalis* and other lepidopteran larvae. Its degradation by hydrolysis and photolysis was recently studied.[86]

Research on bisacylhydrazine-type compounds were published,[87] and steroid-like compounds were investigated[88] for obtaining new insecticides. Non-bisacylhydrazine nonsteroidal EAs are available,[89,90] but none of them developed to commercial insecticide.

Advantages of IGRs

IGRs are nontoxic or slightly toxic (often as low as at g/kg level) to most vertebrates, especially to mammals and birds. Fish may be more susceptible to toxic effects of IGRs. Details of toxicity are presented in *The Pesticide Manual*[28] under the name of each compound separately.

Parasitoid wasps are usually less susceptible to IGRs than to conventional insecticides.[91–95] Moreover, in most (but not all) instances, parasitoid and predatory insects are less affected by IGRs than their respective harmful insect hosts and preys. This seems to be the case also for mites; the pest mite is more susceptible to IGR toxicity than its predatory mite.[53,96] Treatment of livestock with IGRs by feed-through or pour-on methods of application has no notable effect on dung beetles.[42]

The most important advantage of IGRs is that they affect only a relatively narrow range of susceptible species. IGRs are certainly friendlier to the environment and make less damage to non-target species than conventional insecticides that are usually toxic to a wide range of animals, including humans.

Disadvantages of IGRs

The environmental advantages of IGRs (see above) may be disadvantageous economically. The more specific the IGR, the more limited its marketing. Development of an IGR to a commercial product, with permits in many countries, is expensive, and if the expected marketing is limited, a commercial firm would not invest the cost of development.

Although in most instances, parasitoid and predatory insects exhibit low susceptibility to IGRs, this is not always so. Some parasitoid insects may be affected by some IGRs, even when non-IGR insecticides are more toxic than IGRs.[93] IGRs may be harmful to predatory insects.[17,97,98] Also, IGRs may affect other non-target organisms, especially crustaceans.

Besides the direct role of IGRs in parasitoid–pest or predator–pest relations, lethal and sublethal effects of IGRs on beneficial insects, such as pollinators, should also be considered. The sublethal effects of insecticides, including IGRs, on beneficial arthropods were reviewed.[99]

The best-known pollinators are the honey bees, bumblebees, and other bee species. Claims concerning effects of IGRs on bees are contradictory. CSIs, including novaluron, did not affect adult bumblebees (*Bombus terrestris*) but dramatically reduced brood production.[100] Other authors, however, found that novaluron did not affect the brood of this species.[101] An earlier publication reported that some CSIs are toxic to immature stages of bees.[102] Recent studies confirmed that diflubenzuron and novaluron (CSIs) are toxic to the progeny of honey bees and alfalfa leafcutting bees (*Megachile rotundata*).[103–106] Mommaerts and coauthors concluded that fenoxycarb (JHA) and methoxyfenozide (EAs) are safe to bumblebees, whereas pyriproxyfen and kinoprene (JHAs) affect the brood at the maximum field-recommended concentration.[107] Pyriproxyfen was found to be toxic to honey bees in recent studies too.[108,109] In contrast, fenoxycarb was claimed to negatively affect adult emergence in honey bees.[75] An older review on the effect of IGRs on some bees[110] distinguishes between different species and between different doses.

An additional disadvantage is that IGRs are effective only in special stages or instars of the pests: CSIs and EAs before natural molt, and JHAs in the last larval instar. Therefore, IGRs do not prevent some or all damage by the larvae if these larvae are harmful. In extreme instances, JHAs induce supernumerary larval instars[63] with longer-than-normal larval life span and additional damage.

Resistance to IGRs

Insects are capable of developing resistance to any kind of insecticides, and IGRs are not exceptions. Hundreds of scientific publications are devoted to acquired resistance of insects to IGRs.

Several interrelated approaches to IGR resistance may be considered (these do not differ from those to conventional insecticides). One of them is studying field-collected insects; the susceptibility to an IGR of a sample of insects collected in the field is assessed and compared to that of a non-resistant strain maintained in the laboratory or supplied by international or national organizations (WHO, etc.). A much higher dose to affect the field sample than that induces the same effect in the non-resistant strain means that the insects of the field sample are resistant. The effective dose for the field sample divided by that obtained for the non-resistant strain yields the "resistance ratio" (RR), which can be low but can be as high as several thousands. RR values may differ for different populations of the same species within the same country, as exemplified by Kristensen and Jespersen.[111]

Another approach is to expose a non-resistant laboratory strain to an IGR and select the survivors over subsequent generations with increasing doses. The effective dose obtained for the last generation divided by that of the initial dose results in the relevant RR. This reveals the resistance potential of the pest.[111] Obviously, this approach can be combined with the former one by taking an already resistant strain from the field, then expose subsequent generations to increasing doses. The RR for pyriproxyfen resistance of the whitefly, *B. tabaci*, was found to be over 2000[112] and can be as high as 7000.[113]

The third approach is to investigate cross-resistance, meaning that a strain resistant to an IGR shows resistance to other IGRs without formerly being exposed to these IGRs.[111,112] Sometimes, resistance to a conventional, non-IGR insecticide may result in cross-resistance to IGRs.[18]

Other approaches are studies on the inheritance of resistance,[113,114] residual activity,[115] mechanisms of resistance at the molecular level, and gene expressions.[116,117]

Reversal of resistance is studied by maintaining a resistant strain in the laboratory for many subsequent generations, without exposure to the IGR to which the strain is resistant, and assessing the residual resistance of the last generation.[18,19] Reversal of resistance does occur, but it is seldom absolute. Also, when a strain that experienced reversal of resistance is exposed again to the same IGR, it develops resistance more quickly than a strain that is originally non-resistant.

Integrated Pest Management

IPM means employment of multiple control tactics to reduce pest population and usage of chemicals, as well as decrease development of resistance, all with minimal economic cost and environmental damage.[118] The IPM for each pest, in each environment, should be worked out. Efficacy and resistance potential of different insecticides, effects on non-target species, including parasitoids and/or predators of the pest, screening the pest (below a certain level of population density no treatment is needed), and effects on the environment should be considered. Usually IGRs are more efficacious and less harmful to the environment than conventional insecticides. However, this generalization should be verified by carefully gathered actual data. IPM partially solves the problem of resistance, especially resistance to a novel IGR, by alternating IGRs, for example, pyriproxyfen and buprofezin.[51] Such alternation tactics may also include non-IGR insecticides, for example, alternation of novaluron, pyriproxyfen, and neonicotinoids as suggested by Ishaaya and coauthors.[119] The effects of each alternating compound on non-target species and the environment should be studied; the alternating compounds should not induce cross-resistance. In many instances, there is more than one pest harmful to the same object. If so, it is preferable to work out such alternation of IGRs in which one or more compound exert toxic activity on more than just one pest.[15,119]

Obviously, IPM includes not only chemical factors but also biological ones, as exemplified by Mouden et al.[120] and Wilson et al.[121] In the European Union, IPM is compulsory, though its implementation is problematic.[122]

Conclusions

IGRs are insecticides that interfere with insect-specific physiological systems that do not exist in vertebrates, including humans. Consequently, IGRs are either nontoxic or slightly toxic to vertebrates. Within the insect world, IGRs may affect different species differently, even in the same family.[123] The narrow range of susceptibility to IGRs is a great environmental advantage over conventional insecticides, which are usually toxic to a much wider spectrum of species and much more harmful to the environment. However, IGRs may also affect non-target species, parasitoids, and/or predators of the pest, or of other pests, or other beneficial insects, such as pollinators, and even some non-insect species. Therefore, adequate research is needed to estimate the effect of an IGR on one or more pests, beneficial and non-target organisms, and the surrounding environment.

Insects may develop resistance or even cross-resistance to one or more IGRs, which can be prevented, or at least delayed, by IPM. In this respect, IGRs and conventional insecticides are similar.

Despite the advantages of IGRs, there are some disadvantages. Interference with a physiological system means that the insect is susceptible when that physiological system is active. IGRs interfere with production of new cuticle at molt, or disturb metamorphosis, or prevent reproduction. So, the toxic effects of IGRs become overt at the next molt, or in the next or subsequent instars, and meanwhile the pest inflicts additional damage. The ideal targets for IGRs are insects harmless in preadult stages and harmful only as adults (for example, mosquitoes). The belated action is an additional disadvantage; the farmer, or the sanitarian, or anybody who uses IGRs, do not see immediate knockdown of the pest, and doubt, therefore, the efficacy of the IGR.

New IGRs aimed to attack currently susceptible physiological systems are synthesized, tested, and possibly developed to commercial products. However, it is feasible to assume that in the future, additional IGRs and their effects, as well as various approaches attacking other insect-specific physiological or molecular systems, will be discovered and implemented.[124,125]

References

1. Pener, M. P., and T. S. Dhadialla. 2012. An overview of insect growth disruptors; applied aspects. *Adv. Insect Physiol.* 43: 1–162.
2. Pener, M. P. 2013. Insect growth regulators. In *Encyclopedia of Environmental Management*, ed. S. E. Jorgensen, vol. II, 1459–1470. New York: Taylor & Francis.
3. Dhadialla, T. S., A. Retnakaran, and G. Smagghe. 2010. Insect growth- and development-disrupting insecticides. In *Insect Control: Biological and Synthetic Agents*, ed. L. I. Gilbert, and S. S. Gill, 121–184. New York: Elsevier.
4. Subramanian, S., and K. Shankarganesh. 2016. Insect hormones (as pesticides). In *Ecofriendly Pest Management for Food Security*, ed. B. K. Omkar, 613–650. London: Academic Press.
5. Oberlander, H., D. L. Silhacek, E. Shaaya, and I. Ishaaya. 1997. Current status and future perspectives of the use of insect growth regulators for the control of stored product insects. *J. Stored Prod. Res.* 33: 1–6.
6. Collins, D. A. 2006. A review of alternatives to organophosphorus compounds for the control of storage mites. *J. Stored Prod. Res.* 42: 395–426.
7. Alzogaray, R. A., and E. N. Zerba. 2017. *Rhodnius prolixus* intoxicated. *J. Insect Physiol.* 97: 93–113.
8. Axtell, R. C., and T. D. Edwards. 1983. Efficacy and nontarget effects of Larvadex® as a feed additive for controlling house flies in caged-layer poultry manure. *Poult. Sci.* 62: 2371–2377.
9. Anderson, J. F., A. Spandorf, L. A., Magnarelli, and W. Glowa. 1986. Control of house flies in commercial poultry houses in Connecticut. *Poult. Sci.* 65: 837–844.
10. Harris, R. L., E. D. Frazer, and R. L. Younger. 1973. Horn flies, stable flies and house flies: development in feces of bovines treated orally with juvenile hormone analogues. *J. Econ. Entomol.* 66: 1099–1102.
11. Henrick, C. A. 2007. Methoprene. *J. Am. Mosq. Control Assoc.* 23 (supplement 2): 225–239.
12. Mascari, T. M., and L. D. Foil. 2010. Laboratory evaluation of novaluron as a rodent feed-through insecticide against sand fly larvae. *J. Med. Entomol.* 47: 205–209.
13. Mascari, T. M., R. W. Stout, and L. D. Foil. 2012. Evaluation of three feed-through insecticides using two rodent and two sand fly species as models. *J. Am. Mosq. Control Assoc.* 28: 260–262.
14. Pener, M .P., A. Ayali, G. Kelmer, B. Bennettová, V. Němec, M. Rejzek, and Z. Wimmer. 1997. Comparative testing of several juvenile hormone analogues in two species of locusts, *Locusta migratoria migratorioides* and *Schistocerca gregaria*. *Pestic. Sci.* 51: 443–449.
15. Gogi, M. D., R. M. Sarfraz, L. M. Dosdall, M. J. Arif, A. B. Keddie, and M. Ashfaq. 2006. Effectiveness of two insect growth regulators against *Bemisia tabaci* (Gennadius) (Homoptera: Aleyrodidae) and *Helicoverpa armigera* (Hübner) (Lepidoptera: Noctuidae) and their impact on population densities of arthropod predators on cotton in Pakistan. *Pest Manag. Sci.* 62: 982–990.
16. Hattingh, V. 1996. The use of insect growth regulators – Implications for IPM with citrus in southern Africa as an example. *Entomophaga* 41: 513–518.

17. Zotti, M. J., A. D. Grutzmacher, I. H. Lopes, and G. Smagghe. 2013. Comparative effects of insecticides with different mechanisms of action on *Chrysoperla externa* (Neuroptera: Chrysopidae): lethal, sublethal and dose-response effects. *Insect Sci.* 20: 743–752.

18. Reuveny, H., and E. Cohen. 2004. Resistance of the codling moth *Cydia pomonella* (L.) (Lep., Tortricidae) to pesticides in Israel. *J. Appl. Entomol.* 128: 645–651.

19. Wilson, M., P. Moshitzky, E. Laor, M. Ghanim, A. R. Horowitz, and S. Morin. 2007. Reversal of resistance to pyriproxyfen in the Q biotope of *Bemisia tabaci* (Hemiptera: Aleyrodidae). *Pest Manag. Sci.* 63: 761–768.

20. Thoms, E. M., J. E. Eger, M. T. Messenger, E. Vargo, B. Cabrera, C. Riegel, S. Murphree, J. Mauldin, and P. Scherer. 2009. Bugs, baits and bureaucracy: completing the first termite bait efficacy trials (quarterly replenishment of noviflumuron) initiated after adoption of Florida rule, chapter 5E-2.0311. *Am. Entomol.* 55: 29–39.

21. Wright, J. E., and A. Retnakaran. ed. 1987. *Chitin and Benzoylphenyl Ureas.* Series.Entomologica 38. Dordrecht: Dr W. Junk Publisher.

22. Minakuchi, C., and L. M. Riddiford. 2006. Insect juvenile hormone action as a potential target of pest management. *J. Pestic. Sci.* 31: 77–84.

23. Staal, G. B. 1986. Anti juvenile hormone agents. *Annu. Rev. Entomol.* 31: 391–429.

24. Ghoneim, K., and R. F. Bakr. 2018. Physiological activities of anti-juvenile hormone agents against insects and their role devising fourth generation insecticides: a comprehensive review. *Egypt. Acad. J. Biol. Sci. A. Entomol.* 11: 45–138.

25. Dhadialla, T. S., G. R. Carlson, and D. P. Le. 1998. New insecticides with ecdysteroidal and juvenile hormone activity. *Annu. Rev. Entomol.* 43: 545–569.

26. Mordue (Luntz), A. J., E. D. Morgan, and A. J. Nisbet. 2010. Azadirachtin, a natural product in insect control. In *Insect Control: Biological and Synthetic Agents*, ed. L. I. Gilbert and S. S. Gill, 185–203. London: Elsevier, Academic Press.

27. Mordue (Luntz), A. J., E. D. Morgan and A. J. Nisbet. 2010. Addendum: azadirachtin, a natural product in insect control: an update. In *Insect Control: Biological and Synthetic Agents*, ed. L. I. Gilbert, and S. S. Gill, 204–206. London: Elsevier, Academic Press.

28. Turner, J. A. ed. 2018. *The Pesticide Manual*, 18th ed. Aldershot Hampshire: British Crop Production Council.

29. Kelkenberg, M., J. Odman-Naresh, and S. Muthukrishnan. 2015. Chitin is a necessary component to maintain the barrier function of the peritrophic matrix in the insect midgut. *Insect Biochem. Mol. Biol.* 56: 21–28.

30. Cohen, E. 1987. Chitin biochemistry: synthesis and inhibition. *Annu. Rev. Entomol.* 32: 71–93.

31. Cohen, E. 2001. Chitin synthesis and inhibition: a revisit. *Pest Manag. Sci.* 57: 946–950.

32. Muthukrishnan, S., H. Merzendorfer, Y. Arakane, and K. J. Kramer. 2012. Chitin metabolism in insects. In *Insect Molecular Biology and Biochemistry*, ed. L. I. Gilbert, London: Elsevier, Academic Press, 193–235.

33. Boiteau, G., and C. Noronha. 2007. Topical, resudial and ovicidal contact toxicity of three reduced-risk insecticides against the European corn borer, *Ostrinia nubilalis* (Lepidoptera: Crambidae), on potato. *Pest Manag. Sci.* 63: 1230–1238.

34. Merzendorfer, H. 2013. Chitin synthesis inhibitors: old molecules and new developments. *Insect Sci.* 20: 121–138.

35. Matsumura, F. 2010. Studies on the action mechanism of benzoylurea insecticides to inhibit the process of chitin synthesis in insects: a review on the status of research activities in the past, the present and the future prospects. *Pestic. Biochem. Physiol.* 97: 133–139.

36. Doucet, D., and A. Retnakaran. 2012. Insect chitin: metabolism, genomics and pest management. *Adv. Insect Physiol.* 43: 437–511.

37. Sun, R., C. Liu, H. Zhang, and Q. Wang. 2015. Benzoylurea chitin synthesis inhibitors. *J. Agric. Food Chem.* 63: 6847–6865.

38. Weiland, R.T., F. D. Judge, T. Pels, and A. C. Grosscurt. 2002. A literature review and new observations on the use of diflubenzuron for control of locusts and grasshoppers throughout the world. *J. Orthoptera Res.* 11: 43–54.

39. Webb, G. A. 2017. Efficacy of bistrifluron termite bait on *Coptotermes lacteus* (Isoptera: Rhinotermitidae) in southern Australia. *J. Econ. Entomol.* 110: 1705–1712.

40. Liu, G., Q. Zhou, and W. Lu. 2001. The formation of 2-chlorobenzamide upon hydrolysis of the benzoylphenylurea insecticide 1-(2-chlorobenzoyl)-3-(4-chlorophenyl) urea in different water systems. *J. Agric. Food Chem.* 49: 1304–1308.

41. Sukartana, P., G. Sumarni, and S. Broadbent. 2009. Evaluation of chlorfluazuron in controlling the subterranean termite *Coptotermes curvignathus* (Isoptera: Rhinotermitidae) in Idonesia. *J. Trop. For. Sci.* 21: 13–18.

42. Kryger, U., C. Deschodt, A. L. V. Davis, and C. H. Scholtz. 2007. Effects of cattle treatment with fluazuron pour-on on survival and reproduction of the dung beetle species *Onthophagus gazella* (Fabricius). *Vet. Parasitol.* 143: 380–384.

43. Davis, R. M., E. Cleugh, R. T. Smith, and C. L. Fritz. 2008, Use of a chitin synthesis inhibitor to control fleas on wild rodents important in the maintenance of plague, *Yersinia pestis* in California. *J. Vector Ecol.* 33: 278–284.

44. Sadeghi, A., E. J. M. Van Damme, and G. Smagghe. 2009. Evaluation of the susceptibility of the pea aphid, *Acyrthosiphon pisum*, to a selection of novel biorational insecticides using an artificial diet. *J. Insect Sci.* 9: article 65.

45. Méndez N. 2005. Effects of teflubenzuron on larvae and juveniles of the polychaete *Capitella* sp. B from Barcelona, Spain. *Water Air Soil Poll.* 160: 259–269.

46. Belinato, T. A., A. J. Martins, J. B. P. Lima, and D. Valie. 2013. Effect of triflumuron, a chitin synthesis inhibitor, on *Aedes aegypti*, *Aedes albopictus* and *Culex quinquefasciatus* under laboratory conditions, *Parasit. Vectors* 6: article 83.

47. Chen, L., Q. Wang, R. Huang, C. Mao, J. Shang, and F. Bi. 2005. Synthesis and insecticidal evaluation of propesticides of benzoylphenylureas. *J. Agric. Food Chem.* 53: 38–41.

48. Sun, R., M. Lü, L. Chen, Q. Li, H. Song, F. Bi, R. Huang, and Q. Wang. 2008. Design, synthesis, bioactivity and structure–activity relationship (SAR) studies of novel benzoylphenylureas containing oxime ether group. *J. Agric. Food Chem.* 56: 11376–11391.

49. Sun, R., Y. Zhang, F. Bi, and Q. Wang. 2009. Design, synthesis, and bioactivity study of novel benzoylpyridazyl ureas. *J. Agric. Food Chem.* 57: 6356–6361.

50. Zhang, J., X. Tang, I. Ishaaya, S. Cao, J. Wu, J. Yu, H. Li, and X. Qian. 2010. Synthesis and insecticidal activity of heptafluoroisopropyl-containing benzoylphenylurea structures. *J. Agric. Food Chem.* 58: 2736–2740.

51. Horowitz, A. R., G. Forer, and I. Ishaaya. 1994. Managing resistance in *Bemisia tabaci* in Israel with emphasis on cotton. *Pestic. Sci.* 42: 113–122.

52. Zhang, X., X. Liu, F. Zhu, J. Li, H. You, and P. Lu. 2014. Field evaluation of insecticide resistance in the brown plathopper (*Nilaparvata lugens* Stål) in China. *Crop Prot.* 58: 61–66.

53. Stavrinides, M. C., and N. J. Mills. 2009. Demographic effects of pesticides on biological control of Pacific spider mite (*Tetranychus pacificus*) by the western predatory mite (*Galendromus occidentalis*). *Biol. Control* 48: 267–273.

54. Nauen, R., and G. Smagghe. 2006. Mode of action of etoxazole. *Pest Manag. Sci.* 62: 379–382.

55. Herron, G. A., L. K. Wooley, K. I. Langfield, and Y. Chen. 2018. First detection of etoxazole resistance in Australian two-spotted mite *Tetranychus urticae* Koch (Acarina: Tetranychidae) via bioassay and DNA methods. *Austral. Entomol.* 57: 365–368.

56. Demaeght, P., E. J. Osborne, J. Odman-Naresh, M. Grbić, R. Nauen, H. Merzendorfer, R. M. Clark, and T. Van Leeuwen. 2014. High resolution genetic mapping uncovers chitin synthase-1 as the target site of the structurally diverse mite growth inhibitors clofentezine, hexythiazox and etoxazole in *Tetranychus urticae*. *Insect Biochem. Mol. Biol.* 51: 52–63.

57. Tobe, S. S., and B. Stay. 1985. Structure and regulation of the corpus allatum. *Adv. Insect Physiol.* 18: 305–432.

58. Riddiford, L. M. 1994. Cellular and molecular actions of juvenile hormone I. General considerations and premetamorphic actions. *Adv. Insect Physiol.* 24: 213–274.

59. Wyatt, G. R., and K. G. Davey. 1996. Cellular and molecular actions of juvenile hormone. II. Roles of juvenile hormone in adult insects. *Adv. Insect Physiol.* 26: 1–155.

60. Riddiford, L. M. 2008. Juvenile hormone action: a 2007 perspective. *J. Insect Physiol.* 54: 895–901.

61. Jindra, M., S. R. Palli, and L. M. Riddiford. 2013. The juvenile hormone signaling pathway in insect development. *Annu. Rev. Entomol.* 58: 181–204.

62. Sláma, K., M. Romaňuk, and F. Šorm. 1974. *Insect Hormones and Bioanalogues.* Wien: Springer-Verlag.

63. Parthasarathy, R., and S. R. Palli. 2009. Molecular analysis of juvenile hormone analog action in controlling the metamorphosis of the red flour beetle, *Tribolium castaneum. Arch. Insect Biochem. Physiol.* 70: 57–70.

64. Staal, G. B. 1975. Insect growth regulators with juvenile hormone activity. *Annu. Rev Entomol.* 20: 417–460.

65. Palli, S. R. 2009. Recent advances in the mode of action of juvenile hormones and their analogs. In *Biorational Control of Arthropod Pests*, ed. I. Ishaaya, and A. R. Horowitz, 111–129. Dordrecht: Springer.

66. Guerrero, A. and G. Rosell. 2005. Biorational approaches for insect control by enzymatic inhibition. *Curr. Medicinal Chem.* 12: 461–469.

67. Kostyukovsky, M., B. Chen., S. Atsmi, and E. Shaaya. 2000. Biological activity of two juvenoids and two ecdysteroids against three stored product insects. *Insect Biochem. Mol. Biol.* 30: 891–897.

68. Krämer, B., U. Körner, and P. Wolbert. 2002. Differentially expressed genes in metamorphosis and after juvenile hormone application in the pupa of *Galleria. Insect Biochem. Mol. Biol.* 32: 133–140.

69. Suzuki, Y., J. W. Truman, and L. M. Riddiford. 2008. The role of Broad in the development of *Tribolium castaneum*: implications for the evolution of the holometabolous insect pupa. *Development* 135: 569–577.

70. Ramaseshadri, P., R. Farkas, and S. R. Palli. 2012. Recent progress in juvenile hormone analogs (JHA) research. *Adv. Insect Physiol.* 43: 353–436.

71. Wijayaratne, L. K. W., F. H. Arthur, and S. Whyard. 2018. Methoprene and control of stored-product insects. *J. Stored Prod. Res.* 76: 161–169.

72. Arthur, F. H., S. Liu, B. Zhao, and T. W. Phillips. 2009. Residual efficacy of pyriproxyfen and hydroprene applied to wood, metal and concrete for control of stored-product insects. *Pest Manag. Sci.* 65: 791–797.

73. Mittler, T. E., S. G. Nassar, and G. B. Staal. 1976. Wing development and parthenogenesis induced in progenies of kinoprene-treated gynoparae of *Aphis fabae* and *Myzus persicae. J. Insect Physiol.* 22: 1717–1725.

74. Grenier, S., and A.-M. Grenier. 1993. Fenoxycarb, a fairly new insect growth regulator: a review of its effects on insects. *Ann. Appl. Biol.* 122: 369–403.

75. Aupinel, P., D. Fortini, B. Michaud, F. Marolleau, J.-N. Tasei, and J.-F. Odoux. 2007. Toxicity of dimethoate and fenoxycarb to honey bee brood (*Apis mellifera*), using a new *in vitro* standardized feeding method. *Pest Manag. Sci.* 63: 1090–1094.

76. Sullivan, J. J. 2010. Chemistry and environmental fate of fenoxycarb. In *Reviews of Environmental Contamination and Toxicology*, Vol. 202, ed. D. M. Whitacre, 155–184. New York: Springer-Verlag.

77. Sullivan, J. J., and K. S. Goh. 2008. Environmental fate and properties of pyriproxyfen. *J. Pestic. Sci.* 33: 339–350.

78. Streibert H. P., M. L. Frischknecht, and F. Karrer. 1994. Diofenolan – a new insect growth regulator for the control of scale insects and important lepidopterous pests in deciduous fruit and citrus. *Br. Crop Prot. Conf. Pests Dis.* 1: 23–30.

79. Bowers, W. S., T. Ohta., J. S. Cleere, and P. A. Marsella. 1976. Discovery of insect anti-juvenile hormones in plants. *Science* 193: 542–547.

80. Pener, M. P., L. Orshan, and J. De Wilde. 1978. Precocene II causes atrophy of corpora allata in *Locusta migratoria*. *Nature* 272: 350–353.

81. Klowden, M. J. 2013. *Physiological Systems in Insects*, 3rd ed. San Diego: Academic Press.

82. Nakagawa, Y., and H. Sonobe. 2016. Ecdysteroids. In *Handbook of Hormones*, ed. Y. Takei, H. Ando, and K. Tsutsui, 557–559 and e98-1-e98-14. San Diego: Academic Press.

83. Retnakaran, A., P. Krell, Q. Feng, and B. Arif. 2003. Ecdysone agonists: mechanism and importance in controlling insect pests of agriculture and forestry. *Arch. Insect Biochem. Physiol.* 54: 187–199.

84. Smagghe, G., L. E. Gomez, and T. S. Dhadialla. 2012. Bisacylhydrazine insecticides for selective pest control. *Adv. Insect Physiol.* 43: 163–249.

85. Hamaidia, K., F. Tine-Djebbar, and N. Soltani. 2018. Activity of a selected insecticide (methoxyfenozide) against two mosquito species (*Culex pipiens* and *Culiseta longiareolata*): toxicological, biometrical and biochemical study. *Physiol. Entomol.* 43: 315–323.

86. Hu, J-Y., C. Liu, Y.-C. Zhang, and Z.-X. Zheng. 2009. Hydrolysis and photolysis of diacylhydrazines-type insect growth regulator JS-118 in aqueous solutions under abiotic conditions. *Bull. Environ. Contam. Toxicol.* 82: 610–615.

87. Huang, Z., Q. Cui, L. Xiong, Z. Wang, K. Wang, Q. Zhao, F. Bi, and Q. Wang, . 2009. Synthesis and insecticidal activities and SAR studies on novel benzoheterocyclic diacylhydrazine derivatives. *J. Agric. Food Chem.* 57: 2447–2456.

88. Zou, C., G. Liu, S. Liu, S. Liu, Q. Song, J. Wang, Q. Feng, Y. Su, and S. Li. 2018. Cucurbitacin B acts as a potential insect growth regulator by antagonizing 20-hydroxyecdysone activity. *Pest Manag. Sci.* 74: 1394–1403.

89. Dinan, L., Y. Nakagawa, and R. E. Hormann. 2012. Structure–activity relationships of ecdysteroids and non-steroidal ecdysone agonists. *Adv. Insect Physiol.* 43: 251–298.

90. Kitamura, S., T. Harada, H. Hiramatsu, R. Shimizu, H. Miyagawa, and Y. Nakawaga. 2014. Structural requirement and stereospecificity of tetrahydroquinolines as potent ecdysone agonists. *Bioorg. Medic. Chem. Lett.* 24: 1715–1718.

91. Naranjo, S. E., P. C. Ellsworth, and J. R. Hagler. 2004. Conservation of natural enemies in cotton: role of insect growth regulators in management of *Bemisia tabaci*. *Biol. Control* 30: 52–72.

92. Vianna, U. R., D. Prattisoli, J. C. Zanuncio, E. R. Lima, J. Brunner, F. F. Pereira, and J. E. Serrão. 2009. Insecticide toxicity to *Trichogramma pretiosum* (Hymenoptera: Trichogrammatidae) females and effect on descendant generations. *Ecotoxicology* 18: 180–186.

93. Suma, P., L. Zappalà, G. Mazzeo, and G. Siscaro. 2009. Lethal and sub-lethal effects of insecticides on natural enemies of citrus scale pests. *BioControl* 54: 651–661.

94. Wang, Y., C. Wu, T. Cang, L. Yang, W. Yu, X. Zhao, Q. Wang, and L. Cai. 2014. Toxicity risk of insecticides to the insect egg parasitoid *Trichogramma evanescens* Westwood (Hymenoptera: Trichogrammatidae). *Pest Manag. Sci.* 70: 398–404.

95. Khan, M. A., and J. R. Ruberson. 2017. Lethal effects of selected novel pesticides on immature stages of *Trichogramma pretiosum* (Hymenoptera: Trichogrammatidae). *Pest Manag. Sci.* 73: 2465–2472.

96. Döker, I., M. L. Pappas, K. Samaras, A. Triantafllou, C. Kazak, and G. D. Broufas. 2015. Compatibility of reduced-risk insecticides with the non-target predatory mite *Iphiseius degerans* (Acari: Phytoseiidae). *Pest Manag. Sci.* 71: 1267–1273.

97. Cabral, S., P. Garcia, and A. O. Soares. 2008. Effects of pirimicarb, buprofezin and pymetrozin on survival, development and reproduction of *Coccinella undecimpunctata* (Coleoptera: Coccinellidae). *Biocontrol Sci. Technol.* 13: 307–318.

98. He, F., S. Sun, X. Sun, S. Ji, X. Li, J. Zhang, and X. Jiang. 2018. Effects of insect growth-regulator insecticides on the immature stages of *Harmonia axyridis* (Coleoptera: Coccinellidae). *Ecotoxicol. Environ. Safety* 164: 665–674.

99. Desneux, N., A. Decourtye, and J.-M. Delpuech. 2007. The sublethal effects of pesticides on beneficial arthropods. *Annu. Rev. Entomol.* 52: 81–106.
100. Mommaerts, V., G. Sterk, and G. Smagghe. 2006. Hazards and uptake of chitin synthesis inhibitors in bumblebees *Bombus terrestris*. *Pest Manag. Sci.* 62: 752–758.
101. Malone, L. A., C. D. Scott-Dupree, J. H. Todd, and P. Ramankutty. 2007. No sub-lethal toxicity to bumblebees, *Bombus terrestris*, exposed to Bt-corn pollen, captan and novaluron. *New Zealand J. Crop Horticult. Sci.* 35: 435–439.
102. Chandel, R. S., and P. R. Gupta. 1992. Toxicity of diflubenzuron and penfluron to immature stages of *Apis cerena indica* F. and *Apis mellifera* L. *Apidologie* 23: 465–473.
103. Fine, J. D., C. A. Mullin, M. T. Frazier, and R. D. Reynolds. 2017. Field residues and effects of the insect growth regulator novaluron and its major co-formulant *N*-methyl-2-pyrrolidone on honey bee reproduction and development. *J. Econ. Entomol.* 110: 1993–2001.
104. Wade, A., C.-H. Lin, C. Kurkul, E. R. Regan, and R. M. Johnson. 2019. Combined toxicity of insecticides and fungicides applied to California almond orchards to honey bee larvae and adults. *Insects* 10: article 20.
105. Hodgson, E. W., T. L. Pitts-Singer, and J. D. Barbour. 2011. Effects of the insect growth regulator, novaluron on immature alfalfa leafcutting bees, *Megachile rotundata*. *J. Insect Sci.* 11: article 43.
106. Pitts-Singer, T. L., and J. D. Barbour. 2017. Effects of residual novaluron on reproduction in alfalfa leafcutting bees, *Megachile rotundata*. F. (Megachilidae). *Pest Manag. Sci.* 73: 153–159.
107. Mommaerts, V., G. Sterk, and G. Smagghe. 2006. Bumblebees can be used in combination with juvenile hormone analogues and ecdysone agonists. *Ecotoxicology* 15: 513–521.
108. Chen, Y.-W., P.-S. Wu, E.-C. Yang, Y.-S. Nai, and Z. Y. Huang. 2016. The impact of pyriproxyfen on the development of honey bee (*Apis mellifera* L.) colony in field. *J. Asia-Pacific Entomol.* 19: 589–594.
109. Fisher II, A., C. Colman, C. Hoffmann, B. Fritz, and J. Rangel. 2018. The effects of the insect growth regulators methoxyfenozide and pyriproxyfen and the acaricide bifenazate on honey bee (Hymenoptera: Apidae) forager survival. *J. Econ. Entomol.* 111: 510–516.
110. Tasei, J.-N. 2001. Effects of insect growth regulators on honey bees and non-*Apis* bees. A review. *Apidologie* 32: 527–545.
111. Kristensen, M., and J. B. Jespersen. 2003. Larvicide resistance in *Musca domestica* (Diptera: Muscidae) populations in Denmark and establishment of resistant laboratory strains. *J. Econ. Entomol.* 96: 1300–1306.
112. Ishaaya, I., S. Kontsedalov, and A. R. Horowitz. 2005. Biorational insecticides: mechanism and cross-resistance. *Arch. Insect Biochem. Physiol.* 58: 192–198.
113. Horowitz, A. R., K. Gorman, G. Ross, and I. Denholm. 2003. Inheritance of pyriproxyfen resistance in the whitefly *Bemisia tabaci* (Q biotype). *Arch. Insect Biochem. Physiol.* 54: 177–186.
114. Cao, G.-C., and Z.-J. Han. 2015. Tebufenozide resistance is associated with sex-linked inheritance in *Plutella xylostella*. *Insect Sci.* 22: 235–242.
115. Pavan, F., E. Cargnus, G. Bigot, and P. Zandigiacomo. 2014. Residual activity of insecticides applied against *Lobesia botrana* and its influence on resistance management strategies. *Bull. Insectol.* 67: 273–280.
116. Ghanim, M., and S. Kontsedalov. 2007. Gene expression in pyriproxyfen-resistant *Bemisia tabaci* Q biotype. *Pest Manag. Sci.* 63: 776–783.
117. Jia, B., Y. Liu, Y. C. Zhu, X. Liu, C. Gao, and J. Shen. 2009. Inheritance, fitness cost and mechanism of resistance to tebufenozide in *Spodoptera exigua* (Hübner) (Lepidoptera: Noctuidae). *Pest Manag. Sci.* 65: 996–1002.
118. Castle S., and S. E. Naranjo. 2009. Sampling plans, selective insecticides and sustainability: the case for IPM as 'informed pest management'. *Pest Manag. Sci.* 65: 1321–1328.
119. Ishaaya, I., S. Kontsedalov, and A. R. Horowitz. 2003. Novaluron (Rimon), a novel IGR: potency and cross-resistance. *Arch. Insect Biochem. Physiol.* 54: 157–164.

120. Mouden, S., K. F. Sarmiento, P. G. L. Klinkhamer, and K. A. Leiss. 2017. Integrated pest management in western flower thrips: past, present and future. *Pest Manag. Sci.* 73: 813–822.

121. Wilson, L. J., M. E. A. Whitehouse, and G. E. Herron. 2018. The management of insect pests in Australian cotton: an evolving story. *Annu. Rev. Entomol.* 63: 215–237.

122. Matyjaszczyk, E. 2019. Problems implementing compulsory integrated pest management. *Pest Manag. Sci.* doi:10.1002/ps.5357. In Press.

123. Cabrera, P., D. Cormier, and É. Lucas. 2017. Differential sensitivity of an invasive and an indigenous ladybeetle to two reduced-risk insecticides. *J. Appl. Entomol.* 141: 690–701.

124. Smagghe, G., M. Zotti, and A. Retnakaran. 2019. Targeting female reproduction in insects with birational insecticides for pest management: a critical review with suggestions for future research. *Curr. Opin. Insect Sci.* 31: 65–69.

125. Ishaya, I., S. R. Palli, and A. R. Horowitz, eds. 2012. *Advanced Technologies for Managing Insect Pests*. Dordrecht: Springer.

II

COV:
Comparative
Overviews
of Important
Topics for
Environmental
Management

II

7

Biodiversity and Sustainability

Odo Primavesi

Introduction

One of the greatest concerns related to sustainability today is the degradation and switching off of essential ecosystem services caused by anthropogenic activities like land degradation and pollution. These human activities, by producing solid, liquid, gaseous, and radiating (heat, light, radioactive emissions, and others) wastes and contaminants, lead to physical, chemical, and biological pollution, affecting the whole planet. For example, global warming and climate change result from a combination of air pollutants, such as the excess of greenhouse gases (carbon dioxide, methane, nitrous oxide, and others) and the excess of long-wave infrared or heat radiation (>300 W/m^2) from degraded and dry landscapes, and other factors.

Another example is the pollution of water due to poor management of soils (erosion, siltation of water bodies, and contamination with nitrates, phosphates, and other pollutants) or direct discharge of household, laboratory, or industrial waste. Water pollution and air pollution to a great extent contribute to levels of different human sicknesses and death rates in the very contaminated areas. Lack of good management practices regarding the conservation of natural resources (soil, water, air, and biodiversity) and ecosystem services and the lack of adequate management of waste are also very important issues.

In this entry, the goals are to improve the awareness and understanding of the importance of concepts or objectives like biodiversity, environmental health, and sustainability, and to give some suggestions on how the reader may contribute to individual and communal well-being.

Biodiversity

What Is Biodiversity?

Biodiversity or biological diversity is defined by the United Nations Convention on Biological Diversity as "the variability among living organisms from all sources, including terrestrial, marine, and other aquatic ecosystems and the ecological complexes of which they are part; this includes diversity within species, among species, and of natural and altered ecosystems,"[1,2] including species, ecosystem, morphology, gene, and molecular diversity.

Ecosystem means a dynamic complex of plant, animal, and microorganism communities and the inorganic environment, interacting as a functional unity. Different levels of diversity are considered, among individuals, subspecies, species (most useful level), biological communities, and ecosystems. Species richness increases from colder to warmer latitudes. This is also true for the deep-sea species diversity.[1]

The number of species on Earth is estimated to be 5 to 30 million, from which 1.4 to 2 million were identified in a formal system. The majority are invertebrates (1 million; mainly insects and myriapods), followed by microorganisms (5760; e.g., fungi), chelicerates, protists, nematodes, plants (250,000; around 50,000 as trees), molluscs, crustaceans, and vertebrates (19,100 fishes, 9000 birds, 6300 reptiles, 4200 amphibians, and 4000 mammals).

Occurrence of species diversity follows in this decreasing order: tropical and subtropical moist broadleaf forests >>> tropical and subtropical grasslands, savannas, and shrub lands > deserts and xeric shrub lands = tropical and subtropical dry broadleaf forests > mountain grassland and shrub lands > temperate broadleaf and mixed forests > flooded grasslands and savannas = tropical and subtropical coniferous forests > temperate grasslands, savannas, and shrub lands = mangroves = temperate coniferous forests > Mediterranean forests, woodland, and scrub >> boreal forests or taiga > tundra.[1]

The tropics are the home of most of the species. An example is the Amazonian rainforest, where 60% of all life-forms (e.g., 60,000 plant species) reside.[3] In one hectare of Atlantic rainforest in the southern Bahia state in Brazil, there are up to 454 tree species recorded.[4]

The reason for the species richness of the tropics is not well known. Some ideas proposed are the longer time available to develop new species and potentially greater supply of solar energy, allowing more biomass production or more organisms per unit of area.[1]

Furthermore, soils in temperate climates show greater chemical fertility, water-holding capacity, and clay activity, and the cold switch off of biological activity controls these processes. Under tropical conditions, deep soils with mostly low chemical fertility, low water-holding capacity, and low clay activity, and the higher temperatures throughout the year allow for a greater number of interactions of water/ drought × water table depth × nutrient availability × salinity × temperature/altitude/shade × strong rains × wind × fire × plant residues × organic matter content × photoperiod × oxygen (because of faster respiration rates and heat).

Therefore, habitat variability occurs, with specificities settled in by the different plant species, the first component of the food web and net.

Biological diversity is organized in a food web, with plants as base, harvesting sun energy freely available to them, and humans as top of the pyramidal net, where the individuals act as producer or consumer, or as recycler or decomposer.[5] The diversity of litter, defense substances, and root exudates produced by these different plant species, and correlated fauna, need to be chopped up by, e.g., invertebrates and decomposed by a greater number of microorganisms in soil, because of their specificity in producing degradation enzymes. The great recycling activities in soil need to be considered, because of the big importance of organic material as nutrient source for higher plants, as energy source for the microbial activity, and as a factor in improving soil structure and its water-holding capacity. In the tropics, with the great variability of habitats, the diversity of species is the keystone for high biomass yield per unit of area, making the food net of an ecosystem very complex.

Biodiversity reaches the maximum level when the environment offers enough water and energy and low to medium level of nutrients, such as nitrogen and phosphorus. With high levels of nitrogen and phosphorus occurs a very stiff inter- and intraspecific competition, mainly for light, from some more responsive or demanding species, similar to that occurring in high-fertility and very high-fertility soils or in eutrophic water bodies. There are also growth-restrictive conditions like in very low-fertility soils, with very low phosphorus and/or high aluminum content, or saline soils.

What Is the Importance of Biodiversity?

The greatest importance of biodiversity is still the optimized ecosystem services it provides.[1,6–9] Ecosystem services are flows of material, energy, and information on environmental structures (natural capital), which, combined with services, products, and human capital, generate human well-being. Usually, most of these processes could not be substituted in needed scale by any human technology, and if the ecosystem is extinguished, reversion is in general very difficult to realize in an economic way. The best way is conservation, including the costs in the price of products and services. Usually, ecosystem services occur in an imperceptible way, similarly to the involuntary and vital processes in our organism such as pumping of oxygenated blood or breathing.

Biodiversity provides three functional ecosystem services (production, regulation, and cultural) and a support service. Production and supply include mainly food, freshwater, timber, fiber, fuel, energy, genetic resources, medicines, wildlife, and others. Regulation includes maintenance of climate, carbon sequestration, soil conservation, wind and sea wave power, biodegradation and recycling of wastes, biological remediation of soils, and decontamination and cleaning up of water, cleaning up of air, maintaining soil permeability, and others. Cultural services could include cultural and heritage diversity; aesthetic, ethical, medicinal, and health knowledge; inspiration; educational, spiritual, and religious values; leisure; ecotourism; and so on. Support of life is carried out by maintaining a stratospheric oxygen-ozone layer to filter ultraviolet radiation and a greenhouse gas layer to filter infrared sun radiation and to retain partial infrared or heat radiation from surface, as well as by clouds, which reduce sunshine incidence on Earth surface, avoiding the burning of life. It is also carried out by maintaining a long water cycle (rain–interception–infiltration–storage–internal flow–evaporation and transpiration–air humidity–clouds) with distributed soft rains, by stabilizing air temperature and air humidity, and others.[1,2,10]

Biological diversity also provides resilience or a stabilizing effect on the food web, which sustains the human species, when an environmental disturbance occurs.[11]

Nature uses biodiversity to produce the maximum of life and biomass per square meter and year by optimal use of the available sun energy. Biodiversity is also the result of this settlement process of nature, with a great variety of abiotic conditions, mainly in the tropics.

At the same time, nature, by developing a food chain into a complex food net, allows for greater food availability and diversity for the individuals on the top of the food web or food pyramid, such as the human species, and also ensures their sustainability. However, when biodiversity of the food web is disrupted by the establishment of a monoculture (industrial cropping system) and/or when the environment is under subjection of a degradation process, a population outbreak of the more resistant or adapted members of the food web may occur such as the so-called parasites and pathogens.[12]

The soil is one of the most diverse habitats on Earth and contains one of the most diverse assemblages of living organisms, mainly in the humid tropics,[13] due to its plant diversity. In a broader view, it is advisable to consider soil as the undisassociable soil–plant interaction, mainly in the tropics and subtropics. Soil without a permanent living plant cover will lose its main function of harvesting rain and storing resident available freshwater, essential for life and biological production (food, fiber, wood, biofuel, etc.). This interaction will improve the degree of soil biodiversity, including also the rooting system architecture of the so-called weeds. Their rooting structure, as well as that of crops, may be used as a visible indicator of the degree of soil health.[14,15]

In both natural and agricultural ecosystems, the different groups of soil biota interacting with plants and their debris are responsible for, or strongly influence, the soil properties and optimize processes or ecosystem services such as soil genesis, soil structure, carbon, nutrient and water cycles, agrochemical movement or breakdown, plant protection, growth, and production.[5,16]

Soil organisms act in processes of synthesis or production, transformation and decomposition, or consumption of organic material, affecting abiotic and biotic components, transportation, and soil engineering. Therefore, soil biodiversity is a keystone for sustainable agriculture and it could be used as a good indicator of agro-ecosystem or soil health. Soil biodiversity does not necessarily refer to the number of individuals or species, but to the ratio of functional groups,[17,18] and the result or the tool of their activities, such as the presence and intensity of enzymatic activity.[12] It is necessary to remember the importance of soils.

The settlement of terrestrial environments by life was only possible by storing rainwater in rocks that nature developed to permeable soils, and these soils were maintained permeable by a triple-protection layer: plant canopy, litter, and surface rooting system. To succeed in our activities and also to meet our quality of life, it is necessary to maintain, restore, or mimic this structure and the processes involved using artificial technology.

It can be said that the disappearance of several ancient human civilizations with populations concentrated in cities was partially caused by food insecurity due to soil degradation, combined with freshwater shortage due to destruction of forests and soil permeability, therefore reducing the long water cycle, and by lack of sanitation and waste disposal affecting public health.[19] These problems of slowing down or disrupting ecosystem services are at present global in scale, with the emergence of a new problem—global warming and climate change. Besides, there occurs programmed or accidental inclusion of poisons, toxic substances, heavy metals, nitrates, phosphates, hormones, etc.

To have an idea of the current importance of the whole biodiversity, the monetary value of 17 ecosystem services required to sustain life and the biological production capacity of landscapes was estimated to be about $33 trillion/yr, against the $18 trillion of the global gross domestic product (GDP).[9]

What Destroys Biodiversity and Ecosystem Services?

This can be answered by the simple elimination of plants covering the soil and the prevention of their complete regrowth and occupation of soil surface, as well as by turning the soil impervious, by crusting, compaction, pavement, etc.

How Do We Take Care of Biodiversity and Essential Ecosystem Services?

An interesting point is that, instead of considering the natural climax ecosystem as reference for good environmental practices, we use the primary natural ecosystem (rocky landscape) as reference for characteristics we do not want in our agricultural or urban ecosystems. The following are environmental characteristics we have to avoid in our management program: the primary environment has no capacity to store water; it has no biological carrying capacity for higher species; it sustains no food chain nor web; it presents a very short water cycle (rain, evaporation and runoff); and it shows high temperature and air humidity amplitude during the day.

Hence, primary environments are unsuitable for life and production. For example, land and soil degradation will turn life- and production-friendly environmental characteristics into life- and production-unsuitable conditions (with impervious, compacted, dry, hard soil, like a rock), similar to that occurring in primary natural ecosystems.

Considering the growing soil and landscape degradation, for example, due to further erosion and salinity, we need to be aware that the process of desertification[23–25] is a great challenge in dry lands. These dry lands support 44% of all cultivated systems and are the origin of 30% of world's cultivated plants.[26,27]

Thus, first of all, we need to stop the landscape and ecosystem services degradation, by conservation practices. The second step is to recover soils and landscapes by simple harvesting and storing rainwater, and reducing water losses, e.g., by runoff or evaporation in excess, by protecting the soil surface against erosion, and by allowing the growth of a diverse plant cover, to turn the soil permeable and to stabilize air temperature and humidity.

It is necessary to prevent the destruction of the whole plant cover and their debris. The biologically diverse green areas in rural and urban environments are mandatory. Under tropical and subtropical conditions, this means that we have to avoid large-scale areas of pasture or cropland or buildings (cities) without trees. A permanent tree cover is important to maintain the evapotranspiration and windbreak service and to stabilize air temperature and humidity.[8]

Considering that the greatest land-use change, with massive deforestation, is mainly for agricultural purposes (as well as for wood harvest, coal production, and mining activity), it will significantly alter the micro- and mesoclimate in a region, increasing infrared radiation and heating up of lower atmosphere, and because it will promote the significant degradation of other essential ecosystem services due to its scale, agricultural practices based on ecological principles are advisable.

For large areas, the first step should be the practice of conservation agriculture[20] complemented by windbreaks and vaporizing tree cluster. For small areas, agroforestry[21,22] production systems are useful, mainly under tropical and subtropical conditions, considering that the tropics are the engine of global climate dynamics. Without trees, the heat production over terrestrial areas will be greater, the cloud production will be smaller, and the climate dynamics will be faster or stronger, more dangerous.

Examples of successful ecosystem services restoration are those in conservation agriculture[20,28–30] agro-ecological production systems,[31–33] and agroforestry,[34–39] aside from forest management[40] and those biodiversity conservation practices[41,42] that substitute paid environmental services for free ecosystem functions or services, by using ecological principles.

Also, the need to reduce the use of toxic substances, as well as the production of wastes and their random release to the environment, is urgent. Gaseous wastes affect human and biodiversity health, especially when considering the troposphere ozone and acid rain production, the increasing carbon footprint, and the large amount of smog production, brought about by landscape fire, fire from furnaces (used for coal production), or fire from household wood-burning stoves. Liquid waste degrades freshwater, turns it dangerous for human health, and increases freshwater shortages. Solid waste released in large amounts on the landscape will reduce land and water quality and will increase the ecological footprint.

The best indicator of biodiversity that provides adequate ecosystem services is when we have a greater number of plants per unit of area. Plants are associated with fauna and microorganism.[6]

Which Tools Should Be Used to Improve Biodiversity?

Considering the process nature uses to develop a certain site, or to recover a degraded soil or land under fallow, it could be seen that plant diversity is the key tool used. This is because it allows the complementary activity of individuals with different structures, functions, wastes (debris, root exudations, and others), and needs to flourish in one of the different habitats created by the diverse interaction of the abiotic and emergent biotic factors occurring.

Therefore, vegetative techniques to recover the permanent diverse plant cover together with the improvement of environmental legislation and education are in place.

Environmental Health

What Is Environmental Health?

According to the World Health Organization,[43,44] environmental health addresses all the physical, chemical, and biological factors external to a person, and all the related factors impacting behavior. It encompasses the assessment and control of those environmental factors that can potentially affect health. It is targeted towards preventing disease and creating health-supportive environments.

Which Environmental Conditions Will Affect Health?

Ecosystem degradation or disruption can impact on health in a great variety of ways.[43–46] The most important ones are freshwater pollution (lack of sanitation), contamination (heavy metals, hormones, poisons, and excess of medicines), and degradation (siltation and increase of nitrates and phosphates).

Perhaps we need to have a broader view of what is environmental health, considering that persons are also participants of our environment. In a health environment, for example, aside from waterborne diseases due to pollution and acute and chronic respiratory diseases due to air pollution, we need to also consider malnutrition due to food lacking micronutrients, extreme heat and very low air humidity, lack of education and training, low (or no) income, and so forth.

What Is the Importance of Environmental Health?

A healthy environment, with its natural resources and main structures conserved or improved, will allow the ecosystem services to run in order to benefit our well-being. A healthy environment will provide enough clean, fresh water (150 to 200 L/person/day); it will secure food (1500 to 2000 kcal/person/day, without toxic or dangerous substances and contaminants); it will provide clean air with enough humidity (around 10 g water as vapor/m^3 air, or 40%–60% relative humidity in a range of 20–24°C, the comfort range, without solid microparticles, smog, dangerous gases and substances, and inconvenient odor); and it will lead to stabilized temperature.

An excess or low level of air humidity will bring about and increase the occurrence of several diseases.[47] Unprotected surface/soil will produce higher temperature (Figure 1) and therefore greater amount of long-wave infrared radiation, heating up the lower atmosphere. In addition, increasing air temperature will mainly result in decreasing air humidity, because of the increase of atmospheric demand for water, when no water is available to be evaporated or transpired by plants (Figure 2).

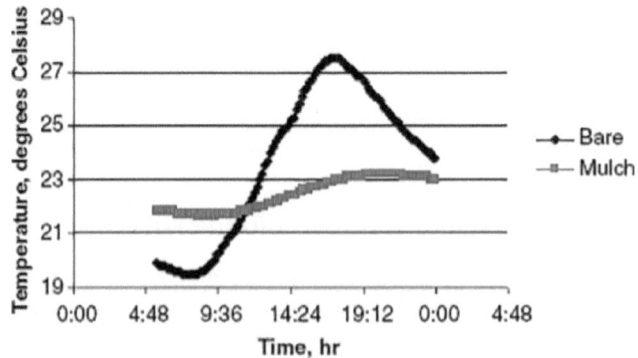

FIGURE 1 Temperature variation (amplitude), at 15 cm depth, in a bare and mulch-covered soil, under tropical conditions (November). The graph shows that it is possible to manage temperature extremes using mulching technique. **Source:** Adapted from Torres (1997) in Primavesi.[8]

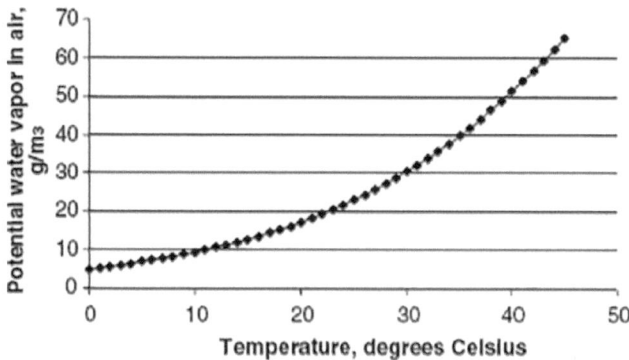

FIGURE 2 Relation between air temperature and potential water demand for air saturation.
Source: Adapted from Addams et al.[52]

Heat in excess, above 30°C, will sharply increase the productivity loss of labor and increase the mistakes or errors in the production line,[48] while polluted and dry air will increase respiratory diseases.[47] Polluted water will cause several gastroenteritis processes.[49–51] Human health degradation will cause problems for the production system, the health care service, and to our well being, with expectation of an earlier death.

What to Do?

The reactivation of the ecosystem services is needed, with the restoration of biodiversity, on a permeable soil, increasing rainwater yield and storage. It is also necessary to reduce waste production, by reduction, reuse, and recycling, as well as by neutralizing and adequate disposal of all harmful materials. We need to use water more efficiently.[52]

Another very important point is to stabilize air temperature (avoiding extremes) and air humidity amplitude, by restoring green evapotranspiring areas, by establishing shade (Table 1) or soil cover (Figure 1), or by managing the surface albedo.[54–62] Nature manages albedo or sunshine reflectivity in function of available liquid water. Water bodies, humid soils, or live plants are darker (with lower albedo) and may absorb more sun energy than dry surfaces, dry leaves (straw), or solid water (snow and ice) fields (with greater albedo).

A notable complementary problem is when fire use is routine in landscape management,[63] especially in dry periods, turning the albedo very low (black surfaces), increasing the heat, and therefore knocking down air humidity and quality, affecting health. Since we have global areas experiencing water shortage[64] [being dry for moderate (3 to 5 months) to long (>6 months) periods], this may result in

TABLE 1 Temperature of Shaded and Unshaded Surfaces in the Tropics

	Temperature		
	Shaded	Unshaded	Increase
Surface	°C		%
Green lawn	32	35	9
Dry lawn	35	52	48
Concrete	37	52	40
Asphalt	37	57	54

Source: Addams et al.[5]

heating up of soils and surfaces, with temperatures above 52°C,[65] and this will produce atmospheric heat in excess (>300 W/ m²),[66] surprisingly where there were no forests.[67]

Trees, due to their darker color, will absorb more sun energy and produce more heat, warming up a cold environment, but when air temperature rises above a certain level, it starts the evapotranspiration process, keeping away the heat in excess. Trees may warm up and cool down an environment, depending on the need. Therefore, it is advisable to make a global effort to plant trees (forests) and manage agroforestry systems to restore ecosystem services such as temperature and air humidity stabilization, as well as to maintain a longer water cycle with more and better- distributed rain.[68–76]

It is necessary to take into account that there are biophysical limits for economic growth (adjusted to nature's biological carrying capacity), which are mainly responsible for pollution as well as for biodiversity and land degradation. From the nine main limitations (biodiversity loss, nitrogen and phosphorus cycle, climate change, acidification of seawater, ozone reduction in stratosphere, freshwater availability, change of soil use, chemical pollution, and aerosol pollution in atmosphere), we did trespass the first three, endangering our health and livelihood.[77]

Indicators of Improvement

The presence of biodiversity developing on permeable soils and the absence of pollution, waste disposal in landscapes, and contaminants result in clean freshwater, soil, air, and food.

Tools to Be Used

Processes and tools based on ecological principles and processes to recover soil permeability (Figures 3 and 4), biodiversity, and ecosystem services need to be used. The keystone of the food web we depend on is plants, especially those with healthy roots. Roots need to be in aerated and humid soil, protected (against temperatures above 33°C, which harms root health), and rich in organic material.

The concept of integrated natural resource management considering the watershed scale, aside from integrated and efficient water, fertilizer, and pest management, is advisable. In relation to waste disposal, pollution, and contamination, environmental technology, eco-technology, and cleaner technology are all in place. In the case of fossil fuel use, the reduction of wasteful use and the substitution by alternative renewable energy sources are advisable. In all cases, improvement in environmental legislation and education is necessary.

FIGURE 3 Crusting of an unprotected prepared seedbed after tropical rainshower, turning soil impervious to rainwater and aeration.
Source: Author's personal archive.

FIGURE 4 Left: permeable healthy soil from a natural ecosystem, plenty of visible roots. Right: the same soil type, compacted, impervious, and dry from the inter row of a sugarcane field after 5 years of continuous intensive cultivation.
Source: Author's personal archive.

Sustainability

What Is Sustainability?

In biology, sustainability means the processes running to maintain biological systems diverse and productive over time.[5] In 1987, the World Commission on Environment and Development established a definition of sustainability, known as the Brundtland Report. It stated that sustainable development should reach the needs of the present without compromising the ability of future generations to meet their own needs. Although this definition has become widely publicized, the term *sustainability* is not limited to one precise definition.[78]

Several authors have discussed the real meaning of sustainability and sustainable development.[16,28,79] Munoz[80] has presented a theoretical model of true sustainability or sustainable development ideas with global to local implications based on the need to balance social, economic, and environmental goals; stakeholders' interests; and issues to induce or determine fairer or more appropriate development solutions, options, and actions.

However, the social component can be seen as part of the environmental component, and the interaction of a health and developed environment (essential ecosystem services running) with educated, trained, organized, and health persons will generate a long-lasting economic component. Hence, the environmental component is the keystone, to sustain the virtual world as well, and only its improvement and quality will allow us to reach a stable social welfare and a sustainable economic profit.

What Is the Importance of Sustainability?

Historically, the first known examples of worry on sustainability, and the establishment of a definition, were from those growing up in the forestry sector, and this is because of wood shortage for a salt mining activity in Germany or for house building and coal production in Japan. Both cases resulted from poor forest management and overexploitation. The question was, "What should we do to maintain a constant wood yield and cash flow for current and future forest owners', managers', and local workers' livelihood?" This question was discussed since 1442 in various regions of Germany, where wood consuming salt-mining activities did occur, and the concept was formulated firstly in 1650 in Saxonia (Germany) and later, independently, in 1666 in some deforested regions of Japan.[81]

The idea was that regenerative living resources, like wood/trees, could only be used/harvested in the same amount of the natural regrowth/recover/refill (with no use of external inputs, like water, energy, or fertilizer), and this by maintaining productivity, vitality, rejuvenation capacity, and biological diversity, in a time span of around 120 years (the time needed by trees to grow for cutting), to avoid natural resource shortage, labor and cash shortage, and livelihood and health problems (mainly related to erosion and flooding).[81]

In 1732, the idea on sustainable use of forest was published for the first time by von Carlowitz, and in 1795, Georg Ludwig Hartig described how to manage a sustainable forest.[81] History shows us also that a sustainable activity may last for 5000 years and more.[82]

Related to sustainable forest management, the current definition is as follows: "The stewardship and use of forests and forest lands in a way, and at a rate, that maintains their biodiversity, productivity, regeneration capacity, vitality and their potential to fulfill, now and in the future, relevant ecological, economic and social functions, at local, national, and global levels, and that does not cause damage to other ecosystems."[40]

Nature teaches us that the real development process of a natural primary environment occurs with the development or restoration of a permeable soil, protected by a diversified vegetation cover with an active rooting system, and the return to soil of diversified organic material, the energy source for the diversified and active soil life.

The soil–vegetation and associated biodiversity interaction could improve the available resident water of a site and also a longer water cycle. The more resident water, the more vegetation and the more permeable the soil, in a growing feedback loop. With greater amounts of available resident water-permeable soil-diversified vegetation and soil life, there is also an improvement in micro- and mesoclimate, with an increase in relative air humidity and a decrease in the maximum temperature and thermal amplitude, characteristic for desert environments.

This friendlier mesoclimate helps more sensible plant and animal species to establish and improves biodiversity, with their additive and emergent characteristics, mainly observed in the humid tropics. Ecology considers desert ecosystems sometimes as sustainable as drier tropical forest ecosystems, due to their richness in biodiversity. Thus, what level of environmental sustainability is desirable? It depends on the biological carrying capacity we want.

The biological carrying capacity represents the concept of primary productivity of an ecosystem, or the rate and amount in which energy is stored by photosynthetic or chemosynthetic activity of the producer organisms as organic substances, food for the food web.[5] The biological carrying capacity also considers the feeding capacity of grazing cattle, or grain equivalent available for humans (4 or 16 persons/ha/yr, with a minimum need of 1000 kcal/day), calculated as available digestive energy or calories per surface unit and year. The biological carrying capacity depends on the recovering capacity of a site (resilience) to produce biomass, after yield, extraction, degradation, or pollution activities.

Hence, considering the exuberant flora and biomass production by the Amazonian forest, the question arises: What is the biological carrying capacity of the around 40% sandy soils (<15% clay content) in the Amazonian basin without that great vegetation and the aggregated or dependent mesoclimate? Something similar to the Sahara? This example shows that the biological carrying capacity may be managed in a certain range by improving the natural resource structural tripod of resident water (harvested and stored rainwater) in permeable soil (organic matter rich) under biodiverse, permanent, alive plant cover and the ecosystem services.

Some agricultural production systems—although their processes result in improvement in environmental characteristics, income, and social inclusion—are not sustainable, because of their great dependence on external inputs [fertilizer, energy, water (including fossil underground water), and technical support]. We could observe, however, that, with time, it is possible to reduce part of this dependence, by switching to more organic and biological processes, after building up a minimum of fertility and organic matter level in soil, and introducing nitrogen-fixing leguminous trees. The goal of such process is to be more efficient and productive while maintaining or improving environment (natural resources) quality.[29]

What Needs to Be Focused on Sustainability?

Considering that the main objective of all human activity, from a global to a local scale, is to promote life and its quality, environmental sustainability will be reached when the biological carrying capacity is adequate to supply the minimal health life requirements of a given human population.

An increase in the biological carrying capacity level will allow a rise in human population density. Instead of this, what occurs now is the destruction of the main natural resource structures and functions (or ecosystem services), a decrease in the biological carrying capacity with an increase in human population density, and an increase in the production of solid, liquid, gaseous, and radiative wastes, thrown randomly in landscapes and marine ecosystems. We are currently watching a global ecological regression process of terrestrial and marine ecosystems, back to conditions unsuitable for human health, livelihood, and life.[8]

We need to reduce losses and wasting of materials, as well as to avoid pollution and contamination of natural resources and products.

Which Indicators Are Usable?

First, we need to reduce our ecological footprint (optimal land use),[83–88] our water footprint (optimal and efficient water use),[89–93] and our carbon footprint (carbon equivalent production in processes and carbon cycle)[94–96] by turning our production systems or life system more efficient and adequate to the global ecosystem carrying capacity. The measure of total sun energy use[97–100] is perhaps the best way to measure and turn the processes more sustainable.

Which Tools Should Be Used?

The following are the main tools we need to use: primarily, knowledge on ecological principles and processes;[5,97] a better understanding of the processes using energy, land, water, and primary products; a retooling of the processes to turn them more efficient, using materials more harmless to the environment and health; and a reduction of waste disposal in ecosystems. All of these are geared towards improving environmental legislation and education.

Conclusion

It is necessary to stop the degradation or regression and pollution processes of terrestrial and marine ecosystems and restore their health. The processes nature uses to develop complex resilient ecosystems with great biological (including human) carrying capacity are known. To conserve, restore, or improve a sustainable and healthy planet with the immense carrying capacity of the human species, biodiversity and ecosystem services should be considered, based on the ecological knowledge of essential natural structures and processes.

Moreover, as in nature, all kinds of wastes need to be reduced and recycled. Our technologies and processes need to maintain, restore, or mimic natural structures and processes to succeed and improve sustainability. A global awareness on the ecological principles and processes for a regional planning of the integrated local participatory activities to restore, maintain, or even improve the environmental health for sustainability of human societies is necessary.

References

1. Millennium Ecosystem Assessment. *Ecosystems and Human Well-Being. Biodiversity Synthesis;* World Resources Institute: Washington, 2005; 1–100. Available at http://www.millenniumassessment.org/documents/document.354.aspx.pdf (accessed March 2010).
2. Millennium Ecosystem Assessment. *Ecosystem and Human Well-Being: Synthesis;* Island Press: Washington, DC, 2005; 1–155. Available at http://www.millenniumassessment.org/en/Synthesis.aspx and http://www.millenniumassessment.org/documents/document.356.aspx.pdf (accessed March 2010).
3. Peneireiro, F.M.; Rodrigues, F.Q.; Brilhante, M.O.; Ludewigs, T. *Apostila do Educador Agroflorestal: Introdução aos Sistemas Agroflorestais—Um Guia Técnico;* Universidade Federal do Acre, Parque Zoobotânico: Rio Branco-AC, Brazil, 1999; 1–76. Available at http://www.agrofloresta.net/artigos/apostila_do_educador_agroflorestal-arboreto.pdf and http://www.slideshare.net/FlaviaCremonesi/apostila-do-educador-agroflorestal-arboreto-1353733 (accessed March 2010) (in Portuguese).
4. Girardi, E.P. *Atlas da Questão Agrária Brasileira;* Unesp: Presidente Prudente-SP, Brazil, 2008; 1–259. Available at http://www4.fct.unesp.br/nera/atlas/configuracao_territorial.htm (accessed March 2010) (in Portuguese).
5. Odum, E.P.; Barrett, G.W. *Fundamentals of Ecology,* 5th. Ed.; Brooks/Cole: New York, 2005; 1–612.
6. Convention on Biological Diversity. COP 6 Decision VI/7. Identification, monitoring, indicators and assessments; UNEP: Montreal, Canada, 2002. Available at http://www.cbd.int/decision/cop/?id=7181 (accessed March 2010).
7. Chivian, E., Ed. Biodiversity: Its importance to human health. *Interim Executive Summary;* Center for Health and Global Environment/Harvard Medical School/WHO/UNDP/UNEP: Harvard, USA, 2003; 1–59. Available at http://chge.med.harvard.edu/publications/documents/Biodiversity_v2_screen.pdf (accessed March 2010).
8. Primavesi, O.; Arzabe, C.; Pedreira, M.S. *Mudanças Climáticas: Visão Tropical Integrada das Causas, Dos Impactos e de Possíveis Soluções para Ambientes Rurais ou Urbanos;* Embrapa Pecuaria Sudeste: Sao Carlos-SP, Brazil, 2007; 1–200 (Embrapa Pecuaria Sudeste. Documentos, 70). Available at http://www.cppse.embrapa.br/080servicos/070publicacaogratuita/documentos/Documentos70.pdf (accessed March 2010) (in Portuguese, with a 3-page Executive Summary in English).
9. Costanza, R.; D'Arge; Groot, R.; Farber, S.; Grasso, M.; Hannon, B.; Limburg, K.; Naeem, S.; O'Neill, R.V.; Paruelo, J.; Raskin, R.G.; Sutton, P.; Belt, M. The value of the world's ecosystem services and natural capital. Nature **1997,** *15* (387, May), 253–260. Partially available at http://myweb.facstaff.wwu.edu/~medlerm/classes/08_09/502/nature-paper.pdf and http://earthmind.net/marine/docs/session2c-on-costanza-global-valuation.ppt (accessed March 2010).
10. Daily, G.C.; Alexander, S.; Ehrlich, P.R.; Goulder, L.; Lub-chenco, J.; Matson, P.; Mooney, H.A.; Postel, S.; Schneider, S.H.; Tilman, D.; Woodwell, G.M. Ecosystem services: Benefits supplied to human societies by natural ecosystems. Issues Ecol. **1997,** *1* (2), 1–18. Available at http://www.esa.org/science_resources/issues/FileEnglish/issue2.pdf (accessed March 2010).
11. Fischer, J.; Lindenmayer, D.B.; Manning, A.D. Biodiversity, ecosystem function, and resilience: Ten guiding principles for commodity production landscapes. Front. Ecol. Environ. **2006,** *4* (2), 80–86. Available at http://people.anu.edu.au/adrian.manning/ten_guiding_principles.pdf (accessed March 2010).

12. Primavesi, A.M. *Manejo Ecologico do Solo: A Agricultura em Regioes Tropicais*; Nobel: São Paulo-SP, Brazil, 1980; 1–541 (in Portuguese).

13. Bunning, S.; Jiménez, J.J. *Soil Biodiversity Portal: Conservation and Management of Soil Biodiversity and Its Role in Sustainable Agriculture*; FAO: Rome, Italy, 2004. Available at http://www.fao.org/ag/agl/agll/soilbiod/default.stm (accessed March 2010).

14. Food and Agriculture Organization. *Biological Management of Soil Ecosystem for Sustainable Agriculture*; Report of the International Technical Workshop: Londrina-PR, Brazil, 2002. FAO: Rome, Italy, 2003; 1–37 (World Soil Resources Report, 101). Available at ftp://ftp.fao.org/docrep/fao/006/Y4810E/Y4810E00.pdf (accessed March 2010).

15. Food and Agriculture Organization. *Integrated Crop Management: An International Technical Workshop Investing in Sustainable Crop Intensification—The Case for Improving Soil Health*; FAO: Rome, Italy, 2008; Vol. 6, 1–139. Available at http://www.fao.org/docrep/012/i0951e/i0951e.pdf (accessed March 2010).

16. Dumanski, J.; Gameda, S.; Pieri, C. *Indicators of Land Quality and Sustainable Land Management: An Annotated Bibliography*; The World Bank: Washington, DC, 1998; 1–126. Available at http://books.google.com.br/books?id=vY3HjKEJitkC&printsec=frontcover&dq=Indicators+of+and+Quality+and+Sustainable+Land+Management:+An+Annotated+Bibliography+dumanski&source=bl&ots=y67bbN2_L-&sig=hveLwdTFCNXfoPEpXFRTYdNRIMo&hl=pt-BR&ei=iIvxS9HQLsX_lgfW-cGzCA&sa=X&oi=book_result&ct=result&resnum=1&ved=0CBoQ6AEwAA (accessed March.2010).

17. Tilman, D.; Cassman, K.G.; Matson, P.A.; Naylor, R.; Polansky, S. Agricultural sustainability and intensive production practices. Nature **2002**, *418*, 671–677. Available at http://pangea.stanford.edu/research/matsonlab/members/PDF/TilmanNaylorMatson2002.pdf (accessed March 2010).

18. Scherer-Lorenzen, M.; Palmborg, C.; Prinz, A.; Schulze, E.-D. The role of plant diversity and composition for nitrate leaching in grasslands. Ecology **2003**, *84* (6), 1539–1552. Abstract available at http://www.jstor.org/pss/3107974 (accessed March 2010).

19. Liebmann, H. *Ein Planet wird unbewohnbar: ein Suenderegister der Menschheit von der Antike bis zur Gegenwart*; R. Piper & Co: Munchen, Germany, 1973; 1–181 (in German).

20. United Nations Environment Programme. *The United Nations Convention to Combat Desertification: A New Response to an Age-Old Problem*; UNDPI: New York, 1997. Available at http://www.un.org/ecosocdev/geninfo/sustdev/desert.htm (accessed March 2010).

21. United Nation Convention to Combat Desertification. GBO-3's Significant Findings Also in What's Unsaid about Landbased Biological Diversity. Press Release, 12/10, 2010. Available at http://www.unccd.int/publicinfo/pressrel/showpressrel.php?pr=press11_05_10 (accessed March 2010).

22. Menne, B.; Bertollini, R. The health impacts of desertification and drought. Down to Earth (Newsletter of the Convention to Combat Desertification) 2000 (14), 4–7. Available at http://www.google.com.br/url?sa=t&source=web&ct=res&cd=1&ved=0CB4QFjAA&url=http%A%2F%2Fwww.unccd.int%2Fpublicinfo%2Fnewsletter%2Fno14%2Fnews14eng.pdf&rct=j&q=The+health+impacts+of+desertification+and+drought&ei=EpnxS5y2KoL6lwf-9Ki0CA&usg=AFQjCNHNGLIRloJFAQZQzDiEUeqHzC9k-Q (accessed March 2010).

23. Food and Agriculture Organization. *Carbon Sequestration in Dryland Soils. Chapter 2. The World's Dry Land* (World Resources Report, 102). FAO/Natural Resources Management and Environment Department: Rome, Italy, 2004; 1-129. Available at http://www.fao.org/docrep/007/y5738e/y5738e06.htm and http://www.fao.org/ag/agl/agll/wrb/soilres.stm (accessed March 2010).

24. United Nation Convention to Combat Desertification. *Only One Earth—Drylands Are Vital*; UNCCD: Bonn, Germany, 2010. Available at http://www.unccd.int/publicinfo/announce/earth_day.php (accessed March 2010).

25. Food and Agriculture Organization. *Conservation Agriculture*. Available at http://www.fao.org/ag/ca/ (accessed March 2010).

26. Food and Agriculture Organization. Realizing the economic benefits of agroforestry: Experience, lessons and challenges. In *State of the World's Forests*; FAO: Rome, Italy, 2005; 88–97. Available at ftp://ftp.fao.org/docrep/fao/007/y5574e/y5574e09.pdf (accessed March 2010).

27. Hailu, M.; Landford, K.; Selvarajah-Jaffery, R.; Vanhoutte, K. (Coord.) *Agroforestry—A Global Land Use.* World Agroforestry Centre: Nairobi, Quenia, 2009; 1–53 (Annual Report 2008–2009). Available at http://www.worldagroforestry.org/downloads/publications/PDFs/B16416.PDF (accessed March 2010).

28. Dumanski, J.; Peiretti, R.; Benites, J.R.; McGarry, D.; Pieri, C. The paradigm of conservation agriculture. Proceedings of the World Association on Soil and Water Conservation, 2006, 58–64. Available at http://www.unapcaem.org/publication/ConservationAgri/ParaOfCA.pdf (accessed March 2010).

29. García-Torres, L.; Martínez-Vilela, A.; Holgado-Cabrera, A.; Gónzalez-Sánchez, E. Conservation agriculture, environment and economic benefits. International Congress of the European Society for Soil Conservation, 2000, 4, Valencia-Spain. Proceedings, 2000; 1–10. Available at http://www.unapcaem.org/publication/ConservationAgri/CA1.pdf (accessed March 2010).

30. Ngandwe, T. Conservation agriculture boosts yields and incomes. SciDev.Net 2006 (January 26). Available at http://www.scidev.net/en/news/conservation-agriculture-boosts-yields-and-incom.html (accessed March 2010).

31. Altieri, M.A. The ecological role of biodiversity in agroecosystems. Agric., Ecosyst. Environ. **1999**, *74,* 19–31. Available at http://comunidades.mda.gov.br/o/1540391 (accessed March 2010).

32. Uphoff, N. *Agroecologically Sound Agricultural Systems: Can They Provide for the World's Growing Populations?* International Research of Food Security, Natural Resource Management and Rural Development: Tropentag, 2005. University of Hohenheim: Stuttgart, Germany, 2005. Available at http://www.tropentag.de/2005/proceedings/node181.html and http://www.tropentag.de/2005/proceedings/node3.html (accessed March 2010).

33. Uphoff, N. Agricultural futures: What lies beyond modern agriculture? Trop. Agric. Assoc., Newsl., London, U.K., **2007**, *27* (3; September), 13–19. Available at http://www.slideshare.net/SRI.CORNELL/0906-agricultural-development-what-comes-after-modern-agriculture (accessed March 2010).

34. Wilkinson, K.M.; Elevitch, C.R. *Integrating Understory Crops with Tree Crops: An Introductory Guide for Pacific Islands*; Permanent Agriculture Resources: Holualoa, Hawaii, 2000; 1–50. Available at http://www.agroforestry.net/pubs/Understory.pdf (accessed March 2010).

35. Elevitch, C.R.; Wilkinson, K.M. *Agroforestry Guides for Pacific Islands*; Permanent Agriculture Resources: Holualoa, Hawaii, 2000; 1–239. Available at http://www.agroforestry.net/afg/ (accessed March 2010).

36. Beetz, A. *Agroforestry Overview*; ATTRA: Fayetteville, Arkansas, 2002; 1–16. Available at http://attra.ncat.org/attra-pub/PDF/agrofor.pdf and http://attra.ncat.org/attra-pub/agroforestry.html (accessed March 2010).

37. Davies, K. *Indian Agroforestry: Some Ecological Aspects of Northeastern India Agroforestry Practices,* 1984. Available at http://www.daviesand.com/Papers/Tree_Crops/Indian_Agroforestry/ (accessed March 2010).

38. Dixon, R.K. Agroforestry systems: Sources or sinks of greenhouse gases? Agroforestry Syst. **1995**, *31* (2), 99–116. Abstract available at http://www.springerlink.com/content/g106873871666629/ (accessed March 2010).

39. Leakey, R.R.B.; Tchoundjeu, Z. Diversification of tree crops: Domestication of companion crops for poverty reduction and environmental services. Exp. Agric. **2001**, *37* (3), 279–296. Available at http://www.wanatca.org.au/acotanc/Papers/Leakey-2/index.htm (accessed March 2010).

40. Convention on Biological Diversity. *A Good Practice Guide: Sustainable Forest Management, Biodiversity and Livelihoods*; CBD: Montreal, Canada, 2009; 1–47. Available at http://www.cbd.int/development/doc/cbd-good-practice-guide-forestry-booklet-web-en.pdf (accessed March 2010).

41. Hopper, K.; Summers, D., Eds. *Protecting the Source: Land Conservation and the Future of America's Drinking Water;* The Trust for Public Land and American Water Works Association: San Francisco, 2004; 1–56 (Water Protection Series). Available at http://earthtrends.wri.org/pdf_library/feature/eco_fea_value.pdf (accessed March 2010).

42. Rand Corporation. *New York City Depends on Natural Water Filtration;* Rand Co.: Santa Monica, California, 2007. Available at http://www.rand.org/scitech/stpi/ourfuture/NaturesServices/sec1_watershed.html (accessed March 2010).

43. World Health Organization. *Environmental Health;* WHO: Geneva, Switzerland, 2010. Available at http://www.who.int/topics/environmental_health/en/ (accessed March 2010).

44. World Health Organization. *Ecosystems and Health;* WHO: Geneva, Switzerland, 2010. Available at http://www.who.int/globalchange/ecosystems/en/ (accessed March.2010).

45. World Health Organization. *Water, Sanitation and Hygiene;* WHO: Geneva, Switzerland, 2010. Available at http://www.who.int/water_sanitation_health/en/ (accessed March 2010).

46. World Health Organization. *Water for Health: Taking Charge;* WHO: Geneva, Switzerland, 2001; 1–40. Available at http://www.who.int/water_sanitation_health/takingcharge/en/ (accessed March 2010).

47. Skuttle. *Impact of Relative Humidity on Air Quality;* Skuttle: Marietta, OH, USA, 2010. Available at http://www.skuttle.com/pdfs/skuttlehumidity.pdf (accessed March 2010).

48. Ciocci, M.V. *Reflexos do excesso de calor na saúde, e na redução da produtividade;* Cabano Engenharia: Belém-PA, Brazil, 2003. Available at http://www.cabano.com.br/excesso_de_calor.htm (accessed March 2010) (in Portuguese).

49. Corocoran, E.; Nellermann, C.; Baker, E.; Bos, R.; Osborn, D.; Savelli, H., Eds. *Sick Water? The Central Role of Wastewater Management in Sustainable Development. A Rapid Response Assessment*; United Nations Environment Programme: Nairobi, Quenia, 2010; 1–88. Available at http://www.cbsnews.com/htdocs/pdf/SickWater_screen.pdf (accessed March 2010).

50. Palaniappan, M.; Gleick, P.H.; Aallen, L.; Cohen, M.J.; Christian-Smith, J.; Smith, C.; Ross, N. *Clearing the Waters: A Focus on Water Quality Solutions*; United Nations Environment Programme: Nairobi, Quenia, 2010; 1–91. Available at http://www.unep.org/PDF/Clearing_the_Waters.pdf (accessed March 2010).

51. United Nations Environment Programme. *World Water Day 2010 Highlights Solutions and Calls for Action to Improve Water Quality Worldwide;* UNEP: Nairobi, Quenia, 2010 (March). Available at http://www.unep.org/Documents.Multilingual/Default.asp?DocumentID=617&ArticleID=6505&l=en&t=long (accessed March 2010).

52. Addams, L.; Boccaletti, G.; Kerlin, M.; Stuchtey, M. *Charting our Water Future: Economic Frameworks to Inform, Decision-Making;* WRG—2030. Water Resources Group/ McKinsey and Company: Chicago, USA, 2009; 1–198. Available at http://www.mckinsey.com/App_Media/Reports/Water/Charting_Our_Water_Future_Full_Report_001.pdf and http://www.nestle.com/InvestorRelations/Events/AllEvents/2030+Water+Resources.htm (accessed March 2010).

53. Gonçalves, C.E.C. *Ruas confortáveis, ruas com vida.: proposição de diretrizes de desenho urbano bioclimático para vias públicas. Av. Juscelino Kubitscheck, Palmas, TO*; UNB: Brasí0lia, Brazil, 2009; 1–149. (MSc Dissertation). Available at http://repositorio.bce.unb.br/bitstream/10482/3901/2/2009_CarlosEduardoCavalheiroGoncalves_pag_82_ate_final.pdf and http://repositorio.bce.unb.br/bitstream/10482/3901/1/2009_CarlosEduardoCavalheiroGoncalves_ate_pag_81.pdf (accessed March 2010).

54. Gash, J.H.C.; Shuttleworth, W.J. Tropical deforestation: Albedo and the surface-energy balance. Clim. Change **1991**, *19* (1–2), 123–133. Partially available at http://books.google.com.br/books?id=XB16EtxyZPUC&pg=PA123&lpg=PA123&dq=Tropical+deforestation:+albedo+and+the+surface-energy+balance&source=bl&ots=2PFRYbSs0l&sig=LOW2DnQ-d4-ZNgthrnFY4BxzYhu0&hl=pt-BR&ei=ccX1S9S3JoL-8Aavm9njCg&sa=X&oi=book_result&ct=result&resnum=5&ved=0CDwQ6AEwBA#v=onepage&q=Tropical%20deforestation%3A%20albedo%20and%20the%20surface-energy%20balance&f=false (accessed March 2010).

55. Gordeau, J. *Albedo;* Environmental Science Published for Everybody Round the Earth—ESPERE; Educational Network on Climate: Cracow, Poland, 2004. Available at http://www.atmosphere. mpg.de/enid/25w.html (accessed March 2010).

56. Russell, R. *Global Warming, Clouds, and Albedo: Feedback Loops;* National Earth Science Teachers Association: Boulder, CO, USA, 2007. Available at http://www.windows.ucar.edu/tour/link=/ earth/climate/warming_clouds_albedo_feedback.html (accessed March 2010).

57. Hartmann, D.L. *Global Physical Climatology;* Academic Press: California, USA, 1994; 1–386 (International Geophysics, v. 56). Partially available at http://books.google.com.br/books?id=Zi 1coMyhlHoC&pg=PA90&lpg=PA90&dq=albedo+straw&source=bl&ots=_SiR7JL67l&sig=8XHC aS4O43ICJR0XkagXhXN_5rs&hl=pt-BR&ei=urRgS9vqMsWtgeipuzYDQ&sa=X&oi=book_res ult&ct=result&resnum=10&ved=0CDgQ6AEwCQ#v=onepage&q=albedo%20 straw&f=false (accessed March 2010).

58. Yang, Z.L. *Physical Climatology: The Global Energy Balance,* 2006; 1–22. Available at http://www. geo.utexas.edu/courses/387H/Lectures/chap2.pdf and http://www.geo.utexas.edu/courses/387H/ default.htm (accessed March 2010).

59. Sharratt, B.S.; Campbell, G.S. Radiation balance of a soil- straw surface modified by straw color. Agron. J. **1994**, *86;* 200–203. Abstract available at http://agron.scijournals.org/cgi/content/ abstract/86/1/200 (accessed March 2010).

60. Triparthi, R.P.; Katiyara, T.P.S. Effect of mulches on the thermal regime of soil. Soil Tillage Res. **1984**, *4* (4), 381–390. Abstract available at http://www.sciencedirect.com/science?_ ob=ArticleURL&_udi=B6TC6-48XDCKB-1B&_user=10&_coverDate=07%2F31%2F1984&_ rdoc=1&_fmt=high&_orig=search&_sort=d&_docanchor=&view=c&_searchStrId=1183152286&_ rerunOrigin=google&_acct=C000050221&_version=1&_urlVersion=0&_userid=10&md5=19d9638 47840c192c30e24092fba1ad2 (accessed March 2010).

61. Heat Island Group. *Cool Roofs;* University of California: Berkeley, USA, 2000. Available at http:// eetd.lbl.gov/HeatIsland/ (at cool roofs) (accessed March 2010).

62. Levinson, R.; Berdahl, P.; Akbari, H. Solar spectral optical properties of pigments—Part II: Survey of common colorants. Sol. Energy Mater. Sol. Cells **2005**, *89* (4), 351–389. Available at http://www.odulo.com/eric/Enviroglobal/email%20documentsfPigments2.pdf (accessed March 2010).

63. European Space Agency. World fire maps now available online in near-real time. ESA Observing the Earth, News, 2006 (May 26). Available at http://www.esa.int/esaEO/SEMRBH9ATME_ index_0.html (accessed March 2010).

64. United States Department of Agriculture. *Soil Moisture Regimes Map.* USDA/Natural Resources Conservation Service/Soil Survey Division/World Soil Resources: Washington, DC, 1999. Available at http://soils.usda.gov/use/worldsoils/mapindex/smr.html (accessed March 2010).

65. Prata, F. *Global Distribution of Maximum Land Surface Temperature Inferred from Satellites;* CSIRO Atmospheric Research, Aspendale: Victoria, 2000; 1–13 (see page 8). Available at http:// www.eoc.csiro.au/associates/aatsr/lst_atlas.pdf (accessed March 2010).

66. The Centre for Australian Weather and Climate Research. *Current Climate Charts: Outgoing Longwave Radiation (OLR) Products—Latest Day and Night-Pass Data;* Bureau of Meteorology Research Centre: Australia, 2009. Available at http://cawcr.gov.au/bmrc/clfor/cfstaff/matw/map-room/index.htm (accessed March 2010).

67. Greenpeace. The World's Last Intact Forest Landscapes (poster A1), 2006. Available at http://www. intactforests.org/pdf.publications/World.IFL.2006.poster_low.pdf and http://www.greenpeace. org/international/campaigns/forests/our-disappearing-forests (accessed March 2010).

68. Natural Environment Research Council. *Satellites Reveal that Green Means Rain in Africa;* ScienceDaily, Sept. 27, 2006. Available at http://www.sciencedaily.com/releases/2006/09/060925064922.htm (accessed March 2010).

69. Coghla, A. More crops for Africa as trees reclaim the desert. New Sci. 2006 (October 14) (Article preview). Available at http://www.newscientist.com/article/dn10293-more-crops-for-africa-as-trees-reclaim-the-desert.html (accessed March 2010).

70. Makarieva, A.M.; Gorshkov, V.G.; Li, B.-L. Conservation of water cycle on land via restoration of natural closed-canopy forests: Implications for regional landscape planning. Ecol. Res. **2006**, *21*, 897–906. Available at http://www.biotic-regulation.pl.ru/offprint/wat_pr1.pdf (accessed March 2010).

71. Makarieva, A.M.; Gorshkov, V.G. Biotic pump of atmospheric moisture as driver of the hydrological cycle on land. Hydrol. Earth Syst. Sci. **2007**, 11, 1013–1033. Available at http://www.biotic-regulation.pl.ru/offprint/hess07.pdf (accessed March 2010).

72. Rowntree, P.R. Review of general circulation models as a basis for predicting the effects of vegetation change on climate (Ch. 8). In *Forests, Climate, and Hydrology: Regional Impacts*; Reynolds, E.R.C., Thompson, F.B., Eds.; The United Nations University: Tokyo, Japan, 1988. Available at http://www.unu.edu/unupress/unupbooks/80635e/80635E00.htm#Contents (accessed March 2010).

73. Sheil, D.; Murdiyarso, D. How forests attract rain: An examination of a new hypothesis. BioScience **2009**, *59* (4), 341–347. Abstract available at http://www.bioone.org/doi/abs/10.1525/bio.2009.59.4.12?journalCode=bisi and http://www.ncriverwatch.org/wordpress/2009/12/14/forests-attract-rain-an-examination-of-a-new-hypothesis/ (accessed March 2010).

74. Bonan, G.B. Forests and climate change: Forcing feedbacks and the climate benefits of forests. Science **2008**, *320* (5882), 1444–1449. Abstract available at http://www.sciencemag.org/cgi/content/abstract/sci;320/5882/1444 (accessed March 2010).

75. Takata, K.; Saito, K.; Yasunari, T. Changes in the Asian monsoon climate during 1700–1850 induced by pre-industrial cultivation. PNAS Online **2009**, *106* (24; June). Available at http://www.nagoya-u.ac.jp/en/pdf/research/activities/090529_hyarc_yasunari.pdf (accessed March 2010).

76. Pearce, F. Rainforests may pump wind worldwide. New Sci. 2009 (2702; April 1), 1–3. Available at http://www.ideastransformlandscapes.org/media/uploads/File/Rainforests%20may%20pump%20winds%20worldwide.pdf (accessed March 2010).

77. Rockström, J.; Steffen, W.; Noone, K.; Persson, A.; Chapin, F.S.; Lambin, E.F.; Lenton, T.M.; Scheffer, M.; Folke, C.; Schellnhuber, H.J.; Nykvist, B.; de Wit, C.A.; Hughes, T.; van der Leeuw, S.; Rodhe, H.; Sörlin, S.; Snyder, P.K.; Costanza, R.; Svedin, U.; Falkenmark, M.; Karlberg, L.; Corell, R.W.; Fabry, V.J.; Hansen, J.; Walker, B.; Liverman, D.; Richardson, K.; Crutzen, P.; Foley, J.A. A safe operating space for humanity. Nature **2009**, *461* (September), 472–475. Abstract available at http://www.nature.com/news/specials/planetaryboundaries/index.html (accessed March 2010).

78. Towers, P.T.; Lumper, K. *Definitions of Sustainability*; The University of Reading: Reading, U.K., 2010. Available at http://www.ecifm.rdg.ac.uk/definitions.htm (accessed March 2010).

79. Food and Agriculture Organization. *Land Quality Indicators and Their Use in Sustainable Agriculture and Rural Development*. Land and Water Bulletin 5, FAO, UNDP, UNEP, World Bank: Rome, 1996; 1–217. Available at http://www.mpl.ird.fr/crea/taller-colombia/FAO/AGLL/pdfdocs/landqual.pdf (accessed March 2010).

80. Munoz, L. Understanding sustainability versus sustained development by means of a WIN Development Model. In *Sustainability Review*; Flint, W., Ed.; Five E's Unlimited: Pungoteague, VA, Canada, 1999 (1; September 6); 1–18. Available at http://theomai.unq.edu.ar/artmunoz001.htm (accessed March 2010).

81. Wikipedia. *Nachhaltigkeit* (Sustainability). Available in German at: http://de.wikipedia.org/wiki/Nachhaltigkeit and http://de.wikipedia.org/wiki/Nachhaltigkeit_%28Forstwirtschaft%29 (accessed March 2010).

82. *Kanshie, T.K. Five Thousand Years of Sustainability? A Case Study on Gedeo Land Use (Southern Ethiopia)*; Kanshie: Wageningen, Netherlands, 2002; 1–295 (Treebook, 5). Available at http://www.treemail.nl/books (accessed March 2010).

83. Chambers, N.; Simmons, C.; Wackernagel, M. *Sharing Nature's Interest: Ecological Footprints as an Indicator of Sustainability;* James and James, Earthscan: London, 2001; 1–200. Partially available at http://en.book2down.com/Sharing-Natures-Interest-Using-Ecological-Footprints-As-An-Indicator-Of-Sustainability/186751 (accessed March.2010).

84. Wikipedia. *List of Countries by Ecological Footprint.* 2008. Available at http://en.wikipedia.org/wiki/List_of_countries_by_ecological_footprint (accessed March 2010).

85. Schaefer, F.; Luksch, U.; Steinbach, N.; Cabeça, J.; Hanauer, J. *Ecological Footprint and Biocapacity: The World's Ability to Generate Resources and Absorb Wastes in a Limited Time Period* (Working paper and studies). Office for Official Publications of the European Communities; Luxembourg, 2006; 1–11. Available at http://epp.eurostat.ec.europa.eu/cache/ITY_OFFPUB/KS-AU-06-001/EN/KS-AU-06-001-EN.PDF (accessed March 2010).

86. Global Footprint Network. *Footprint for Nations. 2009;* Global Footprint Network: Oakland, CA, USA, 2010. Available at http://www.footprintnetwork.org/en/index.php/GFN/page/footprint_for_nations/ (accessed March 2010).

87. World Wide Fund for Nature. *Living Planet Report 2008;* WWF International: Gland, Suisse, 2008; 1–48. Available at http://assets.panda.org/downloads/living_planet_report_2008.pdf (accessed March 2010).

88. Best, A.; Giljun, S.; Simmons, C.; Blobel, D.; Lewis, K.; Hammer, M.; Cavalieri, S.; Lutter, S.; Magguirre, C. *Potential of the Ecological Footprint for Monitoring Environmental Impacts from Natural Resources Use: Analysis of the Potential of the Ecological Footprint and Related Assessment Tools for use in the EU's Thematic Strategy on the Sustainable Use of Natural Resources;* Report to the European Commission, DG Environment, 2008; 1–312. Available at http://ec.europa.eu/environment/natres/pdf/footprint.pdf (accessed March 2010).

89. Waterfootprint Network. *Water Footprint versus Water Scarcity, Self-Sufficiency and Water Import Dependency per Country*; Water Footprint Network/University of Twente: Enschede, Netherlands, 2001. Available at http://www.waterfootprint.org/?page=files/NationalStatistics (accessed March 2010).

90. Waterfootprint Network. *Water Footprints of Crop and Livestock Products (m3/t) for Some Selected Countries (1997–2001)*; Water Footprint Network/University of Twente: Enschede, Netherlands, 2001. Available at http://www.waterfootprint.org/?page=files/Productwaterfootprint-statistics (accessed March 2010).

91. Hoekstra, A.Y.; Chapagain, A.K. Water footprint of nations: Water use by people as a function of their consumption pattern. Water Resour. Manage. **2007**, *21*, 35–48. Available at http://www.waterfootprint.org/Reports/Report18.pdf and http://www.waterfootprint.org/Reports/Hoekstra_and_Chapagain_2007.pdf (accessed March 2010).

92. Hoekstra, A.Y. *Water Neutral: Reducing and Offsetting the Impacts of Water Footprints.* Research report series nr. 28—Value of Water, UNESCO-IHE, 2008; 1–42. Available at http://www.waterfootprint.org/Reports/Report28-WaterNeutral.pdf (accessed March 2010).

93. Hoekstra, A.Y.; Chapagain, A.K.; Aldaya, M.M.; Mekonnen, M.M. *Waterfootprint Manual: State of the Art 2009;* Netherlands: Water Footprint Network: Enschede, Netherlands, 2009; 1–131. Available at http://www.waterfootprint.org/downloads/WaterFootprintManual2009.pdf (accessed March 2010).

94. Wikipedia. *Carbon Footprint.* Available at http://en.wikipedia.org/wiki/Carbon_footprint (accessed March 2010).

95. Wikipedia. *List of Countries by Carbon Dioxide Emission per Capita.* Available at http://en.wikipedia.org/wiki/List_of_countries_by_carbon_dioxide_emissions_per_capita (accessed March 2010).

96. Global Footprint Network. *Carbon Footprint*; Global Footprint Network: Oakland, CA, USA, 2009. Available at http://www.footprintnetwork.org/en/index.php/GFN/page/carbon_footprint/ (accessed March 2010).

97. Odum, H.T. *Environmental Accounting, Emergy and Decision Making*; John Wiley: New York, USA, 1996; 1–370. Partially available at http://dieoff.org/page170.htm (accessed March 2010).

98. Hau, J.L.; Bakshi, B.R. *Promise and Problems of Emergy Analysis*; Ohio State University: Ohio, USA, 2003; 1–13. Available at http://www.che.eng.ohio-state.edu/~bakshi/EcolModel3.pdf (accessed March 2010).

99. Ferreyra, C. Emergy analysis of one century of agricultural production in the rolling pampas of Argentina. Int. J. Agric. Resour. Governance Ecol. **2006**, *5* (2–3), 185–205. Available at http://snre.ufl.edu/graduate/files/publicationsbyalumni/Ferreyra%202006.pdf (accessed March 2010).

100. Martin, J.F.; Tilley, D.R. Accounting for environmental sustainability with emergy analysis. In: *Encyclopedia of Soil Science*; Lal, R., Ed.; Taylor and Francis Group: New York, 2007. Abstract available at http://www.mformaworld.com/smpp/content~db=all~content=a788663399 (accessed March 2010).

8

Biofertilizers

J. R. de Freitas

Introduction

Biofertilizers include microorganisms and their metabolites that are capable of enhancing soil fertility, crop growth, and/or yield. These include both indigenous microbes and microbial inoculants, that is, microorganisms that replace fertilizers or increase a crop's fertilizer use efficiency. Soil microorganisms such as bacteria, ectomycorhiza, arbuscular mycorrhizal fungi, and soil algae, especially the N_2-fixing cyanobacteria have potential as biofertilizers. Nitrogen-fixing inoculants based on *Rhizobium* species were among the first biofertilizers introduced into agroecosystems back in the 19th century. In the 21th century, biofertilizers will become an increasingly important area of research and development.[1] The use of fertilizers and pesticides has increased steadily since the 1970s; consequently, concerns about the impacts of these chemicals on land, air, and water have become significant environmental issues. Biofertilizers provide an alternative to agricultural chemicals as more sustainable and ecologically sound practices to increase crop productivity. Biofertilizer sales forecasts in the United States for the years 2001 and 2006 represent $690 million and $1.6 billion, respectively. Examples of some biofertilizers currently in use worldwide are shown in Table 1.[2]

TABLE 1 Organisms, Mode of Action, Crops, and Producers of Biofertilizers Currently in Use for Agriculture

Type	Mode of Action	Crop	Used in
Rhizobium spp.	N_2 fixation	Legumes	Russia; several countries
Cyanobacteria	N_2 fixation	Rice	Japan; several countries
Azospirillum spp.	N_2 fixation	Cereals	Several countries
Mycorrhizae	Nutrient acquisition	Conifers	Several countries
Penicillium bilaii	P solubilization	Cereals, legumes	Western Canada
Directed compost	Soil fertility	All plants	Several countries
Earthworm	Humus formation	Vegetables, flowers	Cottage industry

Source: Adapted from Tengerdy and Szakacs.[2] Copyright 1998 Elsevier Science.

TABLE 2   Market Price and Potential Price Increments with Yield Increases of 5% and 25% in Selected Vegetable Crops

Vegetable Crop	Market Price ($ per acre)	Price Increments ($ per acre)	
		5% Yield Increase	25% Yield Increase
Carrot	4,520	226	1130
Cauliflower	4,179	209	1045
Celery	10,132	507	2533
Cucumber	3,296	165	824
Lettuce	5,882	294	1471
Tomato	9,966	498	2492

Source: Adapted from USDA.[3]

Market for Biofertilizers

The market potential for biofertilizers includes the high value vegetable industry. A comparison of the base value of various crops and the increased value that can be obtained as the crop yield rises is illustrated in Table 2. Due to high nutrient requirements and high susceptibility to diseases, vegetable growers spend substantial amounts to protect this valuable produce. For example, average broccoli and tomato crops grown in California require ca. $62 and $170 worth of fertilizer and/or fungicide per acre, respectively. When the U.S. government prohibits the use of methyl bromide as a soil fumigant, as anticipated in 2005, development of biological products will be stimulated as an alternative to the use of chemicals.[3]

Mechanisms of Growth Promotion

The mechanisms covered in this entry are those that show commercial market potential; thus, it does not include all modes of action by which biofertilizers promote crop growth. Biofertilizers promote crop growth using several mechanisms with the primary one varying as a function of environmental conditions. Although the mechanisms of commercially available biofertilizers are not always entirely understood, growth promotion has been classified as the result of indirect or direct mechanisms. Indirect plant growth promotion may be associated with the repression of negative effects caused by phytopathogenic organisms, that is, biological control. Conversely, direct growth promotion mechanism may either provide some compound essential to crop development and/or stimulate nutrient uptake. Biofertilizers based on biological control agents *Mycorrhizae* and *Rhizobium* will be discussed in more detail elsewhere in this encyclopedia (e.g., Biological Pest Controls; Mycorrhiza; Rhizobia).

Phytohormones

Production of phytohormones is a commonly noted direct mechanism of plant growth promotion.[4] The nature of growth response may be the result of phytohormone production in the rhizosphere. Phytohormones are produced by many biofertilizers and include a list of plant growth regulators that are important in the plant's metabolism. For example, auxins such as indole-3-acetic acid are known for their ability to stimulate root cell division, differentiation, and promote cell elongation. Other phytohormones such as cytokinin, gibberellin, and ethylene also play key roles in plant development and have been reported to increase the growth of various commercial crops. The horticulture market for biofertilizer products based on gibberellin and other auxins is currently estimated at $600 million per year.

Plant Nutrient Acquisition

Several direct mechanisms are responsible for increased nutrient acquisition.

Biological Nitrogen Fixation (BNF)

Nitrogen (N) is an essential macronutrient, that is, it is the key building block of proteins, thus an indispensable component of the protoplasm of microorganisms, animals, and plants. The supply of biologically available N to agriculture through BNF represents ca. 140×10^6 ton/year, globally.[1] Therefore, BNF represents an economy of millions of dollars. N_2-fixation by free-livingbacteria such as *Azospirillum, Azotobacter, Bacillus,* and *Derxia* species have beenexploited in agricultural systems for many decades and constitute an important source of Ninput into agro-ecosystems.[5] Other BNF associations include the water fern *Azolla* that forms asymbiosis with the heterocystous cyanobacterium *Anabaena azollae.* The *Azolla–Anabaena* system has been used as a biofertilizer in Vietnam and China for rice production and has the potential to supply the entire N requirement (30–50 kg N/ha) for a rice crop during the growing season.[6] Another diazotroph, the N2-fixing actinomycete *Frankia,* forms nodules (actinor- rhizae) in ca. 17 genera of nonlegume wood species with *Alnus* (alder) and the genus *Casuarina* being the most important for forestry and agriculture. Estimates of total N2 fixed range between 50–250 kg/ha/year, depending on the plant species and region. However, in some cases, inoculation with *Frankia* is necessary for nodulation to occur. Ac- tinorrhizal plant species have been successfully inoculated with *Frankia* on a large scale. For example, millions of actinorrhizal trees, especially *Alnus* spp., inoculated with *Frankia* were used in land reclamation programs established in Canada.[7]

Phosphorus Solubilization

Certain microorganisms are very effective in solubilizing phosphorus (P) from insoluble phosphate compounds such as hydroxyapatite through the action of organic acids. Numerous claims have been made about biofertilizers that can enhance plant growth by solubilizing P. A classical example is the bacterium *Bacillus megaterium,* which was formulated into an inoculant under the name of Phosphobacterin in the former Soviet Union. A similar biofertilizer based on P-solubilizing fungi is currently marketed in Canada as JumpStart™ for use on wheat, canola, mustard, and N2-fixing legumes.

Microbial Siderophore Uptake

Iron (Fe) is an important plant micronutrient. Plants assimilate iron by acidifying the rhizosphere and/or secreting phyto-siderophores with subsequent reassimilation of the iron–siderophore complex. However, plants also may benefit from the direct uptake of microbial siderophor–iron complexes. For example, some biofertilizers synthesize siderophores that can solubilize and sequester Fe from soil and provide it to plant cells, thus contributing to the nutrition and development of crops. In fact, studies demonstrate that ferric pseudobactin 358 may stimulate chlorophyll synthesis in carnation and barley.[8]

Other Nutrients

Studies with *Azospirillum* spp. and plant growth-promoting rhizobacteria (PGPR) have demonstrated the ability of these biofertilizers to promote enhancement of nutrient and water uptake into the plant. For example, inoculation of winter wheat seeds with pseudomonad PGPR stimulated the uptake of soil-Fe and fertilizer-^{15}N by winter wheat cultivated in two Canadian soils.[9] Similarly, inoculation of canola seeds with a *Pseudomonas putida* increased phosphate uptake from nutrient solution.[10] In these cases, the authors speculated that plant growth regulators produced by the biofertilizers in the plant's rhizosphere stimulated root development which, in turn, enhanced nutrient acquisition.

Future Research Directions

It is clear that commercial crops can benefit directly from biofertilizers. Certainly, with the development of molecular biology and genetic manipulation of biofertilizers to improve N_2-fixation, rhizosphere competence and ability to be used together with specific chemicals, will contribute to an integrated strategy to reduce the total amount of chemicals used in agriculture. Although biofertilizer products are currently available on the market, consistency is still the major factor that limits their use. Elucidation of mechanisms, development of stable formulations, effective delivery systems, and field demonstration of effective biofertilizers will definitely improve reliability and enhance their use as commercial biofertilizers.

References

1. Killham, K. *Soil Ecology*; Cambridge University Press: Cambridge, U.K., 1994; 242.
2. Tengerdy, R.P.; Szakács, G. Perspectives in agrobiotechnology. J. Biotechnol. **1998**, *66*, 91–99.
3. USDA. USDA Economics and Statistics System. National Agricultural Statistics Service; Cornell University. http://mann77.mannlib.cornell.edu/reports/nassr/fruit/pvg-bban/vegetables_annual_summary-01.16.98 (accessed June 1999).
4. Glick, B.R. The enhancement of plant growth by freeliving bacteria. Can. J. Microbiol. **1995**, *41*, 109–117.
5. Pankhurst, C.E.; Lynch, J.M. The role of soil microbiology in sustainable intensive agriculture. Adv. Plant Pathol. **1995**, *11*, 230–247.
6. Zuberer, D.A. Biological Dinitrogen Fixation: Introduction and Nonsymbiotic. In *Principles and Applications of Soil Microbiology*; Sylvia, D.M., Fuhrmann, J.J., Hartel, P.G., Zuberer, D.A., Eds.; Prentice Hall: Upper Saddle River, NJ, 1998; 295–321.
7. Périnet, P.; Brouillette, J.G.; Fortin, J.A.; Lalonde, M. Large scale inoculation of actinorrhizal plants with frankia. Plant Soil **1985**, *87*, 175–183.
8. Duiff, B.J.; de Kogel, W.J.; Bakker, P.A.H.M.; Schipper, B. Significance of Pseudobactin 358 for the Iron Nutrition of Plants. In *Improving Plant Productivity with Rhizosphere Bacteria*; Ryder, M.H., Stephens, P.M., Bowen, G.D., Eds.; Third International Workshop on Plant-Growth Promoting Rhizobacteria, Adelaide, Australia, March 7–11, CSIRO Division of Soils: South Australia, 1994; 142–144.
9. De Freitas, J.R.; Germida, J.J. Growth promotion of winter wheat by fluorescent pseudomonads under growth chamber conditions. Soil Biol. Bioch. **1992**, *24*, 1127–1135.
10. Lifshitz, R.; Kloepper, J.W.; Kozlowiski, M.; Simonson, C.; Carlson, J.; Tipping, E.M.; Zaleska, I. Growth promotion of canola (rapeseed) seedlings by a strain of *pseudomonas putida* under gnotobiotic conditions. Can. J. Microbiol. **1987**, *33*, 390–395.

9

Ecosystems: Large-Scale Restoration Governance

Shannon Estenoz,
Denise Vienne,
and Alka Sapat

Introduction

Large-scale ecosystem restoration involves layers of complexity including, at the most fundamental level, ecological cause/effect relationships that may not be well understood, immediately observable, or measurable. Ecological complexity alone ensures that ecosystem restoration will be characterized by difficulty and uncertainty. However, in the context of large-scale ecosystem restoration, ecological complexity is compounded by other complexities including, for example, governance systems and regulatory frameworks not designed to approach restoration at the ecosystem scale and complex political interactions associated with resource consumption, private property rights, and economic development. Restoration leaders have created new governance structures, alternatively termed "collaborative resource governance" (CRG) or "collaborative environmental management," designed to work through and around such layered complexity.[1] While these structures vary in form and scope across restoration programs, there are common threads among them that include the scale and complexity of problems they are designed to address, emphasis on locally tailored solutions, the allowance for experimentation and dynamic adaption, and the importance of information sharing and collaborative problem solving at multiple overlapping levels.[2] The main focus of this entry is on the nature of CRG as a means of managing complexity and cross-jurisdictional issues in large-scale ecosystem restoration. In doing so, we review the evolution of these approaches in environmental and ecosystem management and the characteristics, challenges, and opportunities that such approaches present for governance and ecosystem restoration. More specifically, we discuss collaborative governance structures that have been adopted to manage two of the largest ecosystem restoration programs in the country, Gulf Coast ecosystem restoration and Everglades restoration.[3,4] Similar institutional mechanisms, i.e., ecosystem restoration taskforces, were put in place for both these ecosystems. These two programs present an opportunity for ascertaining whether synergies can be created in an ecosystem restoration program by cross-pollinating engagement at many levels with another restoration program.

We begin our discussion of this topic by examining how collaborative governance has emerged as one of the predominant mechanisms for addressing complex policy problems. Next, we note the evolution of environmental management in the United States toward large-scale ecosystem management through employment of collaborative governance. Although ecosystem restoration projects increasingly rely on this approach for governance and implementation, they are extremely diverse as a reflection of their political, administrative, and environmental contexts. Thus, the following section attempts to derive important lessons gleaned from these specific projects. Then, CRG will be defined as a distinct type of governance model that exhibits several general characteristics. A specific example of CRG, the ecosystem restoration task force model is the focus of the final sections. The Florida Everglades and the Gulf of Mexico, two of the world's largest ecosystem restoration projects that employ the task force model, will be discussed and compared. The conclusion elaborates how experience with these models can serve as the basis for future comparative research and as a guide for governance structures and environmental management.

Governance Models: Collaborative Environmental Management

Since the 1980s, collaborative governance has emerged as a predominant paradigm to resolve various policy issues, and this shift has been part of a gradual and broader movement in both the theory and praxis of public administration. Public managers have increasingly been addressing complex problems through networks, strategic partnerships, alliances, coalitions, contractual relationships, committees, consortia, and councils; these mechanisms have been used to develop collaborative mechanisms across jurisdictions, governments, and sectors. Collaborative governance approaches are thus increasingly being used in response to public demands and to deal with boundary-spanning policy problems that cannot be addressed by traditional regulatory approaches relying on single agencies and jurisdictions.[5,6]

Governance models for natural resource management have mirrored and, in some instances, even led to the gradual shift in collaborative governance, due to the emergence of various forms of partnerships and civic environmentalism that emerged to deal with environmental problems.[7,8] Reasons for the emergence of collaborative environmental management are many, overlapping, and varied. Some types of collaborative environmental governance emerged as a result of federal cutbacks, leading to civic environmentalism and citizens pushing for environmental changes.[8] For the most part, though, forms of collaborative environmental management came about as a result of the problems stemming from command-and-control measures, which often resulted in protracted conflict or failure to achieve key objectives. This led to calls for change from bureaucratic, adversarial, technology-based regulatory approaches, which were the basis for many environmental policies in the 1970s and early 1980s, to "results-based" and voluntary approaches to regulation.[9] Similar to other policy problems, calls for more collaborative measures also stemmed from the complexity of environmental problems that were not effectively addressed by conventional rule-based, top-down, and hierarchical approaches. The latter were seen as problematic in that they did not allow for more democratic forms of participation, stymied potential innovation, were ineffective in addressing multimedia environmental hazards and those stemming from nonpoint sources of pollution, and relied on unrealistic models of administrative and individual rationality.[7,9–11] To deal with these cross-jurisdictional problems, there were calls for more holistic and integrated approaches to deal with ecosystems such as large watersheds and forests. The adoption of more holistic approaches through collaborative environmental management grew during the 1990s particularly the Clinton administration; during this period, the ecosystem approach was greatly expanded both to better administer large restoration projects (such as the Everglades and the Northwest Forest plans) and to address smaller ecosystem and habitat conservation planning.[12]

Collaborative environmental management takes many forms: some entail collaboration between multiple agencies only, while others are characterized by the presence of multiple stakeholders such as actors in the private and nonprofit sectors. In some cases of collaborative environmental management, government has led the efforts; in others, government has encouraged it through grants and other incentives; and, at times, the efforts at collaborative management have been pioneered by other stakeholders such as citizens or nonprofit groups.[13] What distinguishes these collaborative approaches from earlier forms of environmental management is the movement away from command-and-control policies toward those that are both more inclusive of and seek to incentivize public participation. The search for mutually beneficial policy solutions by encouraging broad participation from local stakeholders, underscoring the importance of consensus and voluntary approaches, and building trust-based policy networks are defining characteristics of collaborative environmental approaches. In addition, collaborative environmental governance also involves improvements in scientific understanding of how ecological processes affect resource outcomes across artificial jurisdictional and political boundaries[14] and careful scientific monitoring to allow for managerial adaptation as necessary.[12]

Since collaborative environmental governance is an institutional mechanism for natural resource management, its development and success may depend on the set of institutional rules applied in ecosystem management. As noted by Elinor Ostrom, successful institutions tend to have certain design principles, such as rules adapted to local circumstances, clearly defined resource boundaries, information about resource variability, monitoring and sanction mechanisms, and local conflict resolutions forums.[15] While these principles may not ensure success, they could be contributory factors, as noted in our discussion about CRG structures adopted for large-scale ecosystem restoration.

In the next section, we discuss how collaborative governance approaches have been used in ecosystem management, which has seen a shift from place-based to species-based conservation.

Collaborative Governance Approaches and Ecosystem Management

Over the past 30 years, place-based and species-based approaches to environmental conservation have been giving way to more holistic and collaborative approaches to ecosystem management.[16] The earlier strategies involved prescriptive, or command-and-control, regulatory approaches that targeted discrete, identifiable sources of environmental damage.[17] Traditional regulatory approaches target the "low-hanging fruit" with respect to environmental problems, but they are insufficient for resource management and the restoration of ecosystems that are subject to multiple interdependent and interacting conditions, lacking clear or easily discernable sources of damage.[17] Collaborative ecosystem governance is a response to seemingly intractable environmental problems that do not fall neatly within the domain of any one governmental jurisdiction or agency.[1] Comparatively speaking, there is a great deal of variation in how such collaborative governing bodies are convened, their constellation of participants, the type and pattern of interactions, and the ecosystem problems they manage.

On the surface, these diverse institutional arrangements appear to be distinct problem-specific environmental responses because they conform to the particular resource systems they are designed to manage. However, collaborative ecosystem governance models share common characteristics that set them apart as a distinctive alternative to the traditional legal and institutional approaches relied upon for environmental management.[2] Furthermore, since they are characteristically more flexible, participatory, deliberative, and heterarchical forms of organization, they are considered to be more responsive, legitimate, and effective. Yet, little is known about the preconditions for their effectiveness or the ultimate consequences for democratic processes.[16]

The concern for democratic process is not inconsequential, and it is twofold. In the most general sense, the nature of these new governance structures precludes direct lines of political accountability. The decision-making process encompasses a diverse range of public and private actors along with nonelected administrators, and it supersedes jurisdictional authority. This not only threatens political accountability but also can be an impediment to collaboration to the extent that it undermines the political autonomy of participants. For example, governors are likely to be faced with choosing which to subordinate—the political will of their state or the will of the collaborative governance structure. In the case of interstate water compacts, this dilemma lies at the heart of legal conflicts. Despite commitment to interstate water agreements, legal recourse is often the only viable mechanism for conflict resolution when public officials are trapped between the terms of their compact agreements and evolving demands of their constituencies.[18] Given the long-term nature of these governance models, the challenge is to address both the issue of democratic process and strategies for maintaining broad-based public support and the political will necessary to sustain and inform them.

Opportunities and Challenges of Large-Scale Ecosystem Restoration

There are numerous challenges presented by collaborative ecosystem governance that stem from the complexity inherent both in the nature of large-scale ecosystem restoration and in the governance models themselves. Conceptualizing resource management as ecosystem management expands the range of considerations to include multiple competing uses and users, nonpoint sources of damage, nonlinear environmental reactions, and uncertain interactive effects and threshold points. The corresponding governance models are no less complex because they involve designing governing arrangements to accommodate specific environmental issues that are often framed by competing economic, political, and social contexts. As such, they necessarily overlap political jurisdictions and the functional boundaries of agencies, and they encompass multiple levels of governmental decisionmakers, regulatory bodies, and nongovernmental participants. Sustaining active cooperation, maintaining public and political support, and securing sufficient commitment resources necessary for uncertain and changing environmental problems over an indeterminate time frame are just some of the practical challenges. At the same time, embracing postsovereign governing arrangements, "rolling-rule" regulatory models, and adaptive management techniques represent paradigmatic shifts that entail philosophical challenges, which depend on more fundamental changes.[16,17,19,20] Reliance on intersovereign agreements and sovereign-dominated approaches has not been an adequate match for the scale of ecosystem problems particularly when they merely result in lowest-common-denominator agreements on uses and measures.[19] Nevertheless, expanding the range of issues addressed and the number of parties involved in governance raises questions about the role of government in facilitating effective decisions operating outside direct political control and in the "shadow of hierarchy."[16] In addition to overcoming philosophical and practical challenges, there is the additional challenge of designing the appropriate governance structure given the context and the particular ecosystem problem to be addressed.[20]

   The propensity to focus on the challenges overlooks important opportunities presented by large-scale ecosystem governance. Collaborative environmental management offers a highly dynamic and participative approach to policy implementation that integrates localized knowledge and scientific learning into future decisionmaking.[2] Adaptive management approaches are more conducive to experimentation required in conditions of uncertainty, allowing for creativity, flexibility, and learning.[17] Broad-based participation improves the public and political buy-in necessary for credibility and legitimacy, the commitment and pooling of resources, and long-term collaboration.[20] Moreover, collaborative governance models offer an opportunity to address long-recognized challenges presented by the decentralization and devolution of government and related concerns about democratic control associated with the increasing discretion afforded networks of administrators and nongovernmental actors in performing functions of the state. The state has an important role to play in preventing dysfunction and facilitating these governance models by ensuring democratic processes and integrating technical and normative considerations toward effective implementation.[16] Distinct from networks of policy actors, public–private partnerships, or devolved federal responsibilities, CRG is a much narrower, issue-specific, model of democratic governance.

Collaborative Resource Governance

Collaborative resource governance can be defined as a diverse group of public and private stakeholders working together to address shared problems that extend beyond their individual capacities.[1,16] Unlike other policy configurations common in the literature, such as public–private partnerships, policy networks that seek to influence government, or networks of government actors that informally coalesce around "wicked problems," they are distinctive problem-solving, polyarchic governance models focused on complex ecological problems that overwhelm the capacity of the sovereign state.[17] Although examples of CRG vary in scale, focus, and structure, they are an increasingly important focus of research for the purpose of ascertaining factors that determine their effectiveness, the specific role of the state, and their level of success in managing large-scale ecological problems.[20,1]

The form a collaborative governance model assumes depends on environmental, political, and economic contexts. Comparative research is beginning to identify specific factors that contribute to how they function, as well as emergent concerns. For the most part, the diversity of these initiatives belies similarities among them. Some defining characteristics of CRG are its hybrid public-private structure, the scale and complexity problems they are designed to address, emphasis on locally tailored solutions, the allowance for experimentation and dynamic adaption, and the importance of information sharing and collaborative problem solving at multiple overlapping levels.[2] Success, on the other hand, is highly dependent on how they are organized, funded, and governed.[21] Integrating science into the decision-making process has also been seen as critically important, necessitating broad-based participatory processes that preclude science from being manipulated, or trumped, by the political process.[20] Process is just as important as structure for ensuring flexibility and adaptability in order to incorporate new scientific information and to make necessary course corrections. Despite the dependence on nongovernmental actors and decentralized and fragmented approaches, there is a critical role for government in providing definitional guidance, participatory incentives, and enforcement capabilities.[16] One common problem is that most initiatives severely underestimate the expense (financial, time, and personnel investment) of collaborative approaches to ecosystem management relative to traditional regulatory approaches.[20] Examples of CRG programs are indicative of specific lessons learned.

The Chesapeake Bay Program is generally acknowledged as one of the oldest and most organizationally successful of these programs despite questionable improvements in water quality.[17] It exemplifies the importance of having reliable mechanisms in place in order to facilitate participation, public outreach, and the integration of scientific information into decisionmaking.[20] Diffuse governance that is not contained by traditional political authority requires special effort not only to capitalize on the respective strengths of participants but also to cultivate legitimate and credible processes.

The California Bay-Delta Program (CALFED) has demonstrated the importance of generating broad-based scientific input, instituting internal and external peer review of scientific proposals, and devising conceptual models for communicating scientific information to decision makers and stakeholders. The CALFED has also shown the benefit of dedicating facilitators and planners to ensure vertical integration and secure long-term funding. However, problems did emerge with CALFED; for instance, several independent reports criticized its governance structure, stakeholders moved water management decisions outside the CALFED process using environmental litigation, the executive director and lead scientist resigned in 2004, and faced with mounting criticism, the CALFED Bay-Delta Authority voted to disband itself in 2005.[14]

The Comprehensive Everglades Restoration Plan (CERP) is an example of how clearly defined problems and agreement over the urgency of an environmental issue can coalesce political and financial support. However, CERP is also an example of how political tensions between state and federal levels can impede cooperation. These tensions stem in part from prematurely determining CERP's organizational structure and possible alternative solutions for Everglades restoration prior to scientific involvement, which constrained options by favoring methods preferred by the United States Army Corps of Engineers (USACE).[20] The CERP has demonstrated the benefit of adaptive monitoring in the early stages of new strategies and also, for the sake of long-term goals, the need for clarifying performance measures and indicators.

Since the 2010 oil spill, the Louisiana Coastal Area Ecosystem Restoration Program (LCA) has transitioned to the task-force model followed by CERP. However, problems with its previous model are instructive. The LCA, like CERP, experienced state-federal tensions for differing reasons but which were also exacerbated by restrictive control under the USACE. Options available to the LCA were severely constrained by political pressures by powerful stakeholders. Still, the LCA's structure and processes were not adequately designed to facilitate broad-based participation, generate public buy-in, or integrate science in ongoing decision-making.[20] The result was an excessive focus on local symptoms to the exclusion of root causes tied to resource practices in the Mississippi Ohio–Missouri river basin.

Lastly, the Glen Canyon Adaptive Management Program exemplifies the potential for adaptive management in these new governance models.[20] Adaptive management emphasizes learning, adjustment, and the acceptance of rolling rules rather than grounding in static regulatory or managerial approaches. However, there is still much misunderstanding about these approaches that requires education in order to facilitate a transition from traditional, static approaches to adaptive methods of regulation and localized policy implementation.[17] The experiences of these CRG programs are indicative of jurisdictional issues relating to ecosystem management that do not easily correspond with traditional forms governance or regulatory approaches.

Cross-Jurisdictional Issues

Jurisdictional issues are of particular importance to CRG because the geographic dimensions of ecosystem restoration defy traditional political and functional boundaries, presenting new challenges for environmental regulation and management. The geographic scale of an ecosystem means that it often will encompass some combination of municipalities, counties, states, regions, and nations.[17] Similarly, resource users and sources of environmental degradation are not contained by any particular jurisdiction or subsumed under any one political authority. Thus, problems neither fall neatly within the control of any particular authority nor the functional realm of any particular agency. Nevertheless, the magnitude of environmental problems exceeds the capacity, resources, and expertise of any one governmental entity.

Traditional regulatory methods are not equipped to deal with uncertainty, complexity, and continuous change. They are appropriate for targeting point-source problems but cannot account for the diversity presented by numerous local circumstances.[16] Legal scholars are examining the implications of CRG for the future of environmental regulation and the use of adaptive rolling rules more suitable to uncertainty and change.[16,17,19] Similarly, the decentralized and fragmented governance structure necessitated by the nature and magnitude of resource problems presents an enormous challenge for management. There are numerous public and private parties involved, and addressing problems of this scope depends on long-term commitment and public support. This requires managing collaborative activities of numerous parties from multiple jurisdictions and diverse functional backgrounds that often have competing interests. The challenge is not only to coordinate participants but also to sustain cooperation. Although CALFED demonstrated the importance of broad participation and "bottom-up" policy approaches, the problems that emerged highlighted the importance of having a clear direction and identifiable goals. Authority may be diffuse, but leadership is imperative in order to maintain momentum, negotiate compromises among competing interests, and translate and communicate across functions (e.g., political decisionmakers and scientists, scientists and the public, and across agency missions and cultures). Adaptive management is appropriate for CRG because it allows for experimentation, learning, adaptation, and course corrections.[17,20] Adaptive management represents a paradigm shift, and possibly a hurdle, for politicians, administrators, and regulators used to operating within defined jurisdictions. As researchers consider the list of factors that determine the success or failure of CRG as a general approach to large-scale ecosystem restoration, it is helpful to highlight one specific CRG structure that has been both in place for a considerable period of time and recently replicated. As such, may provide important and unique opportunities to increase understanding of CRG.

Focus: Ecosystem Restoration Task Force Model

The restoration of the Gulf of Mexico and the restoration of the Florida Everglades are two of the largest and most complex ecosystem restoration programs in the world.[22–24] The highest level of intergovernmental coordination for both programs occurs under the auspices of intergovernmental task forces created by the federal government. This model warrants closer consideration because although the two task forces differ somewhat in constitution and scope, there are enough similarities between them to make the case that they are two examples of a single intergovernmental coordination and governance structure. This "duplication" of a specific CRG structure in two large-scale ecosystem restoration programs provides an opportunity to observe the successes and challenges for this institutional design and structure over time in a way that cannot occur when focused solely on a structure that is unique to a specific ecosystem. In addition, the model warrants closer consideration because of its longevity in the case of the Everglades and, in the case of the Gulf Coast, its relevance to the most current and high profile issues in large-scale ecosystem restoration.

South Florida Ecosystem Restoration Task Force

The origins of the South Florida Ecosystem Restoration Task Force (SFERTF) can be traced to the fractured state of intergovernmental relationships that existed in the late 1980s due to contentious litigation between the federal government and the state of Florida over degraded water quality in the Everglades.[23] The litigation, which began in 1988 and continues today, produced a settlement agreement in 1992 that, for the time being, opened a window of opportunity to improve intergovernmental relations in order to begin addressing a broader range of Everglades restoration issues.[23,25] At the beginning of the Clinton administration, then secretary of the U.S. Department of the Interior, Bruce Babbitt, exploited the litigation lull and ushered in a new era of collaboration on Everglades restoration issues, which included the first incarnation of the SFERTF, created by Bruce Babbitt.[23,25]

Membership of this original SFERTF was limited to six federal agencies, but 3 years later, Congress created a new task force (also called the SFERTF) and expanded its membership to include nonfederal representatives.[23] The new Task Force was to be chaired by the Secretary of the Department of the Interior and was to include the secretaries of the Commerce, Army, Agriculture, and Transportation departments, the Administrator of the Environmental Protection Agency (EPA), and the attorney general. Congress allowed these Presidential Cabinet members to designate appointees to represent them; however, the statute requires that designees be at the assistant secretary or equivalent level of authority.[4] The SFERTF also includes representation for the state of Florida, the Miccosukee Tribe of Florida, the Seminole Tribe of Florida, the South Florida Water Management District, and local governments. In the same legislation, Congress directed the Secretary of the Army to develop a "... proposed comprehensive plan for the purpose of restoring, preserving and protecting the South Florida ecosystem" [(4) §§ 528(b)(1)(A)(i)-528(f)(3)].

Initially, the SFERTF was charged with a number of coordination and oversight responsibilities during the restoration plan development phase; however the statute did not include a sunset provision to dissolve or transform the SFERTF at the completion of that phase. On the contrary, over the decade that followed, the Congress wove SFERTF oversight into the implementation phase of restoration, as did federal regulations governing the restoration program that were developed in 2002.[26,27] The responsibilities of the SFERTF include consultation with the Secretary of the Army; coordination of restoration policy, strategy, priorities, and programs; the exchange of information among task force members; facilitation of intergovernmental dispute resolution; coordination of restoration science;

the support of implementing agencies; the coordination of financial reporting and budget requests; and reporting biennially to Congress on its own activities and on the progress of restoration efforts and results.[4]

Gulf Coast Ecosystem Restoration Task Force

In the spring of 2010, the Deepwater Horizon oil spill became one of the worst man-made environmental disasters in American history.[22] In total, an estimated 4.9 million gallons of oil was released into the Gulf of Mexico, resulting in short- and long-term environmental and economic impacts that may not be fully quantified or understood for years to come.[22] The emergency response to the Deep-water Horizon crisis required significant intergovernmental coordination to address rescue and recovery, well closure, and cleanup.[3,22] During the response period, U.S. President Barack Obama ordered the Secretary of the Navy, Ray Mabus, to develop a vision for moving from response to recovery and restoration in the Gulf of Mexico.

The so-called "Mabus Report" was released in September 2010 and took an expansive view of Gulf restoration that went beyond ecological damage caused by the oil spill and included consideration of the broad range of ecosystem challenges in the Gulf and Gulf Coast that preceded the Deepwater Horizon crisis.[22] Secretary Mabus issued a set of recommendations, which in part focused on short- and long-term intergovernmental coordination in the recovery and restoration effort. The secretary recommended that Congress establish the Gulf Coast Recovery Council to coordinate federal, state, and tribal restoration and recovery actions and to coordinate with and support activities conducted under the Natural Resources Damage Assessment (NRDA) process.[22] However, the Secretary also recommended the immediate designation of a lead federal restoration agency and the immediate creation of the Gulf Coast Restoration Task Force. The executive branch of government could carry out these recommendations without Congressional action. Mabus suggested that the task force initiate the development of a restoration and recovery strategy for the Gulf and pointed out that if Congress acted to establish the recommended council, it could subsume the task force.[22]

A month after the release of the Mabus Report, President Barack Obama created the Gulf Coast Ecosystem Restoration Task Force (GCERTF) by Executive Order.[3] The President's order states that "[t]o effectively address the damage caused by the BP Deepwater Horizon Oil Spill, address the [sic] long-standing ecological decline, and begin moving toward a more resilient Gulf Coast ecosystem, ecosystem restoration is needed" (Section 1). The Executive Order specifies federal membership of the GCERTF as including "senior officials" [Section 2 (a)(1)] of the Departments of Defense (Army Civil Works), Justice, Interior, Agriculture, Commerce, and Transportation; the EPA; the White House Offices of Management and Budget and Science and Technology Policy; and the White House Councils on Environmental Quality and Domestic Policy. The task force includes representatives of the five Gulf coast states and has the authority to add representatives of affected tribal governments.[3] The GCERTF was created as an advisory body to coordinate intergovernmental restoration efforts, support the NRDA process, present to the President a strategy for Gulf restoration, coordinate scientific research, engage stakeholders and the public, and report to the President biennially on the progress of the restoration strategy.[3]

Common Threads between the SFERTF and GCERTF

The similarities of structure and scope are evident between these two governance bodies. Every federal agency represented on the SFERTF is also represented on the GCERTF, a reflection of the significant jurisdictional and geographic overlap between the Everglades and the Gulf Coast; both include state representation, and both are responsible for strategic planning, the coordination of intergovernmental activities and science, stakeholder engagement, and biennial reporting on restoration progress.[3,4] Both task forces are administered by senior executives of the federal government who supervise full-time staff dedicated exclusively to task force administration.

More than 700 miles of Gulf coastline and 100% of the Everglades ecosystem are located in the state of Florida, and in fact, the Greater Everglades Ecosystem includes significant Gulf Coast resources such as the Caloosa-hatchee Estuary.[24,28] This shared geography not only ensures a great deal of intergovernmental cross-pollination at both the political and staff levels, but it also ensures high levels of cross-pollination among stakeholders, nongovernmental organizations, scientists, journalists, and elected officials involved in both programs. Operational or "day-to-day" collaboration, above and beyond the collaboration that flows from formal governance structures like the GCERTF and SFERTF, is an important factor in many large-scale ecosystem restoration programs.[1] Over time, the cross-pollination occurring between the Gulf Coast and Everglades restoration programs may have synergistic effects on operational collaboration for both. These two programs may help operationalize the premise that collaboration among ecosystem restoration efforts creates synergies, efficiencies, and benefits across programs, a premise central to the mission of efforts such as America's Great Waters Coalition.[29]

Conclusion

As discussed in this entry, large-scale ecosystem restoration, as an approach to solving complex, multidimensional environmental problems, presents challenges of scale, causality, and jurisdiction that can overwhelm the capacity of a single state, jurisdiction, or authority. Configurations of governmental, private sector, and nongovernmental actors have produced a variety of collaborative arrangements in response to addressing high levels of complexity. While models vary across restoration programs and tend to pursue strategies tailored to specific environmental problems, there are identifiable common features among models, such as the integration of adaptive management principles, stakeholder engagement processes, and the incorporation of science in decisionmaking. The government's comprehensive approach to ecosystem restoration in the aftermath of the Deepwater Horizon oil spill may be an indication that CRG is an approach to environmental management that is in its ascendancy. The Gulf ecosystem restoration program borrows its model of collaborative governance from a model that has been in place in the Everglades restoration program for 16 years. This is an opportunity for comparative research on collaborative resource management that can inform future applications of CRG about specific challenges and opportunities relating to large-scale ecosystem restoration. Importantly, future research must also account for several issues that are beyond the scope of this entry. For example, a nuanced understanding of the political, administrative, and logistical challenges requires an in-depth understanding of the histories and the legal challenges confronted in each of these cases, as well as assessments of the level of success in meeting respective environmental goals.

References

1. Gerlack, A.; Heikkla, T. Comparing collaborative mechanisms in large-scale ecosystem governance. Nat. Resour. J. **2006**, *46* (Summer), 657–707.
2. Karkkainen, B. Collaborative Ecosystem Governance: Scale, Complexity, and Dynamism, available at http://www.law.virginia.edu/lawweb/lawweb2.nsf/0/2ba27078dc464a84852569700060de96/$FILE/HDOCSscalecomplex.pdf.2002 (accessed September 2011).
3. Obama, B. Executive Order 13554 of October 5, 2010, Establishing the Gulf Coast Ecosystem Restoration Task Force. Federal Register. Vol. 75 (No. 195), Presidential Documents; 2010.
4. *Water Resources Development Act*, Pub. L. No. 104–303, 110 Stat. Section 528 3767–3773; 1996.
5. Agranoff, R.; M. McGuire. *Collaborative Public Management: New Strategies for Local Governments*; Georgetown University Press: Washington, DC, 2003.
6. O'Leary, R.; Bingham, L., Eds. *The Collaborative Public Manager: New Ideas for the 21st Century*; Georgetown University Press: Washington, DC, 2009.
7. Koontz, T.M.; Thomas, C.W. What do we know and need to know about the environmental outcomes of collaborative management? Pub. Admin. Rev. **2006**, *66*, 111–121.

8. John, DeWitt. *Civic Environmentalism: Alternatives to Regulation in States and Communities*; CQ Press: Washington, DC, 1994.

9. Durant, R.F.; Fiorino, D.; O'Leary, R., Eds. *Environmental Governance Reconsidered: Challenges, Choices, and Opportunities*; The MIT Press: Cambridge, MA, 2004.

10. National Academy of Public Administration. *Resolving the Paradox: EPA and the States Focus on Results*; NAPA: Washington, DC, 1997.

11. O'Leary, R.; Durant, R.F.; Fiorino, D.; Weiland, P.S. *Managing for the Environment: Understanding the Legal, Organizational, and Policy Challenges*; Jossey-Bass: San Francisco, 1999.

12. Vig, N.J.; Kraft, M., Eds. *Environmental Policy: New Directions for the 21st Century*; CQ Press: Washington, DC, 2010.

13. Koontz, T.M.; Steelman, T.A.; Carmin, J.; Korfmacher, K.S.; Mosely, C.; Thomas, C.W. *Collaborative Environmental Management: What Roles for Government?* Resources for the Future: Washington, DC, 2004.

14. Lubell, M.; Segee, B. Conflict and cooperation in natural resource management. In *Environmental Policy: New Directions for the 21st Century*; Vig, N., Kraft, M., Eds.; CQ Press: Washington, DC, 2010; 171–196.

15. Ostrom, E. *Governing the Commons: The Evolution of Institutions for Collective Action*; Cambridge University Press: New York, 1990.

16. Gunningham, N. The new collaborative environmental governance: The localization of regulation. J. LawSoc. **2009**, *36* (1), 145–166.

17. Ruhl, J. Regulation by adaptive management—Is it possible? Minn. J. Law, Sci., Tech. **2006**, *7* (1), 21–57.

18. Heikkila, T.; Schlager, E.; Davis, M. The role of cross-scale institutional linkages. In common pool resource management: Assessing interstate river mangement. Policy Stud. J. **2011**, *39* (1), 121–145.

19. Karkkainen, B. Post-sovereign environmental governance. Global Environ. Polit. **2004**, *4* (1), 72–96.

20. Van Cleve, A.; Simenstad, C.; Geotz, F.; Mumford, T. Application of the "best available science" in ecosystem restoration: Lessons learned from large-scale restoration project efforts in the USA. Tech. Rep. **2004**, *1* (May), 1–29.

21. Wiley, H.; Canty, D. *Regional Environmental Initiatives in the United States: A Report to the Puget Sound Shared Strategy*; Evergreen Funding Consultants: Seattle, Washington, 2003, available at http://www.sharedsalmonstrategy.org/files/Final_regional%20initiatives.pdf (accessed October 2011).

22. Mabus, R. America's Gulf Coast: A Long Term Recovery Plan After the Deepwater Horizon Oil Spill 2010, available at http://www.restorethegulf.gov (accessed November 2011).

23. Salt, T.; Langton, S.; Doyle, M. The Challenges of Restoring the Everglades Ecosystem. In *Large-scale Ecosystem Restoration: Five case Studies from the United States*; Doyle, M., Drew, C., Eds.; Island Press: Washington, DC, 2008.

24. Gulf Coast Ecosystem Restoration Task Force. *Gulf of Mexico Regional Ecosystem Restoration Strategy* (Preliminary); 2011.

25. Grunwald, M. *The Swamp: The Everglades, Florida and the Politics of Paradise*; Simon and Schuster:New York, 2006.

26. *Water Resources Development Act*, Public Law 106–541, 114 Stat. Section 601; 2000; 2680–2693.

27. *Programmatic Regulations for the Comprehensive Everglades Restoration Plan*, Final Rule 33 CFR Part 385; 2002.

28. South Florida Ecosystem Restoration Task Force. *Strategy and Biennial Report*; July 2008–June 2010.

29. National Wildlife Federation. America's Great Waters Coalition, available at http://www.nwf.org/Wildlife/What-We-Do/Waters/Great-Waters-Restoration/Great-Waters-Coalition.aspx (accessed November 2011).

10

Ecosystems: Planning and Trade-offs

Ioan M. Ciumasu,
Keith Culver,
Mihai Costica,
and Jean-Paul
Vanderlinden

Introduction

This chapter presents an updated and extended version of prior work, while retaining an earlier definition and overview of the notion of ecosystem planning—and the related prospect of reconciliation of disciplinary tensions and realities slowing progress in understanding of the merits of ecosystem planning.[1] This chapter extends the original discussion of natural landscapes and urban systems to the potentially positive roles of ecosystem planning in communities of actors and stakeholders dealing with the concurrent effects of urbanization and climate changes.

Humans have a long history of modifying their environment to suit their own perceived needs. Science and technology provide a certain, limited capacity to anticipate the behavior of ecosystems and plan a chosen set of desired ecosystem responses. However, because ecosystems are complex systems, i.e., governed by nonlinear behavior via self-organization and thresholds, true planning activities are bound to have a very limited relevance and effect. In various ecosystem contexts, some planning may or may not be possible—depending on the type and scale of the intended ecosystem responses. Therefore, it is important to define the scope of ecosystem planning. The term ecosystem planning is very rarely used in the literature, and this formulation might well be regarded as arriving with negative connotations likely to hamper the future use of the concept in disciplinary contexts. However, its simplicity and straightforwardness have a certain appeal for environmental managers in their search for a synthetic vocabulary to help them in the necessary integration of knowledge from many fields and disciplines. The term "ecosystem planning" may, for this reason, be usefully adopted in practice, albeit in epistemologically controversial ways. An eventual adoption of the term by the growing community of problem-solving-oriented environmental managers will require some substantial definition clarification efforts and responses to epistemological difficulties.

Context

Any observer of the use of the expression "ecosystem planning" in the extant literature will see it frequently used as shorthand for "ecosystem management planning," most notably in situations of management of environmental resources. For example, silvicultural practices often employ the term "planning" in the context of forest management, sometimes referring to "forest planning" or even "ecological forest planning."[2] Such contexts are typically described in terms of managerial objectives such as "multi-objective managerial planning," to which descriptors such as "ecological" or "ecosystem" are often applied to account for the natural processes involved.[3] In this context, planning seems to refer to procedures applied to attain some desired, pre-established objectives in terms of exploitation of resources. One can observe that, despite occasional claims of the contrary, the literature tends to recognize that planning refers to some empirical knowledge-related procedure to obtain a benefit *from* ecosystems, not a procedure to *make* ecosystems. Yet, this distinction is not always sufficiently clear, with certain confusions persisting in the authors' understanding of the fundamental differences between the aspiration to plan and the nature of ecosystems. Ecosystem planning is often mentioned in contexts that imply manipulating ecosystems to the purpose of obtaining desired outcomes. This fact, together with the lack of agreement on what ecosystem planning is supposed to designate, is a warning that more work is needed to clarify the concept and the practice. This entry takes on this objective and retains the original definition as baseline for further developments and reviews.

To various extents, properties of ecosystems may be changed. At one end of the spectrum (gradient) of human intervention in ecosystems is the domain of conservation biology, which mainly aims at protecting–preserving a given ecosystem (or components of it) as is.[4,5] At the other end of the gradient is "ecological engineering," which essentially pursues objectives related to "ecological reconstruction" or "ecosystem restoration" of heavily degraded areas or even aims at developing new ecosystems.[6] The term ecosystem planning incorporates an inbuilt contradiction between the self-organizing, nonlinear character of ecosystems[7,8] and the linear, human-organizing character of planning. Despite this, the incontestable global reality is that most terrestrial and water ecosystems are under some form of modification by humans and thus subject to management involving both natural ecosystem dynamics and planning. Even though nature conservation's original aim was the prioritized protection of pristine and near-pristine environments, and ecology seeks to understand natural processes, i.e., those undisturbed by human interventions, we have to deal every day, and in most situations, with ecosystems that are heavily influenced by human choice and action. Indeed, given the fact of global changes, we cannot just continue to talk about "untouched" or "wild" nature. The term ecosystem planning might, therefore, gain widespread use, in diverse circumstances, notwithstanding the epistemological difficulties it carries along.

Definition

In order to reconcile the nature of ecosystems with human planning reflexes and with the need for a practically useful, integrative understanding of the term "ecosystem planning," we propose a working definition: *ecosystem planning refers to the human activity of anticipating and inducing the generation of a set of ecosystem goods and services (EGS), within the limits allowed by the intrinsic dynamics of ecosystems.*

For our purposes, EGS are "the capacity of natural processes and components to provide goods and services that satisfy human needs, directly or indirectly." EGS involve the translation of ecological complexity (structures and processes) into a more restricted set of ecosystem functions that ground the provision of goods and services that are valued by humans. Admittedly, the term "ecosystem function" has been subject to conflicting interpretations, sometimes referring to internal functioning of an ecosystem (e.g., energy and matter fluxes, food chains, and food webs) and sometimes relating to the benefits derived by humans from the properties and processes of ecosystems (e.g., food production).[9,10]

Such semantic hesitations illustrate or perhaps reproduce the epistemological difficulties that tend to be encountered at the human/nature interface. Indeed ecosystem planning is implicit in the literature concerned with Coupled Human and Natural Systems (CHANS).[11] In the definition above, ecosystem planning is explicitly ascribed to, and constrained by, the limited human capacity to influence ecosystems. This attribute should be read as the crux of the concept and a key idea of this work piece.

Dynamics intrinsic to ecosystems, thresholds and nonlinearity, are explained by the laws of physics, and ecosystems can be understood as complex systems—with the support of complexity science. Ecosystem properties are subject to some limited influence, but they are not responsive to human planning imperatives. To use a somewhat less precise but more intuitive illustration, an ecosystem (as a multi-individual system, in terms of system biology) is analogous to a human body (an individual system) in that it can be trained and improved, and a certain number of its parts may be changed, but it cannot be changed as a system—any attempt to do so will cross organizational and functional thresholds that are steps towards its destruction. The system has its own character. It can be enlarged, squeezed, accelerated, or made more productive or more beautiful—but its essential characteristics as a system cannot be changed. Ecosystem planning is, therefore, like education of a human individual: it can aim at reaching the utilization of its maximum potential, but it cannot change its genetic endowment.

Term Uses

Viewed in isolation and out of context, the term ecosystem planning makes little sense. Its meaning and value become apparent only once it is considered in association with other types of planning and with attention to the concept's carrying of a specific understanding of intrinsic ecosystem dynamics.

Our working definition has several consequences and lessons. One is that the term ecosystem planning captures only partially the relation between humans and ecosystems. Employment of unduly broad or general meanings is harmful because they would allow (or inadvertently convey) the mistaken assumption that human action can compensate for ecosystem destruction by sheer planning. In reality, once destroyed (e.g., by overexploitation), the capacity of ecosystems to provide EGS (or "ecosystem carrying capacity" in the ecological economics vocabulary) cannot be simply restored by further planning. Carrying capacity is grounded in the ecosystem's intrinsic properties, and these develop during the long-term ecological history of the place. In economically equivalent terminology, the carrying capacity represents the "natural capital" (K_N), or, differently put, the ensemble of EGS extractable from a given ecosystem. Unlike the physical (K_P) and the human capital (K_H), natural capital is irreplaceable since it is dependent on ecological thresholds.

At this point, we should recall that one fundamental characteristic of biological–ecological systems dynamics is their irreversibility, meaning that once an internal threshold is crossed, an essential set of characteristics of that ecosystem are irremediably lost because passage across biological–ecological thresholds cannot be reversed. This is a consequence of the laws of thermodynamics and can be taken either as a source of ultimate ecological pessimism or simply a stark warning that the sustainability of human civilization allows no major mistake. In any case, the recognition of this reality is a foundation stone for many research threads that are bundled together in the heterodox tradition of ecological economics and the related concept of bioeconomy.[12–17] In the second half of the 19th century, Ernst Haeckel, while trying to understand the consequences of the Darwinian multifaceted struggle for survival, coined the term "ecology" to denote "the body of knowledge concerning the economy of nature." But this term only became popular in the second half of the 20th century after the introduction of the notion of "ecosystem" in 1935 by the British plant ecologist A.G. Tansley to describe the system of interactions between a biocoenosis (i.e., a community of living organisms) and their biotope (i.e., their environment), followed by the introduction of the notion of "ecosystem ecology" in 1953 by the American brother ecologists Eugene Odum and Howard T. Odum in the aftermath of the development of a slurry of mathematical models of interactions between species in the first half of the 20th century.[18–23]

Indeed, there is an even older notion which deserves renewed attention in the context of ecosystem planning and sustainability practice: that of geonomy (or geonomie), a term introduced by the Romanian biologist–ecologist and oceanographer Grigore Antipa, a student of Ernst Haeckel and the father of modern dioramas currently used in the museums of natural sciences. Antipa first used "geonomy" in 1909 to describe a system of rational management of natural resources within the watershed of Danube River and the Black Sea.[24] French juridical expert, economist, and urban planner Maurice-François Rouge in turn used the concept of geonomy to explain the organization of physical space and the structures and equilibria that determine the land uses from the triple perspective of geography, sociology, and economy.[25] This concept is known in English as "geonomics," a term which American economic geographer Ray Hugues Whitbeck used in 1926 to redefine human economy in its biophysical context of planet Earth.[26–28]

The concept of geonomics emerged initially from the holistic perspective of evolutionary ecology and then later that of economic geography. Both departure points lead to a shared insight which is now central to sustainability studies, namely that we must cease conceiving of humanity as distinct from or in opposition to nature, and instead recognize humanity as part of nature. This insight may be stated in more formal terms as the recognition that any economic system is a subsystem depending for its existence on a larger social system, in turn a subsystem of a natural (biophysical) system on which its existence depends.[29,30] The concept is roughly equivalent to—but is larger than, and includes—what is now known in the literature as the CHANS evoked in the previous subchapter. It is *prima facie* plausible to regard all of these terms as elements of the domain of sustainability studies—also known as sustainability science.[31,32] Yet when considered in light of the aim of ecosystem planning to achieve long-term sustainability, the term geonomy holds the special potential to structure a more coherent body of knowledge—i.e., more like a conception of a traditional discipline—than the idea of "sustainability studies" which stands as a broad reference to a collection of contributions to problem definition which nonetheless lack *inter alia* the internal consistency characteristic of recognized disciplines. This is a unique potential to foster systematic and comprehensive approaches to a problem domain. Use of such a unifying term would rally those ready to add to the early days of problem analysis a complementary conceptual picture of the landscape to be inhabited by an array of solutions. This is much more important than usually acknowledged in complex projects where agreements on values and knowledge are a prerequisite for project success.[33]

The three forms of capital mentioned above form together the total capital stock,[34] recognizing as a crucial constraint that natural capital cannot be replaced by the other forms of capital, and that local increases of the total capital stock available may occur while the utility of that stock decreases. Thus, when the natural capital on which human activity depends is depleted irreversibly, the value of the other two major forms of capital also decreases—they are rendered unsustainable by the destruction of the underpinning natural capital.

As argued by Costanza et al.[35] in a widely cited article, it is possible to imagine generating human welfare without natural capital/EGS in artificial "space colonies," but this possibility is still too remote and impractical for all but conceptual experimentation with the bounds of possibility. One additional, more practically relevant way to conceive the value of EGS is to determine what it would cost to replicate them in a technologically produced, artificial biosphere. Past experience with manned space missions and with Biosphere 2 (a vivarium greenhouse, i.e., a materially closed ecological system, built on 1.27 ha between 1987 and 1991) in Arizona indicates that this would be an exceedingly complex and expensive project. By contrast, Biosphere 1 (the Earth) is incomparably more efficient as least-cost provider of life support services for humans. Nevertheless, given the extent of constant human impact on biosphere, Earth's biosphere is maintained in equilibrium between its natural dynamics and the human factor (exerted through human technology), a situation which is likely to also characterize any sustainable future of humanity on Earth. To take this into account, it has been proposed that we should give great attention to the conditions required for a sustainable (or unsustainable) technobiosphere.[36] Nonetheless, greenhouse-based approaches enhanced with computer-controlled sensor systems (model laboratory terrestrial ecosystems called "ecotrons") can provide insights about impact of forcing of biogeochemical cycles (e.g., CO_2) and climate changes (e.g., temperature) on ecosystems[37–40] and thus support science-based planning of human interventions on ecosystems at planetary and local scales—aimed at mitigation of, and adaptation to, climate and other environmental changes.

Discussions on costs open, of course, the issue of potential valuation of EGS in a market. The concept of "payment for ecosystem (goods and) services" (PES/PEGS) remains controversial.[41,42] Consideration of an EGS-inclusive economy is nonetheless useful to the extent that it brings attention to the importance of incorporating into a comprehensive analysis such previously marginal concepts like "intangible values" (social and intellectual aspects—they are also connected with ecosystems) and "externalities" (ecosystem resources usually taken as "given" and not factored in by the current, neoclassical economics). In a review of this increasingly urgent matter, as current human consumption of EGS outstrips ecosystem production capacities, Baggethun et al.[42] have shown that the recent advances towards monetization of EGS indicate the slow progress of the concept. While monetization has helped draw policy and economic attention to ecosystems and EGS, it has also absorbed it into the logic of the aging yet still dominant neoclassical theory of economics in ways that deprive the concept of many of its initial virtues related to ecological and social values. Reviewers of the literature may observe that the incomplete incorporation of EGS into mainstream policy and decision-making is attributable to the prevalence of an approach that remains essentially disciplinary. EGS was repackaged or translated from ecology to current economics, along the way losing many of the virtues of EGC. Successful incorporation of EGS into policy decision-making will require a problem-driven (rather than a tool- or discipline-driven) transdisciplinary approach capable of synthesizing tools, skills, and methodologies. As their review suggests—reflecting wider opinions in the literature—valuation needs not always amount to an exclusively monetary valuation. This recommendation is naturally applicable to ecosystem planning in the sense discussed here.

Planning efficiency is nonetheless dependent on financial costs; EGS and PES represent essential advancements towards a solid grounding and meaningful use of the concept of ecosystem planning. However, as the initial concept has become diluted by disciplinary habits of subdivision and specialization, current developments and achievements are limited in their scope and practical utility. They are usefully and positively regarded as important initial forays into the conditions of sustainability.[43] On the socioeconomically transformative journey that is the transition to sustainability, ecosystem planning needs to factor in how much exploitation of EGS ecosystems can bear. In other words, how much of the carrying capacity of ecosystems can be consumed by humans without risking the collapse of the ecosystem carrying capacity. In this sense, ecosystem planning can draw support from the concept of "ecological footprint," an accounting tool for assessing the natural capital and its degree of depletion at various scales—individuals, cities, nations, or the entire planet. This approach uses estimates of consumption of natural resources for food, shelter, transportation, personal care, pollution absorption capacity, and other uses of the natural capital and compares it with the carrying capacity of ecosystems. This is one highly relevant tool for overall assessment of ecosystem planning: a value of less than 1.0 indicates that the ecological footprint remains below the carrying capacity of the ecosystems (at the scale taken into consideration), which means that ecosystem exploitation is within the limits of sustainability. On the contrary, a value greater than 1.0 indicates that a person, city, nation, or the entire humanity lives beyond the natural support capacity limits of the ecosystems/biosphere.[44–47] Values that surpass 1.0 are of utmost importance because any excess represents "eating up" the regenerative capacity of the ecosystems/biosphere, with the consequence that the overall carrying capacity decreases. At the planetary scale, it appears that humanity is already close to this tipping point or already beyond it, and the possibility of collapse is very real and a reason for accelerating and improving our efforts to understand and respond to this possibility.

The solutions for an effective transition to sustainability are still to be developed. At epistemic and moral levels, one option is to aim for a steady-state economy. Such a steady state would amount to a kind of plateau of development—as permitted by the Earth's carrying capacity—and would maintain this constant value in the future.[48] Another approach is to commit to "degrowth," meaning a collective decision of humanity to consume less, especially in those affluent countries that consume beyond the ecosystem carrying capacity available to them.[49] This idea is likely to encounter major social acceptability obstacles, certainly in developed countries, and especially in developing countries. The degrowth proposal appears to express a chosen rejection of the excessive monetization of intangible assets (social and cultural values) and ecosystem services. However, the grounds to be considered for sustainable development are first and foremost biophysical: natural limits cannot be overcome as they are thermodynamic realities already known for decades—since the origin of ecological economics.[50] The relevance of ecosystem planning relates to the immutable character of the physical laws: planning for a sustainable equilibrium of humans with ecosystems may be a matter of social decision-making and acceptance, but the consequences of those decisions are not. In other words, ecosystem planning will need to help humans accomplish sustainable development (and thus, survive). If this fails, collapse will follow. This is not science fiction: rather, it is a prediction grounded in history as civilizations have in fact overused their natural resources and faced collapse. At the planetary scale, this event is the natural outcome in the absence of a downward adjustment in the ecological footprint/carrying capacity balance.[51-53]

Taking this emergency seriously, Daily et al.[54] propose a conceptual frame for operationalizing the relation between humans and ecosystems, conceived as a never-ending circuit of five links (rendered in all caps here) and types of relations between them (rendered in italics here): ECOSYSTEMS → *biophysical models* → SERVICES → *economic and cultural models* → VALUES → *information* → INSTITUTIONS → *incentives* → DECISIONS → *actions and scenarios* → ECOSYSTEMS. Scenarios are developed for an applied case in Hawaii, called The Natural Capital Project, which tries to develop a scientific basis and connect it with policy and finance mechanisms, aiming to incorporate natural capital into resource- and land-use decisions on local and larger scales. This can be regarded as a useful example of current attempts to link ecosystem conservation with development and to render EGS concepts useful in policy and business. Yet, at the same time, we are far from the desirable situation when this would become mainstream practice. In the light of these considerations, the relevance of the term ecosystem planning remains ambiguous in the literature dedicated to the theory and applications of environmental/ecosystem management.

Mac Nally et al.[55] delimit ecosystem planning in a more cautious manner. By "ecosystem-based planning," they mean planning activities for biodiversity conservation purposes. Often, the expression "conservation planning" is used to the same effect.[5] Margules and Pressey[56] identify different stages of biodiversity conservation planning in relation to ecosystem services. Conservation biology (a term designating the aim of biodiversity and ecosystem conservation) is gradually converging with the concept of EGS.[57-59] Sometimes, an equally cautious alternative concept and expression are used, that of "ecosystem-based management," where planning activities are discussed in terms of the constraints imposed by ecosystem properties and dynamics in general and in specific contexts and case studies. In an extensive book on the matter,[60] Randolph uses terminology including "environmental protection," "land conservation," "environmental management," "environmental land use management," and "ecosystem management," as well as "land-use planning," "environmental planning," and "habitat conservation planning," to systematically describe the intricacies of human–nature interactions.

Among the rationales behind a potentially useful concept of "ecosystem planning," one can count earlier efforts to integrate environment planning and development via ecosystem approaches. Thus, the necessity to extend planning activities from human-created environment and modified environments to the natural environment arises from the expansion of human activities themselves. Therefore, the ideal situation of environmental planning would be one where it was not needed. This applies *mutatis mutandis* to ecosystem planning as well. The point we want to make here is that ecosystem planning may reveal itself to be not a goal per se, but a necessary practice towards an aspiration that is not yet concretized into a specific, readily measurable, or quantifiable goal. Implemented well, it will embody a managerial compromise between human activity expansion and the need to control, limit, and minimize the impact of human activities on the carrying capacity of biophysical systems.

One common feature is evident throughout the literature surveyed above, namely, the avoidance of a direct association between "ecosystem" and "planning." Instead, authors seem to be searching for new concepts to describe those actual situations where "ecosystem" and "planning" coincide. The authors surveyed are generally preoccupied with selection of the most apt definitions and ascription of the most appropriate meanings, hoping that the reality of human–nature interactions is being neither hastily misconceived nor ignored. For example, expressions such as "environmental planning" or "land-use planning" bear a lesser epistemological burden than "ecosystem planning." Such terms are general enough to avoid asserting an unnecessarily specific, contentious relation between planning and ecosystems, yet they clearly do include reference to sheer planning of use/occupancy of natural resources, especially in the case of water or land resources. Instead of being merely a matter of semantics, the expression ecosystem planning cannot be a useful concept unless it is associated with a clear distinction between employing knowledge about ecosystems and the effects of potential human action on ecosystems. The concept of ecosystem planning must help illuminate the border area between what humans know and can do about environment and ecosystems. Various disciplines and approaches can contribute to such a new understanding of "ecosystem planning." However, the lack of a compact body of literature makes the epistemological and technical reviewing process a highly demanding endeavor—a potential subject for future examinations.

In practice, however, understanding of the relation between humans and ecosystems may not always precede norms and decisions, as social rules and behaviors may not wait for detailed scientific clarifications. Kagan[61] examines ecosystem planning from a legal perspective, using ecosystem planning when referring to human activities related to environmental management and decision process over land/resources use, seemingly unaware of the ecological implications of putting together the words "ecosystem" and "planning." The main point of his perspective, however, is that it reflects a situation where ecosystems are being viewed, consciously or not, as something "out there," exterior to social matters. This is not very different from the currently dominant neoclassical view in economics, according to which natural resources and social matters are external to the economic processes and associated accounts, in the sense that they are "externalities"—their values or influences need not be accounted for in setting and operating a core process towards a desired outcome. All those narrow perspectives, however, are now being challenged by the widening community of practitioners for sustainable development. Nicholas Stern, the author of the homonymous report on the cost of climate changes due to greenhouse gas emissions, has arguably provided the best synthesis of the man–nature relations in a conference at the Royal Economic Society in 2007: "Climate change is a result of the greatest market failure the world has seen. The evidence on the seriousness of the risks from inaction or delayed action is now overwhelming. We risk damage on a scale larger than the two world wars of the last century. The problem is global and the response must be collaboration on a global scale." The more urgent it becomes, the great the need to reassemble disciplinary perspectives into a common understanding of the relations between natural, social, political, and economic realms.

At the borders between legislation, natural sciences, economy, and social sciences, one often-used terminology is "forest planning," meaning, in silvicultural practices, planning for forest ecosystem exploitation. The literature on forest management is among the earliest to acknowledge the tension between the deterministic approach of forest exploitation planning and the uncertainties intrinsic to ecosystem dynamics. Given the importance of forests as resource generators for human populations, uncertainty was "accommodated" in forest exploitation planning[62] in terms of managerial approaches that accounted for disturbances and other unpredictable, less deterministic phenomena. In this sense, forests represent a good case study for the relation between the management of EGS, ecosystems, and socioeconomic systems, where natural resource management planning is often seen as akin if not identical to "ecosystem planning." An extensive overview on the matter is beyond the scope of this entry. Some examples, however, may be illustrative. Exploitation of ecosystems is obviously related to the property bearing lands and waters. In Central and Eastern Europe, for example, Romania has served as a natural experiment in this regard. The recovering of the individual property rights to forests (lost at the time of military imposition of communism by the Soviet Union, when 23% of the country's forests were in private hands) in a context of a still-recovering economy and society has led to massive forest overexploitation. Large areas of forests were clear-cut in the 1990s for the immediate purpose of selling the wood (120,000 ha, almost half of the first wave of forest restitutions). Obviously, firm property rights are necessary for effective management of forests, because they provide marketability, i.e., options for future planning conservation of EGS and various uses—timber and other biomass, recreation, carbon sequestration, water retention and decontamination, and so forth. These property rights are therefore not in question, but their consequences are, in this perilous situation where the postcommunist socioeconomic transition has left a management and regulatory vacuum. Fortunately, the last decades have seen a gradual socioeconomic recovery of the country and improvements in the definition of the property rights, giving rise to remarkable improvements in forest management. Entrepreneurial activities are now related to uses of forests, based on the (1) right to access; (2) (resource) withdrawal right; (3) management right; (4) exclusion right; and (5) alienation right.[63] Along the way, a process of reforestation is taking place in the entire eastern half of the European Union. The region's forests are now recovering from some of the losses suffered during the previous years of harsh transition, with various countries at various levels of reforestation—now matching the levels of socioeconomic convergence with the more developed western European countries.[64] In such situations, ecosystem planning bears the full weight of socioeconomic contexts.

A related example that is relevant to ecosystem planning is linked to the ecological reconstruction that may follow a cycle of ecosystem overexploitation and partial ecosystem recovery. Reforestation may be difficult on highly degraded and low-accessible lands, which may become prone to soil erosion. However, knowledge of natural processes may provide the means to plan for revegetation of degraded slopes. To remain within our example from Central Europe, past experiences show how the natural vegetation succession can be used as a tool. Thus, field experiments with planting a certain shrub like the common sea-buckthorn (*Hippophae rhamnoides*) on degraded slopes in Romania have led to fast revegetation with the shrub as quasi-exclusive species. In later stages, this monospecific vegetation allows an evolution of the structure of the soil and the advent of other species (grasses and tree seedlings) to grow under its shade. In a next stage, oak (*Quercus*) and beech (*Fagus*) tree juveniles, as well as other shrubs, have grown taller and shaded the sea-buckthorn and gradually replaced it as new forest, which, even if it is characteristic of the area, would not have grown there because the land was too degraded and the conditions were suboptimal prior to the establishment of the sea-buckthorn in the first place.[65] Application of such lessons from the field, however, is only possible under favorable socioeconomic conditions. Recent advances in the country and the region may reopen the possibility of using older field ecological experiments. In such situations, ecosystem planning means planning for a desired favoring of certain natural processes, with the purpose of halting the loss of natural capital and the recovery of certain ecosystem functions and services.

A fresh question now arises: to what extent can we take advantage of our knowledge to manipulate natural processes towards recovery, albeit partial, of some initial ecosystem structures and functions? The discipline of ecological engineering is one way to explore answers to this question. Its original principles have been updated to reflect the improving understanding of the limited planning powers of humans upon ecosystems.[6,66] (1) It is based on ecosystems' self-organization capacity; (2) it can be a field test of ecological theory; (3) it relies on integrated system approaches; (4) it conserves nonrenewable energy; and (5) it supports biological conservation. Such developments seem to indicate broad acceptance that planning, and for that matter "ecosystem planning," must incorporate the idea of adaptive management, both at the scales of landscapes and those of local settlements and despite epistemological uncertainties.[67,68]

Landscape Planning

The first and most obvious application of a future-oriented, broadly accepted concept of ecosystem planning is in the area of landscape planning. However, many authors in this area may understand ecosystem planning as "the process of land-use decision-making that considers organisms and processes that characterize the ecosystem as a whole." In other words, ecosystem planning is used as shorthand for ecosystem-based planning of land uses across landscapes. Much of the literature deploys a concept that may have a certain utility in its own right, albeit narrow, yet it is often used in epistemically irregular and unstudied constructions. Musacchio et al.[69] use the expression "landscape ecological planning (LEP)." There are major challenges in integrating the concept of EGS in landscape planning and, for that matter, also in the concept of ecosystem planning as defined in this entry, simply because there is as yet no coherent and integrated approach for practical applications of EGS in planning, management, and decision-making in general.[53,70] Even in the hypothetical case where the literature will start displaying decreasing pluralism (which is not necessarily desirable) and increasing use of a shared concept (which is in principle desirable)—even a cluster concept admitting of many conceptions—it remains practically the case that there are no developments in the literature associated with substantive changes in practice. This stands in interesting contrast to other experiments regarding the place of individual human flourishing in the context of larger systems, such as the "living wage" movement's experiment in Finland with a minimum guaranteed income.[71,72] Humans are unquestionably engaged in large-scale experiments regarding new ways of living together in conditions distributing available resources with attention to the conditions of human dignity; yet those experiments are not substantially engaging the scope and scale of the sustainability challenge—which warrants special attention from researchers and managers.

Recent research programs have tried to tackle this obstacle. For example, De Groot et al.[73] have proposed a framework that involves a chain of five major links: (1) understanding and quantifying how ecosystems provide services; (2) valuing ecosystem services; (3) use of ecosystem services in trade-off analysis and decision-making; (4) use of ecosystem services in planning and management; and (5) financing sustainable use of ecosystem services. Inevitably, such a succession of actions bears the intellectual virtue of a coherent synthesis of what we can imagine to be eventually done for successfully approaching the desired equilibrium between humans and nature. But it also carries a certain naivety with respect to what may be possible—for example, it sees ecosystems as "still poorly understood" and proposes more detailed quantification as the best way to achieve a better grasp of how ecosystems function and provide EGS, as if it is only a matter of time until we get it and equally a matter which must be resolved more fully than at present if meaningful downstream action is to be taken. This way of approaching the issue, however, seems to ignore (1) the serious epistemic hurdles involved in seeking total, comprehensive understanding complex systems, especially with a reductionist mindset and (2) the practical feasibility of such an endeavor in terms of available time and resources. To give only one example, tens of thousands of chemicals exist on the market,[74] and thousands of new chemicals are introduced to the market (and released into the broader environment) every year, and there is no chance that ecotoxicological test batteries could ever be developed for each of them to "better grasp" how ecosystems function, not to mention the inherently limited relevance of each bioassay. In general, it is thought that different patterns of urbanization have a direct impact on ecosystem dynamics through feedback mechanisms and complex interactions between urban activities, land/water uses, and environmental changes.[75] Some integrative, multitier studies can be done to account for changes in ecosystem parameters in space and time, for some specific indicators, for example, water quality along a river basin, upstream and downstream from a city that is a source of water pollution.[76–78] Notably, urbanization facilitates the transition of rivers from transporters to transformers of carbon.[79]

But even those are very far from being sufficient for understanding the state of ecosystems and the quality status of EGS. What seems to be missing in the five-link type of framework proposed in the analysis cited above is the acknowledgment of the facts that we only have a limited amount of time and material resources to be allotted to advancing towards the equilibrium between humanity and the ecosystems in which it is embedded, and that those resources are themselves dependent upon our mankind–ecosystems relation that we want to address. We simply cannot plan everything we would like to. As such, the mentioned paper takes a tool-based approach rather than a problem-based approach, being concerned more with understanding than understanding sufficiently for remedial action. Apart from this difficulty, and to be fair to the mentioned authors, the framework provided by De Groot et al.[73] does appear to us to have an essentially "color-test" value for the concept of ecosystem planning proposed hereby. Under the definition of ecosystem planning employed in their paper, the chain of actions proposed by the abovementioned team at Wageningen University can actually be understood as applied to "ecosystem planning," as a useful organizer of what needs to be done by humans within the limits of the possible. Without the definition above, the same chain of actions proposed by De Groot et al. allows "everything and nothing" in a progressively increasing demand for finite. Thus, mapping landscape functions for planning the management of EGS is potentially a smart and justifiable approach, on the condition that it is integrated in a management process that leads to less destruction of EGS and, thus, interrupts the series of blind actions upon ecosystems.

A necessary feature of "ecosystem planning," as a synthetic concept, must be simplicity in the sense of operational readiness. Proposed methods should maintain sensitivity to the complexity of the phenomenon but avoid creating supplementary complications, since those would exacerbate the problem by postponing practical action. Ecosystem planning should recognize limits of human knowledge and actions and seek genuinely integrative solutions, i.e., simple yet responding to a complexity of pressures. Humanity has not managed to integrate the fragmented knowledge we have produced so far into a coherent whole, and the tension between humanity and ecosystem continues to increase. We cannot realistically hope to solve this problem by producing more piles of disparate (and mutually untranslatable disciplinary) knowledge. Ecosystem planning, if effective, will need to consist in solutions that are relevant (adaptable or instructive) to situations as various as floodplains, polders, warm and cold deserts, tropical and boreal forests, marine and fluvial ecosystems, and so forth.

Although the literature explicitly using the concept of ecosystem planning in a substantial way is scarce, there is a large and eclectic body of knowledge on implicit ecosystem planning attempts. In a study on the potential overlap between EGS and conservation policies in the coast eco-region of central California, Chan et al.[58] have extracted a set of key insights into the convergence of conservation and EGS planning: (1) both suitability of sites and demand from citizens are main drivers in what ecosystem planning can do for protecting biodiversity and EGS—near cities demand is higher but suitability is lower; (2) spatial scales may vary for optimum of biodiversity conservation or/and different types of EGS—pollination services result from a variety of small locations, while water quality from entire river basins; (3) population centers yield tensions in planning between estimated value and demand—for example, a low-value/high-demand situation can occur near cities; (4) data are usually lacking, and thorough research and analysis are necessary; (5) it must involve multidisciplinary and transdisciplinary teams, for the integration of theoretical and empirical expertise from diverse fields; (6) efficiency is conditioned by considering both trade-offs and side effects of biodiversity conservation and EGS conservation—often trade-offs are the most efficient way, and often side effects reveal points of common relevance for various goals of planning.

While the concept of EGS appears to prioritize or privilege focus on human–ecosystem relations, challenges for future applications involving the concept of ecosystem planning can be related (without being equivalent) to situations as various as land-use planning; degraded landscape management, salt-marshes and river regulation, flood prevention and management of river catchments, management of freshwater quality, silviculture (with related concepts like "ecosystem stewardship planning," "forest stewardship plans," and "forest stewardship"), and planning for carbon sequestration in terrestrial and water ecosystems. In the current context of increasing need for mitigation and adaptation to climate change, forms of ecosystem planning have been imagined at planetary scale. Thus, geo-engineering approaches propose that the planetary ocean should undergo fertilization with limiting substances (e.g., iron ions) to allow faster carbon sequestration in phytoplankton biomass. This, however, involves large-scale, unpredictable risks that require thorough investigation in a race against time.[80]

Urban Planning

In spite of some controversy, the concept of ecosystem is appropriate in the study of cities.[81] Rees and Wackernagel[44] have proposed estimations of the ecological footprint of cities and show why cities cannot be sustainable per se, even while they are actually central to sustainability. Cities can only be considered sustainable together with their hinterland, from where they draw their resources and on which they depend. In the current era of globalization, cities have become part of a planetary network of cities, with a "common hinterland," which is the Earth's biosphere. Cities play preeminent roles in the global exchanges of information and matter energy and have an immediate impact on, and are immediately affected by, the entire biosphere. Ecological footprint estimates allow cities to track the necessary resources and the plan for development according to this. Consequently, urban development planning for sustainability will require evaluations of the ecological demand of a given city (disaggregated by type of activities) and comparison with available ecosystem resources. In the same logic, ecological deficits may determine the way the city may develop, if it intends to remain sustainable.[46] Urban ecosystem services can be identified and described in urban areas just as as in nonurban areas.[82] When proposing an integrated planning tool for sustainable cities, Rotmans et al. implicitly refer to "ecosystem planning." In their paper, the concept of environmental capital seems to be equated with natural capital, as part of a wider planning for sustainable city.[83]

Niemelä describes urban ecology in the context of the relation between ecology and urban planning, where urban planning is a type of land-use planning.[84] He makes the essential observation that urban nature has been regarded in prior studies as a true field experiment about human impact on ecosystems. This resonates with the wider preoccupations with the place of urban areas within the landscapes, for example, with the degrees of vegetation cover[85] or habitat patch corridors to avoid excessive habitat fragmentation in highly urbanized areas and to allow landscape connectivity between local plant/animal populations as part of the wider metapopulations (network of local populations) within landscapes. In a previous communication concerned with the place of cities as a human social construct within local ecosystems, we have described the city in terms of an ecosystem disturbance.[86] Applying current ecological theory—the intermediate disturbance hypothesis (IDH) and island biogeography theory (IB)—we have identified an ecological taxonomy of cities and city areas as a function of disturbance intensity in a local biogeographical context: small city/city periphery (low disturbance intensity), medium-sized city/city near-center (intermediate disturbance intensity), and large city/city center (high disturbance intensity). Within this framework, any urban unit can be described on a gradient of ecological disturbance and in the local biogeographical context, with biodiversity theories and indices serving as proxies for the state and dynamics of ecosystems and ecosystems–humans dynamics.[87,88]

An abundant urban biodiversity literature has emerged providing the means for use of urban ecology-biodiversity studies in urban planning and for planning for biodiversity conservation in urban areas.[89–92] Further literature points to the transformative impact of urbanization, particularly the reduction of native habitat and the creation of new habitats, and ecosystem homogenization.[93,94] These complex aspects must be accounted for by urban/ecosystem planning. In effect, city planning emerges as a form of planning the impact (types and intensity of disturbance) of humans upon ecosystems. The management of urban areas per se and the management of city hinterlands become a matter of planning for extraction of EGS for use in cities and human settlements in general. The place of the city within the surrounding ecosystems is, therefore, a matter of "ecosystem planning."

At the level of immediate impacts, urban planning may require intra-urban or peri-urban planning, which may involve a diversity of terrestrial/water ecosystems, coastal ecosystems, and inclusion of the effects of climate changes, grazing, and agricultural land. City planners must face the concomitant emergence of new knowledge domains and a plethora of new challenges. The central feature of urban ecology is that "cities are emergent phenomena of local-scale, dynamic interactions among socioeconomic and biophysical forces" that give rise "to a distinctive ecology and to distinctive forcing functions."[95]

The role of a city as disturbance of the local ecosystems is the result of a complex dynamic involving major concerns as disparate as urban energy systems[96] and global change,[97] solid waste and wastewater,[98,99] and infrastructure reliability engineering,[100] to cite only a few. All this requires effective community coordination, based on social multi-criteria conception and evaluation of urban sustainability policies.[101–103] In this context, the use of some multiple criteria decision aid (MCDA) methodologies seems to be unavoidable for any urban and associated ecosystem planning, and effective coordination is yet another step further. The best candidate to providing an organic integration of all these issues and perspective is the systemic view of sustainability, according to which any economic system is a subsystem of a social system, itself a subsystem of a biophysical system.[29,30] However, translation from such ecological–economic insight to management strategies, actually tried and revised, seems to be waiting.

Elsewhere, starting from the systemic conception of sustainability, we have proposed an integrative approach for problem structuring and identifying sustainable/unsustainable management scenarios for solid wastes.[104] The method consists in translating the nested inclusion relationship between economy, society, and ecosystems into an ordered set of sustainability filters to be respected in managerial practices and proposed solutions *in illo ordine:* ecological sustainability filter (EcSF), social sustainability filter (SoSF), and economic sustainability filter (EnSF). Successfully passing all filters (i.e., meeting threshold values and indicators negotiated/listed in the technical definition of each filter) indicates that a given policy or solution proposed for the stated purpose (e.g., solving the problem of an old landfill) is consistent with sustainable development. This would be considered a sustainable scenario of action, under the conditions and knowledge available at the given place and time. Under this framework, the role of ecosystem planning can be recognized at the priority level of ecosystem carrying capacity (as reflected in EcSF) prior to action being taken towards further exploitation of EGS. Alternative methods can be identified to assess sustainability of solutions proposed for urban landfills, like the so-called sustainability potential analysis (SPA), and extension of the bioecological potential analysis (BEPA) that was originally developed for local and regional landscape analyses,[105] a reminder that any urban planning is part of planning for an equilibrium between humans and the natural support systems.

Trade-offs

One idea unites all discussions about ecosystem planning, namely the need for trade-offs. For example, in the case of both landscapes and waterscapes, river drainage basins (catchment areas) are considered necessary because they represent the minimal unit by which natural forces (starting with gravity) unite to shape the surface of the planet as understood in geological and ecological terms. The European Union's Directive 2000/60/EC, the so-called Water Framework Directive (WFD) legally requires each EU member state to have river basin management plans as a common EU-wide tool for implementing a drainage basin-based approach to integrated water resources management. Although this approach has both advantages and disadvantages, it is considered a best trade-off between the need for locally optimized performance of technical approaches to natural resource management and the need for EU-wide comparisons, which requires a common understanding and coordination, which in turn requires a commonly agreed set of procedures between over 300 methods for monitoring water quality across Europe.[106,107] Creating a common methodological body required an intercalibration exercise (which is an example of agreement on knowledge) at the scale of the European Union. So planning across large areas requires coordination which requires agreements on values and knowledge. While the WFD is an example of agreement on values achieved at political level, the intercalibration exercise is an example of agreement on knowledge achieved at the level of experts in water physics, chemistry, and biology but directed towards managerial objectives aiming at obtaining a set of EGS bundled in terms of "ecologically clean water status."

In the case of human communities, cities can be regarded as the best candidate scale and unit for modeling and testing states and processes in the transition to sustainability, because cities present the best trade-off between system-level relevance and managerial feasibility. The city is large enough to incorporate all fundamental functions that exist in a human system, yet small enough to still allow an intelligent agent to have a holistic overview so that action can be coordinated. Lower scales, for example, city districts, are too small because they are too specialized to be representative for an entire local community (e.g., as a residential area or industrial area or commercial area). Higher scales, for example, regional or national scales, are typically too large and administered through administrative divisions like specialized departments (ministries, agencies, and offices). Systemic changes necessary in the transition to sustainability may be subject to experimentation at smaller scales, and it will be cheaper and easier than at whole-city scales, but the results will not be relevant for how whole human systems work. At scales larger then cities, experimentation is not feasible without major disruptions from which the communities may be unable to recover.

In terms of ecosystem planning, the various instances of urban green and blue infrastructure (green areas and surface water bodies) represent planning trade-offs between the natural dynamics of ecosystems and the local needs of cities, notably mitigation of urban heat island effects and flood risks.[108,109] Thus, planting trees (to create a canopy cover and dissipate heat through biological processes) can help limit the heat island effect during the day, but other factors become important for temperature relief during the night (which is important for health, as human body needs to recover during the night): reducing soil sealing (impervious surfaces).[110] At the same time, tree canopy per se and that of particular species may be part of the local culture, which adds a social component that must be taken into account in urban ecosystem planning; failing to do so can result in strong rejection, civil disobedience, and reversal of the measures. This is what happened in 2013 in the city of Iasi, Romania, where a number of lime trees (*Tilia* sp., a species common in temperate climates of Europe and with strong cultural resonance in the city and in the country) from the city's historical center have been fell for sanitary-security (related to the old age of the trees) and esthetic reasons (related to the desire to clear up facades of historic buildings and respecting the initial plans of architects), but this managerial measure was executed in a manner that was perceived as an unacceptable bureaucratic imposition—namely, without public consultation. In addition, the lime trees were replaced with Japanese pagoda tree (*Sophora japonica*)—a world-popular tree for urban forests, one that is resistant to urban pollution and heat island effect and fits with the good practice of high diversity of genera and species of urban trees[111] but which was not part of the city's *spiritus loci* (spirit of the place). Less than 2 years later, following public revolt against what was perceived as a savage massacre against a local living emblem, and after an online local referendum which showed an overwhelming support of 94% for lime trees and in spite of the large sunken cost involved, *Sophora* was dug out and replanted elsewhere in the city while the city center was replanted with lime trees as predominant species.

Such situations show that a complex balance between factors must be reached through adaptive planning of land uses and vegetation structure in each sector of the city, within the limits of what is known at a certain moment in time and, most importantly, within the local cultural frame. Often this requires combining historic and democratic traditions, certain species bearing strong symbolism codifying common cultural references and social values that reflect the local history and details of the relations between people and nature in certain biographic provinces and cultural regions of the world.

Perspectives

It is not clear yet whether the term ecosystem planning will acquire a widely recognized meaning in its own right. Nevertheless, the concept often appears to function as a convenient stylistic contraction employed in the description of complex, multilevel, multidiscipline-fed issues and topics related to environmental management. Should the term enter mainstream terminology, it will need to overcome some major epistemic difficulties presently overlooked while the term is used as an aspirational gesture unconstrained by specific operational meaning and associated specific performance measures. A variety of methodologies are expected to be developed in the future, and further redefinitions of the term are likely. No matter how methodological reflections and terminological development occur, however, it remains the case that all reflection and working terms must face the fundamental tension between human-centered planning and human-independent characteristics of ecosystems. Ecosystems are an autonomous part of the environment, with their own, conspicuously nonlinear, dynamics. Planning regarding these dynamics cannot be made fully operational, if at all, by managerial task subdelegation let alone semantic reduction. The fact that planning may work effectively in certain sectors of environmental management does not translate into effective planning at the level of ecosystems. Rather, it typically means that (1) the concept of environmental planning has been limited to engineering- and management-tractable issues; (2) it has benefitted a certain capacity of environment as a whole (i.e., ecosystem included) to absorb disturbances by human activity; and (3) environmental planning needs further refining to clarify the limitations of planning and account for the reality of nonlinear behavior of ecosystems.

In the current context of vast and fast environmental changes, which are calling for coherent responses from science and management, certain knowledge domains may assimilate the concept of ecosystem planning. Notably, conservation biology aims at more effective nature protection, going well beyond the traditional focus on biodiversity conservation, with planning emerging as a necessary component of any comprehensive approach (for a review and discussion on the matter, see Reyers et al.[112]). The imperative of a global transition to sustainable development commands a profound transformation of the economy and society, which entails effective knowledge integration and management. A recent global survey that included experts in both social and natural sciences and that was carried out by the International Council for Science and the International Social Science Council has revealed a set of five grand challenges that need to be tackled within the coming years[113] and which can be summarized in four points as follows: (1) higher management-relevant capacity to anticipate environmental changes; (2) coordinated observation of environmental changes; (3) comprehension and anticipation of disruptive environmental changes; and (4) social–institutional transformation for the transition to sustainable development. As discussed here, ecosystem planning is a focal concept for current knowledge integration and management and, despite some challenging epistemic questions, has the potential to amalgamate into a core vocabulary of sustainability. Among the multiple approaches and contributions towards sustainability, landscape and urban planning emerge as particularly effective platforms for intellectual developments involving ecosystem planning.

References

1. Ciumasu, I.M.; Buzdugan, L.; Stefan, N.; Culver, K. Ecosystems: Planning. In *Encyclopedia of Environmental Management*, Jorgensen, S.E. (ed.), 4 Volume Set, pp. 632–642. Boca Raton: CRC Press, 2012.
2. Zagas, T.D.; Raptis, D.I.; Zagas, D.T. Identifying and mapping the protective forests of southeast Mt. Olympus as a tool for sustainable ecological and silvicultural planning, in a multi-purpose forest management framework. *Ecol. Eng.* **2011**, *37*, 286–293.
3. Palahi, M.; Trasobares, A.; Pukkala, T.; Pascual, L. Examining alternative landscape metrics in ecological forest planning: A case for capercaillie in Catalonia. *For. Syst.* **2004**, *13* (3), 527–538.
4. Soulé, M.E. What is conservation biology? *BioScience* **1985**, *35* (11), 727–734.
5. Naidoo, R.; Balmford, A.; Ferraro, P.J.; Polasky, S.; Ricketts, T.H.; Rouget, M. Integrating economic costs into conservation planning. *Trends Ecol. Evol.* **2006**, *21* (12), 681–687.
6. Mitsch, W.J.; Jorgensen, S.E. *Ecological Engineering: An Introduction to Eco Technology*. John Wiley & Sons: New York, 1989.
7. Koch, E.W.; Barbier, E.B.; Silliman, B.R.; Reed, D.J.; Perillo, G.M.E.; Hacker, S.D.; Granek, E.F.; Primavera, J.H.; Muthiga, N.; Polasky, S.; Halpern, B.S.; Kennedy, C.J.; Kappel, C.V.; Wolanski, E. Non-linearity in ecosystem services: Temporal and spatial variability in coastal protection. *Front. Ecol. Environ.* **2009**, *7* (1), 29–37.

8. Walther, G.R. Community and ecosystem responses to recent climate change. *Philos. Trans. R. Soc. B* **2010**, *365* (1549), 2019–2024.

9. De Groot, R.S. *Functions of Nature: Evaluation of Nature in Environmental Planning, Management and Decision Making.* Wolters-Noordhoff: Groningen, 1992.

10. De Groot, R.S.; Wilson, M.A.; Boumans, R.M.J. A typology for the classification, description and valuation of ecosystem functions, goods and services. *Ecol. Econ.* **2002**, *41*, 393–408.

11. Kandziora, M.; Burkhard, B.; Müller, F. Interactions of ecosystem properties, ecosystem integrity and ecosystem service indicators—A theoretical matrix exercise. *Ecol. Indic.* **2013**, *28*, 54–78.

12. Georgescu-Roegen, N. Inequality, limits and growth from a bioeconomic viewpoint. *Rev. Soc. Econ.* **1977**, *35* (3), 361–375.

13. Daly, H.E. On Nicholas Georgescu-Roegen's contributions to economics: An obituary essay. *Ecol. Econ.* **1995**, *13* (3), 149–154.

14. Gowdy, J.; Mesner, S. The evolution of Georgescu-Roegen's bioeconomics. *Rev. Soc. Econ.* **1998**, *56* (2), 136–156.

15. Mayumi, K., Giampietro, M.; Gowdy, J.M. Georgescu-Roegen/daly versus solow/stiglitz revisited. *Ecol. Econ.* **1998**, *27* (2), 115–117.

16. Mayumi, K. *The Origins of Ecological Economics: The Bioeconomics of Georgescu-Roegen.* Routledge: London, 2002.

17. Kleidon, A. Life, hierarchy, and the thermodynamic machinery of planet Earth. *Phys. Life Rev.* **2010**, *7* (4), 424–460.

18. Haeckel, E. *Generelle Morphologie der Organismen. Allgemeine Grundzüge der organischen Formen-Wissenschaft, mechanisch begründet durch die von C. Darwin reformirte Descendenz-Theorie,* Berlin: G. Reimer, Vol. 2. 1966.

19. Tansley, A.G. The use and abuse of vegetational concepts and terms. *Ecology* **1935**, *16* (3), 284–307.

20. Tansley, A.G. British ecology during the past quarter-century: The plant community and the ecosystem. *J. Ecol.* **1939**, *27* (2), 513–530.

21. Odum, E.; Odum, H.T. *Fundamentals of Ecology.* XV. B. Saunders Co.: Philadelplhia, 1953.

22. Tauber, A.I. The immune system and its ecology. *Philos. Sci.* **2008**, *75* (2), 224–245.

23. Egerton, F.N. History of ecological sciences, part 47: Ernst Haeckel's ecology. *The Bull. Ecol. Soc. Am.* **2013**, *94* (3), 222–244.

24. *Encyclopédie Roumaine,* édition de 1900, t. *II,* p. 528.

25. Gohier, J. Un cours à l'EPHE, la géonomie de Maurice François Rouge. *Ann. Rech. Urb.* **1988**, *37* (1), 94–97.

26. Whitbeck, R.H. A science of geonomics. *Ann. Assoc. Am. Geogr.* **1926**, *16* (3), 117–123.

27. Finch, V.C. Ray Hughes Whitbeck 1871–1939. *J. Geogr.* **1939**, *38* (6), 252–253.

28. Floyd, B. On reversing the image of a simplistic geography: The task of socio-economic geography (geonomics). *J. Geogr.* **1971** *70* (2), 84–90.

29. Giddings, B.; Hopwood, B.; O'Brien, G. Environment, economy and society: Fitting them together into sustainable development. *Sustain. Dev.* **2002**, *10* (4), 187–196.

30. Gowdy, J.; Erickson, J.D. The approach of ecological economics. *Camb. J. Econ.* **2005**, *29*, 207–222.

31. Brandt, P.; Ernst, A.; Gralla, F.; Luederitz, C.; Lang, D.J.; Newig, J.; Reinert, F., Abson, D.J.; Von Wehrden, H. A review of transdisciplinary research in sustainability science. *Ecol. Econ.* **2013**, *92*, 1–15.

32. Lam, J.C., Walker, R.M.; Hills, P. Interdisciplinarity in sustainability studies: A review. *Sustain. Dev.* **2014**, *22* (3), 158–176.

33. Castle, D.; Culver, K. Getting to 'no': The method of contested exchange. *Sci. Public Policy* **2013**, *40* (1), 34–42.

34. Barbier, E.B. *Natural Resources and Economic Development.* Cambridge University Press: Cambridge, 2005.

35. Costanza, R.; D'Arge, R.; De Groot, R.; Farber, S.; Grasso, M.; Hannon, B.; Limburg, K.; Naeem, S.; O'Neill, R.V.; Paruelo, J.; Raskin, R.G.; Sutton, P.; Van Den Belt, M. The value of the world's ecosystem services and natural capital. *Nature* **1997**, *387* (6630), 253–260.

36. Turner, D.P. Global vegetation monitoring: Toward a sustainable technobiosphere. *Front. Ecol. Environ.* **2011**, *9* (2), 111–116.

37. Lawton, J.H.; Naeem, S.; Woodfin, R.M.; Brown, V.K.; Gange, A.; Godfray, H.J.C.; Heads, P.A.; Lawler, S.; Magda, D.; Thomas, C.D.; Thompson, L.J. The Ecotron: A controlled environmental facility for the investigation of population and ecosystem processes. *Philos. Trans. R. Soc. Lond., B, Biol. Sci.* **1993**, *341* (1296), 181–194.

38. Lawton, J.H. The Ecotron facility at Silwood Park: The value of "big bottle" experiments. *Ecology* **1996**, *77* (3), 665–669.

39. Jones, T.H.; Thompson, L.J.; Lawton, J.H.; Bezemer, T.M.; Bardgett, R.D.; Blackburn, T.M.; Bruce, K.D.; Cannon, P.F.; Hall, G.S.; Hartley, S.E.; Howson, G. Impacts of rising atmospheric carbon dioxide on model terrestrial ecosystems. *Science* **1998**, *280* (5362), 441–443.

40. Lange, M.; Eisenhauer, N.; Sierra, C.A.; Bessler, H.; Engels, C.; Griffiths, R.I.; Mellado-Vázquez, P.G.; Malik, A.A.; Roy, J.; Scheu, S.; Steinbeiss, S. Plant diversity increases soil microbial activity and soil carbon storage. *Nat. Commun.* **2015**, *6*, 6707.

41. Redford, K.H.; Adams, W.M. Payment for ecosystem services and the challenge of saving nature. *Conserv. Biol.* **2009**, *23* (4), 785–787.

42. Gómez-Baggethun, E.; De Groot, R.; Lomas, P.L.; Montes, C. The history of ecosystem services in economic theory and practice: From early notions to markets and payment schemes. *Ecol. Econ.* **2010**, *69*, 1209–1218.

43. Daily, G.C.; Matson, P.A. Ecosystem services: From theory to implementation. *Proc. Natl. Acad. Sci. U. S. A.* **2008**, *105* (28), 9455–9456.

44. Rees, W.; Wackernagel, M. Urban ecological footprints: Why cities cannot be sustainable—and why they are a key to sustainability. *Environ. Impact Assess. Rev.* **1996**, *16* (4–6), 223–248.

45. Wackernagel, M.; Onisto, L.; Bello, P.; Callejas Linares, A.; Lopez Falfan, I.S.; Mendez Garcia, J.; Suarez Guerrero, A.I. National natural capital accounting with the ecological footprint concept. *Ecol. Econ.* **1999**, *29*, 375–390.

46. Wackernagel, M.; Kitzes, J.; Moran, D.; Goldfinger, S.; Thomas, M. The ecological footprint of cities and regions: Comparing resource availability with resource demand. *Environ. Urbanization* **2006**, *18* (1), 103–112.

47. Goldfinger, S.; Wackernagel, M.; Galli, A.; Lazarus, E.; Lin, D. Footprint facts and fallacies: A response to Giampietro and Saltelli (2014) "Footprints to Nowhere". *Ecol. Indic.* **2014**, *46*, 622–632.

48. Czech, B.; Daily, H.E. The steady state economy: What it is, entails, and connotes. *Wildl. Soc. Bull.* **2004**, *32* (2), 598–605.

49. Martínez-Alier, J.; Pascual, U.; Vivien, F.D.; Zaccai, E. Sustainable de-growth: Mapping the context, criticisms and future prospects of an emergent paradigm. *Ecol. Econ.* **2010**, *69*, 1741–1747.

50. Georgescu-Roegen, N. The steady state and ecological salvation: A thermodynamic analysis. *BioScience* **1977**, *27* (4), 266–270.

51. Liu, J.; Diamond, J. China's environment in a globalizing world. *Nature* **2005**, *435*, 1179–1186.

52. Costanza, R.; Graumlich, L.; Steffen, W.; Crumley, C.; Dearing, J.; Hibbard, K.; Leemans, R.; Redman, C.; Schimel, D. Sustainability or collapse: What can we learn from integrating the history of humans and the rest of nature? *Ambio* **2007**, *36* (7), 522–527.

53. Costanza, R.; de Groot, R.; Braat, L.; Kubiszewski, I.; Fioramonti, L.; Sutton, P.; Farber, S.; Grasso, M. Twenty years of ecosystem services: How far have we come and how far do we still need to go? *Ecosys. Serv.* **2017**, *28*, 1.

54. Daily, G.C.; Polasky S.; Goldstein, J.; Kareiva, P.M.; Mooney, H.A.; Pejchar, L.; Ricketts, T.H.; Salzman, J.; Shallenberger, R. Ecosystem services in decision making: Time to deliver. *Front. Ecol. Environ.* **2009**, *7*, 21–28.

55. Mac Nally, R.; Bennett, A.F.; Brown, G.W.; Lumsden, L.F.; Yen, A.; Hinkley, S.; Lillywhite, P.; Ward, D. How well do ecosystem-based planning units represent different components of biodiversity? *Ecol. Appl.* **2002**, *12* (3), 900–912.

56. Margules, C.R.; Pressey, R.L. Systematic conservation planning. *Nature* **2000**, *405*, 243–253.

57. Kremen, C. Managing ecosystem services: What do we need to know about their ecology? *Ecol. Lett.* **2005**, *8*, 468–479.

58. Chan, K.M.A.; Shaw, M.R.; Cameron, D.R.; Underwood, E.C.; Daily, G.C. Conservation planning for ecosystem services. *PLoS Biol.* **2006**, *4* (6), 2138–2152.

59. Egoh, B.; Rouget, M.; Reyers, B.; Knight, A.T.; Cowling, R.M.; Van Jaarsveld, A.S.; Weltz, A. Integrating ecosystem services into conservation assessments: A review. *Ecol. Econ.* **2007**, *63*, 714–721.

60. Randolph, J. *Environmental Land Use Planning and Management*. Island Press: Washington, DC, 2004.

61. Kagan, R.A. Political and legal obstacles to collaborative ecosystem planning. *Ecol. Law Q.* **1997**, *24*, 871–876.

62. McCarthy, M.A.; Burgman, M.A. Coping with uncertainty in forest wildlife planning. *For. Ecol. Manage.* **1995**, *74*, 23–36.

63. Nichiforel, L.; Schenz, H. Property rights distribution and entrepreneurial rent-seeking in Romanian forestry: A perspective of private forest owners. *Eur. J. For. Res.* **2011**, *130* (3), 369–381.

64. Taff, G.N.; Müller, D.; Kuemmerle, T.; Ozdeneral, E.; Walsh, S.J. Reforestation in Central and Eastern Europe after the breakdown of socialism. *Reforesting Landscapes* **2010**, *10*, 121–147.

65. Stefan, N. Contributii la studiul sindinamicii asociatiei Hyppophaetum *rhamnoides* in Subcarpatii de Curbura. *Academia Româna—Memoriile Sectiilor Stiintifice* **1991**, seria IV, tom *XIV* (1), 223–233.

66. Mitsch, W.J.; Jorgensen, S.E. Ecological engineering: A field whose time has come. *Ecol. Eng.* **2003**, *20*, 363–377.

67. Lessard, G. An adaptive approach to planning and decisionmaking. *Landscape Urban Plann.* **1998**, *40*, 81–87.

68. Williams, B.K. Adaptive management of natural resources—Framework and issues. *J. Environ. Manage.* **2011**, *92* (5), 1346–1353.

69. Musacchio, L.R.; Coulson, R.N.; Robert, N. Landscape ecological planning process for wetland, waterfowl, and farmland conservation. *Landscape Urban Plann.* **2001**, *56*, 125–147.

70. ICSU, UNESCO, UNU. Ecosystem change and human wellbeing. Research and Monitoring Priorities Based on the Findings of the Millennium Ecosystem Assessment, 2008, Paris, International Council for Science, ISBN 978-0-930357-67-2, available at http://www.icsu.org/icsu-asia/news-centre/news/archive-2006-2010/ICSUUNESCO-UNU_Ecosystem_Report.pdf (accessed December 12, 2011).

71. Koistinen, P.; Perkiö, J. Good and bad times of social innovations: The case of universal basic income in Finland. *Basic Income Stud.* **2014**, *9* (1–2), 25–57.

72. Marchal, S.;Van Mechelen, N. A new kid in town? Active inclusion elements in European minimum income schemes. *Soc. Policy Adm.* **2017**, *51* (1), 171–194.

73. De Groot, R.S.; Alkemade, R.; Braat, L.; Hein, L.; Willemen, L. Challenges in integrating the concept of ecosystem services and values in landscape planning, management and decision making. *Ecol. Complexity* **2010**, *7*, 260–272.

74. Gustavsson, M.B.; Hellohf, A.; Backhaus, T. Evaluating the environmental hazard of industrial chemicals from data collected during the REACH registration process. *Sci. Total Environ.* **2017**, *586*, 658–665.

75. Alberti, M. Maintaining ecological integrity and sustaining ecosystem function in urban areas. *Curr. Opin. Environ. Sustain.* **2010**, *2* (3), 178–184.

76. De Pauw, N.; Heylen, S. Biotic index for sediment quality assessment of watercourses in Flanders, Belgium. *Aquat. Ecol.* **2001**, *35*, 121–133.

77. Neamtu, M.; Ciumasu, I.M.; Costica, N.; Costica, M.; Bobu, M.; Nicoara, M.N.; Catrinescu, C.; Becker van Slooten, K.; De Alencastro, L.F. Chemical, biological, and ecotoxicological assessment of pesticides and persistent organic pollutants in the Bahlui River, Romania. *Environ. Sci. Pollut. Res.* **2009**, *16*, S76–S85.

78. Von der Ohe, PC.; De Deckere, E.; Prüß, A.; Muñoz, I.; Wolfram, G.; Villagrasa, M.; Ginebreda, A.; Hein, M.; Brack, W. Toward an integrated assessment of the ecological and chemical status of European river basins. *Integr. Environ. Assess. Manage.* **2009**, *5*, 50–61.

79. Smith, R.M.; Kaushal, S.S. Carbon cycle of an urban watershed: Exports, sources, and metabolism. *Biogeochemistry* **2015**, *126* (1–2), 173–195.

80. Ciumasu, I.M.; Costica, M.; Secu, C.V.; Gurjar, B.R.; Ojha, C.S.P. Adapting to climate change. In *Greenhouse Gas Emissions and Climate Change*; Surampalli, R.Y., Zhang, T., Ojha, C.S.P., Gurjar, B.R., Eds. ASCE Press – American Society of Civil Engineers: Reston, Virginia, 2013.

81. Pickett, S.T.; Grove, J.M. Urban ecosystems: What would Tansley do? *Urban Ecosyst.* **2009**, *12* (1), 1–8.

82. Bolund, P.; Hunhammar, S. Ecosystem services in urban areas. *Ecol. Econ.* **1999**, *29*, 293–301.

83. Rotmans, J.; Van Asselt, M.; Vellinga, P. An integrated planning tool for sustainable cities. *Environ. Impact Assess. Rev.* **2000**, *20*, 265–276.

84. Niemelä, J. Ecology and urban planning. *Biodiversity Conserv.* **1999**, *8*, 119–131.

85. Pauleit, S.; Duhme, F. Assessing the environmental performance of land cover types for urban planning. *Landscape Urban Plann.* **2000**, *52*, 1–20.

86. Ciumasu, I.M.; Culver, K. The city as disturbance of the local ecosystems. ISEE2011—Advancing Ecological Economics—Theory and Practice. *The 9th International Conference of the European Society of Ecological Economics*, Istanbul, June 14–17, 2011.

87. Marzluff, J.M. Island biogeography for an urbanizing world: How extinction and colonization may determine biological diversity in human-dominated landscapes. *Urban Ecosyst.* **2005**, *8*, 157–177.

88. Kattel, G.R.; Elkadi, H.; Meikle, H. Developing a complementary framework for urban ecology. *Urban For. Urban Gree.* **2013**, *12* (4), 498–508.

89. Wittig, R. The origin and development of the urban flora of Central Europe. *Urban Ecosyst.* **2004**, *7*, 323–339.

90. Dearborn, D.C.; Kark, S. Motivations for conserving urban biodiversity. *Conserv. Biol.* **2010**, *24* (2), 432–440.

91. Tzoulas, K.; James, P. Making biodiversity measures accessible to non-specialists: An innovative method for rapid assessment of urban biodiversity. *Urban Ecosyst.* **2010**, *13*, 113–127.

92. MacGregor-Fors, I.; Morales-Perez, L.; Schondube, J.E. Does size really matter? Species–area relationships in human settlements. *Diversity Distrib.* **2011**, *17*, 112–121.

93. McKinney, M.L. Urbanization as a major cause of biotic homogenization. *Biol. Conserv.* **2006**, *126*, 247–260.

94. Kühn, I.; Klotz, S. Urbanization and homogenization—Comparing the floras of urban and rural areas in Germany. *Biol. Conserv.* **2006**, *127* (3), 292–300.

95. Alberti, M.; Marzluff, J.M.; Shulenberger, E.; Bradley, G.; Ryan, C.; Zumbrunnen, C. Integrating humans into ecology: Opportunities and challenges for studying urban ecosystems. *BioScience* **2003**, *53* (12), 1169–1179.

96. Manfred, M.; Caputo, P.; Costa, G. Paradigm shift in urban energy systems through distributed generation: Method and models. *Appl. Energy* **2011**, *88*, 1032–1048.

97. Grimm, N.B.; Faeth, S.H.; Golubiewski, N.E.; Redman, C.L.; Wu, J.; Bai, X.; Briggs, J.M. Global change and the ecology of cities. *Science* **2008**, *319* (5864), 756–760.

98. Dyson, B.; Chang, N.B. Forecasting municipal solid waste generation in a fast-growing urban region with system dynamics modeling. *Waste Manage.* **2005**, *25*, 669–679.

99. Muga, H.E.; Mihelcic, J.R. Sustainability of wastewater treatment technologies. *J. Environ. Manage.* **2008**, *88*, 437–447.

100. Zio, E. Reliability engineering: Old problems and new challenges. *Reliability Eng. Syst. Saf.* **2009**, *94*, 125–141.
101. Lahdelma, R.; Saminen, P.; Hokkanen, J. Using multicriteria methods in environmental planning and management. *Environ. Manage.* **2000**, *26* (6), 595–605.
102. Munda, G. Social multi-criteria evaluation for urban sustainability policies. *Land Use Policy* **2006**, *23*, 86–94.
103. Ling, C.; Hanna, K.; Dale, A. A template for integrated community sustainability planning. *Environ. Manage.* **2009**, *44*, 228–242.
104. Ciumasu, I.M.; Costica, M.; Costica, N.; Neamtu, M.; Dirtu, A.C.; De Alencastro, L.P.; Buzdugan, L.; Andriesa, R.; Iconomu, L.; Stratu, A.; Popovici, O.A.; Secu, C.V.; Paveliuc- Olariu, C.; Dunca, S.; Stefan, M.; Lupu, A.; Stingaciu-Basu, A. ; Netedu, A.; Dimitriu, R.I.; Gavrilovici, O.; Talmaciu, M.; Borza, M. Complex risks from old urban waste landfills—A sustainability perspective from Iasi, Romania. *J. Hazard., Toxic Radioact. Waste* **2012**, *16* (2), 158–168.
105. Lang, D.J.; Scholz, R.W.; Binder, C.R.; Wiek, A.; Stäubli, B. Sustainability Potential Analysis (SPA) of landfills—A systemic approach: Theoretical considerations. *J. Cleaner Prod.* **2007**, *17*, 1628–1638.
106. Birk, S.; Bonne, W.; Borja, A.; Brucet, S.; Courrat, A.; Poikane, S.; Solimini, A.; Van De Bund, W.; Zampoukas, N.; Hering, D. Three hundred ways to assess Europe's surface waters: An almost complete overview of biological methods to implement the Water Framework Directive. *Ecol. Indic.* **2012**, *18*, 31–41.
107. Poikane, S.; Zampoukas, N.; Borja, A.; Davies, S.P.; van de Bund, W.; Birk, S. Intercalibration of aquatic ecological assessment methods in the European Union: Lessons learned and way forward. *Environ. Sci. Policy* **2014**, *44*, 237–246.
108. Demuzere, M.; Orru, K.; Heidrich, O.; Olazabal, E.; Geneletti, D.; Orru, H.; Bhave, A.G.; Mittal, N.; Feliu, E.; Faehnle, M. Mitigating and adapting to climate change: Multi-functional and multi-scale assessment of green urban infrastructure. *J. Environ. Manage.* **2014**, *146*, 107–115.
109. Žuvela-Aloise, M.; Koch, R.; Buchholz, S.; Früh, B. Modelling the potential of green and blue infrastructure to reduce urban heat load in the city of Vienna. *Clim. Change* **2016**, *135* (3–4), 425–438.
110. Ziter, C.D.; Pedersen, E.J.; Kucharik, C.J.; Turner, M.G. Scale-dependent interactions between tree canopy cover and impervious surfaces reduce daytime urban heat during summer. *Proc. Natl. Acad. Sci.* **2019**, *116* (15), 7575–7580.
111. Santamour Jr, F.S. Trees for urban planting: Diversity uniformity, and common sense. In *The Overstory Book: Cultivating Connections with Trees*; C. Elevitch, Ed., pp. 396–399. Permanent Agriculture Resources, 2004.
112. Reyers, B.; Roux, D.J.; Cowling, R.M.; Ginsburg, A.E.; Nel, J.L.; O'Farrell, P. Conservation planning as a transdisciplinary process. *Conserv. Biol.* **2010**, *24* (4), 957–965.
113. Reid, W.V.; Chen, D.; Goldfarb, L.; Hackmann, H.; Lee, Y.T.; Mokhele, K.; Ostrom, E.; Raivio, K.; Rockström, J.; Schellnhuber, H.J.; Whyte, A. Environment and development. Earth system science for global sustainability: Grand challenges. *Science* **2010**, *330* (6006), 916–917.

101. ...

102. ...

103. ...

11

Natural Enemies: Conservation

Cetin Sengonca

Introduction

After the successful utilization of two methods of biological control, classical biological control (importation and establishment of exotic natural enemies against either exotic or native pests) and augmentation of natural enemies (either inundative or inoculative releases of mass reared natural enemies), the third method—conservation and enhancement—has become more and more important during recent years.[1,2] Conservation of natural enemies is probably the most important concept in the practice of biological control and, fortunately, is one of the easiest to understand and readily available to growers. Most authors consider conservation as an environmental modification to protect and enhance natural enemies.[3] This definition will be the main subject of this entry.

Natural enemies of arthropod pests, also known as biological control agents, include predators and parasitoids that occur in all production systems from commercial fields to backyards where they have adapted to the local environment and target pests. Their conservation is generally noncomplicated and cost-effective.[4] With relatively little effort the activities of these natural enemies can be observed. Natural control agents are a major factor in controlling agricultural pests and need to be considered when making pest management decisions. Today, therefore, the conservation of natural enemies is considered inseparable from enhancement and together they represent a successful biological control method.

Avoid Harmful Practices of Pesticides

Pesticides used in agriculture are not only killing target pests but they also can have direct effects on natural enemies by killing them, or indirectly by eliminating their hosts or preys and causing them to starve. In contrast, the conservation concept of natural enemies attempts to avoid the application of particularly broad-spectrum, highly disruptive pesticides. Applying selective or specific and beneficially safe pesticides may contribute much toward preserving natural enemies. Pesticide selectivity to beneficial arthropods has been broadly classified into two forms. The first of these is physiological selectivity, that is, pesticides are less toxic to natural enemies than to their target pest when applied at the recommended rate. The second form is ecological selectivity that pertains to the means and domains in which pesticides are used. Systemic pesticides killing leaf-feeding herbivores, for example, may have little or no effect on the many natural enemies that have contact only with the leaf surface. In some cases, pesticides can be successfully integrated into pest management systems with little or no detrimental effect on natural enemies, and this trend is likely to increase substantially in the future. Nowadays, the pesticide industry places increasing emphasis on the development of beneficially safe and environmentally friendly pesticides that exhibit greater selectivity for natural enemies and have minimal environmental impacts. In the same way, governmental regulatory agencies increasingly consider the adverse effects of pesticides on natural enemies in their registration process for pesticides, reflecting the growing concern over negative effects on beneficial insects. Despite these important steps and the great progress that has been made, the latent effects of pesticides and the impact of "cocktail applications" on natural enemy populations are still not fully understood. Further research and implementation of research results are urgently needed.

Alternatively, when selective pesticides are unavailable, recommended conservation tactics usually involve exact timing of pesticide applications. Careful forecasting and observation of the occurrence and growth of pest populations can substantially reduce the number of pesticide applications. Forecasting systems should be based on a defined economic threshold for each pest, considering also the presence of natural enemies.[5] Another approach is the selective placement of pesticides in agricultural fields. Limiting pesticide application only to infested parts of the field will reduce costs and also conserve natural enemies. An ideal alternative approach is the use of microbial insecticides, such as commercially available *Bacillus thuringi- ensis* and fungal and viral products that have little adverse effects on natural enemies and the environment.

Habitat and Environmental Manipulation

Another form of natural enemy conservation is habitat and environmental manipulation. The agricultural landscape is currently so intensively managed that the species diversity of many natural habitats has disappeared or become endangered. A similar reduction can be observed among natural enemies too. It has been strongly suggested by many experts that natural enemies also can be conserved by simply encouraging vegetational diversity of the agroecosystem. In this context, hedgerows, cover crops, strips inside and bordering fields, and even in-field balks provide important refuges for parasitoids and predators of many pest species. There they find and benefit from safe shelters, sources of pollen and nectar, and also alternative prey or hosts in case of food scarcity in cultivated fields. And thus, at such sites, a long-lasting and self-regulating biocoenosis will develop. A higher acceptance of weeds in agricultural crops may also increase the efficacy of natural enemies. Similarly, mixed plantings, for example, Umbelliferae, mustard, and *Phacelia tanacetifolia*,[6] growing weed strips even within fields, and providing flowering field borders significantly increase habitat diversity. *P. tanacetifolia* has been cultivated widely between crops in the production system in Germany for more than a decade. At the same time this will provide shelters and alternative food sources for natural enemies. Another important concept in conservation by habitat manipulation is that of connectivity. Natural or less disturbed habitats are often scattered and isolated within the agricultural landscape. Connecting these habitats, for example, by hedgerows and woods, will establish a continuous network of corridors allowing movement of natural enemies between fields.

Experiments have shown that a constant population of natural enemies can be established and conserved by releasing their prey or hosts during periods of scarcity, for example, releasing the red mite, *Tetranychus urticae* Koch, to support the establishment of its predatory mite, *Phytoseiulus persimilis* Athias-Henriot, in cucumber and bean cultures widely in greenhouses in middle Europe.[7] Similarly, distributing sterilized *Eupoecelia ambiguella* Hb. eggs between the two generations of this lepidopteran pest preserved its egg parasitoid *Trichogramma semblidis* (Auriv.) in the Ahr valley in Germany. A classical example is the black scale, *Saissetia oleae* (Oliv.), which interrupts its development for a short period during summer in the hot arid areas of central California. Planting irrigated oleander plants adjacent to citrus orchards allowed the black scale a continuous development. As a result, this enabled its specific parasitoid *Metaphycus helvolus* (Comp.) to maintain its population, particularly during the hot summer months.[7] In another example, it has been found that preservation of nettles, *Urtica* spp., an important host plant of *Aglais urticae* L., can enhance the efficacy of *T. semblidis* on the second generation of *E. ambiguella*. The reason is that the parasitoid maintains its population on this alternative host during the two nonoverlapping generations of *E. ambiguella*.[8]

The manipulation of some simple cultural measures can also conserve natural enemies. A famous example is strip harvesting hay alfalfa, allowing mobile natural enemies to disperse from cut strips to half-grown strips. Similarly, *Trissolcus vasilievi* (Mayr) and *T. semistriatus* Nees, two important egg-parasitoids of the sunn pest, *Eurygaster integriceps* Put., were successfully conserved in Turkey by growing shade trees in hot arid areas, providing shade and thus suitable climatic conditions for these parasitoids.

Overwintering and Shelter Sites

Natural enemies build up considerably high population densities during summer periods, but then suffer from lack of overwintering or shelter sites and unfavorable climatic conditions during winter. As a result of these detrimental environmental factors, extreme low entomophagous arthropod densities are often present the following year. This permits pest populations to explode and in consequence requires more pesticide applications. In contrast, by preserving existing or providing artificial hibernation sites or shelters, natural enemies can be conserved during overwintering periods. For example, planting of trees and perennial bunch grasses near agricultural sites, the use of burlap or cloth trees, and wrap and stones provided as hiding places at overwintering time allow coccinellid lady beetles higher survival rates during hibernation. In the same way "trunk traps" and "trap bands" can be used as artificial overwintering sites for predatory bugs and lace-wings,[9] and felt belts can be wrapped around the trunks of fruit trees and vines for the predatory mite *Typhlodromus pyri Scheuten*. The green lacewing, *Chrysoperla carnea* (Stephens), overwinters as an adult in barns, roof trusses, houses, and under the bark of trees, where mortality rates during hibernation in middle European climatic conditions may still reach 60%–90%. By using specially designed, simple wooden shelters (hibernation boxes) the overwintering mortality was reduced to only 4%–8%.[10] On this ground, these hibernation boxes are now being accepted and commonly used by farmers, gardeners, and also environmental protectionists in Germany and Switzerland.

Future Concerns

The majority of pest problems in agriculture are due to the elimination of natural enemies by the indiscriminate and intensive use of pesticides. Improper habitat manipulation and mismanagement of ecosystems have further intensified this problem by reducing the available flora and fauna. Conservation and enhancement of natural enemies is the easiest and least costly method of biological control offering solutions to most pest problems without harming and disturbing the natural ecosystem. Unfortunately, however, this field has received little attention and very little investment has been made in research. There is a serious need for research into the areas of direct conservation and enhancement of natural enemies during the vegetation period and hibernation. In a self-regulating mechanism focusing on conservation and enhancement, natural enemies can keep agricultural pests below their economic threshold and help to reduce the number and frequency of pesticide applications. Furthermore, integration of conservation and enhancement of natural enemies into existing IPM programs[11] will lead to a more sustainable and cost-effective agriculture system.

See also Cosmetic Standards, pages 152–154; Conservation of Biological Controls, pages 138–140; Augmentative Controls, pages 36–38; Biological Controls, pages 57–60; pages 61–63; pages 64–67; pages 68–70; pages 71–73; pages 74–76; pages 77–80; pages 81–84.

References

1. Ehler, L.E. Conservation Biological Control: Past, Present and Future. In *Conservation Biological Control*; Barbosa, P., Ed.; Academic Press: New York, 1998; 1–8.
2. Bugg, R.L.; Pickett, C.H. Introduction: Enhancing Biological Control–Habitat Management to Promote Natural Enemies of Agricultural Pests. In *Enhancing Biological Control*; Pickett, C.H., Bugg, R.L., Eds.; University of California Press: Berkeley, 1998; 1–23.
3. DeBach, P. *Biological Control of Insect Pests and Weeds*; Chapman and Hall: New York, 1964; 844.
4. Barbosa, P. Agroecosystems and Conservation Biological Control. In *Conservation Biological Control*; Barbosa, P., Ed.; Academic Press: New York, 1998; 39–59.
5. Sengonca, C. Conservation and enhancement of natural enemies in biological control. Phytoparasitica **1998**, *26* (3), 187–190.
6. Sengonca, C.; Frings, B. Einfluss von phacelia tanacetifolia auf schaedlings- und nuetzlingspopulation in Zuckerruebe. Pedobiologia **1988**, *32* (5/6), 311–316.
7. Krieg, A.; Franz, J.M. *Lehrbuch der biologischen Schaedlingsbekaempfung*; Verlag Paul Parey: Berlin, 1989; 302.
8. Schade, M.; Sengonca, C. Foerderung des traubenwickler- eiparasitoiden Trichogramma semblidis (Auriv.) (Hym., Trichogrammatidae) durch bereitstellung von ersatzwirten an brennesseln im weingebiet ahrtal. Vitic. Enol. Sci. **1998**, *53* (4), 157–161.
9. Beane, K.A.; Bugg, R.L. Natural and Artificial Shelter to Enhance Arthropod Biological Control Agents. In *Enhancing Biological Control*; Pickett, C.H., Bugg, R.L., Eds.; University of California Press: Berkeley, 1998; 240–253.
10. Sengonca, C.; Frings, B. Enhancement of green lacewing Chrysoperla carnea (Stephens) by providing artificial facilities for hibernation. Turk. Entomol. Derg. **1989**, *13* (4), 245–250.
11. *CRC Handbook of Pest Management in Agriculture*; Pimentel, D., Ed.; CRC Press: Boca Raton, FL, 1991; *2*, 757.

12

Pests: Landscape Patterns

F. Craig Stevenson

Pest and Landscape Patterns

Small Patches

Certain areas within a field may be more conducive to pest survival, establishment, and development relative to other areas. The size and proximity of localized infestations with a given field differs among the three major pest groups. Past research has shown that weed and disease patches generally have a radius of about 25 m (if you assume that they have a circular shape), but may be as large as 100 m.[1–3] In the case of wild oat *(Avena fatua* L.), the patch may be associated areas that have higher soil water and nutrient availability (see Figure 1). Weeds such as perennial sowthistle *(Sonchus arvensis* L.), with seeds adapted to wind dissemination, may occur in areas of higher elevation where seeds are trapped as they are blown across a field. Foliar disease patches generally are associated with low wet areas of fields or sheltered field margins that have higher canopy humidity and lower wind speeds. Landscape patterns for insects, however, are less persistent and harder to predict because of rapid reproductive rates and pest movement.[4] To a lesser extent, this same level of complexity also occurs for polycyclic diseases that easily become airborne. However, the overwintering phases in northern agricultural regions and

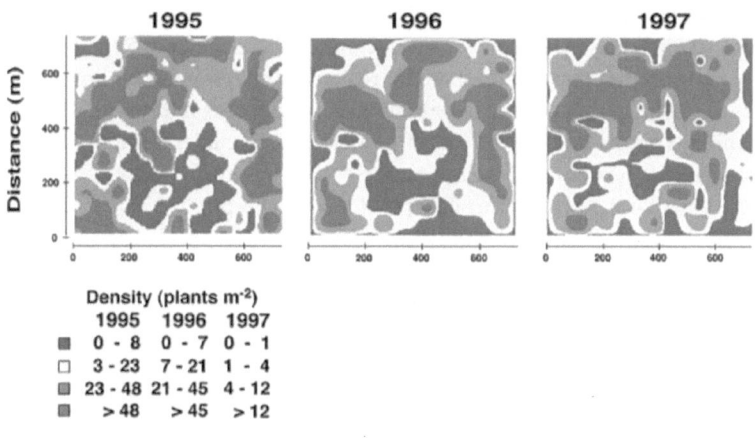

FIGURE 1 Landscape pattern of wild oat in a 65 ha field.
Source: Thomas.[6]

nonflying phases of insects may be associated with areas of a field with a more friable soil structure (e.g., grasshoppers) or other soil and microclimatic factors. In addition, for insects and to a lesser extent diseases, landscape patterns are complicated pest–predator interactions that can affect infestation of that pest in the given space in the next growing season. For example, the landscape pattern of nematodes has been shown to be partly a function of nematophagous fungi.[5] Therefore, the proximity of areas more or less conducive for the pest is ultimately dependent on the topography of the field and spatial variation of microclimatic conditions, especially for those pests with restricted mobility.

Larger Scale Landscape Patterns

These patterns are seen where larger areas of a field, or numerous fields, are being infested by a pest. These types of landscape patterns generally are dependent on external factors that move the pest from a smaller patch to adjacent areas. Climatic processes, such as wind, that blow weed seeds, disease spores, and insects can rapidly increase the area of pest infestation. Farming practices such as tillage and harvest operations also cause larger scale landscape patterns. For example, a study showed that *Polygonum* spp. of weeds tended to vary with the direction of tramlines up to distances of 635 m, a distance about 600 m more than that observed in the direction perpendicular to tramlines.[1] The spatial distribution of pests can be influenced by tillage practices and harvest operations that move crop residues and soil across a field. For example, nematodes are spread by cultivation from initial infestation foci.[5] Also, grain combine harvesters and shank-type tillage implements can move weed seeds and other reproductive structures across a field quickly.

Temporal Stability of Landscape Patterns

Time tends to complicate and/or obscure distinct landscape patterns for pests within a field. Varying climatic patterns (rainfall, temperature, etc.), within and among growing seasons can have a profound effect on the ability of pests to survive, establish, and develop. This complexity, interacting with polycyclic and multigeneration reproductive strategies within a growing season and factors moving pests across or between fields, poses a major hurdle to a holistic understanding of pests and landscape patterns. Extensive field research combined with predictive modeling may be a fruitful avenue for such a complex phenomenon.

Future Developments

The advent of global positioning systems (GPS), satellite imagery, and geographic information systems (GIS) software have heightened awareness and provided insight into landscape patterns and pests. Ultimately, these technologies may allow for the site-specific management of pests in accordance with their landscape patterns.[7] For example, spatially referenced maps could be linked to GPS on pesticide applicators to reduce the total amount of pesticide applied, thus improving economic returns and resulting in farming systems less dependent on pesticides. Our current understanding clearly shows that pest infestations often occur in patterns across a field, however, a great deal more effort will be necessary to provide information to accurately predict where pests will occur and affect crop production most extensively. These challenges will be especially difficult considering the highly variable and dynamic nature of current climatic conditions and farm management systems.

References

1. Nordbo, E.; Christensen, S.;Kristensen, K.;Walter, M. Patch spraying of weed in cereal crops. Asp. Appl. Biol. **1994**, *40,* 325–334.
2. Zadoks, J.C.; van den Bosch, F. On the spread of plant disease: a theory on foci. Ann. Rev. Phytopathol. **1994**, *32,* 503–521.

3. Zanin, G.; Berti, A.; Riello, L. Incorporation of weed spatial variability into the weed control decision-making process. Weed Res. **1998**, *38*, 107–118.

4. Hassell, M.P.; Comins, H.N.; May, R.M. Spatial structure and chaos in insect population dynamics. Nature **1991**, *353*, 255–258.

5. Webster, R.; Boag, B. Geostatistical analysis of cyst nematodes in soil. J. Soil Sci. **1992**, *43*, 583–595.

6. Thomas, A.G. *Agriculture and Agri-Food Canada*; Saskatoon, SK, Canada, 1999, unpublished data.

7. *The State of Site Specific Management for Agriculture*; Sadler, E.J.,Pierce,F.J.,Eds.; American Society of Agronomy, Inc., Crop Science Society of America, Inc., Soil Science Society of America, Inc.: Madison, WI, 1997; 423.

III

CSS: Case Studies of Environmental Management

III

13

Biological Control of Vertebrates: Myxoma Virus and Rabbit Hemorrhagic Disease Virus as Biological Controls for Rabbits

Peter Kerr and
Tanja Strive

Myxomatosis

Myxomatosis is a generalized, lethal disease of European rabbits classically characterized by gross skin swellings, sometimes termed tumors or myxomas, which appear mucinous on cut sections, mucopurulent blepharoconjunctivitis, swollen face and head, thickened, edematous, often drooping ears and grossly swollen perineum. In acute infections, death occurs 8–12 days after infection or 3–5 days after development of obvious clinical signs. The case fatality rate (CFR) can approach 100%. Myxomatosis was not a disease of European rabbits in their natural range; it is caused by myxoma virus (MYXV), which originally circulated in the South American jungle rabbit or tapeti (*Sylvilagus menensis* also called

Sylvilagus brasiliensis). In this species, MYXV causes a relatively inconsequential cutaneous fibroma at the site of inoculation. Myxomatosis is an example of an emerging disease that was only seen when a susceptible host species, the European rabbit, was imported into South America where MYXV was circulating.

Myxoma Virus

MYXV is a poxvirus (family: *Poxviridae*; subfamily: *Chordopoxvirinae*; genus: *Leporipoxvirus*; species: *Myxoma virus*). Like other chordopoxviruses, MYXV is a large brick-shaped particle (286 × 230 × 75 nm). The type virus (Lausanne strain) has a double-stranded DNA genome of 161.8 kb encoding 158 unique genes, 12 of which are duplicated in terminal inverted repeats of 11.5 kb at each end of the genome. The virus replicates in the cytoplasm of infected cells and so must encode the necessary enzymes for DNA replication and RNA transcription. Like other large DNA viruses, many genes encode virulence factors which suppress host immune responses or antiviral mechanisms (Cameron et al. 1999, Kerr 2012, Kerr et al. 2015). In the natural host, these virulence factors are probably essential for local suppression of immune responses, thus maintaining virus in the fibroma for sufficient time to allow transmission. However, in European rabbits, the combination of these immune suppressive factors and the ability of the virus to replicate in lymphocytes and monocytes of this species allow the virus to disseminate from the site of initial infection and cause generalized systemic disease with lymphoid tissue destruction and profound immunosuppression (Best et al. 2000, Best and Kerr 2000, Jeklova et al. 2008). MYXV is passively transmitted on the mouthparts of biting arthropods such as mosquitoes or fleas that probe through the virus-rich skin lesions in search of a blood meal. The virus does not replicate in the arthropod vector. In European rabbits, transmission may also occur via virus in conjunctival or nasal discharges.

The Use of Myxoma Virus as a Biological Control for Rabbits in Australia

The use of MYXV to control rabbits was suggested as early as 1919 (Fenner and Ratcliffe 1965) and some early experiments were done in Australia. However, it was not until the 1930s that systematic studies were undertaken, initially in Britain and subsequently in Australia, to examine the lethality for rabbits, the host range, and the likely impact of MYXV on other Australian species (Martin 1936, Bull and Dickinson 1937). Field studies in Australia were undertaken in 1937–1939 in enclosures on Wardang Island off the coast of South Australia and subsequently, in the summer of 1940–1941, in an enclosure in low-rainfall country on the mainland where native stickfast fleas (*Echidnophaga* spp.) were present as a vector (Bull and Mules 1944). These studies were followed by field releases in the same area. The virus had a substantial impact on local rabbit populations when fleas were present and predators such as foxes were removed to prevent them from killing sick rabbits. However, it did not spread beyond the study sites and was regarded as unlikely to control rabbits on any scale (Bull and Mules 1944).

Recognizing the need for mosquitoes as vectors, experimental studies were undertaken in 1950 in higher rainfall districts in the Murray River valley near Albury on the border between New South Wales (NSW) and Victoria and near Gunbower in north-central Victoria (Myers 1954). The virus appeared to spread initially but then died out at the release sites. However, myxomatosis reappeared in December of that year at one of the sites and simultaneously at other locations along the Murray River and at locations up to 650 km away (Ratcliffe et al. 1952, Myers 1954). Transmission was driven by mosquitoes, particularly *Culex annulirostris*, which were in high numbers due to flooding in the northern Murray-Darling river system and the vast numbers of rabbits which were completely susceptible to this novel pathogen. The scale of the epizootic was unprecedented with estimates of spread covering an area of 1600 km south to north and 1750 km east to west (Ratcliffe et al. 1952). The virus trickled on in local outbreaks through the winter and then reappeared in epizootics the following summer once mosquitoes emerged. Over the next 3 years, MYXV spread, either naturally or by deliberate introduction, wherever rabbits had established and there were enough mosquitoes or other vectors to transmit the virus.

Myxomatosis reduced rabbit numbers by perhaps 90% to 99% with CFRs due to the released virus (termed the standard laboratory strain; SLS) estimated at 99.8% (Fenner et al. 1953, Myers et al. 1954). However, in some local epizootics, the estimated CFR was as low as 40% with surviving rabbits identified by scarring and the presence of serum antibodies to MYXV (Fenner et al. 1953).

Rabbits that recover from myxomatosis are probably immune for life, with immunity potentially boosted by reexposure to the virus. In addition, recovered females will transfer maternal antibodies across the placenta to kittens, which may provide limited passive protection in the early weeks of life (Fenner and Marshall 1954, Kerr 1997).

It soon became apparent that slightly attenuated strains of MYXV were occurring in the field (Fenner et al. 1953, Myers et al. 1954, Marshall et al. 1955). While still highly lethal, with CFRs of 90%–99%, these strains had a selective advantage because they allowed infected rabbits to survive for a slightly longer time, thus providing more opportunity for mosquito transmission of the virus (Marshall et al. 1955, Fenner et al. 1956).

Field strains of virus were subsequently classified into five virulence grades (1–5) based on CFR, survival times, and clinical signs of small groups of infected laboratory rabbits (Fenner and Marshall 1957). Grade 3 viruses, with CFRs of 70%–95% and survival times of 17–28 days, became the predominant strains in the field within 4 years (Figure 1) and remained so for the next 30 years after which there are no systematic data available (Fenner and Ross 1994). However, both highly virulent and highly attenuated strains can still be recovered from the field (Kerr et al. 2017, unpublished data).

With such high CFRs, MYXV was exerting very high selection pressure on the rabbit population and the emergence of somewhat attenuated viruses probably increased the frequency with which rabbits

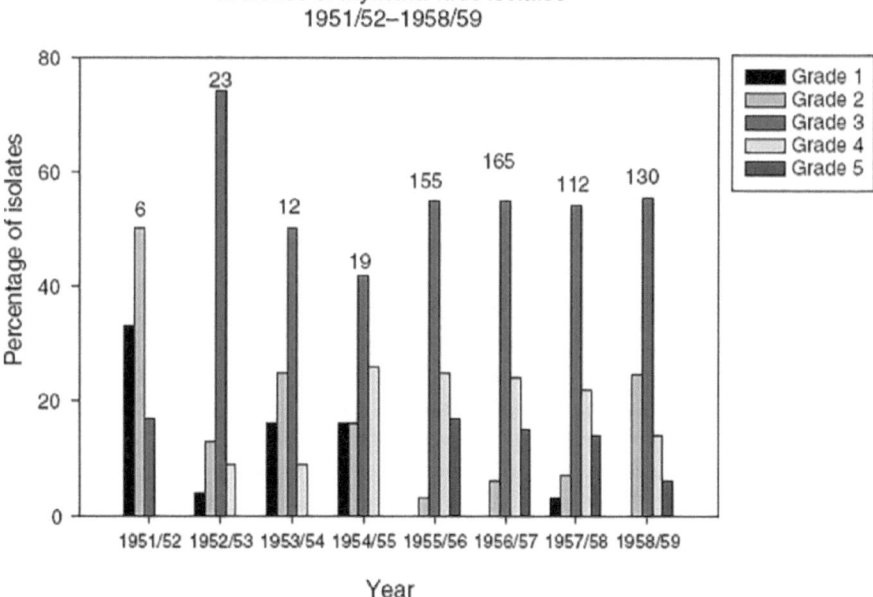

FIGURE 1 Virulence changes in Australian isolates of MYXV from 1950–1959. Virulence of MYXV field isolates is shown as the percentage of isolates made each season classified into grades 1–5. The number of isolates tested each year is shown above the bars. Data for 1951/1952–1957/1958 are from Marshall and Fenner (1960) and data for 1958/1959 are from Fenner and Woodroofe (1965). Virulence was classified as: Grade 1 average survival time (AST): ≤ 13 days; CFR: 99.5%; Grade 2 AST: > 13 ≤ 16 days; CFR: 99%; Grade 3 AST > 16 ≤ 28 days; CFR: 70%–95%; Grade 4: AST > 28 ≤ 50 days; CFR: 50%–70%; Grade 5 AST not applicable; CFR < 50%.

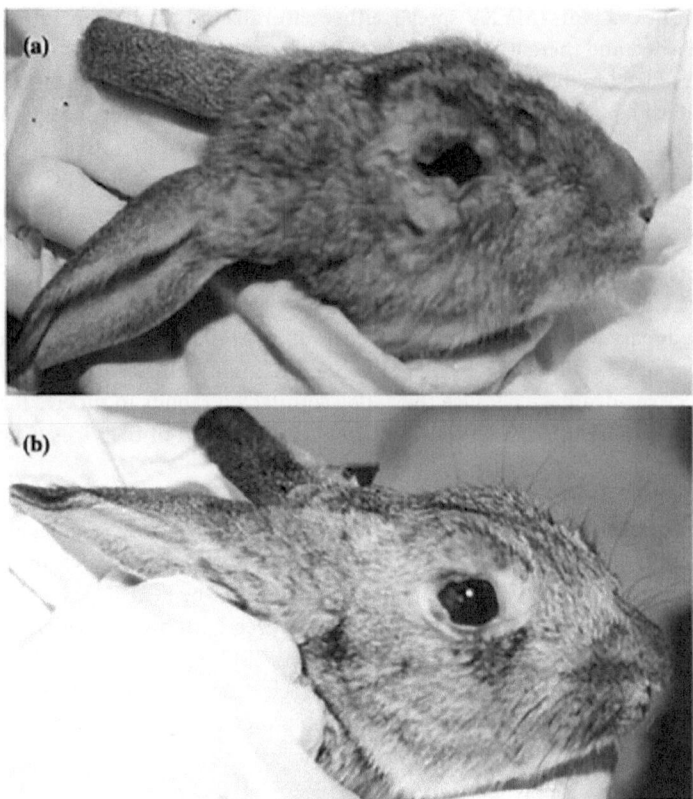

FIGURE 2 Genetic resistance to MYXV in Australian wild rabbits. (a) A wild rabbit infected 10 days previously with the grade 1 virulent SLS MYXV introduced into Australia in 1950. The rabbit shows relatively mild clinical signs of myxomatosis. (b) A wild rabbit 20 days after infection with the same virus. This rabbit has virtually fully recovered. SLS originally killed essentially 100% of infected rabbits with an average survival time of ≤13 days. (Figures are from Best SM and Kerr PJ (2000) Virology 267, 36–48, Elsevier publishing, with permission.)

with alleles conferring some resistance to myxomatosis survived infection (Figure 2) (Alves et al. 2019). In addition, high environmental temperatures enhanced survival of rabbits infected with moderately attenuated virus strains (Marshall 1959, Marshall and Douglas 1961).

A dramatic example of the selection for resistance was provided by a study on rabbits from Lake Urana in NSW. Within 7 years of natural selection by annual epizootics of myxomatosis, the CFR in these rabbits, infected under laboratory conditions with the KM13 strain of MYXV, dropped from 90% to 26% (Marshall and Fenner 1958, Marshall and Douglas 1961).

The emergence of attenuated virus strains and genetically resistant rabbits in the decade from 1950–1960 reduced the impact of myxomatosis. However, there is no doubt that the disease continued to suppress rabbit populations especially in the agricultural areas of Australia where myxomatosis, in occasional widespread epizootics and more regular local outbreaks, complemented conventional control measures. In an experiment where myxomatosis was experimentally controlled, rabbit numbers increased by 8–12 fold in a region where myxomatosis had previously been regarded as not contributing to rabbit control (Parer et al. 1985).

In the early 1990s, 40 years after the initial epizootic, it was estimated that rabbit numbers in the agricultural areas of Australia were on average fluctuating around 5% of their pre-myxomatosis numbers (Fenner and Ross 1994, Williams et al. 1995). However, in the low rainfall rangeland and semiarid areas, where conventional management was more difficult or nonexistent, populations were about 25%

of pre-1950 numbers and population irruptions were occurring (Fenner and Ross 1994, Williams et al. 1995). The rabbit breeding cycle and climatic variation means that there will be large fluctuations in population sizes particularly in the arid zones where the natural decline in rabbit populations due to drought may be as much as 99%, but it is also in these areas that rabbits have their greatest impact on preventing regeneration of vegetation (Williams et al. 1995, Cooke 2012).

Since 1995/1996, the impact of myxomatosis on rabbit numbers has been more difficult to discern because of the presence of rabbit hemorrhagic disease (RHD), which may also alter the timing of epizootics of myxomatosis by removing the seronegative young rabbits needed to sustain an outbreak (Mutze et al. 2002, 2010a). Although difficult to quantify, it is likely that myxomatosis continues to suppress rabbit numbers across Australia at essentially no economic cost.

The Introduction of Novel Insect Vectors to Australia to Enhance Biological Control by Myxomatosis

MYXV in Australia was largely vectored by mosquitoes with some local spread by direct contact and native stickfast fleas. In dry seasons or arid zones, there might be no transmission because there was little or no mosquito breeding. There were areas such as Tasmania, the northern tablelands of NSW, the western district of Victoria, or south-western West Australia where mosquito activity was low (Williams et al. 1995) with a consequent lower impact of myxomatosis.

In Britain and parts of Europe, the main vector of MYXV, following its release in 1952 (see below), was the European rabbit flea (*Spilopsyllus cuniculi*), however, this parasite had not been introduced to Australia with the rabbit. Rabbit fleas are present on rabbits all year round and move freely between rabbits, thus potentially allowing transmission of MYXV when mosquitoes are absent or in low numbers.

S. cuniculi was imported to Australia in 1966 and released from quarantine in 1968 after rather minimal testing for environmental impact (Sobey and Menzies 1969). The fleas were widely released throughout Australia, and in many regions, this produced a drop in rabbit numbers that was comparable to the earlier impact of myxomatosis (Cooke 1983, King et al.1985, Mutze et al. 2002, Cooke et al. 2013). The overall result appears to have been an increase in the efficacy of myxomatosis despite attenuated field strains of virus and genetic resistance in rabbits. In particular, fleas could drive epizootics when mosquito activity was low. *S. cuniculi* was also introduced into rabbit populations on the subantarctic Macquarie Island and the Kerguelen archipelago.

Despite its initial establishment, *S. cuniculi* tended to die out in rabbit populations in the hot dry areas of Australia (annual rainfall <200 mm) (Cooke 1984). To overcome this limitation, the Spanish rabbit flea *Xenopsylla cunicularis*, which is adapted to hot, arid conditions, was introduced (Cooke 1990). This flea was released from quarantine in 1993 after studies on species specificity showed that native animals were not a preferred host. Large-scale breeding was undertaken and over one million pupae or newly emerged fleas were subsequently released at some 500 sites (Fenner and Fantini 1999, Saunders et al. 2010). Unfortunately, due to the emergence of RHD in 1995, virtually no follow-up studies were done on the impact of this flea on the epidemiology of myxomatosis although there were reports of unseasonable outbreaks of myxomatosis (Fenner and Fantini 1999). More recent anecdotal evidence indicates that this flea remains established in rabbit populations.

Introduction of More Virulent Strains of Myxoma Virus to Australia

From the time of the first epizootic of myxomatosis in Australia, MYXV was deliberately introduced into rabbit populations. Preparations of SLS were made available for field release, and in NSW and later in other states, a field isolate derived from SLS known as the Glenfield strain was used. The distinctive

Lausanne strain of MYXV (used for release in Europe, see below) was widely used in release campaigns in Australia from the 1970s particularly in association with *S. cuniculi* release (Fenner and Ross 1994). Lausanne and Glenfield are more virulent than SLS when tested in genetically resistant wild rabbits although the virulence of the three viruses is indistinguishable in laboratory rabbits. Releases of Lausanne continued until the late 1990s even though it was recognized that released viruses were outcompeted by field strains of virus and at best might create local epizootics (Myers et al. 1954, Marshall et al. 1955, Fenner et al. 1957, Berman et al. 2006). Virulent virus is no longer produced for release in Australia, and releases of MYXV are not occurring (rabbit hemorrhagic disease virus (RHDV) is produced for release). Proposals for new introductions of virulent strains of MYXV have been made (Henzell et al. 2008, Mutze et al. 2010a), although it is difficult to see how these would compete in the field based on previous releases of virulent virus. For example, despite its widespread release in Australia, no circulating strains of MYXV derived from the Lausanne strain have been identified (Saint et al. 2001, Kerr et al. 2019).

Myxomatosis in Europe and Other Regions

The Lausanne strain of MYXV was released in France in June 1952 by a landholder seeking to control rabbits. The virus established and spread, killing an estimated 90% of wild rabbits in France (Fenner and Ross 1994). From France, MYXV spread to the rest of Europe wherever rabbits were present in the wild and into domestic rabbitries. It was illegally introduced into Britain in 1953 and spread through the British wild rabbit population with estimates of population reduction as high as 99% (Fenner and Ross 1994). As in Australia, there was emergence of both attenuated virus strains and genetically resistant rabbits as the virus and its new host coevolved, although this seems to have happened more slowly in Britain (Fenner and Ross 1994).

In Europe, the rabbit is a keystone species in some ecosystems and a valued game animal for hunting. Collapse of rabbit populations, initially due to myxomatosis and later due to RHD, has added to pressure on endangered predators such as the Spanish imperial eagle (*Aquila adalberti*) and Iberian lynx (*Lynx pardinus*) (Rogers et al. 1994). In 2018, a widespread epizootic of myxomatosis occurred in hares (*Lepus granatensis*) in the Iberian Peninsula potentially signaling a species shift (Dalton et al. 2019).

MYXV was also introduced into Chile in 1954 (and later into Argentina) to control feral rabbit populations. It has also been released onto the subantarctic Macquarie Island and Kerguelen Archipelago.

Rabbit Hemorrhagic Disease as a Biological Control for European Rabbits

RHD was first described in 1984 as a novel disease outbreak in domestic Angora rabbits imported into China from Germany. It subsequently spread to, or emerged in, Europe, the Americas, Korea, and other parts of the world causing substantial losses in domestic and wild rabbit populations (Cooke and Fenner 2002).

In its original form, RHD typically presented as a sudden onset of high mortality in subadult and adult rabbits. Kittens less than 4–6 weeks of age were generally unaffected. In the peracute form of RHD, rabbits often died with no premonitory signs and some still had food in their mouths. More normally, the disease manifested as a short period of malaise with fever and elevated respiratory rate for up to 24 hours prior to death, which might be accompanied by ataxia, posterior paresis, convulsions and coma, hematuria, and a bloody discharge from the nostrils. Death is due to fulminant hepatitis complicated by disseminated intravascular coagulation with hemorrhages in lungs, kidneys, and other tissues. Subacutely affected rabbits that survive for a few days may show signs of jaundice.

Rabbit Hemorrhagic Disease Virus

RHD is caused by a calicivirus (Family: *Caliciviridae*; genus: *Lagovirus*; species: *Rabbit hemorrhagic disease virus*; RHDV). RHDV is a small non-enveloped virus approximately 40 nm in diameter with icosahedral symmetry. It has a single-stranded positive-sense RNA genome of 7.4 kb and replicates in the cytoplasm of infected cells; it cannot be routinely isolated in cell culture.

Two distinct genotypes of the virus now circulate: viruses derived from the original RHDV (here referred to as RHDV1) and a more recently emerged genotype, which will be referred to as RHDV2. A new nomenclature for the lagoviruses has been proposed with a single species *Lagovirus europaeus* that subsumes RHDV and the related *European brown hare syndrome virus* (EBHSV) each of which would become a genogroup. Genogroups are then further subdivided into genotypes and variants based on phylogenetic relationships. In this proposal, RHDV1 is genotype G1.1 and RHDV2 becomes genotype G1.2 (Le Pendu et al. 2017).

RHDV probably evolved from an apathogenic or mildly pathogenic rabbit calicivirus (RCV). These viruses replicate in the small intestine (Hoehn et al. 2013) and are widely distributed in domestic and wild rabbits in Europe and Australia and probably other parts of the world where farmed, feral, or wild European rabbits occur (Capucci et al. 1996, Cooke et al. 2004, Marchandeau et al. 2005, Forrester et al. 2007, Strive et al. 2009, Bergin et al. 2009, Jahnke et al. 2010, Strive et al. 2010, Marchandeau et al. 2010, Le Gall Reculé et al. 2011b, Nicholson et al. 2017). RCVs are closely related to RHDV, induce cross-reactive antibodies and, in some cases, appear to provide cross-protection to RHDV.

Following the initial spread of RHDV1, an antigenic variant termed RHDVa emerged in Europe (Capucci et al. 1998) and appeared to outcompete RHDV. The more divergent RHDV2 emerged in 2010 and appears to have evolved independently (Le Gall Reculé et al. 2011a, Silvério et al. 2018). RHDV2 overcomes the age resistance to disease in young kittens (Dalton et al. 2014, Neimanis et al. 2018) and has a broader host range spilling over from rabbits to cause disease in multiple species with some evidence of circulation in hares (Puggioni et al. 2013, Camarda et al. 2014, Hall et al. 2017, Neimanis et al. 2018). RHDV1 and a recombinant of RHDV1 have also been retrospectively demonstrated in Iberian hares with clinical signs of RHD (Lopes et al. 2014, Lopes et al. 2017).

Recombination of RHDV1, RHDV2, and low virulence RCVs is a significant feature of the ongoing evolution of these viruses (Lopes et al. 2015, Silvério et al. 2018, Hall et al. 2018).

Infection with RHDV can occur by most routes of inoculation, but infection via the oropharynx is probably most common in the field. RHDV replicates to very high titres in the liver causing massive hepatic necrosis. Virus RNA is present in liver, spleen, mesenteric lymph nodes, bile, thymus, urine, blood, and feces (Shien et al. 2000). In experimental studies, where relatively high doses of virus are used, death generally occurs within 2–3 days of infection. The CFR in susceptible rabbits following infection with RHDV1 is over 90%.

Kittens less than 4 weeks of age inoculated with RHDV1 support virus replication and can transmit virus (Lenghaus et al. 1994, Mikami et al. 1999, Matthaei et al. 2014) but generally do not develop clinical disease and are subsequently immune to reinfection. This age-related resistance is gradually lost between 4 and 9 weeks of age (Robinson et al. 2002b). However, kittens infected with RHDV2 succumb to disease with high CFRs.

The titres of RHDV in the blood are sufficiently high for fleas and mosquitoes to passively transmit the virus; it does not replicate in these insects (Lenghaus et al. 1994). However, the duration of viremia prior to death is very short so it is unlikely that biting insects play a significant role in transmission. In contrast, carrion-feeding flies play a major role in transmission by feeding on carcasses and passively shedding virus in feces and regurgita (Asgari et al. 1998, McColl et al. 2002a, Schwensow et al. 2014).

Rabbits that have recovered from RHD are probably immune for life to the infecting antigenic type with potential boosting by natural reexposure to the virus (Cooke et al. 2002). Immune females will transfer protective antibodies across the placenta to their kittens. These passive maternal antibodies can protect kittens for up to 11 weeks after birth depending on the titre of antibody in the doe. Importantly, kittens that are challenged with RHDV while passive antibody titres are waning may become infected but survive and be subsequently immune (Robinson et al. 2002b). It is uncertain how much immunological cross-protection occurs between RHDV1 and RHDV2. Anecdotally, vaccination against RHDV1 seems poorly protective against RHDV2 (Dalton et al. 2012) and limited field studies showed little cross-protection (Peacock et al. 2017). However, in a study with experimentally infected rabbits, some cross-protection appeared to occur (Calvete et al. 2018).

RHDV is relatively resistant to environmental degradation, and infective virus can persist in carcasses for at least 20 days at 22°C and perhaps as long as 3 months, although this would be affected by the climatic conditions (McColl et al. 2002b, Henning et al. 2005). Thus, RHDV has two modes of transmission: direct contact between susceptible and infected animals shedding virus and environmental contamination from carcasses and flies that feed on infected carcasses. Predators such as foxes that feed on infected rabbits or scavenge carcasses do not support virus replication but may also pass infectious RHDV in their feces and potentially spread the virus.

Introduction of RHDV to Australia as a Biological Control Agent

The potential of RHDV as a biological control for European rabbits was quickly recognized following epizootics in wild rabbits in Spain (Cooke and Fenner 2002), and in 1991, the Czech 351 strain of RHDV was imported into the high-security Australian Animal Health Laboratory at Geelong for evaluation. Studies of virulence, pathogenesis, and transmission in laboratory and wild rabbits of different ages were undertaken (Lenghaus et al. 1994). Species-specificity testing indicated that rabbits were the only susceptible species of those tested. Approval was then given for the virus to be evaluated in pens on Wardang Island 5 km off the coast of South Australia; the same location that had been used nearly 60 years earlier for trials with MYXV.

Testing on Wardang Island commenced in March 1995. Despite careful security, including insecticide treatments, RHDV escaped from the island and appeared in rabbit populations on the adjacent mainland in October 1995 (Cooke and Fenner 2002). It was probably transmitted by blowflies (*Calliphora* spp.) or bushflies (*Musca vetustissima*), which had not been previously suspected as vectors. Contingency plans to contain the virus by eradicating rabbits in the area were put into action, but it quickly became obvious that the virus was not just spreading between warrens but was making large jumps to new populations hundreds of kilometres away (Cooke and Fenner 2002). This long-distance spread, together with the escape from Wardang Island, clearly implicated flying insects in dissemination of the virus (Cooke 2002).

Following this accidental release of RHDV, regulatory inquiries, including public submissions, and further species-specificity testing suggested that the virus was unlikely to infect other species and it was registered as a pest control agent (Cooke and Fenner 2002). This allowed government agencies to produce virus for deliberate release as an inoculation for rabbits or for distribution on carrot baits.

Impact of RHDV as a Biological Control

Within 12 months of its escape, RHDV was reported in all the mainland states of Australia. As with myxomatosis 45 years earlier, the impact on rabbit populations was dramatic with some populations being reduced by 95% (Mutze et al. 1998, Saunders et al. 1999, Story et al. 2004, Bruce and Twigg 2004). This was particularly so in the low rainfall regions of Australia. In the more temperate regions, the impact was mixed. Generally, populations were considerably reduced, but at some sites, there was little or no impact following RHD (Saunders et al. 1999, Henzell et al. 2002, Bruce and Twigg 2005, Richardson et al. 2007).

In the subsequent years, there was a gradual reduction in the impact of RHDV on rabbit populations but numbers were still at lower levels than pre-RHD (Mutze et al. 2002, 2014). In southern West Australia, RHDV disappeared and was successfully reintroduced (Bruce and Twigg 2005). Reduction in rabbit populations in some arid regions was strongly associated with recovery of endangered mammals (Pedler et al. 2016). RHDV was also released on Macquarie Island in 2011 as part of a successful rabbit eradication campaign.

Emergence and Release of Novel RHDV Strains

In 2014, an exotic RHDVa strain was detected in south-eastern Australia (Mahar et al. 2018b); it is not known how this virus entered Australia. The virus is a recombinant between an RCV-like and an RHDVa virus most closely related to a virus from China. In 2015, routine surveillance detected an exotic RHDV2 strain in south-eastern Australia (Hall et al. 2015). Phylogenetic studies indicated that it had been circulating for some months prior to detection; this virus spread across the rabbit-infested areas of Australia largely replacing RHDV1 (Mahar et al. 2018a). A novel recombinant between this virus and RCV has also been reported (Hall et al. 2018). These RHDV2 strains also spill over into introduced European hares causing some lethal infections (Hall et al. 2017). Studies with the originally isolated Australian RHDV2 virus showed that it caused similar disease processes in kittens and adults, as had already been shown in Europe and, in this way, was quite distinct from RHDV1 (Neave et al. 2018, Neimanis et al. 2018). It appears to have substantially reduced rabbit populations (Mutze et al. 2018, Ramsey et al. 2020) and, as of 2019, was the dominant virus in the environment. It is possible that the ability of RHDV2 to kill very young rabbits has interfered with the spread of RHDV1 by removing a susceptible cohort from the population. Combined with the ability to overcome immunity to RHDV1, this has likely given RHDV2 a selective advantage, which may wane as the breeding female population becomes dominated by survivors of RHDV2, which will pass passive antibody protection to their offspring.

From 2010 onwards, a large research program in Australia examined other globally occurring variants of RHDV for their potential to augment the impact of circulating RHDV and slow the growth in rabbit populations. An RHDVa strain from Korea was selected because it overcame partial cross-protection conferred by recent infection with benign RCV (see below) and had a better ability to infect genetically resistant wild rabbits than other strains. This virus (termed K5) was released in a nationwide campaign in March 2017 (Strive and Cox 2019).

Although K5 achieved a local reduction in rabbit populations at some release sites, there is so far no evidence that it has established in the rabbit population. This is likely due to the competing RHDV2. K5 remains available as a commercial product for rabbit control but probably only has impact as a local biocide rather than disseminating widely. Whether it provides additional genetic potential for recombination with other strains remains to be seen.

Adult wild rabbits are likely to have survived RHD and, therefore, be immune to further challenge with the same antigenic type (McPhee et al. 2009). This means that the virus will have its main impact on the susceptible kittens and subadults each breeding season. However, depending on the timing of epizootics, the impact of RHD may be reduced by age resistance, passively transferred maternal antibodies, and potentially by cross-protection from apathogenic caliciviruses (see below) (McPhee et al. 2009, Mutze et al. 2010a, 2010b, Cox et al. 2017, Cooke et al. 2018). The co-circulation of two antigenic types of RHDV that can at least partially overcome existing immunity may be altering the epidemiology in complex ways (Dalton et al. 2012, Peacock et al. 2017).

Interference with RHDV by Rabbit Caliciviruses

The impact of RHD was substantially less in the higher rainfall, more temperate areas of Australia (Henzell et al. 2002, Cooke 2002, McPhee et al. 2002). This may be due, at least in part, to an apathogenic RCV (Rabbit Calicivirus Australia-1; RCV-A1) (Strive et al. 2009, Strive et al. 2010). Infection with RCV-A1 induces antibodies that cross-react with RHDV (Nagesha et al. 2000, Cooke et al. 2002, Robinson et al. 2002a). Recent infection with RCV-A1 provides some protection from experimental challenge with

RHDV (Strive et al. 2013). However, this protection is relatively short-lived and appears to be independent of the titres of cross-reacting antibody. In addition to attenuating RHDV infection, previous exposure to RCV has been suggested to reduce the rate of RHDV transmission at a population level (Cooke et al. 2018).

As already noted, similar caliciviruses are widely distributed in farmed and wild European rabbits around the world. Recombination between RCV and RHDV, in particular exchange of the nonstructural and structural genes, generating new variants appears to be relatively common (Lopes et al. 2015, Hall et al. 2018).

Impact of RHDV in New Zealand

RHDV was illegally imported into New Zealand from Australia and released into rabbit populations in October 1997. Following its release, RHDV spread over the rabbit infested areas of New Zealand. Decrease in rabbit populations ranged from 90% in some areas to less than 20%, essentially no effect, in others (Sanson et al. 2000, Parkes et al. 2002, Norbury et al. 2002, Henning et al. 2006). Cross-reacting antibodies to RCV and possible protection from RHDV were also observed in New Zealand wild rabbits (O'Keefe et al. 1999, Nicholson et al. 2017). As in Australia, once the initial impact on the naïve population had passed, there was a gradual recovery in some rabbit populations. However, the rate of increase has been much slower in the presence of RHD than occurred following conventional controls (Parkes et al. 2008). Predation, particularly by cats, may be locally important in maintaining suppression in low density populations in the presence of RHD (Reddiex et al. 2002, Henning et al. 2008). In 2018, RHDV2 emerged in New Zealand via an unknown entry pathway. In the same year, New Zealand also released K5 as an additional tool for rabbit control. However, the effects of RHDV2 and K5 on New Zealand rabbit populations have yet to be quantified.

What Is the Future for Biological Control with RHDV?

Adult rabbits experimentally infected with RHDV1 generally only survive for 2–3 days. Based on the myxomatosis model, it might be expected that selection would occur for virus that allowed the rabbit to survive for longer and shed virus for longer. There has been no systematic study of RHDV virulence in the way MYXV was studied. However, there is no compelling epidemiological evidence to suggest that the virus is attenuating either in Australia or Europe and some data showing the emergence of strains of higher virulence compared to the released virus (Elsworth et al. 2014). The avirulent RCVs identified so far are clearly not derived from RHDV (Capucci et al. 1996, Strive et al. 2009, Jahnke et al. 2010, Mahar et al. 2016).

It is probable that RHDV is not attenuating because high virulence, leaving carcasses with extremely high titres of virus as a source of contamination for insect vectors or for direct transmission, is an effective transmission strategy that offsets the advantages of shedding for longer (Elsworth et al. 2014, Schwensow et al. 2014). By causing such rapid death, the virus almost completely avoids the adaptive immune response (Cooke and Berman 2000), which may also reduce selective pressure. In addition, very young rabbits can be infected but are resistant to disease caused by RHDV1 and, thus, could maintain virulent virus without selection pressure. However, this may be of limited importance since in most populations kittens will have passive maternal antibodies which would prevent infection or limit shedding during this period of age resistance (Robinson et al. 2002b). The emergence of RHDV2 has also considerably altered the role of kittens in RHDV epidemiology.

It has been reported that some populations of wild rabbits in Australia exhibit resistance to infection with RHDV (Saunders et al. 2010, Elsworth et al. 2012, Schwensow et al. 2017). Similar observations have been made in Europe (Fouchet et al. 2009). Studies on the mechanism of resistance have focused on the expression of complex carbohydrate molecules called histo-blood group antigens (HBGAs) on the surface of epithelial cells (Guillon et al. 2009). Binding of RHDV to these molecules appears important in the infection process although other cell-surface receptors must also be utilized since HBGAs are not expressed on hepatocytes.

Rabbits that do not express particular HGBAs on intestinal epithelium are significantly overrepresented in populations following epizootics of RHDV suggesting a strong selection advantage (Guillon et al. 2009, Nyström et al. 2011). Experimentally, such rabbits were significantly more likely to survive oral challenge with low doses of virus although they still became infected, and challenge with higher doses of virus was lethal (Nyström et al. 2011). Regulation of HBGA expression in rabbits is complex and not fully understood (Nyström et al. 2015).

Different strains of RHDV utilize different HGBAs for attachment suggesting that there is likely to be a complex and ongoing selection in rabbit populations for expression of particular HBGAs (and potentially other receptors) and in virus populations for recognition of different HBGAs. It should not be assumed that this is the only mechanism of resistance that might emerge. Indeed, other factors such as major histocompatibility molecules type I (MHC-I) have been implicated in genetic studies (Schwensow et al. 2017). This ongoing host–pathogen coevolution may have unpredictable consequences for viral virulence and biological control (Fouchet et al. 2009).

The emergence of RHDV2 may also be altering the ongoing host–pathogen dynamic in unpredictable ways. One could envisage a scenario of alternating dominance of RHDV1 and RHDV2 variants depending on the degree of immunological cross-protection. As with myxomatosis, RHD is likely to continue to suppress rabbit populations at essentially no economic cost but with gradually diminishing impact. Economic modeling has suggested that the combined impact of myxomatosis and RHD has accrued benefits to agriculture of 70 billion Australian dollars (at 2011 values) (Cooke et al. 2013). Population suppression also means that other natural processes such as predation, parasitism, and other disease agents have more impact.

Bibliography

Alves, J. M., M. Carneiro, J. Y. Cheng, A. Lemos de Matos, M. M. Rahman et al. 2019. Parallel adaptation of rabbit populations to myxoma virus. *Science* 363:1319–1326.

Asgari, S., J. R. E. Hardy, R. G. Sinclair, and B. D. Cooke. 1998. Field evidence for mechanical transmission of rabbit haemorrhagic disease virus (RHDV) by flies (Diptera: Calliphoridae) among wild rabbits in Australia. *Virus Res.* 54:123–132.

Bergin, I. L., A. G. Wise, S. R. Bolin, T. P. Mullaney, M. Kiupel et al. 2009. Novel calicivirus identified in rabbits, Michigan, USA. *Emerg. Infect. Diseases* 15:1955–1962.

Berman, D., P. J. Kerr, R. Stagg, B. H. van Leeuwen, and T. Gonzalez. 2006. Should the 40-year-old practice of releasing virulent myxoma virus to control rabbits (*Oryctolagus cuniculus*) be continued? *Wildl. Res.* 33:549–556.

Best, S. M., S. V. Collins, and P. J. Kerr. 2000. Coevolution of host and virus: cellular localization of myxoma virus infection of resistant and susceptible European rabbits. *Virology* 277:76–91.

Best, S. M., and P. J. Kerr. 2000. Coevolution of host and virus: the pathogenesis of virulent and attenuated strains of myxoma virus in resistant and susceptible European rabbits. *Virology* 267:36–48.

Bruce, J. S., and L. E. Twigg. 2004. Rabbit haemorrhagic disease virus: serological evidence of a non-virulent RHDV-like virus in south-western Australia. *Wildl. Res.* 31:602–612.

Bruce, J. S., and L. E. Twigg. 2005. The reintroduction, and subsequent impact of rabbit haemorrhagic disease virus in a population of wild rabbits in south-western Australia. *Wildl. Res.* 32:139–150.

Bull, L. B., and C. G. Dickinson. 1937. The specificity of the virus of rabbit myxomatosis. *J CSIR* 10:291–294.

Bull, L. B., and M. W. Mules. 1944. An investigation of *myxomatosis cuniculi* with special reference to the possible use of the disease to control rabbit populations in Australia. *J. CSIR* 17:79–93.

Calvete, C., M. Mendoza, A. Alcaraz, M. P. Sarto, M. P. Jiménez-de-Bagüéss et al. 2018. Rabbit haemorrhagic disease: cross-protection and comparative pathogenicity of G1.2/RHDV2/b and G1.1/RHDV lagoviruses in a challenge trial. *Vet Microbiol.* 219:87–95.

Camarda, A., N. Pugliese, P. Cavadini, E. Circella, L. Capucci et al. 2014. Detection of the new emerging rabbit haemorrhagic disease type 2 virus (RHDV2) in Sicily from rabbit (*Oryctolagus cuniculus*) and Italian hare (*Lepus corsicanus*). *Res. Vet. Sci.* 97:642–645.

Cameron, C., S. Hota-Mitchell, L. Chen, J. Barrett, J-X. Cao et al. 1999. The complete DNA sequence of myxoma virus. *Virology* 264:298–318.

Capucci, L., F. Fallacara, S. Grazioli, A. Lavazza, M. L. Pacciarini et al. 1998. A further step in the evolution of rabbit hemorrhagic disease virus: the appearance of the first consistent antigenic variant. *Virus Res.* 58:115–126.

Capucci, L., P. Fusi, A. Lavazza, M. L. Pacciarini, and C. Rossi. 1996. Detection and preliminary characterization of a new calicivirus related to rabbit haemorrhagic disease virus but non pathogenic. *J. Virol.* 70:8614–8623.

Cooke, B. D. 1983. Changes in the age-structure and size of populations of wild rabbits in South Australia, following the introduction of European rabbit fleas, *Spilopsyllus cuniculi* (Dale), as vectors of myxomatosis. *Aust. Wildl. Res.* 10:105–120.

Cooke, B. D. 1984. Factors limiting the distribution of the European rabbit flea *Spilopsyllus cuniculi* (Dale) (Siphonoptera) in inland South Australia. *Aust. J. Zool.* 32 (4):493–506.

Cooke, B. D. 1990. Notes on the comparative reproductive biology and the laboratory breeding of the rabbit flea *Xenopsylla cunicularis* Smit (Siphonaptera: Pulicidae). *Aust. J. Zool.* 38:527–534.

Cooke, B. D. 2002. Rabbit haemorrhagic disease: field epidemiology and the management of wild rabbit populations. *Rev. Sci. Tech. Off. Int. Epiz.* 21:347–358.

Cooke, B. D. 2012. Rabbits: manageable environmental pests or participants in new Australian ecosystems? *Wildl. Res.* 39:279–289.

Cooke, B. D., and D. Berman. 2000. Effect of inoculation route and ambient temperature on the survival time of rabbits, *Oryctolagus cuniculus* (L.) infected with Rabbit haemorrhagic disease virus. *Wildl. Res.* 27:137–142.

Cooke, B. D., J.-L. Chapuis, V. Magnet, A. Lucas, and J. Kovaliski. 2004. Potential use of myxoma virus and rabbit haemorrhagic disease virus to control feral rabbits in the Kerguelen Archipelago. *Wildl. Res.* 31:415–420.

Cooke, B. D., P. Chudleigh, S. Simpson, and G. Saunders. 2013. The economic benefits of the biological control of rabbits in Australia, 1950–2011. *Aust. Econ. Hist. Rev.* 53:1–17.

Cooke, B. D., R. P. Duncan, I. McDonald, J. Liu, L. Capucci et al. 2018. Prior exposure to non-pathogenic calicivirus RCV-A1 reduces both infection rate and mortality from rabbit haemorrhagic disease in a population of wild rabbits in Australia. *Transbound. Emerg. Dis* 65:e470–e477.

Cooke, B. D., and F. Fenner. 2002. Rabbit haemorrhagic disease and the biological control of wild rabbits, *Oryctolagus cuniculus*, in Australia and New Zealand. *Wildl. Res.* 29:689–706.

Cooke, B. D., S. McPhee, A. J. Robinson, and L. Capucci. 2002. Rabbit haemorrhagic disease: does a pre-existing RHDV-like virus reduce the effectiveness of RHD as a biological control in Australia. *Wildl. Res.* 29:673–682.

Cox, T., J. Liu, R. van de Ven, and T. Strive. 2017. Different serological profiles to co-occurring pathogenic and nonpathogenic caliciviruses in wild European rabbits (*Oryctolagus cuniculus*) across Australia. *J. Wildl. Dis.* 53:472–481.

Dalton, K. P., J. M. Martin, I. Nicieza, A. Podadera, D. de Llano et al. 2019. Myxoma virus jumps species to the Iberian hare. *Transbound. Emerg. Dis.* 66(6):2218–2226.

Dalton, K. P., I. Nicieza, J. Abrantes, P. J. Esteves, and F. Parra. 2014. Spread of new variant RHDV in domestic rabbits on the Iberian Peninsula. *Vet Microbiol.* 169:67–73.

Dalton, K. P., I. Nicieza, A. Balseiro, M. A. Muguerza, J. M. Rossell et al. 2012. Variant Rabbit Haemorrhagic Disease Virus in young rabbits in Spain. *Emerg. Inf. Dis.* 18:2009–2012.

Elsworth, P., B. D. Cooke, J. Kovaliski, R. Sinclair, E. C. Holmes et al. 2014. Increased virulence of rabbit haemorrhagic disease virus associated with genetic resistance in wild Australian rabbits (*Oryctolagus cuniculus*). *Virology* 464–465:415–423.

Elsworth, P. G., J. Kovaliski, and B. D. Cooke. 2012. Rabbit haemorrhagic disease: are Australian rabbits (*Oryctolagus cuniculus*) evolving resistance to infection with Czech CAPM 351 RHDV? *Epidemiol. Infect.* 140:1972–1981.

Fenner, F., M. F. Day, and G. M. Woodroofe. 1956. Epidemiological consequences of the mechanical transmission of myxomatosis by mosquitoes. *J. Hyg. (Camb.)* 54:284–303.

Fenner, F., and B. Fantini. 1999. *Biological Control of Vertebrate Pests. The History of Myxomatosis – An Experiment in Evolution.* New York: CAB International.

Fenner, F., and I. D. Marshall. 1954. Passive immunity in myxomatosis of the European rabbit (*Oryctolagus cuniculus*): the protection conferred on kittens by immune does. *J. Hyg. (Camb.)* 52:321–336.

Fenner, F., and I. D. Marshall. 1957. A comparison of the virulence for European rabbits (*Oryctolagus cuniculus*) of strains of myxoma virus recovered in the field in Australia, Europe and America. *J. Hyg. (Camb.)* 55:149–191.

Fenner, F., I. D. Marshall, and G. M. Woodroofe. 1953. Studies in the epidemiology of infectious myxomatosis of rabbits. I. Recovery of Australian wild rabbits (*Oryctolagus cuniculus*) from myxomatosis under field conditions. *J. Hyg. (Camb.)* 51:225–244.

Fenner, F., W. E. Poole, I. D. Marshall, and A. L. Dyce. 1957. Studies in the epidemiology of infectious myxomatosis of rabbits. VI. The experimental introduction of the European strain of myxoma virus into Australian wild rabbit populations. *J. Hyg. (Camb.)* 55:192–206.

Fenner, F., and F. N. Ratcliffe. 1965. *Myxomatosis.* Cambridge: Cambridge University Press.

Fenner, F., and J. Ross. 1994. Myxomatosis. In *The European Rabbit. The History and Biology of a Successful Colonizer*, edited by H. V. Thompson and C. M. King, 205–240. Oxford: Oxford University Press.

Fenner, F., and G. M. Woodroofe. 1965. Changes in the virulence and antigenic structure of strains of myxoma virus recovered from Australian wild rabbits between 1950 and 1964. *Aust. J. Exptl. Biol. Med. Sci.* 43:359–370.

Forrester, N. L., R. C. Trout, and E. A. Gould. 2007. Benign circulation of rabbit haemorrhagic disease virus on Lambay Island, Eire. *Virology* 358:18–22.

Fouchet, D., J. Le Pendu, J.-S. Guitton, M. Guiserix, S. Marchandeau et al. 2009. Evolution of microparasites in spatially and genetically structured host populations: the example of RHDV infecting rabbits. *J. Theor. Biol.* 257:212–227.

Guillon, P., N. Rovoen-Clouet, B. le Moullac-Valdye, S. Marchandeau, and J. Le Pendu. 2009. Association between expression of the H histo-blood group antigen, α1,2 fucosyltransferases polymorphism of wild rabbits, and sensitivity to rabbit haemorrhagic disease virus. *Glycobiology* 19:21–28.

Hall, R. N., J. E. Mahar, S. Haboury, V. Stevens, E. C. Holmes et al. 2015. Emerging rabbit hemorrhagic disease virus 2 (RHDVb) in Australia. *Emerg. Inf. Dis.* 21:2276–2278.

Hall, R. N., J. E. Mahar, A. J. Read, R. Mourant, M. Piper et al. 2018. A strain-specific multiplex RT-PCR for Australian rabbit haemorrhagic disease viruses uncovers a new recombinant virus variant in rabbits and hares. *Transbound. Emerg. Dis* 65:e444–e456.

Hall, R. N., D. E. Peacock, J. Kovaliski, J. E. Mahar, R. Mourant et al. 2017. Detection of RHDV2 in European brown hares (*Lepus europaeus*) in Australia. *Vet. Record* 180:121.

Henning, J., J. Meers, P. R. Davies, and R. S. Morris. 2005. Survival of rabbit haemorrhagic disease virus (RHDV) in the environment. *Epidemiol. Infect.* 133:719–730.

Henning, J., D. U. Pfeiffer, P. R. Davies, J. Meers, and R. S. Morris. 2006. Temporal dynamics of rabbit haemorrhagic disease virus infection in a low-density population of wild rabbits (*Oryctolagus cuniculus*) in New Zealand. *Wildl. Res.* 33:293–303.

Henning, J., D. U. Pfeiffer, P. R. Davies, M. A. Stevenson, and J. Meers. 2008. Mortality patterns over 3 years in a sparse population of wild rabbits (*Oryctolagus cuniculus*) in New Zealand, with an emphasis on rabbit haemorrhagic disease (RHD). *Eur. J. Wild. Res.* 54:619–626.

Henzell, R. P., B. D. Cooke, and G. D. Mutze. 2008. "The future biological control of pest populations of European rabbits, *Oryctolagus cuniculus.*" *Wildl. Res.* 35:633–650.

Henzell, R. P., R. B. Cunningham, and H. M. Neave. 2002. Factors affecting the survival of Australian wild rabbits exposed to rabbit haemorrhagic disease. *Wildl. Res.* 29:523–542.

Hoehn, M., P. J. Kerr, and T. Strive. 2013. In situ hybridisation assay for localisation of rabbit calicivirus Australia-1 (RCV-A1) in European rabbit (*Oryctolagus cuniculus*) tissues. *J. Virol. Meth.* 188:148–152.

Jahnke, M., E. C. Holmes, P. J. Kerr, J. D. Wright, and T. Strive. 2010. Evolution and phylogeography of the non-pathogenic calicivirus RCV-A1 in wild rabbits in Australia. *J. Virol.* 84:12397–12404.

Jeklova, E., L. Leva, J. Matiosovic, K. Kovarcik, H. Kudlackova et al. 2008. Characterisation of immuno-suppression in rabbits after infection with myxoma virus. *Vet. Microbiol.* 129:117–130.

Kerr, P.J. 1997. An ELISA for epidemiological studies of myxomatosis: persistence of antibodies to myxoma virus in European rabbits (*Oryctolagus cuniculus*). *Wildl. Res.* 24:53–65.

Kerr, P. J. 2012. Myxomatosis in Australia and Europe: a model for emerging infectious diseases. *Antiviral Res.* 93:387–415.

Kerr, P. J., I. M. Cattadori, J. Liu, D. G. Sim, J. W. Dodds et al. 2017. Next step in the ongoing arms race between myxoma virus and wild rabbits in Australia is a novel disease phenotype. *Proc. Natl. Acad. Sci. USA.* 114:9397–9402.

Kerr, P. J., J-S. Eden, F. Di Giallonardo, D. Peacock, J. Liu, et al. 2019. Punctuated evolution of myxoma virus: rapid and disjunct evolution of a recent viral lineage in Australia. *J. Virol.* 93:e01994–18.

Kerr, P. J., J. Liu, I. M. Cattadori, E. Ghedin, A. F. Read et al. 2015. Myxoma virus and the leporipoxviruses: an evolutionary paradigm. *Viruses* 7:1020–1061.

King, D. R., A. J. Oliver, and S. H. Wheeler. 1985. The European rabbit flea, *Spilopsyllus cuniculi*, in south-western Australia. *Aust. Wildl. Res.* 12:227–236.

Le Gall Reculé, G., F. Zwinglestein, S. Boucher, B. Le Normand, G. Plassiart et al. 2011a. Detection of a new variant of Rabbit Haemorrhagic Disease Virus in France. *Vet. Record* 168:137–138.

Le Gall Reculé, G., F. Zwinglestein, M. P. Fages, S. Bertagnoli, J. Gelfi et al. 2011b. Characterisation of a non-pathogenic and non-protective infectious rabbit lagovirus related to RHDV. *Virology* 410:395–402.

Le Pendu, J., J. Abrantes, S. Bertagnoli, J.-S. Guitton, G. Le Gall Reculé et al. 2017. Proposal for a unified classification system and nomenclature of lagoviruses. *J. Gen. Virol.* 98:1658–1666.

Lenghaus, C., H. Westbury, B. Collins, N. Ratnamohan, and C. Morrissy. 1994. Overview of the RHD project in the Australian Animal Health Laboratory. In *Rabbit Haemorrhagic Disease: Issues in Assessment for Biological Control*, edited by R. K. Munro and R. T. Williams, 104–129. Canberra: Bureau of Resource Sciences.

Lopes, A. M., K. P. Dalton, M. J. Magalhães, F. Parra, P. J. Esteves et al. 2015. Full genome analysis of new variant rabbit haemorrhagic disease virus revealed multiple recombination events. *J. Gen. Virol.* 96:1309–1319.

Lopes, A. M., S. Marques, E. Silva, M. J. Magalhães, A. Pinheiro et al. 2014. Detection of RHDV strains in the Iberian hare (*Lepus granatensis*): earliest evidence of rabbit lagovirus cross-species infection. *Vet. Res.* 45:94.

Lopes, A. M., D. Silvério, M. J. Magalhães, H. Areal, P. C. Alves et al. 2017. Characterization of old RHDV strains by complete genome sequencing identifies a novel genetic group. *Scientific Reports* 7:13599.

Mahar, J. E., R. N. Hall, D. Peacock, J. Kovaliski, M. Piper et al. 2018a. Rabbit hemorrhagic disease virus 2 (RHDV2; G1.1) is replacing endemic strains of RHDV in the Australian landscape within 18 months of its arrival. *J. Virol.* 92:e01374-17.

Mahar, J. E., L. Nicholson, J-S. Eden, S. Duchêne, P. J. Kerr et al. 2016. Benign rabbit caliciviruses exhibit evolutionary dynamics similar to those of their virulent relatives. *J. Virol.* 90:9317–9329.

Mahar, J. E., A. J. Read, X. Gu, N. Urakova, R. Mourant et al. 2018b. Detection and circulation of a novel rabbit hemorrhagic disease virus in Australia. *Emerg. Inf. Dis.* 24:22–31.

Marchandeau, S., S. Bertagnoli, Y. Leonard, H. Santin-Janin, B. Peralta et al. 2010. Serological evidence for the presence of non-pathogenic rabbit haemorrhagic disease virus-like strains in rabbits (*Oryctolagus cuniculus*) of the Kerguelen archipelago. *Polar Biol.* 33:985–989.

Marchandeau, S., G. Le Gall-Recule, S. Bertagnoli, J. Aubineau, G. Botti et al. 2005. Serological evidence for a non-protective RHDV-like virus. *Vet. Res.* 36:53–62.

Marshall, I. D. 1959. The influence of ambient temperature on the course of myxomatosis in rabbits. *J. Hyg. (Camb.)* 57:484–497.

Marshall, I. D., and G. W. Douglas. 1961. Studies in the epidemiology of infectious myxomatosis of rabbits. VIII. Further observations on changes in the innate resistance of Australian wild rabbits exposed to myxomatosis. *J. Hyg. (Camb.)* 59:117–122.

Marshall, I. D., A. L. Dyce, W. E. Poole, and F. Fenner. 1955. Studies in the epidemiogy of infectious myxomatosis of rabbits. IV. Observations of disease behaviour in two localities near the northern limit of rabbit infestation in Australia, May 1952 to April 1953. *J. Hyg. (Camb.)* 53:12–25.

Marshall, I. D., and F. Fenner. 1958. Studies in the epidemiology of infectious myxomatosis of rabbits. V. Changes in the innate resistance of wild rabbits exposed to myxomatosis. *J. Hyg. (Camb.)* 56:288–302.

Marshall, I. D., and F. Fenner. 1960. Studies in the epidemiology of infectious myxomatosis of rabbits. VII. The virulence of strains of myxoma virus recovered from Australian wild rabbits between 1951 and 1959. *J. Hyg. (Camb.)* 58:485–488.

Martin, C. J. 1936. Observations on *Myxomatosis cuniculi* (Sanarelli) made with a view to the use of the virus in the control of rabbit plagues. *CSIR Australia Research Bulletin, No. 96*:1–28.

Matthaei, M., P. J. Kerr, A. J. Read, P. Hick, S. Haboury et al. 2014. Comparative quantitative monitoring of rabbit haemorrhagic disease viruses in rabbit kittens. *Virol. J.* 11:109.

McColl, K. A., J. C. Merchant, J. Hardy, B. D. Cooke, A. Robinson et al. 2002a. Evidence for insect transmission of rabbit haemorrhagic disease virus. *Epidemiol. Infect.* 129:655–633.

McColl, K. A., C. J. Morrissy, B. J. Collins, and H. A. Westbury. 2002b. Persistence of rabbit haemorrhagic disease virus in decomposing rabbit carcases. *Aust. Vet. J.* 80:298–299.

McPhee, S. R., D. Berman, A. Gonzales, K. L. Butler, J. Humphrey et al. 2002. Efficacy of a competitive enzyme-linked immunosorbent assay (cELISA) for estimating prevalence of immunity to rabbit haemorrhagic disease virus (RHDV) in populations of Australian wild rabbits (*Oryctolagus cuniculus*). *Wildl. Res.* 29:635–647.

McPhee, S. R., K. L. Butler, J. Kovaliski, G. Mutze, L. Capucci et al. 2009. Antibody status and survival of Australian wild rabbits challenged with rabbit haemorrhagic disease virus. *Wildl. Res.* 36:447–456.

Mikami, O., J. H. Park, T. Kimura, K. Ochial, and C. Itakura. 1999. Hepatic lesions in young rabbits experimentally infected with rabbit haemorrhagic disease virus. *Res. Vet. Sci.* 66:237–242.

Mutze, G., P. I. Bird, J. Kovaliski, D. Peacock, S. Jennings et al. 2002. Emerging epidemiological patterns in rabbit haemorrhagic disease, its interactions with mxyomatosis, and their effects on rabbit populations in South Australia. *Wildl. Res.* 29:577–590.

Mutze, G., P. Bird, D. Peacock, N. De Preu, J. Kovaliski, B. D. Cooke et al. 2014. Recovery of South Australian rabbit populations from the impact of rabbit haemorrhagic disease. *Wildl. Res.* 41:552–559.

Mutze, G., B. Cooke, and P. Alexander. 1998. The initial impact of rabbit hemorrhagic disease on European rabbit populations in South Australia. *J. Wildl. Dis.* 34:221–227.

Mutze, G., N. De Preu, T. Moody, D. Koerner, D. McKenzie et al. 2018. Substantial numerical decline in South Australian rabbit populations following the detection of rabbit haemorrhagic disease virus 2. *Vet. Rec.* 182(20):574.

Mutze, G., J. Kovaliski, K. Butler, L. Capucci, and S. McPhee. 2010a. The effect of rabbit population control programmes on the impact of rabbit haemorrhagic disease in south-eastern Australia. *J. App. Ecol.* 47:1137–1146.

Mutze, G., R. Sinclair, D. Peacock, J. Kovaliski, and L. Capucci. 2010b. Does a benign calicivirus reduce the effectiveness of rabbit haemorrhagic disease virus (RHDV) in Australia? Experimental evidence from field releases of RHDV on bait. *Wildl. Res.* 37:311–319.

Myers, K. 1954. Studies in the epidemiology of infectious myxomatosis of rabbits. II. Field experiments, August-November 1950, and the first epizootic of myxomatosis in the Riverine plain of south-eastern Australia. *J. Hyg. (Camb.)* 52:47–59.

Myers, K., I. D. Marshall, and F. Fenner. 1954. Studies in the epidemiology of infectious myxomatosis of rabbits. III. Observations on two succeeding epizootics in Australian wild rabbits on the Riverine plain of south-eastern Australia. *J. Hyg. (Camb.)* 52:337–360.

Nagesha, H. S., K. A. McColl, B. J. Collins, C. J. Morrissy, L. F. Wang et al. 2000. The presence of cross-reactive antibodies to rabbit haemorrhagic disease virus in Australian wild rabbits prior to the escape of virus from quarantine. *Arch Virol* 145:749–757.

Neave, M. J., R. N. Hall, N. Huang, K. A. McColl, P. Kerr et al. 2018. Robust innate immunity of young rabbits mediates resistance to rabbit hemorrhagic disease caused by *Lagovirus europaeus* G1.1 but not G1.2. *Viruses* 10:512.

Neimanis, A., U. Larsson Pettersson, N. Huang, D. Gavier-Widén, and T. Strive. 2018. Elucidation of the pathology and tissue distribution of *Lagovirus europaeus* G1.2/RHDV2 (Rabbit haemorrhagic disease virus 2) in young and adult rabbits (*Oryctolagus cuniculus*). *Vet. Res.* 49:46.

Nicholson, L. J., J. E. Mahar, T. Strive, T. Zheng, E. C. Holmes et al. 2017. Benign rabbit calicivirus in New Zealand. *Appl. Environ. Microbiol.* 83:e00090–17.

Norbury, G., R. Heyward, and J. Parkes. 2002. Short-term ecological effects of rabbit haemorrhagic disease in the short-tussock grasslands of the South Island, New Zealand. *Wildl. Res.* 29:599–604.

Nyström, K., J. Abrantes, A. M. Lopes, B. Le Moullac-Vaidye, S. Marchandeau et al. 2015. Neofunctionalization of the Sec1 α1,2fucosyltransferase paralogue in leporids contributes to glycan polymorphism and resistance to rabbit haemorrhagic disease. *PLoS Pathog* 11:e1004759.

Nyström, K., G. Le Gall Reculé, P. Grassi, J. Abrantes, N. Ruvoën-Clouet et al. 2011. Histo-blood group antigens act as attachment factors of rabbit hemorrhagic disease virus infection in a virus strain-dependent manner. *PLoS Pathog* 7:e1002188.

O'Keefe, J. S., J. E. Tempero, M. X. J. Motha, M. F. Hansen, and P. H. Atkinsona. 1999. Serology of rabbit haemorrhagic disease virus in wild rabbits before and after release of the virus in New Zealand. *Vet. Microbiol.* 66:29–40.

Parer, I., D. Conolly, and W. R. Sobey. 1985. Myxomatosis: the effects of annual introductions of an immunizing strain and a highly virulent strain of myxoma virus into rabbit populations at Urana, New South Wales, Australia. *Aust. Wildl. Res.* 12:407–424.

Parkes, J. P., B. Glentworth, and G. Sullivan. 2008. Changes in immunity to Rabbit haemorrhagic disease virus, and in abundance and rates of increase of wild rabbits in Mackenzie Basin, New Zealand. *Wildl. Res.* 35:775–779.

Parkes, J. P., G. L. Norbury, R. P. Heyward, and G. Sullivan. 2002. Epidemiology of rabbit haemorrhagic disease (RHD) in the South Island, New Zealand, 1997–2001. *Wildl. Res.* 29:543–555.

Peacock, D., J. Kovaliski, R. Sinclair, G. Mutze, A. Iannella et al. 2017. RHDV2 overcoming RHDV immunity in wild rabbits (*Oryctolagus cuniculus*) in Australia. *Vet. Record.* doi:10.1136/vr.104135.

Pedler, R. D., R. Brandle, J. L. Read, R. Southgate, P. Bird et al. 2016. Rabbit biocontrol and landscape-scale recovery of threatened desert mammals. *Conserv. Biol.* 30:774–782.

Puggioni, G., P. Cavadini, C. Maestrale, R. Scivoli, G. Botti et al. 2013. The new French 2010 *Rabbit haemorrhgic disease virus* cases an RHD-like disease in the Sardinian Cape hare (*Lepus capensis mediterraneus*). *Vet. Res.* 44:96.

Ramsey, D. S. L., T. Cox, T. Strive, D. M. Forsyth, I. Stuart et al. 2020. Diverse impacts of two novel strains of rabbit haemorrhagic disease virus on wild rabbit populations in Australia. *J. App. Ecol.* 57:630–641.

Ratcliffe, F. N., K. Myers, B. V. Fennessy, and J. H. Calaby. 1952. Myxomatosis in Australia. A step towards the biological control of the rabbit. *Nature* 170:7–19.

Reddiex, B., G. J. Hickling, G. L. Norbury, and C. M. Frampton. 2002. Effects of predation and rabbit haemorrhagic disease on population dynamics of rabbits (*Oryctolagus cuniculus*) in North Canterbury, New Zealand. *Wildl. Res.* 29:627–633.

Richardson, B. J., S. Philips, R. A. Hayes, S. Sindhe, and B. D. Cooke. 2007. Aspects of the biology of the European rabbit (Oryctolagus cuniculus) and rabbit haemorrhagic disease virus (RHDV) in coastal eastern Australia. *Wildl. Res.* 34: 398–407.

Robinson, A. J., P. D. Kirkland, R. I Forrester, L. Capucci, B. D. Cooke et al. 2002a. Serological evidence for the presence of a calicivirus in Australian wild rabbits, *Oryctolagus cuniculus*, before the introduction of rabbit haemorrhagic disease virus (RHDV): its potential influence on the specificity of a competitive ELISA for RHDV. *Wildl. Res.* 29:655–652.

Robinson, A. J., P. T. M. So, W. J. Muller, B. D. Cooke, and L. Capucci. 2002b. Statistical models for the effect of age and maternal antibodies on the development of rabbit haemorrhagic disease in Australian wild rabbits. *Wildl. Res.* 29:663–671.

Rogers, P. M., C. P. Arthur, and R. C. Soriguer. 1994. The rabbit in continental Europe. In *The European Rabbit. The History and Biology of a Successful Colonizer*, edited by H. V. Thompson and C. M. King, 22–63. Oxford: Oxford University Press.

Saint, K. M., N. French, and P. Kerr. 2001. Genetic variation in Australian isolates of myxoma virus: an evolutionary and epidemiological study. *Arch. Virol.* 146:1105–1123.

Sanson, R. L., H. V. Brooks, and G. W. Horner. 2000. An epidemiological study of the spread on rabbit haemorrhagic disease virus amongst previously non-exposed rabbit populations in the North Island of New Zealand. *N. Z. Vet. J.* 48:105–110.

Saunders, G., D. Choquenot, J. McIlroy, and R. Packwood. 1999. Initial effects of rabbit haemorrhagic disease on free-living rabbit (*Oryctolagus cuniculus*) populations in central-western New South Wales. *Wildl Res* 26:69–74.

Saunders, G., B. Cooke, K. McColl, R. Shine, and T. Peacock. 2010. Modern approaches for the biological control of vertebrate pests: an Australian perspective. *Biol. Control* 52:288–295.

Schwensow, N., B. Cooke, J. Kovaliski, R. G. Sinclair, D. Peacock et al. 2014. Rabbit haemorrhagic disease virus: persistence and adaptation in Australia. *Evolutionary Adaptations* 7:1056–1067.

Schwensow, N. I., H. Detering, S. Pederson, C. Mazzoni, R. Sinclair et al. 2017. Resistance to RHD virus in wild Australian rabbits: comparison of susceptible and resistant individuals using a genomewide approach. *Mol. Ecol.* 26:4551–4561.

Shien, J. H., H. K. Shieh, and L. H. Lee. 2000. Experimental infections of rabbits with rabbit haemorrhagic disease virus monitored by polymerase chain reaction. *Res. Vet. Sci.* 68:255–259.

Silvério, D., A. M. Lopes, J. Ferriera-Melo, M. J. Magalhães, P. Monterroso et al. 2018. Insights into the evolution of the new variant rabbit haemorrhagic disease virus (G1.2) and the identification of novel recombinant strains. *Transbound. Emerg. Dis.* 65:983–982.

Sobey, W. R., and W. Menzies. 1969. Myxomatosis: the introduction of the European rabbit flea *Spilopsyllus cuniculi* (Dale) into Australia. *Aust. J. Sci.* 31:404–405.

Story, G., D. Berman, R. Palmer, and J. Scanlan. 2004. The impact of rabbit haemorrhagic disease on wild rabbit (*Oryctolagus cuniculus*) populations in Queensland. *Wildl. Res.* 31:183–193.

Strive, T., and T. Cox. 2019. Lethal biological tools for landscape-scale mitigation of rabbit impacts in Australia. *Aust. Zool.* doi:10.7882/AZ.2019.016.

Strive, T., P. Elsworth, J. Liu, J. D. Wright, J. Kovaliski et al. 2013. The non-pathogenic Australian rabbit calicivirus RCV-A1 provides temporal and partial cross-protection to lethal rabbit haemorrhagic disease virus infection which is not dependent on antibody titres. *Vet. Res.* 44:51.

Strive, T., J. Wright, J. Kovaliski, G. Botti, and L. Capucci. 2010. The non-pathogenic Australian lagovirus RCV-A1 causes a prolonged infection and elicits partial cross-protection to rabbit haemorrhagic disease virus. *Virology* 398:125–134.

Strive, T., J. D. Wright, and A. J. Robinson. 2009. Identification and partial characterisation of a new lagovirus in Australian wild rabbits. *Virology* 384:97–105.

Williams, K., I. Parer, B. Coman, J. Burley, and M. Braysher. 1995. *Managing Vertebrate Pests: Rabbits. Canberra: Bureau of Resource Sciences*. Australian Government Publishing Services.

Cabbage Disease Ecology and Management

Anthony P. Keinath,
Marc A. Cubeta,
and David B.
Langston, Jr.

Introduction

Cabbage (*Brassica oleracea* 'Capitata Group') has long been cultivated as an important vegetable crop and a source of fiber, minerals, and vitamins, particularly during cold seasons in temperate climates. More recently, cabbage and other cruciferous vegetables (members of the Brassicaceae) have been recognized as important sources of dietary chemoprotective phytochemicals. Cabbage and other brassica vegetables were grown on 2.4 million hectares worldwide in 2018 (Food and Agriculture Organization, 2020). Cabbage is a productive vegetable based on biomass per area of cultivation, which averaged 28.7 ton/ha in 2018. However, many diseases, particularly those caused by bacteria and fungi, affect the production of this crop. This chapter focuses on six diseases of worldwide importance in cabbage production. These diseases also affect other cole crops, i.e., vegetables derived from *B. oleracea*, including broccoli, Brussels sprouts, cauliflower, collard, kale, and kohlrabi, and other genetically related cruciferous vegetables, such as Chinese cabbage, mustard, rutabaga, and turnip. Emphasis will be placed on pathogen life cycle stages susceptible to management techniques. Control measures will be presented in an integrated pest management (IPM) framework.

Major Diseases and Pathogen Ecology

Black Rot

Black rot is caused by the bacterium *Xanthomonas campestris* pathovar *campestris*. *X. campestris* pv. *raphani* also has been isolated from necrotic spots on cabbage seedlings (Lange et al. 2016). Because this bacterium can be seedborne, black rot occurs in most areas of the world where cabbage and other crucifers are grown. The pathogen produces V-shaped chlorotic and necrotic lesions starting at the margins of leaves, but it also causes wilting of plants if it reaches the vascular system in the stem (systemic infection). Blackening of the leaf veins is a helpful diagnostic symptom. The pathogen survives in infested crop debris and usually survives only for a few months in soil in the absence of host debris.

Clubroot

Clubroot is caused by the protozoan-like organism *Plasmodiophora brassicae*. This soilborne organism is an obligate biotroph, completing its unique life cycle within the root cells of crucifers. Infected root cells enlarge and divide to produce the diagnostic swollen, club-like roots. The pathogen produces resting spores in the clubs that persist in soil for at least 10 years after they decay. Isolates of *P. brassicae* differ in host range, and races have been found that are pathogenic on the few resistant cultivars of cabbage. Resting spores are extremely resistant to desiccation and easily spread with wind-blown soil and irrigation water.

Black Spot, Dark Leaf Spot

Five species of *Alternaria*, but primarily *A. brassicae* and *A. brassicicola*, infect cabbage and other crucifers. *A. brassicicola* has a higher optimal temperature for growth, sporulation, and spore germination (20°C–30°C) than *A. brassicae* (18°C–24°C). Both fungi can be seedborne and airborne but do not survive apart from infested host debris in soil. Infested debris left on the soil surface can be a significant source of pathogen spores for up to 12 weeks after harvest (Humpherson-Jones 1989). Seedborne inoculum can lower seed germination and vigor but usually is not damaging to seedlings.

Downy Mildew

Crucifer downy mildew is caused by the oomycete *Hyaloperonospora parasitica*. This biotroph produces airborne sporangia on the underside of leaves and oospores inside infected leaves, roots, and stems. The pathogen is believed to survive as dormant oospores in roots and stems and on living hosts in frost-free regions. Cabbage is affected by downy mildew particularly during the seedling and heading growth stages. High relative humidity, dew, and fog are favorable for infection.

Watery Soft Rot, Sclerotinia Stem Rot, White Mold

Sclerotinia sclerotiorum has a wide host range but is especially damaging to cabbage, because it not only infects the head in the field but also can cause decay in transit and storage. Infection occurs primarily at head maturity when wrapper leaves shade the soil, providing a cool, moist environment that favors the pathogen. This fungus produces airborne sexual spores (ascospores) that infect plants, and soilborne survival structures (sclerotia) cause infection when they germinate near a plant. The closely related species *Sclerotinia minor* also has been reported on brassica vegetables.

Wirestem

Wirestem, a postemergence disease, is caused by the soilborne fungus *Rhizoctonia solani* anastomosis groups (AGs) 2-1 and 4 (Budge and Shaw 2009). In soils cropped repeatedly to crucifers, AG 2-1 predominates. At low pathogen levels, wirestem is more prevalent or severe than preemergence damping-off. Seedlings may be killed by wirestem when lesions girdle stems. Older plants may be killed later or be stunted and fail to produce a marketable-sized head. Root rot also occurs when infection is severe but is absent when discrete stem lesions are the only symptoms.

Control

General Control Principles

Exclusion. It is extremely important to prevent contamination of clubroot-free land by excluding the pathogen. Movement of field-grown transplants, soil, and equipment from clubroot-infested fields or farms should be avoided (Hwang et al. 2014). Growers in clubroot-free areas should avoid purchasing field-grown transplants or equipment from infested areas. Before purchase, greenhouse-grown transplants should be inspected for symptoms of downy mildew, black rot, and black spot.

Eradication. Testing seed for the fungal pathogen *Plenodomus* (=*Phoma*) *lingam* has greatly reduced outbreaks of black leg disease of crucifers. Cruciferous weeds should be eradicated to eliminate sources of the pathogens causing black rot, downy mildew, and clubroot (Table 1). Cruciferous ornamentals also can be infected by the same species of *Alternaria*, *Hyaloperonospora*, *Plasmodiophora*, and *Xanthomonas* that infect cabbage. The parasitic fungus *Paraconiothyrium minitans*, the active ingredient in the biofungicide Contans®, infects and destroys sclerotia of *S. sclerotiorum* and *S. minor*.

Avoidance. Cabbage crops should be rotated to a different field or plot each season to avoid contact with infested debris remaining from previous crops. Using a shallow planting depth for transplants reduces wirestem incidence, because the susceptible hypocotyl is not directly in contact with *Rhizoctonia*-infested soil. Avoid wounding plants to prevent black rot, bacterial soft rot, and watery soft rot.

Resistance. Host plant resistance is widely available in green (white) and red cabbage for yellows (caused by *Fusarium oxysporum* f. sp. *conglutinans*). Some hybrid cultivars have partial resistance to black rot that restricts lesions to the wrapper leaves. A few cabbage cultivars (mostly red cabbage) have moderate resistance to *Alternaria*. Cabbage cultivars available in the United States are susceptible to clubroot, downy mildew, white mold, and wirestem.

Protection. Seed treatment with fungicides is very effective in preventing damping-off caused by species of *Pythium* and *R. solani*. Protectant fungicides are effective against black spot and downy mildew and used for managing black rot, clubroot, and wirestem with varying degrees of success. Potassium phosphite is an alternative fungicide that is effective against downy mildew. The fungicides boscalid and penthiopyrad are registered in the United States to control *Sclerotinia* on cole crops.

TABLE 1 Management Practices for Common Diseases of Cabbage

Disease	Plant-Resistant Cultivars	Use Healthy Seed or Transplants	Eliminate Weeds	Avoid Wounding	Bury Crop Residue	Rotate with Nonhosts	Apply Protectant Fungicide or Bactericide
Black spot	+	+	+	–	+	+	+
Bacterial soft rot	–	–	–	+	–	–	–
Black leg	–	+	–	–	+	+	+
Black rot	+	+	+	+	+	+	–
Clubroot	–	+*	+	–	–	–	+/–
Downy mildew	–	+	+	–	+	+	+
Yellows	+	–	–	–	–	–	–
Sclerotinia stem rot	–	–	+	+	+	–	+/–
Damping-off	–	–	–	–	–	–	+
Wirestem	–	+*	–	–	+	–	+

+ = Practice can be used to manage the disease; – = Practice is ineffective or inappropriate, based on pathogen life cycle;
+/– = Practice may be useful under certain conditions.
* The pathogen is not seedborne but can be spread on infected, field-grown transplants.

Therapy. The only measure to control cabbage diseases postinfection is the application of systemic fungicides for downy mildew.

Examples of Integrated Disease Management

Cabbage scouting guides have been developed and are useful for surveying production fields for diseases, insects, and weeds. Controlling weeds, especially ragweed (*Ambrosia artemisiifolia*), can reduce the incidence of watery soft rot. Ascospores of *Sclerotinia* can also infect ragweed flowers that fall onto cabbage leaves and heads because flower parts provide nutrients for the pathogen (Dillard and Hunter 1986). Lastly, it is important to control flea beetles (*Phyllotreta cruciferae*), which carry spores of *Alternaria brassisicola* on their bodies and in their frass and transmit conidia while feeding.

Managing Seedborne Pathogens

Seed should originate from seedlots that have tested negative for the presence of the black leg and black rot pathogens. Hot water seed treatment is useful to control seedborne black rot bacteria, provided the water temperature is monitored carefully so it remains at 50°C for 25 minutes to avoid reducing seed germination and quality. Minimize leaf wetness periods when producing transplants in glasshouses, because of the ease of spreading pathogens. Apply protectant fungicides to seed crops to prevent infection of seed by *Alternaria*.

Managing Soilborne Pathogens

Soil fumigants generally are not used against soilborne pathogens in cabbage production because of the high cost, although they may be used to disinfest seedbeds and suppress clubroot. Field-grown transplants may be sources of the clubroot and wirestem pathogens and spread them to noninfested fields. Because of this risk, transplants should be produced in soilless mixes in greenhouses when possible. Do not plant any cruciferous cover crops or vegetables in fields before or after cropping to cabbage. Use monocots as rotation crops, because *R. solani* AG-4 and *S. sclerotiorum* have wide host ranges among dicotyledonous crops. The resting spores of the clubroot organism cannot be eradicated by rotation. Instead, liming soil to raise the pH above 7.2 with calcium oxide or hydrated lime prevents infection of roots.

Managing Foliar Pathogens

Diseases caused by foliar pathogens, such as *Alternaria* and *Xanthomonas*, can be managed with crop rotation during the period when infested host debris is decaying in affected fields, because these foliar pathogens of cabbage do not survive longer than 1 or 2 years in soil, respectively. Disk and bury or compost unmarketable cabbage heads. Apply protectant fungicides as needed based on environmental conditions and host susceptibility. Black spot can be reduced by increasing row width and plant spacing to promote air circulation that dries leaves, because *Alternaria* spp. require relatively long periods of leaf wetness for infection (a minimum of 5–9 hours).

Conclusion

The diseases dark spot, downy mildew, watery soft rot, and wirestem often can be managed successfully using a combination of biological, chemical, and cultural control measures. The biological and cultural methods listed in Table 1 are amenable to organic production systems. Management of black rot and clubroot remains challenging. In the future, resistance to downy mildew and improved resistance to black rot may be available in cabbage cultivars. It may be possible to transfer downy mildew resistance from broccoli to cabbage using molecular genetics methods. Additional molecular genetics research is needed to clarify the identity of races of the downy mildew and clubroot organisms.

Acknowledgments

Technical Contribution No. 5100 of the Clemson University Experiment Station. We thank Richard Morrison, Sakata Seed Company, and J. Powell Smith, Clemson University, for reviewing this chapter.

References

Budge, G.E., and Shaw, M.W. 2009. Characterization and origin of infection of *Rhizoctonia solani* associated with *Brassica oleracea* crops in the UK. *Plant Pathol.* 58:1059–1070.

Dillard, H.R., and Hunter, J.E. 1986. Association of common ragweed with Sclerotinia rot of cabbage. *Plant Dis.* 70 :26–28. Available at http://www.fao.org/statistics/en/.

Humpherson-Jones, F.M. 1989. Survival of *Alternaria brassicae* and *Alternaria brassicicola* on crop debris of oilseed rape and cabbage. *Ann. Appl. Biol.* 115:45–50.

Hwang, S.F., Howard, R.J., Strelkov, S.E., Gossen, B.D., and Peng, G. 2014. Special Issue: Management of clubroot (*Plasmodiophora brassicae*) on canola (*Brassica napus*) in western Canada. *Can. J. Plant Pathol.* 36:49–65.

Lange, H. W., Tancos, M. A., Carlson, M. O., and Smart, C. D. 2016. Diversity of *Xanthomonas campestris* isolates from symptomatic crucifers in New York State. *Phytopathology* 106:113–122.

15

Natural Enemies and Biocontrol: Artificial Diets

Simon Grenier and
Patrick De Clerq

Introduction

Arthropod parasitoids and predators used in biological control strategies are at present mainly produced on natural or alternative hosts or prey. However, their large-scale production may be more convenient and cost-effective when using artificial diets/media. Studies aiming at the successful development of arthropod parasitoids and predators under artificial conditions have started a long time ago, but the practical use of insects and mites grown on artificial diets is still in its infancy. Besides their use for the production of natural enemies, artificial media may be valuable tools for physiological and behavioral studies of entomophagous arthropods due to a simplification of their environment. Different types of artificial diets with or without insect additives can support the development and/ or reproduction of natural enemies. Successes have been achieved for several species of parasitoids and predators but these have mainly been restricted to an experimental level. Comparisons of the performances of artificially vs. naturally reared natural enemies (as quality control) have primarily been conducted in the laboratory, and only very rarely in the field. The promising results achieved in recent years open up new prospects for natural enemy producers.

Artificial Diets for Predators and Parasitoids

The culture of entomophagous insects and mites involves rearing not only of the host/prey, but often also of the host's/prey's plant food, and thus requires a tritrophic level system. Different steps were taken to try to reduce the production line for entomophagous arthropods. The complete line comprises plant growing, host/prey rearing, and parasitoid/predator rearing. The simplified line includes the use of artificial diets instead of plants for the phytophagous host/prey, or of factitious hosts/prey that are easier to rear in the laboratory than the natural food (e.g., eggs of *Ephestia kuehniella* or *Sitotroga cerealella*, larvae of *Galleria mellonella* or *Tenebrio molitor*). The ultimate reduction of the production line consists only of an artificial diet for direct parasitoid/predator rearing. Mass rearing entomophagous insects on artificial media, first suggested 60 years ago, holds the promise to increase the ease and flexibility of insect production, including automation of procedures, and to reduce cost. The early and subsequent efforts at developing artificial diets have extensively been reviewed.[1–3] The basic qualitative nutritional requirements of parasitoids and predators are similar to those of free-living insects. But the very fast growth of some parasitoids such as tachinid larvae requires a perfectly well-balanced diet[4] to minimize intermediate metabolism and toxic waste product accumulation.

Essentially, two types of artificial diets can be distinguished: Those including and those excluding insect components. The availability of media without insect components offers a greater independence from insect hosts/prey, even if in some countries insect components are cheap and easily available by-products, e.g., from silk production in Asia or South America.[5] In diets containing insect additives, such varied components as hemolymph, body tissue extract, bee brood extract or powder, egg juice, or homogenate of the natural host have been used. Products of insect cell culture have also been incorporated into diets as host factors. The composition of most media for in vitro rearing of *Trichogramma* egg parasitoids is based on lepidopterous hemolymph.[6] Media for the tachinid fly *Exorista larvarum*, the chalcid wasp *Brachymeria intermedia*, and the ichneumonid wasp *Diapetimorpha introita* contain various insect components. Bee extracts or bee brood have been commonly added in diets for predatory coccinellids.[1,5] Only few diets devoid of insect additives are composed of ingredients that are fully chemically defined in their composition and structure. Besides proteins or protein hydrolysates, most of such diets contain crude or complex components, e.g., hen's egg yolk, chicken embryo extract, calf serum, cow's milk, yeast extract or hydrolysate, meat or liver extract, or plant oils. Beef or pork meat and liver have extensively been used as basic components of diets for feeding coccinellids and several predatory heteropterans.[1,5]

Successes and Failures with Artificial Diets

Both biochemical and physical aspects determine the success of an artificial diet. Artificial diets should be nutritionally adequate to support development and reproduction of an insect and should be formulated in such a manner that the medium is easily recognized and accepted for feeding or oviposition; the food should be readily ingested, digested, and absorbed.[7] For parasitoids, the diet must also allow the growing larvae to satisfy other physiological needs like respiration and excretion without diet spoiling. The best results on artificial media were obtained with idiobiontic parasitoids such as egg or pupal parasitoids and with polyphagous predators. Different tachinid species were also successfully grown in vitro, but the koinobiontic Hymenoptera appear the most difficult group to be reared in vitro, probably because of a close relationship with their living host that supplies them with crucial growth factors. Ectoparasitoids are generally easier to culture in vitro than endoparasitoids for which the diet is also the living environment of the immature stages.[2] Several predatory insects have been reared for successive generations on artificial diets, including heteropterans (e.g., *Geocoris punctipes, Orius laevigatus, Podisus maculiventris*), coccinellids (e.g., *Coleomegilla maculata, Harmonia axyridis*), and chrysopids (e.g., *Chrysoperla carnea, Chrysoperla rufilabris*).[3]

Artificial rearing of natural enemies has mostly remained at an experimental level, and the practical experience with natural enemies produced in artificial conditions has remained quite limited. Wasps of the genus *Trichogramma* reared on factitious host eggs are the most common agents used worldwide in biological control in many field crops and forests. In China, *Trichogramma* spp. and *Anastatus* spp. produced on a large scale in artificial host eggs have been released on thousands of hectares of different crops with a parasitization rate above 80%, leading to an effective pest control level equal to that of naturally reared parasitoids.[5] In the U.S.A., field tests with encouraging first results were conducted using the pteromalid parasitoid *Catolaccus grandis* reared for successive generations on artificial diet for the control of the cotton boll weevil *Anthonomus* grandis.[5] Since the late 1990s, biocontrol companies in the U.S.A. and Europe have started producing a number of natural enemies (partially) on artificial diets.

Quality Control of Natural Enemies Produced on Artificial Diets

Long-term rearing on artificial diets could lead to genetic bottleneck effects inducing high selection pressure on the entomophages and possible reduction of their effectiveness. Periodic population renewals from nature may circumvent this drawback. The use of natural enemies in augmentative biological control requires a reliable mass production of good quality insects. Therefore, quality control is a key element for the efficiency and the long-term viability of biological control. The quality control procedures developed for in vivo production of entomophages could be recommended as a first approach for in vitro production.[8] Many parameters can be used as quality criteria. Size, weight, life cycle duration, survival rate, and especially fecundity, longevity, and predation/parasitization efficiency are the most relevant characters.[5] Besides its value as a quality criterion, the biochemical composition (based upon carcass analyses) of the insects produced on artificial diets may be a powerful tool for improving the composition and performance of the diets through the detection of excess or deficiency of some nutrients. Often, different criteria are closely linked; hence, the quality control process may be simplified if one easily measured parameter can be used to predict another one that is more complex or time consuming to determine (e.g., fecundity). Arguably, excellent field performance of the artificially produced natural enemy against the target pest remains the ultimate quality criterion. However, quality assessments of artificially reared natural enemies have mostly been performed at a laboratory scale or in semifield conditions, and only rarely so in practical field conditions.

Conclusions

At present, rearing systems using natural or factitious foods remain the only effective way for industrial production of most entomophagous insects and mites. However, success achieved for a restricted number of species of parasitoids (e.g., *Trichogramma* spp., *Exorista larvarum, Catolaccus grandis)* and predators (e.g., *Orius* spp., *Geocoris punctipes, Chrysoperla* spp., *Harmonia axyridis)* has prompted producers to increasingly incorporate artificial diets into their mass rearing systems. Further behavioral and physiological investigations may lead to significant improvements in artificial rearing through a better knowledge of the host-parasitoid and predator-prey relationships. Besides an easier mechanization of the production line, the use of artificial diets opens new possibilities for preimaginal conditioning of parasitoids/predators to targeted hosts/prey by adding specific chemicals in their food. Artificial diets also seem the only way of mass rearing for some middle-sized egg parasitoids (Encyrtidae, Eulophidae, Eupelmidae, Scelionidae to name a few) that are promising pest control agents but are unable to develop normally in the small lepidopteran substitution host eggs commonly used nowadays *(Ephestia kuehniella, Sitotroga cerealella).*

References

1. Thompson, S.N. Nutrition and culture of entomophagous insects. Annu. Rev. Entomol. **1999**, *44*, 561–592.
2. Grenier, S.; Greany, P.D.; Cohen, A.C. Potential for mass release of insect parasitoids and predators through development of artificial culture techniques. In *Pest Management in the Subtropics: Biological Control—A Florida Perspective;* Rosen, D., Bennett, F.D., Capinera, J.L., Eds.; Intercept: Andover, U.K., 1994; 181–205.
3. Thompson, S.N.; Hagen, K.S. Nutrition of entomophagous insects and other arthropods. In *Handbook of Biological Control: Principles and Applications;* Bellows, T.S., Fisher, T.W., Eds.; Academic Press: San Diego, CA, 1999; 594–652.
4. Grenier, S.; Delobel, B.; Bonnot, G. Physiological interactions between endoparasitic insects and their hosts— Physiological considerations of importance to the success of in vitro culture: an overview. J. Insect Physiol. **1986**, *32* (4), 403–408.

5. Grenier, S.; De Clercq, P. Comparison of artificially vs. naturally reared natural enemies and their potential for use in biological control. In *Quality Control and Production of Biological Control Agents: Theory and Testing Procedures*; van Lenteren, J.C., Ed.; CABI Publishing: Wallingford, U.K., 2003; 115–131.

6. Grenier, S. Rearing of *Trichogramma* and other egg parasit- oids on artificial diets. In *Biological Control with Egg Para- sitoids*; Wajnberg, E., Hassan, S.A., Eds.; CAB International: Wallingford, U.K., 1994; 73–92.

7. Cohen, A.C. *Insect Diets—Science and Technology*; CRC Press: Boca Raton, U.S.A., 2003.

8. van Lenteren, J.C.; Hale, A.; Klapwijk, J.N.; van Schelt, J.; Steinberg, S. Guidelines for quality control of commercially produced natural enemies. In *Quality Control and Production of Biological Control Agents: Theory and Testing Procedures*; van Lenteren, J.C., Ed.; CABI Publishing: Wallingford, U.K., 2003; 265–303.

IV

DIA: Diagnostic Tools: Monitoring, Ecological Modeling, Ecological Indicators, and Ecological Services

IV

16

Animals: Toxicological Evaluation

Vera Lucia S.S.
de Castro

Introduction

Developmental toxicants may be a chemical, microorganism, physical agent, or deficiency state that alters the morphology or a physiological process of a developing organism, both before and after birth. Exposure of the developing embryo or fetus to some environmental agents like gamma irradiation and thalidomide is known to produce anatomical anomalies leading to in utero death or structural birth defects. Developmental toxicology studies the causes, mechanisms, manifestation, and prevention of developmental deviations produced by developmental toxicants. Several environmental agents are established as causing developmental toxicity in humans, while many others are suspected of causing developmental toxicity in humans on the basis of data from experimental animal studies.

Nowadays, all the possible manifestations of developmental toxicity (death, structural abnormalities, growth alterations, and behavioral and functional deficits) are of concern, while in the past there has been a tendency to consider only malformations and death as end points of concern. Developmental toxicity usually results from prenatal exposures to toxicants experienced by the mother, but it can also result from paternal exposures, e.g., in rural workers exposed to pesticides. Effects can include birth defects, reduced body weight at birth, growth and developmental retardation, organ toxicity, death, abortion, and functional dysfunctions. It can also lead to behavioral deficits that become manifest as the organism develops since the chemicals can impair postnatal development up to pubertal development. Further considerations related to the efforts to assemble an internationally harmonized source of common nomenclature for use in describing observations of fetal and neonatal external, visceral, and skeletal abnormalities can be found in Makris.[1]

In May 2001, more than 90 nations adopted the Stockholm Convention on Persistent Organic Pollutants (POPs), with significant contributions from non-governmental organizations, trade unions, and private companies. The Stockholm Convention is a global treaty created to protect human health and the environment from chemicals that remain intact in the environment for long periods, become widely distributed geographically, accumulate in the fatty tissue of humans and wildlife, and have adverse effects to human health or to the environment. Exposure to POPs can lead to serious health effects, including certain cancers, birth defects, dysfunctional immune and reproductive systems, greater susceptibility to disease, and even diminished intelligence. Given their long-range transport, no one government acting alone can protect is citizens or its environment from POPs. In response to this global problem, the Stockholm Convention entered into force in 2004, requiring parties to take measures to eliminate or reduce the release of POPs into the environment. The convention is administered by the United Nations Environment Programme and is based in Geneva, Switzerland. The substances covered initially are eight pesticides (aldrin, chlordane, DDT, dieldrin, endrin, heptachlor, mirex, and toxaphene), two industrial chemicals (hexachlorobenzene and polychlorinated biphenyls), and two POP by-products (dioxins and furans). In 2009, the Conference of the Parties decided to undertake a work program to provide guidance to parties on how best to restrict and eliminate nine newly listed POPs and invited parties to support work on the evaluation of alternatives and other work related to the restriction and elimination of these new POPs [α-hexachlorocyclohexane, ß-hexachlorocyclohexane, lindane, pentachlorobenzene, perfluorooctane sulfonic acid (PFOS), PFOS salts, perfluorooctane sulfonyl fluoride, tetrabromodiphenyl ether, and pentabromodiphenyl ether (commercial pentabromodiphenyl ether)] (http://chm.pops.int/default.aspx).[2]

Exposure of the developing embryo or fetus to some environmental agents like gamma irradiation and thalidomide is known to produce anatomical anomalies leading to in utero death or structural birth defects, commonly termed teratogenesis. Perhaps less well appreciated is that such environmental exposures also can cause functional disorders that persist postnatally and into adult life. The spectrum of such postnatal consequences is growing, and more recently is thought to include disorders of the immune system, brain function, obesity, and diseases such as diabetes and cancer, to name a few.[3]

Importance of Oxidative Stress

Xenobiotics such as phenytoin and benzo[a]pyrene can be bioactivated by enzymes like the cytochromes P450 (CYPs); however, the developing embryo and fetus have relatively low levels of most CYP isozymes. The xenobiotic bioactivation to a free radical intermediate by enzymes associated with peroxidase activities within the embryo or fetus can be a critical determinant of teratogenesis. In the developing embryo and fetus, enhanced formation of reactive oxygen species by xenobiotics may adversely alter development by oxidatively damaging cellular lipids, proteins, and DNA, and/or by altering signal transduction. The postnatal consequences may include an array of birth defects, postnatal functional deficits, and diseases.[3]

Also, estrogen-like substances, such as several organochlorine pesticides, have been demonstrated to induce defeminization, miscarriages, malformations, and transplacental carcinogenesis. This variety of effects results from the interference of these substances with the metabolism of steroid and protein hormones, therefore altering a whole spectrum of complex developmental functions.[4]

Importance of Toxic Effects on Nervous System Development

Only about 200 chemicals out of more than 80,000 registered with the U.S. Environmental Protection Agency (EPA) have undergone extensive neurotoxicity testing, and many chemicals found in consumer goods are not required to undergo any neurodevelopmental testing. The cumulative effects of co-contaminants and the difficulties in analyzing biomarkers of exposure in human tissues have complicated comprehensive risk assessment. Furthermore, population-based studies that measure subtle effects on neurobehavioral outcomes are challenging to interpret and costly to conduct.[5]

The effects observed following developmental exposure may be different both quantitatively and qualitatively from adult exposure because of the potential to affect processes in the developing child that have no parallel process in adults. Central nervous system development consists of a series of processes that occur in sequence and are dependent on each other, such that interference with one stage may also affect later stages of development. This makes the timing of a potential environmental neurotoxicant a critical parameter in the risk for subsequent neurologic effects. The sequence includes proliferation, migration, differentiation, synaptogenesis, apoptosis, and myelination.

There are multiple windows of vulnerability during which environmental exposures can interfere with normal development. For many developmental toxicants, there is a spectrum of adverse outcome depending on dose timing of exposure, maternal and fetal susceptibility, and interactions with other environmental factors.[6]

Therefore, environmental exposure to a toxic agent that affects synaptogenesis, such as lead, may affect brain areas differentially depending on timing of exposure. Since different brain areas develop on different time lines during prenatal and postnatal life, an environmental neurotoxic agent may produce impairment in different functional domains depending on the time of exposure. For example, the same exposure at different points in development could result in an adverse effect on motor systems versus memory or executive functions. Similarly, exposures at different concentrations or for different lengths of time could potentially produce differential effects. Therefore, the constellation of observed effects should not be expected to be the same in different children exposed to the same neurotoxic agent.[7]

Concern on Developmental Impairment in Children

The findings of the some studies imply that children's exposure to pesticides may bring about impairments in their neurobehavioral development. The long-term neurotoxicity risks caused by prenatal exposures to pesticides are sometimes unclear. Effective control of exposure is complicated by variable exposure sources and variable contaminant levels in food and environment. This awareness has also been extended to effect(s) of toxic contaminants on breastfeeding women and their children.[8]

The information deriving from epidemiological studies indicate a need to increase awareness among people and children exposed to pesticides about the association between the use of pesticides and neurodevelopmental impairments. There are modest epidemiological evidence on occupational exposures of female workers to industrial chemicals and the consequences in regard to the child's neurodevelopment. The majority of the occupational studies identified aimed to assess organic solvents and organophosphate pesticide effects in the offspring, and neurobehavioral impairments were reported. In some reports, however, the evidence suffers from a variety of shortcomings and sources of imprecision. These problems would tend to cause an underestimation of the true extent of the risks. Due to the vulnerability of the brain during early development, a precautionary approach to neurodevelopmental toxicity needs to be applied in occupational health.[9,10]

With increasing evidence of the high prevalence of pesticide use and the considerable risk it poses to children, it is of concern that there has been little research into the health implications of household pesticide use. Children are exposed to pesticides in various ways, not only environmentally, but also through food and through use of pesticides at home and in public areas. The exposure depends on a large variety of factors related to chemical characteristics and use, and children's activities. In spite of its potential health and environmental risks and contribution to agribusiness, the use of agricultural chemicals for yard care has not been well studied. The probability that a household chooses a mix of do-it-yourself and hired applications of synthetic chemicals increases with income, age, and the presence of preschoolers.[11]

Also, children who live in farming communities are furthermore exposed to both agricultural and household pesticides. Farmworkers bring home pesticide residues on their clothing, boots, and skin, placing other household members at risk, particularly children.[12] Recent studies on in utero exposure to the organochlorine pesticide dichlorodiphenyltrichloroethane and its breakdown product, dichlorodiphenyltrichloroethane, indicate that exposure is associated with poorer infant (6 mo and older) and child neurodevelopment depending on the end point evaluated. Research on organophosphate pesticide exposure and neurodevelopment also suggests some negative association of exposure and neurodevelopment at certain ages. About abnormal reflexes in neonates and in young children (2–3 years), two separate studies observed an increase in maternally reported pervasive developmental disorder with increased levels of organophosphate exposure.[13]

For example, children whose mothers worked in the flower industry during pregnancy scored lower on communication and fine motor skills, and had higher odds of having poor visual acuity, compared with children whose mothers did not work in the flower industry during pregnancy, after adjusting for potential confounders. These facts showed that maternal occupation in the cut-flower industry during pregnancy may be associated with delayed neurobehavioral development of children aged 3–23 mo. However, possible hazards associated with working in the flower industry during pregnancy include pesticide exposure, exhaustion, and job stress.[14]

The increased emphasis on children's exposures to pesticides and other organic pollutants has led to a surge in recent years in the number of research studies aimed at this specific susceptible population. Continuous strong investment in research plus strong preventive action by the government is required for further progress in environmental pediatrics and for better control of the diseases caused in children by environmental toxic exposures. This research will have high costs and demanding long-term multiyear studies.[15]

Developmental Animal Experimental Studies

Before 1960, governmental recommendations for the assessment of chemical effects on the reproductive cycle involved limited animal testing. During the early 1960s, the thalidomide disaster evidenced, on the one hand, the greater vulnerability of the embryo and fetus, and on the other, that the complexity of the mother-child unity warranted special consideration. Thalidomide was used as a sedative drug in the 1950s, also in pregnant women. Later, it was found that it could cause malformations in newborns of mothers who ingested the drug during the sensitive period.[16] This disaster fostered the establishment of formal laboratory animal-testing procedures for assessing fetal development.

In contrast to most other toxicological tests, developmental studies are usually required in rodent and nonrodent subjects. One of the reasons for this requirement is the thalidomide disaster. When the developmental toxicity of thalidomide was studied in experimental animals, large interspecies differences were found in effective doses and in the types of effects. However, the discrimination between direct and indirect (i.e., as a consequence of maternal toxicity) developmental effects was often doubtful, and is one of the factors that could explain the apparent differences between species.[16]

Despite the limitations, animals can be useful predictors of chemical hazards to humans. A specific animal model might be chosen for any conjunction of widely varying reasons. Accessibility of embryos, cost of acquiring or maintaining animals, availability of genomic analyses or probes, and/or close similarity to human physiology might factor in the design of a laboratory experiment. Growth and development are compressed into a shorter period in animals, which makes interpretation of animal testing inherently more difficult. Each experimental species has its own advantages and the use of laboratory animal models is based on diverse practical grounds arising from an assumption of generalizability across species. The conjunction of evolutionary and developmental biology shows that the timing and sequence of early events in brain development are remarkably conserved across mammals.[17] During mammalian development, the fetal organism is exposed to its own gonadal hormones, placental steroids, and maternal hormones that may cross the placental barrier.[18]

Manifestations of developmental toxicity observed in humans are not always reproduced in experimental animals, and in general there is at least one experimental species that mimics the types of effects seen in humans. The fact that every species may not react in the same way could be due to species-specific differences in critical periods, differences in timing of exposure, metabolism, developmental patterns, placentation, or mechanisms of action.[6]

The vast majority of the laboratory studies on the developmental effects of environmental contaminants use the pregnant or lactating animal as the conduit to deliver the contaminant to the developing offspring. The traditional approach in developmental toxicology adopts a linear perspective for the interpretation of the effects of contaminants on the offspring as a particular one is given to the mother in order to expose the fetus in utero or postnatally via lactation. Variations of this approach may manipulate time of exposure or use cross-fostering strategies to separate the effects of in utero from those of lactational exposure.

Maternal behavior is not always monitored during the treatment period, even though there is evidence of the effects of these compounds on adult behaviors, particularly on behaviors that are sensitive to hormonal manipulations. As is the case with all mammals, maternal care in female rodents comprises very specific behaviors that help ensure the survival of the offspring by providing nourishment, warmth, sensory stimulation, cleaning, and protection. Maternal behavior begins even before parturition as the dam builds a nest in order to provide warmth and protection for the coming offspring. The fact that many developmental outcomes are determined or modulated by the amount and quality of maternal care raises the question of the importance of possible changes in maternal behavior in determining the consequences of their exposure during early development.[19] There are some guidelines proposed in the literature in order to evaluate the maternal behavior and its effects on litter development.[20,21]

Reproductive and Developmental Protocols

Developmental toxicology bioassays are designed to identify agents with the potential to induce adverse effects and include dose levels that induce maternal toxicity. In reproduction toxicity studies, the determination of the high dose is important since changes in body weight are often used as an index of toxicity. The highest dose level should be chosen with the aim to induce some parental toxicity (e.g., clinical signs, decreased body weight gain, not more than 10%) and/or evidence of toxicity in a target organ. Consequently, the knowledge on maternal (and to some degree, paternal) toxicity is important as a natural limiter to prevent underdosing. A comparison between doses causing effects in adults and offspring can also be used, although a direct comparison is difficult since the level of observation applied in offspring is often much higher than in adults. It is also useful in order to obtain information on the influence of pregnancy and/or lactation on the susceptibility to a test compound. However, if dosing was high enough to cause maternal toxicity, these doses often also cause some effects in the offspring.[22]

Prenatal developmental toxicity studies are designed to provide general information concerning the effects of exposure to the pregnant test animal on the developing organism. Although exposures during a typical guideline prenatal developmental toxicity study are designed to include either the entire period of gestation or limited species-specific gestation periods, developmental end points are considered to be an integral concern in the assessment of potential health effects from continuous lifetime exposures to a toxicant. Pregnancy and fetal development are thus considered to represent a potentially susceptible life stage that should be considered in lifetime or chronic assessments. If studies and information on reproductive or developmental toxicity are absent in a health-effect assessment, specific uncertainty factors (e.g., database) may be applied to the final point of departure used in risk calculations. It is also well established, however, that developmental toxicity may occur in response to single exposure, such as during specific developmental windows of susceptibility. Whereas this circumstance does not influence the relevance of typical guideline developmental studies to the evaluation of chronic or lifetime assessments of health effects, it does signify that developmental end points observed in these repeated dose studies are relevant in health-effect assessments of shorter-term exposures, including acute exposures.[23] Results from developmental toxicology bioassays have significant predictive value in identifying potential health risks to the human embryo/fetus.[24]

Data Quality Control of Experimental Studies

To accomplish good experimental planning, some points before the beginning of the study should be observed for adequate data interpretation in view of the experimental delineation.[25,26] Animals to be used in laboratory research should experience an acceptable welfare (for ethical reasons) and show normal behavioral and physiological reaction patterns to guarantee the quality of research.[27] In this sense, the performing laboratory should maintain a historical control database to track any changes in the data over time in the animals and/or in the equipment. The value of historical data depends on its quality and its reliability at the side of contemporary controls.

International Protocols

Different international organizations have developed protocols for testing the reproductive and developmental toxicity of chemicals. The information produced by them, related to toxic effects produced by environmental pollutants, is very useful worldwide. Besides, new protocols are being evaluated viewing to reduce the number of animals used and to improve the predictability for human health hazards identification.

Developmental study methodology has been extensively reviewed and evaluated over the last 25 years. This has included the conduct of a number of meetings and collaborative studies involving experts from academic, industry, regulatory, and public interest groups. For example, in recent years, the International Life Sciences Institute (ILSI), under a cooperative agreement with the EPA, established a working group of scientists from government, industry, and academia, to discuss developmental neurotoxicity test protocol ending with a public workshop in which occasion the conclusions of the working group were presented.[28]

Food Quality Protection Act

Developmental and reproductive toxicity testing protocols such as those recommended by the EPA, Food and Drug Administration, and Organization for Economic Cooperation and Development are useful for characterizing toxicity in developing animals and for assessing risks to children that might arise from in utero and postnatal exposures.[7] However, there is a global interest in reducing, refining, and replacing (3Rs) the use of animals in research.[29]

The Food Quality Protection Act (FQPA) of 1996 enactment by U.S. Congress amended the Federal Insecticide, Fungicide, and Rodenticide Act and the Federal Food, Drug, and Cosmetic Act by fundamentally changing the way the EPA regulates pesticides. The major requirements of the FQPA include stricter safety standards, especially for infants and children, and a complete reassessment of all existing pesticide tolerances. They include an additional safety factor to account for developmental risks and incomplete data when considering a pesticide's effect on infants and children, and any special sensitivity and exposure to pesticide chemicals that infants and children may have (http://www.epa.gov/opp00001/regulating/laws/fqpa/).

The FQPA mandated that all pesticides in the United States undergo re-registration with a focus on reducing cumulative risk of exposure to pesticides sharing a common mode of action. Enforcement of FQPA has resulted in the modification of use patterns and removal (or pending removal) of many organophosphate insecticides that had previously seen wide use.[30]

The FQPA also requires the EPA to consider the cumulative effects of exposure to pesticides having a common mechanism of toxicity, considering for a cumulative risk assessment the exposure to all chemicals that act by a common mechanism of toxicity, as well as the exposure to each chemical via various routes and sources in an aggregate risk assessment. To support the grouping of different chemicals together for purposes of cumulative risk assessment, there must be sufficient evidence to support a common adverse effect that is associated with a common mechanism of action in specific target tissues. However, the criteria that are required to establish a common mechanism of toxicity with a specific toxic effect have not always been achieved for various pesticides as the common mechanism of toxicity of organophosphorus and carbamate insecticides (inhibition of acetylcholinesterase activity) that can be associated with adverse effects (cholinergic signs of intoxication). For example, a determination of common mechanism of toxicity in mammals is complicated by the number of potential biological target sites and effects expressed by various pyrethroid insecticides on these targets. Probably, the differences of action on neuronal ion channels among the pyrethroid insecticides contribute to the diversity of neurologic and behavioral manifestations of acute toxicity that are evident in the whole animal.[31]

European Union's Reach Legislation

REACH is the European Community Regulation on chemicals and their safe use. It deals with the Registration, Evaluation, Authorization, and Restriction of Chemicals. The law entered into force on June 1, 2007. The REACH regulation places greater responsibility on industry to manage the risks from chemicals and to provide safety information on the substances. Manufacturers and importers are required to gather information on the properties of their chemical substances, which will allow their safe handling, and to register the information in a central database in Helsinki (http://ec.europa.eu/environment/chemicals/reach/reach_intro.htm). It transfers responsibility for risk assessment from government to the manufacturers and importers, and includes downstream uses in the registration and management process. It introduces authorization and restriction procedures for the most hazardous chemicals and creates a new European Chemicals Agency. The legal permission to market products is conditional on the firms testing them for toxicity. If firms do not provide data required by the program, their products will not be permitted to enter (or remain in) the market. This program holds some promise for detecting developmental toxicants before they enter commerce and cause adverse effects. Whether this will work for subclinical neurotoxic and other developmental effects depends on tests the European Union requires. The REACH testing strategy is to require fewer tests for products produced in lesser amounts and to require more tests and more detailed tests as the production volume increases.[32]

In particular, large numbers of industrial chemicals are unlikely to be tested under this paradigm, and there is no specific requirement for developmental neurotoxicological testing under the new European Union law governing chemical regulation (REACH), which was passed in 2006 and went into effect in 2007. In response to the need for broader screening for developmental neurotoxicity, efforts are under way to develop additional developmental neurotoxicity screening paradigms. Additional efforts will be needed to focus on identifying possible chemical class- specific targets and biomarkers of effect, and on ways to differentiate normal variability in response from changes that are adverse.[33]

Perspectives

There is a need to expand risk assessment paradigms to evaluate exposures relevant to children from preconception to adolescence, taking into account the specific susceptibilities at each developmental stage. Risk assessment approaches for exposures in children must be linked to life stages. Establishing causal links between specific environmental exposures and complex, multifactorial health outcomes is difficult and challenging, particularly in children. For children, the stage in their development when the exposure occurs may be just as important as the magnitude of exposure. Very few studies have characterized exposures during different developmental stages. Some examples of health effects resulting from developmental exposures include those observed prenatally and at birth (e.g., miscarriage, stillbirth, low birth weight, birth defects), in young children (e.g., infant mortality, asthma, neurobehavioral and immune impairment), and in adolescents (e.g., precocious or delayed puberty). Emerging evidence suggests that an increased risk of certain diseases in adults (e.g., cancer, heart disease) can result in part from exposures to certain environmental chemicals during childhood. Advancing technology and new methodologies now offer promise for capturing exposures during these critical windows. This will enable investigators to detect conceptions early and estimate the potential competing risk of early embryonic mortality when considering children's health outcomes that are conditional on survival during the embryonic and fetal periods.[34]

Furthermore, emerging technologies such as gene expression, electrical activity measurements, and meta-bonomics have been identified as promising tools for evaluating neurotoxicity. In a combination with other assays, the in vitro approach could be included into a developmental neurotoxicological intelligent testing strategy to speed up the process of developmental neurotoxicity evaluation mainly by initial prioritization of chemicals with developmental neurotoxicity potential for further testing. Also, emerging nano/microtechnologies (piezoelectric spotting and microcontact printing of different biomolecules to create protein microarrays) can be used to promote cell differentiation and make the model suitable for developmental neurotoxicity screening.[35]

Developmental Toxic Effects on Nontarget Organisms in Environment

In addition to their potential role as human reproductive toxicants, pesticides are also implicated in reproductive failure of wildlife species exposed to pesticide sprays and residues. One main example of the ecological consequences of teratogenic pesticides is related to the organochlorines, which, besides inducing malformations in embryos, cause calcification problems in eggshells and impair reproduction in several wild bird species. This problem still remains in numerous areas, due to organochlorine residue accumulation through food chains and to wild populations exposed to organochlorine-contaminated sites.[36] Also, organochlorine pesticides can be maternally transferred to the developing eggs of alligators. This maternal exposure is associated with reduced clutch success and increased embryonic mortality.[37]

Toxicity assays are available for the evaluation of pesticide impact on wildlife or their surrogates, and many compounds have been shown to cause teratogenesis in fish, amphibian, avian, and mammalian species. Some chemicals that have been detected in the environment may be disrupting of both target and non-target systems in exposed populations of wildlife and fish.[38] Environmental compounds can also interfere with the endocrine systems of wildlife. Surface waters are the main sinks of endocrine disrupters, which are mainly of anthropogenic origin. Thus, aquatic organisms, especially lower vertebrates such as fish and amphibians, are the main potential targets for endocrine disrupters at direct or indirect risk via ingestion and accumulation of endocrine disrupters, direct exposure, or via the food chain. The impact of these compounds on reproductive biology can be mediated through four principal mechanisms of action: estrogenic, anti-estrogenic, androgenic, and anti-androgenic.[39]

Furthermore, population studies have revealed disruptions in crustacean growth, molting, sexual development, and recruitment that are indicative of environmental endocrine disruption. However, environmental factors other than pollution (i.e., temperature, parasitism) also can elicit these effects, and definitive causal relationships between endocrine disruption in field populations of crustaceans and chemical pollution is generally lacking.[40]

Amphibians are considered reliable indicators of environmental quality, in particular due to their biphasic life (aquatic and terrestrial) and semipermeable skin. These vertebrates are sensitive to a great number of pollutants dispersed in the environment, such as pesticides, heavy metals, and polychlorinated biphenyls. Field studies on frogs from polluted and reference sites have provided information on the effects of chronic exposure to contaminants.[41]

Despite the fact that almost all environmental chemical exposure is to mixtures, the current understanding of environmental health risks is based almost entirely on the evaluation of chemicals studied in isolation. Consequently, it is essential to develop and validate methods to accurately predict effects of endocrine-disrupting mixtures beyond the individual exposure to a single chemical, in order to protect humans and wildlife from the risk associated with potentially cumulative effects of these mixtures.[36,42] Its focus should be on the biological system or the target tissue rather than on the mechanism of toxicity or even a single signaling pathway.[43]

Although the importance of multiple stressors is widely recognized in aquatic ecotoxicology, pesticide mixture studies pose some major challenges such as experimental design difficulties (e.g., near-insurmountable factorial complexity for large numbers of chemicals), poorly understood pathways for chemical interaction, potential differences in response among species, and the need for more sophisticated statistical tools for analyzing complex data.[44]

Conclusion

The ability of a species to reproduce successfully requires the careful orchestration of developmental processes during critical time points, particularly the late embryonic and early postnatal periods. Standard developmental toxicology bioassays are designed to identify agents with the potential to induce adverse effects. Government agencies that regulate the use of pesticides and various industrial chemicals evaluate the toxicity of these agents to the developing embryo/fetus as an integral part of the testing protocol used to assess potential dangers to the public. However, an adequate experimental delineation is important to the data interpretation.

Humans and other non-target organisms are exposed to a mixture of chemicals. Various chemicals may target the organism's development during the same critical developmental period. Although developmental studies have been historically conducted on a chemical-by-chemical basis, the interest on considering cumulative risks of chemicals is growing. Chemicals as pesticides represent a risk for the reproduction and development of children and non-target organisms of terrestrial and aquatic ecosystems.

References

1. Makris, S.L.; Solomon, H.M.; Clark, R.; Shiota, K.; Barbellion, S.; Buschmannf, J.; Emag, M.; Fujiwarah, M.; Grotei, K.; Hazeldenj, K.P.; Hewk, K.; Horimotol, M.; Ooshimam, Y.; Parkinsonn, M.; Wiseo, L.D. Terminology of developmental abnormalities in common laboratory mammals (version 2). Reprod. Toxicol. **2009**, *28*, 371–434.
2. Rodan, B.D.; Pennington, D.W.; Eckley, N.; Boethlin, R.S. Screening for persistent organic pollutants: Techniques to provide a scientific basis for POPs criteria in international negotiations. Environ. Sci. Technol. **1999**, *33*, 3482–3488.
3. Wells, P.G.; McCallum, G.P.; Chen, C.S.; Henderson, J.T.; Lee, C.J.J.; Perstin, J.; Preston, T.J.; Wiley, M.J.; Wong, A.W. Oxidative stress in developmental origins of disease: Teratogenesis, neurodevelopmental deficits, and cancer. Toxicol. Sci. **2009**, *108* (1), 4–18.
4. Chernoff, N. The reproductive toxicology of pesticides. In *Toxicology of Pesticides: Experimental, Clinical and Regulatory Perspectives;* Costa, L.G., Galli, C.L., Murphy, S.D., Eds.; NATO ASI-Cell Biology; Springer-Verlag: Berlin, Germany, 1987; H13, 109–123.
5. Miodovnik, A. Environmental neurotoxicants and developing brain. Mt. Sinai J. Med. **2011**, *78*, 58–77.
6. U.S. EPA. *Guidelines for Developmental Toxicity Risk* Assessment; U.S. Environmental Protection Agency, Risk Assessment Forum: Washington, D.C.; EPA/600/FR-91/001, 1991.
7. Mendola, P.; Selevan, S.G.; Gutter, S.; Rice, D. Environmental factors associated with a spectrum of neurodevelopmental deficits. Mental retardation and developmental disabilities. Res. Rev. **2002**, *8*, 188–197.

8. Lakind, J.S.; Berlin, C.M.; Naiman, D.Q. Infant exposure to chemicals in breast milk in the United States: What we need to learn from a breast milk monitoring program. Environ. Health Perspect. **2001**, *109*, 75–88.

9. Jurewicz, J.; Hanke, W. Prenatal and childhood exposure to pesticides and neurobehavioral development: Review of epidemiological studies. Int. J. Occup. Med. Environ. Health **2008**, *21* (2), 121–132.

10. Julvez, J.; Grandjean, P. Neurodevelopmental toxicity risks due to occupational exposure to industrial chemicals during pregnancy. Ind. Health **2009**, *47*, 459–468.

11. Templeton, S.R.; Zilberman, D.; Yoo, S.J.; Dabalen, A.L. Household use of agricultural chemicals for soil-pest management and own labor for yard work. Environ. Resour. Econ. **2008**, *40*, 91–108.

12. Strong, L.L.; Starks, H.E.; Meischke, H.; Thompson, B. Perspectives of mothers in farmworker households on reducing the take-home pathway of pesticide exposure. Health Educ. Behav. **2009**, *36* (5), 915–929.

13. Rosas, L.G.; Eskenazi, B. Pesticides and child neurodevelopment. Curr. Opin. Pediatr. **2008**, *20* (2), 191–197.

14. Handal, A.J.; Harlow, S.D.; Breilh, J.; Lozoff, B. Occupational exposure to pesticides during pregnancy and neurobehavioral development of infants and toddlers. Epidemiology **2008**, *19* (6), 851–859.

15. Landrigan, P.J.; Miodovnik, A. Children's health and the environment: An overview. Mt. Sinai J. Med. **2011**, *78*, 1–10.

16. Janer, G.; Slob, W.; Hakkert, B.C.; Vermeire, T.; Piersma, A.H. A retrospective analysis of developmental toxicity studies in rat and rabbit: What is the added value of the rabbit as an additional test species? Regul. Toxicol. Pharmacol. **2008**, *50*, 206–217.

17. Clancy, B.; Finlay, B.L.; Darlington, R.B.; Anand, K.J.S. Extrapolating brain development from experimental species to humans. NeuroToxicology **2007**, *28*, 931–937.

18. Gore, A.C. Developmental programming and endocrine disruptor effects on reproductive neuroendocrine systems. Front. Neuroendocrinol. **2008**, *29*, 358–374.

19. Cummings, J.A.; Clemens, L.G.; Nunez, A.A. Mother counts: How effects of environmental contaminants on maternal care could affect the offspring and future generations. Front. Neuroendocrinol. **2010**, *31*, 440–451.

20. Champagne, F.; Francis, D.; Mar, A.; Meaney, M. Variations in maternal care in the rat as a mediating influence for the effects of environment on development. Physiol. Behav. **2003**, *79*, 359–371.

21. Slamberová, R.; Charousová, P.; Pometlová, M. Maternal behavior is impaired by methamphetamine administered during pre-mating, gestation and lactation. Reprod. Toxicol. **2005**, *20*, 103–110.

22. Buschmann, J. Critical aspects in reproductive and developmental toxicity testing of environmental chemicals. Reprod. Toxicol. **2006**, 22, 157–163.

23. Davis, A.; Gift, J.S.; Woodall, G.M.; Narotsky, M.G.; Foureman, G.L. The role of developmental toxicity studies in acute exposure assessments: Analysis of single-day vs. multiple-day exposure regimens. Regul. Toxicol. Pharmacol. **2009**, *54*, 134–142.

24. Chernoff, N.; Rogers, E.H.; Gage, M.I.; Francis, B.M. The relationship of maternal and fetal toxicity in developmental toxicology bioassays with notes on the biological significance of the "no observed adverse effect level". Reprod. Toxicol. **2008**, *25* (2), 192–202.

25. Slikker, Jr., W.; Acuff, K.; Boyes, W.; Chelonis, J.; Crofton, K.; Dearlove, G.; Li, A.; Moser, V.; Newland, C.; Rossi, J.; Schantz, S.; Sette, W.; Sheets, L.; Stanton, M.; Tyl, S.; Sobotka, T. Behavioral test methods workshop. Neurotoxicol. Teratol. **2005**, *27*, 417–427.

26. Festing, M.; Altman, D. Guidelines for the design and statistical analysis of experiments using laboratory animals. ILAR J. **2002**, *43* (4), 244–258.

27. Olsson, I.A.S.; Westlund, K. More than numbers matter: The effect of social factors on behaviour and welfare of laboratory rodents and non-human primates. Appl. Anim. Behav. Sci. **2007**, *103*, 229–254.

28. Tyl, R.W.; Crofton, K.; Moretto, A.; Moser, V.; Sheets L. P.; Sobotka, T.J. Identification and interpretation of developmental neurotoxicity effects. A report from the ILSI Research Foundation/ Risk Science Institute expert working group on neurodevelopmental endpoints. Neurobehav. Toxicol. **2008**, *30*, 349–381.

29. Matthews, E.J.; Kruhlak, N.L.; Benz, R.D.; Contrera, J.F. A comprehensive model for reproductive and developmental toxicity hazard identification: I. Development of a weight of evidence QSAR database. Regul. Toxicol. Pharmacol. **2007**, *47*, 115–135.

30. Jones, V.P.; Steffan, S.A.; Hull, L.A.; Brunner, J.F.; Biddinger, D.J. Effects of the loss of organophosphate pesticides in the US: Opportunities and needs to improve IPM Programs. Outlooks Pest Manage. **2010**, *21* (4), 161–166.

31. Weiner, M.L.; Nemec, M.; Sheets, L.; Sargent, D.; Breckenridge, C. Comparative functional observational battery study of twelve commercial pyrethroid insecticides in male rats following acute oral exposure. NeuroToxicology **2009**, *30S*, S1–S16.

32. Cranor, C. The legal failure to prevent subclinical developmental toxicity. Basic Clin. Pharmacol. Toxicol. **2008**, *102*, 267–273.

33. Raffaele, K.C.; Rowland, J.; May, B.; Makris, S.L.; Schumacher, K.; Scarano, L.J. The use of developmental neurotoxicity data in pesticide risk assessments. Neurotoxicol. Teratol. **2010**, *32*, 563–572.

34. Louis, G.B.; Damstra, T.; Díaz-Barriga, F.; Faustman, E.; Hass, U.; Kavlock, R.; Kimmel, C.; Kimmel, G.; Krishnan, K.; Luderer, U.; Sheldon, L. Principles for evaluating health risks in children associated with exposure to chemicals. Environ. Health Criteria **2006**, *237*, 1–327.

35. Bal-Price, A.K.; Hogberg, H.T.; Buzanska, L.; Lenas, P.; van Vliet, E.; Hartung T. In vitro developmental neurotoxicity (DNT) testing: Relevant models and endpoints NeuroToxicology **2010**, *31*, 545–554.

36. Hotchkiss, A.K.; Rider, C.V.; Blystone, C.R.; Wilson, V.S.; Hartig, P.C.; Ankley, G.T.; Foster, P.M.; Gray, C.L.; Gray, L.E. Fifteen years after "Wingspread"—Environmental endocrine disrupters and human and wildlife health: Where we are today and where we need to go. Toxicol. Sci. **2008**, *105* (2), 235–259.

37. Rauschenberger, R.H.; Wiebe, J.J.; Buckland, J.E.; Smith, J.T.; Sepulveda, M.S.; Gross, T.S. Achieving environmentally relevant organochlorine pesticide concentrations in eggs through maternal exposure in *Alligator mississippiensis.* Mar. Environ. Res. **2004**, *58*, 851–856.

38. Ankley, G.T.; Brooks, B.W.; Huggett, D.B.; Sumpter, J P. Repeating history: Pharmaceuticals in the environment. Environ. Sci. Technol. **2007**, *41*, 8211–8217.

39. Kloas, W.; Lutz, I. Amphibians as model to study endocrine disrupters. J. Chromatogr. A **2006**, *1130*, 16–27.

40. LeBlanc, G.A. Crustacean endocrine toxicology: A review. Ecotoxicology **2007**, *16*, 61–81.

41. Falfushinska, H.I.; Romanchuk, L.D.; Stolyar O.B. Different responses of biochemical markers in frogs (*Rana ridibunda*) from urban and rural wetlands to the effect of carbamate fungicide. Comp. Biochem. Physiol. C **2008**, *148*, 223–229.

42. Jia, Z.; Misra H.P. Developmental exposure to pesticides zineb and/or endosulfan renders the nigrostriatal dopamine system more susceptible to these environmental chemicals later in life. NeuroToxicology **2007**, *28*, 727–735.

43. Rider, C.V.; Furr, J.R.; Wilson, V.S.; Gray, Jr., L.E. Cumulative effects of in utero administration of mixtures of reproductive toxicants that disrupt common target tissues via diverse mechanisms of toxicity. Int. J. Androl. **2010**, *33*, 443–462.

44. Laetz, C.A.; Baldwin, D.H.; Collier, T.K.; Hebert, V.; Stark, J.D.; Scholz, N.L. The synergistic toxicity of pesticide mixtures: Implications for risk assessment and the conservation of endangered pacific salmon. Environ. Health Perspect. **2009**, *117* (3), 348–353.

17

Bioindicators for Sustainable Agroecosystems

Joji Muramoto
and Stephen
R. Gliessman

Introduction

The growth of larger-scale monocultures that heavily depend on use of synthetic pesticides, chemical fertilizers, and fossil fuel-based agricultural machinery has significantly increased crop yield. However, this has also brought about both direct and indirect negative consequences on ecological (e.g., soil erosion, nitrate and pesticide contamination of groundwater, loss of agrobiodiversity), economic (e.g., increase of production and marketing costs and decrease of the net income of farmers), and social (e.g., loss of agricultural communities due to farm consolidations) sustainability of agriculture.[1,2] Especially, agricultural intensification is one of the major drivers of global biodiversity loss as a result of associated habitat fragmentation, land conversion, and agrochemical applications.[3,4] Since the 1970s, and more so after the 1980s, therefore, various alternative farming practices have been developed in pursuit of sustainable agriculture.[5–12] Farmers around the world have adopted these practices to varying degrees, but evaluation of their success needs to be conducted locally. To implement sustainable agricultural practices and policies, managers and policy makers need tools for assessing changes in agroecosystems in various time and spatial scales. Numerous indicators for agricultural sustainability including physicochemical, socioeconomic, and biological indicators, or bioindicators, have been developed.[13–17] Bioindicators are important because they are direct measures of the desired outcome, i.e., sustained or increased biodiversity. They also are living, dynamic, and active indicators, often responding quickly to the way farming is carried out. As evidence regarding the role of biodiversity in maintaining agroecosystem structure and function accumulates, its role as a bioindicator increases.

This chapter will first briefly review the background, definitions, concepts, and history of bioindicators. Then, we discuss the current status of developing bioindicators with two case studies: Europe and Latin America (shaded coffee systems). The former case represents temperate agroecosystems with 4000 years of farming history. The second case characterizes upland tropical agroecosystems known to occur in regions of the world with some of the highest, yet most threatened, biodiversity. Furthermore, examples of recent studies on use of aquatic organisms as bioindicators for agricultural sustainability and an example of the participatory approach in implementing agrobiodiversity policy in Japan are presented.

Agricultural Sustainability and Agrobiodiversity

Although many definitions of sustainable agriculture exist, most of them address ecological, economic, and social goals.[18,19] From an ecological perspective, sustainable farming practices (1) maintain their natural resource base; (2) rely on minimal artificial inputs from outside the farm system; (3) manage pests and diseases through internal regulating mechanisms; and (4) allow the recovery from the disturbances caused by cultivation and harvest.[20] Ecological characteristics of an agroecosystem, such as diversity, trophic structure, energy flow, nutrient cycles, population-regulating mechanisms, stability, and resilience, are metrics to help determine if a particular farming practice, input, or management decision is sustainable.[2]

Functions of biodiversity in agroecosystems, or agrobiodiversity, are foundations of sustainable farming practices. Ever since agriculture began some 10,000–12,000 years ago, biodiversity has allowed farming systems to evolve by providing genetic resources including edible plants, crop species, and livestock species.[21] However, diversities of crops and livestock have been decreasing rapidly since the introduction of monocultures of high-yielding cultivars and industrialization.[22] Insect and plant biodiversity in an agroecosystem provide multiple ecological services for agriculture. Examples include pollination,[23,24] insect pest management by intercropping,[25] and disease control by mixed planting of multiple cultivars.[26] Soil biodiversity, even though perhaps fewer than 10% of the species have been identified, is vital to soil fertility, decomposition of organic matter, soil structure, and soil health.[27,28] With higher diversity, there is greater microhabitat differentiation; there are more opportunities for coexistence and beneficial interference between species; there are more possible kinds of beneficial interactions between herbivores and their predators; there is more efficient use of resources of soil, water, and light; and there is reduced risk for the farmer. Examples of farming practices that can enhance agrobiodiversity and agroecosystem sustainability are listed in Table 1. Studies show that species and taxonomic groups respond to varying degrees to environmentally friendly management.[29,30]

Bioindicators

Heink and Kowarik[31] proposed an all-encompassing definition of indicators: "An indicator in ecology and environmental planning is a component or a measure of environmentally relevant phenomena used to depict or evaluate environmental conditions or changes or to set environmental goals. Environmentally relevant phenomena are pressures, states, and responses as defined by OECD (2003)"[32] (see Figure 1).

Bioindicators are based on biota that serve as indicators of the quality of the environment, the biotic component, or human impacts within an ecosystem.[34] The concept of bioindicators has been around for a long time—for example, the use of canaries to detect deadly carbon monoxide and methane gas buildup in mines has a long history. In the United Kingdom, it had been used since 1911 until it was completely replaced with gas detectors in 1986.[35] Other early uses of bioindicators include aquatic organisms to evaluate water quality in the United States[36] and the melanic form of moths to detect air pollution in the United Kingdom.[37,38] The number of publications about environmental indicators including bioindicators has been increasing since the 1980s.[39] The inaugural issues of the journals *Ecological Indicators* and *Environmental Bioindicators* (changed to *Environmental Indicators* in 2010) were published in 2001 and 2006, respectively. At the 2002 Johannesburg World Summit on Sustainable Development, representatives from 190 countries committed to achieving a significant reduction of the

TABLE 1 Farming Practices That Can Enhance Agrobiodiversity and Agroecosystem Sustainability

Habitat diversification	
Spatial	Temporal
Intercropping	Rotations
Trap crops	Fallow
Hedgerows	Cover crops
Shelterbelts	
Windbreaks	
Agroforestry	
Mosaic landscape	
Organic amendment applications	
Compost, organic mulch	
Green manure	
Conservation or minimum tillage	
Biological pest management	
No or reduced use of pesticides, fungicides, herbicides, and fumigants	
Use of beneficial insects	
Anaerobic soil disinfestation[a]	
Physical pest management	
Solarization	
Flooding	
Plant resistance	

[a] A biological method to control a range of soilborne pathogens using a principle of acid fermentation in anaerobic soil. See Rosskopf et al.[33] for more details.

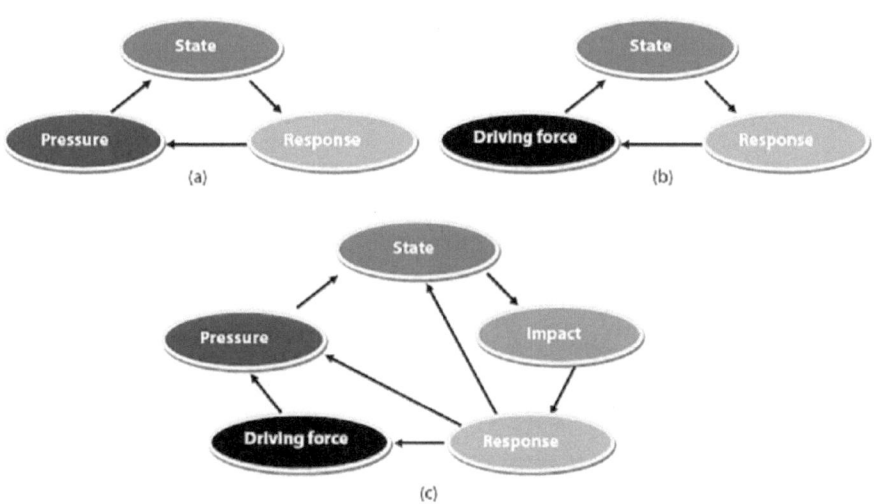

FIGURE 1 Varying causal chain frameworks. (a) PSR (pressure–state–response), (b) DSR (driver–state–response), and (c) DPSIR (driving force–pressure–state–impact–response). (Niemeijer and de Groot.[54])

current rate of biodiversity loss at the global, regional, and national levels by 2010.[40] Since then, the use of bioindicators for evaluating the status and functions of ecosystems and sustainability of agroecosystems has risen considerably,[41,42] yet the majority of recent studies of bioindicators has focused on the detection of pollution.[39]

Successful bioindicators should not only have biological and methodological relevance but also societal relevance because indicators require long-term monitoring to separate real change from natural fluctuations. Features of bioindicators are listed in Table 2. Inventory and classification are the foundations for developing indicators.[43] Developing an indicator then involves several steps: define objectives; determine end uses; construct indicators; determine norms and thresholds; and test sensitivity, probability, and usefulness.[44] Compared to abiotic indicators, however, developing a bioindicator has greater challenges for several reasons. First, biotic parameters are highly variable both temporally and spatially. There is a paucity of information on temporal and spatial variability of most natural species populations. Second, due to high variability, it costs more to collect additional replications to achieve the same statistical power in data. Standardized sampling methods for bioindicators are often lacking. Databases for bioindicators are critically limited especially in developing countries[45] although number of such studies are increasing.[46,47] Finally, it is more difficult with bioindicators to define background levels, norms, and thresholds.[42]

Nevertheless, demand for bioindicators as evaluative tools at diverse levels is increasing. Some of these are policy driven and others are market driven. For example, international organizations, such as the Organisation for Economic Co-operation and Development (OECD)[48–50] and the

TABLE 2 Features of Bioindicators for Environmental and Ecological Health Assessment

Biological relevance	• Provides early warning
	• Exhibits changes in response to stress
	• Changes are measured on appropriate time scale
	• Intensity of changes related to intensity of stressors
	• Change occurs when effect is real
	• Changes are biologically important and occur early enough to prevent catastrophic effects
	• Changes can be attributed to a cause
	• Changes indicate effects on ecosystem services
	• Can be used as sentinels for humans
Methodological relevance	• Easy to use in the field
	• Can be used by nonspecialists
	• Easy to analyze and interpret data
	• Measures what it is supposed to measure
	• Useful to test management questions
	• Can be used for hypothesis testing
	• Can be conducted in a reasonable time
	• Does not require expensive or complicated equipment
	• Easily repeatable with little training
Societal relevance	• Of interest to the public
	• Of interest to regulators and public policy makers
	• Easily understood by the public
	• Methods transparent to the public
	• Measures related to environment, ecological integrity, and human health
	• Cost-effective
	• Adds measurably to other indicators
	• Complements other indicators

Source: Modified from Burger.[34]

European Union (EU),[51] promote development of standardized bioindicators as a policy to compare biodiversity worldwide. Among commercial sectors, bioindicators have been used to certify "environmentally sound" products, such as migratory birds for shade-grown coffee.[52] Ecologists are using bioindicators and various other indicators at different levels of organization to evaluate the health of ecosystems (Table 3) and make recommendations for future management.[53,54] Bioindicators are often integrated into the frameworks of sustainability indicators of agroecosystems.[55,56]

TABLE 3 Usefulness of Indicators at Different Biological Levels of Organization to Ecological Health

Ecological Level	Type of Indicator	Ecological Health
Individual	Contaminant levels	Used to evaluate health of individuals
	Lesions	For evaluation of risk to higher-level consumers
	Disease	As an indicator of health of its foods, including prey
	Tumors	
	Infertility	
	Growth	
	Longevity	
	Reproduction	
	Age of reproduction	
	Hormonal balance	
	Proper development and maturation	
Population	Reproductive rates	Used to evaluate health of populations of species, particularly endangered or threatened species
	Growth rates	
	Survival rates	For comparison among populations
	Movements	For temporal comparisons
	Population genetics related to the breadth of the gene bank	Sources of resistance and pressure of natural selection
Community	Foraging guilds	Measures health of species using the same niche, such as colonial birds nesting in a colony or foraging animals such as dolphins and tuna
	Breeding guilds (groups of related species)	
	Predator–prey interactions	Indicates relationship among different species within guilds or assemblages
	Competitive interactions	
	Pathogen–host, pest–host relationships	For spatial and temporal comparisons
	Species richness	For evaluating efficacy of management options
Ecosystem	Decomposition rates	Measure changes in relative presence of species, how fast nutrients and energy will become available, how fast nutrients in soil will no longer be available, how much photosynthesis is occurring
	Erosion rates	
	Primary productivity	
	Energy transfer	Examines overall structure of the ecosystem in terms of the relationships among trophic levels
	Nutrient flow	
	Relationship among different trophic levels	For evaluating efficacy of management options
	Biomass	
	Energy flow	
Landscape	Relative amounts of different habitats	Measures dispersion of different habitat types, indicates relative species diversity values
	Patch size	Measures the difference among habitats
	Corridors between habitat types or different ecosystems	Measures distribution of corridors and refugia within the landscape
	The extent of uniform genetics	Also can measure the relationship between developed and natural areas for evaluating the importance of specific ecosystems within the landscape

Source: Modified from Burger.[34]

Bioindicator Development in European Union Agroecosystems

The importance of agro-environmental indicators has been highlighted by the EU[51] where current agricultural policies aim to increase multifunctionality of agricultural production. Intensive studies on bioindicators have been conducted in European countries. Bioindicators demonstrated to be sensitive to farm management intensities in European and some other agroecosystems are listed in Table 4. Generally, it is observed among invertebrate species that with less intensive management, there are more specialists and less generalists (as a result of ecological succession), greater biodiversity, and greater resilience.[42] An example of a bioindicator based on these correlations is European spiders; habitat preferences of spiders, particularly the ratio of "pioneer species (mostly Linyphiidae)" versus "wolf spiders (Araneae: Lycosidae)," can be a sensitive indicator for the assessment of farming intensity.[56] Many bioindicators listed in Table 4, however, have critical use limitations due to technically complex sampling methods and greater temporal and spatial variability. Although special instruments are required, recent advancements in molecular techniques may make some highly sensitive bioindicators such as nematodes much more accessible.[57,58]

TABLE 4   Examples of Potential Bioindicators for Sustainability of Farming Practices in European and Other Agroecosystems

Bioindicator	Parameter	Comments	References
Arthropods			
Ground beetles (Carabidae)	Abundance	• Sensitive to management intensity but needs intensive data collection	[91–96]
Spiders (Araneae)	Habitat preferences Percent pioneer species	• Highly sensitive to management intensity and database is available on ecological characteristics of central European spiders	[97–105]
Hoverflies (Syrphidae)	Percent stenotopic species	• Diversity of landscape structure adjacent to the field enhances species numbers	[106–111]
Pollinators	Individuals Populations Ecological guilds	• Environmental stress brought about by pesticides and habitat modification reduce pollinators	[112–114]
Arthropod community	Abundance of key species	• Diverse arthropod communities are affected by landscape diversity and environmental stress	[98,115–116]
Soil fauna			
Ants	Diversity Community composition	• Good indicator for rangeland monitoring and land-use changes	[117–122]
Earthworm	Biomass Species number Ecological guilds	• Suitable indicator for soil structure or compaction, tillage practice, pesticides, and other toxic chemicals	[123–131]
Collembola Protozoa Nematode Micro-arthropods Mites Soil animal	Physiotype Biodiversity Trophic index Maturity index Food web	• Highly sensitive to management intensity but time consuming and special skills are required for identification. Recent advancement in molecular approach makes these more accessible	[58,132–157]
Soil microbiota			
Soil enzymes (e.g., glucosinases for cellulose decomposition, phenol oxidases for lignin decomposition)	Activities	• Moderately sensitive to management intensity and relatively easy to measure but moderately variable both temporally and spatially	[58,158–160]

(Continued)

TABLE 4 (*Continued*) Examples of Potential Bioindicators for Sustainability of Farming Practices in European and Other Agroecosystems

Bioindicator	Parameter	Comments	References
Soil protists (e.g., algae, amoeba, diatom)	Species diversity Photosynthetic activity	• Less studied but sensitive to tillage, pesticides, and fertilizers	[161–163]
Microbial communities	Composition Functional diversity PLFA[a] profiles qPCR[b]	• Moderately sensitive to management intensity, but special skills and equipment are required and difficult to interpret	[58,164–166]
Functional groups	Mycorrhizae Nitrification Root pathogens	• Highly sensitive to management intensity, but special skills and instruments are required. DNA-based approaches (e.g., TRFLP[c], qPCR) became popular	[58,167–171]
Microbial activity	Soil respiration Mineralization Multiple substrate-induced respiration (MSIR)	• Relatively easy to measure but highly variable both temporally and spatially	[58,159,160,171]
Microbial biomass	C, N, and P biomass	• Relatively easy to measure but highly variable both temporally and spatially	[159,160]
		Birds	
Terrestrial birds Farmland birds	Species abundance of focal species and taxonomic composition at landscape scales	• Bird communities can be used to evaluate the effect of agriculture on surrounding ecosystems and hydrology and to explore sustainable land-use scenarios at regional scales	[62,172–175]
		Plants	
Higher plants	Numbers of "characteristic" species, weeds, functional groups, and endangered species Cover of litter in vegetation Diversity Evenness indices Habitat age	• Capable of being integrated into sophisticated floristic diversity at the habitat scale but requires intensive data collection	[176–180]

[a] PLFA, phospholipid fatty acid.
[b] qPCR, quantitative polymerase chain reaction.
[c] TRFLP, terminal restriction fragment length polymorphism.

To practically implement agro-environmental policy in the EU, efforts have been made to develop relatively easy-to-measure surrogate indicators (e.g., length of borders, farm size, area managed with organic farming).[42] Another practical bioindicator is a list of indicator plant species to evaluate species richness of a farm. A total of 28 indicator flower species for meadows and pastures that can be easily identified by local farmers were selected in Baden-Württemberg, Germany. Agro-environmental payments are granted to farms that have at least 4 of these 28 indicator species in all of the meadows and pastures on the farm.[59] In designing more efficient agro-environmental schemes, advantages of result-oriented remuneration (e.g., payment towards species-rich meadows) over action-oriented remuneration (e.g., payment towards manuring and mowing once per year) have been discussed.[60] It has been further recognized that the preservation of biodiversity is only possible through the (re)establishment of a mosaic of habitat patches at the landscape level. To meet this need, Geographical Information System

(GIS)-based landscape-oriented indicators have been examined.[61] GIS approaches are powerful tools for special analysis of bioindicators such as studies on correlations of bioindicator distributions (e.g., birds) and environmental and social factors at a landscape level.[62] Landscape context can have significant effect on biodiversity.[63,64]

To select objective, broad-scale, and unbiased indicators, conceptual frameworks for agroenvironmental indicators have been proposed. Varying causal chain frameworks such as pressure–state–response (PSR), driving force–state–response (DSR),[65] and driving force–pressure–state–impact–response (DPSIR)[66] have been developed (Figure 1). Enhanced DPSIR (eDPSIR) was proposed as a way to provide improved conceptual guidance in indicator selection based upon the DPSIR approach, systems analysis, and causal networks.[54] Sustainability Assessment of Farming and the Environment (SAFE), a hierarchical framework for assessing the sustainability of agricultural systems in Belgium, was created as a consistent and comprehensive framework of principles, criteria, and indicators.[55] These frameworks integrate bioindicators as a component of sustainability indicators of agroecosystems.

Biodiversity has multiple dimensions, and it is difficult to measure by a single indicator. Feest et al. proposed the "biodiversity quality" approach using multiple indicators (e.g., species richness, evenness/dominance, density/population, relative biomass, and species conservation value index) in combination to assess the balance between a range of indices and their relative magnitude.[67,68] Four different types of indicators of biodiversity change were suggested as a tool to facilitate indicator development (Figure 2).[69] With a goal of stable or increasing populations in all species associated with agricultural landscapes, Butler et al. developed a cross-taxonomic index for quantifying the health of farmland biodiversity by which the detrimental impacts of agricultural change to a broad range of taxonomic groupings can be assessed.[70]

For nationwide soil monitoring in the United Kingdom, a framework in selecting soil bioindicators for balancing scientific and technical opinion to assist policy development was established.[58]

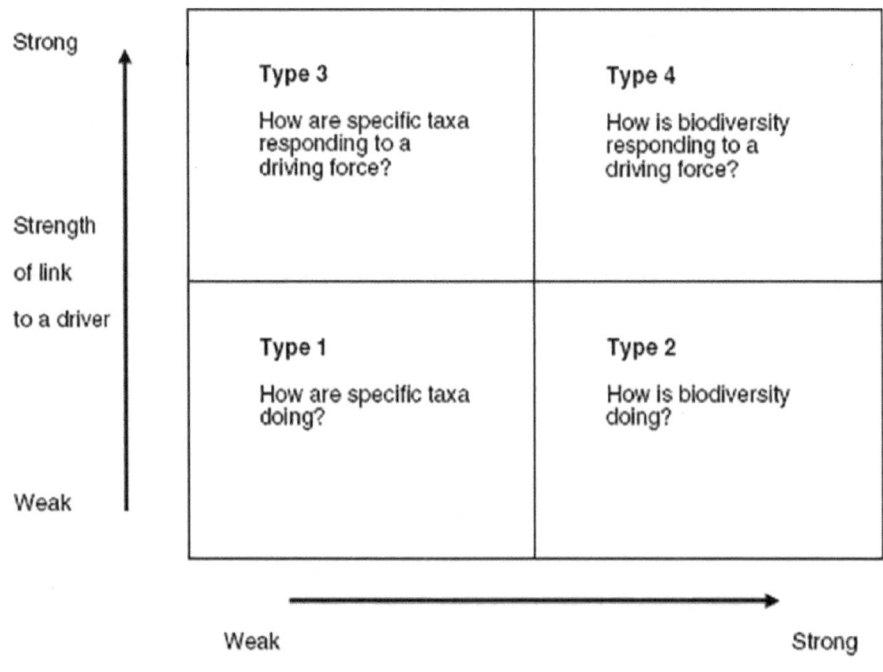

FIGURE 2   Typology of biodiversity change indicators. (van Strien et al.[69])

This semi-objective approach using "logical sieve" yielded 17 bioindicators that cover a range of genotypic-, phenotypic-, and functional-based indicators for different trophic groups out of 183 candidate bioindicators. This framework allows transparency in the decision-making process in selecting soil bioindicators as well as flexibility in including other indicators depending on priorities of the monitoring. The need for unambiguous and broad definitions of terms such as *indicators* and *biodiversity* was addressed for improving communication among interdisciplinary researchers, policy makers, and stakeholders.[71] To make indicators relevant to potential users and have long-term public support, the necessity of stakeholder participation in developing ecological indicators was also emphasized.[72]

Bioindicator Development in Shaded Coffee in Latin America

Coffee is the second most traded commodity in the world after petroleum and forms the principal economic activity of more than 20 million people in farming communities throughout much of the developing world.[73] Traditionally grown in the understory of forest cover or planted shade trees, throughout the 1970s and 1980s, many coffee farmers adopted more modern production practices, planting higher-yielding varieties in full sun, eliminating shade trees, and increasing pesticide and fertilizer applications.[74] The loss of shade cover, and associated biodiversity, has led to many environmental problems such as soil erosion, loss of water capture and recharge ability, and contamination from the excessive fertilizer and pesticide use associated with sun-grown coffee.

Recent research has documented the high levels of agrobiodiversity and ecosystem services (such as pollination, soil conservation, and natural pest control) associated with diverse shade coffee production.[75,76] From this research, several very important bioindicators are being developed. The number of tree species in the shade canopy is an important indicator of both conservation potential of coffee plantations and increased options for farmer livelihood alternatives.[77–79] The species richness of ants and insect feeding birds are greater in shade coffee than in sun coffee, especially ground-foraging ants that act as important predators and birds that are bark gleaners and leaf surface foragers. It also appears that the higher diversity of predaceous ants restricts the development of pest ants such as the fire ant *Solenopsis* sp. (Hymenoptera), a very common pest in open landscapes and sun-grown coffee. Associated species such as orchids in the shade tree canopy can also be important bioindicators. Local farmers can be trained to recognize orchids, demonstrating how the development of bioindicators of sustainability can be an accessible and local methodology that adds value to more sustainable farming practices. Orchids have an intrinsic value from a conservation perspective and, at the same time in this region of northern Nicaragua, have added an attractive value for an emerging agro-ecotourism industry associated with shade-grown coffee landscapes.[80]

Aquatic Organisms as Bioindicators

Recent studies showed that a range of aquatic organisms that inhabit water bodies in agroecosystems or in watersheds including agroecosystems can be used as bioindicators for evaluating sustainability of agroecosystems or their impacts on the environment. A study in organic and conventional paddy fields in Costa Rica found that some aquatic macroinvertebrates (*Baetis* sp., *Fallceon* sp., *Leptohyphes* sp., *Tricorythodes* sp., *Farrodes* sp., *Phyllogomphoides* sp., *Hydroptila* sp., *Mayatrichia* sp., *Neotrichia* sp., *Oxyethira* sp., *Nectopsyche* sp.1, *Nectopsyche* sp.2, and *Oecetis* sp.) were found only in organic paddy fields and can be used as sensitive bioindicators of water quality in these agroecosystems.[81]

Rhinella arenarum, a toad species, inhabiting ponds associated with agroecosystems, showed a smaller egg diameter, a higher frequency of abnormal eggs, and a higher mortality of oviposition compared to those inhabiting a reference pond without agricultural influences.[82] The δ15N signature of large, long-lived, freshwater mussels (Mollusca: Bivalvia: Unionidae) was successfully used as an integrated bioassessment tool for tracking agricultural nitrogen inputs into watersheds.[83]

Improving Awareness of Farmers and Consumers

Often, farmers themselves are not fully aware of the biodiversity in their farms. Moreover, to economically sustain environmentally friendly farming practices, recognition and support from consumers are necessary. To implement sustainable agricultural policy, therefore, improved awareness of both farmers and consumers on agrobiodiversity is required.

An example of such activity is the participatory biodiversity inventory in paddy rice ecosystems in Japan, where paddy fields occupy ~50% of the cultivated lands of the country. Japanese paddy agroecosystems contain a diverse group of organisms: Recent inventory found 5668 species including birds (189 spp.), fish (143 spp.), reptiles and amphibian (61 spp.), arthropods (1726 spp.), and plants (2075 spp.).[84] Since the 1960s, however, many native species inhabiting paddy fields have decreased in abundance mainly due to nontarget effects of pesticide applications and the construction of concrete ditches. Consequently, 3 mammal species, 15 insect species, 5 bird species, 27 species of fish, 3 amphibian species, 3 mollusks, and 26 plant species are listed as endangered species inhabiting paddy fields.[84] Nongovernmental organizations (NGOs), scientists, farmers, the general public, and administrators collaborated to create nationwide inventories of biodiversity in paddy agroecosystems and proposed a broad range of bioindicators (237 animal species[85] and 223 plant species[86]) to enhance the awareness of both farmers and consumers on paddy field biodiversity (Figure 3).

FIGURE 3   A portion of Japanese poster for improving farmer and consumer awareness of agrobiodiversity in paddy fields. Central figure indicates a bowl of rice being equivalent to 3000–4000 grains of rice, 3 stubs of rice plants, and 35 tadpoles in a paddy. This can be interpreted as eating a bowl of rice supports 35 tadpoles in a paddy. Numbers of species are averages from the national survey of agrobiodiversity in paddy fields in Japan. Clockwise from the upper left, the species listed are *Daphnia pulex* (water flea), *Pantala flavescens* (Globe Skimmer), *Sympetrum frequens* (Autumn Darter), *Triops longicaudatus* (longtail tadpole shrimp), *Branchinella kugenumaensis* (fairy shrimp), *Hyla japonica* (Japanese tree frog), *Rhabdophis tigrinus* (tiger keelback), *Misgurnus anguillicaudatus* (Dojo Loach), *Oryzias latipes* (Medaka), *Rana nigromaculata* (Dark-spotted frog), *Rana rugosa* (Japanese wrinkled frog), and *Ceriagrion melanurum* (damselfly). (Une.[181])

Conclusions

Indicators represent a compromise between scientific knowledge of the moment and simplicity of use.[44] Compared to assessment systems for natural ecosystems,[43,87] the current status of developing bioindicators for sustainability of farming practices appears to still be in its early stages but is making good progress as seen in increased number of studies from places outside of Europe. For using bioindicators to implement sustainable agricultural policy, however, we need not only more research on the science of bioindicators but also better awareness on agrobiodiversity among farmers and consumers.

Future studies on bioindicators for sustainable farming practices should address the following: (1) the importance of stability and reproductive potential of not only pest and beneficial species but also other species typical of agroecosystems;[42,70,88] (2) standardization of sampling methods;[42] (3) expansion of databases and improvement of the statistical techniques to minimize potential bias of indicators;[68,89] (4) improved understanding of the mechanisms linking the status of biodiversity, Earth system processes, human decisions, and ecosystem services impacting human welfare through interdisciplinary studies;[52,89] (5) use of participatory approaches in the processes of developing bioindicators;[72,90] (6) development of multiple sets of bioindicators tailored to different end users (general public, farmers, policy makers, and scientists);[42] and (7) constructing a hierarchical system that integrates different types of bioindicators and ecological, social, cultural, and economic indicators to evaluate sustainability of farming practices.[55].

References

1. Pretty, J. Agricultural sustainability: Concepts, principles and evidence. *Philos. Trans. R. Soc., B* **2008**, *363* (1491), 447–465.
2. Gliessman, S.R. *Agroecology: The Ecology of Sustainable Food Systems*, 3rd ed.; CRC Press/Taylor and Francis: Boca Raton, FL, 2015.
3. Green, R.E.; Cornell, S.J.; Scharlemann, J.P.W.; Balmford, A. Farming and the fate of wild nature. *Science* **2005**, *307* (5709), 550–555.
4. Tscharntke, T.; Klein, A.M.; Kruess, A.; Steffan-Dewenter, I.; Thies, C. Landscape perspectives on agricultural intensification and biodiversity—Ecosystem service management. *Ecol. Lett.* **2005**, *8* (8), 857–874.
5. International Federation of Organic Agriculture Movements (IFOAM). *Towards a Sustainable Agriculture: Papers*; Wirz: Aarau, Switzerland, 1978.
6. Douglass, G.K. *Agricultural Sustainability in a Changing World Order*; Westview Press: Boulder, CO, 1984.
7. Jackson, W.; Berry, W.; Colman, B. *Meeting the Expectations of the Land: Essays in Sustainable Agriculture and Stewardship*; North Point Press: San Francisco, CA, 1984.
8. Dover, M.J.; Talbot, L.M. *To Feed the Earth: Agro-ecology for Sustainable Development*; World Resources Institute: Washington, DC, 1987.
9. Altieri, M.A. Beyond agroecology: Making sustainable agriculture part of a political agenda. *Am. J. Altern. Agric.* **1988**, *3* (4), 142–143.
10. Andow, D.A.; Hidaka, K. Experimental natural history of sustainable agriculture: Syndromes of production. *Agric. Ecosyst. Environ.* **1989**, *27*, 447–462.
11. MacRae, R.J.; Hill, S.B.; Henning, J.; Mehuys, G.R. Agricultural science and sustainable agriculture—A review of the existing scientific barriers to sustainable food-production and potential solutions. *Biol. Agric. Hort.* **1989**, *6* (3), 173–219.
12. Soule, J.D.; Piper, J.K. *Farming in Nature's Image: An Ecological Approach to Agriculture*; Island Press: Washington, DC, 1992.
13. Liverman, D.M.; Hanson, M.E.; Brown, B.J.; Merideth, R.W.J. Global sustainability toward measurement. *Environ. Manage.* **1988**, *12* (2), 133–144.

14. Jansen, D.M.; Stoorvogel, J.J.; Schipper, R.A. Using sustainability indicators in agricultural land use analysis: An example from Costa Rica. *Neth. J. Agric. Sci.* **1995**, *43* (1), 61–82.

15. Food Agriculture Organization of the United Nations (FAO). Land Quality Indicators and Their Use in Sustainable Agriculture and Rural Development: Proceedings of the Workshop; FAO: Rome, Italy, 1997.

16. MAFF U.K. *Toward Sustainable Agriculture: A Pilot Set of Indicators*; MAFF Publications: London, UK, 2000.

17. Bell, S.; Morse, S. Sustainability Indicator: Measuring the Immeasurable, 2nd Ed.; Earthscan: London, UK, 2008.

18. Spedding, C.R.W. Agriculture and the Citizen, 1st Ed.; Chapman and Hall: London, UK and New York, 1996.

19. Ikerd, J.E. Rethinking the first principles of agroecology: Ecological, social, and economic. In Sustainable Agroecosystem Management: Integrating Ecology, Economics, and Society; Bohlen, P.J., House G.J., Eds.; CRC Press/Taylor & Francis: Boca Raton, FL, 2009; 41–52.

20. Edwards, C.A. Sustainable agricultural practices. In Encyclopedia of Pest Management; Pimentel, D., Ed.; Marcel Dekker/Taylor & Francis: Boca Raton, FL, 2002; 812–814.

21. Hillel, D.; Rosenzweig, C. The role of biodiversity in agronomy and agroecosystem management in the coming century. In Sustainable Agroecosystem Management: Integrating Ecology, Economics, and Society; Bohlen, P.J., House, G.J., Eds.; CRC Press/Taylor & Francis: Boca Raton, FL, 2009; 167–193.

22. Thrupp, L.A. The importance of biodiversity in agroecosystems. *J. Crop Improv.* 2004, *11* (1/2), 315–338.

23. Kearns, C.A.; Inouye, D.W.; Waser, N.M. Endangered mutualisms: The conservation of plant–pollinator interactions. *Annu. Rev. Ecol. Syst.* **1998**, *29*, 83–112.

24. Buchmann, S.L.; Nabhan, G.P. *The Forgotten Pollinators*; Island Press/Shearwater Books: Washington, DC, 1996.

25. Ogol, C.; Spence, J.R.; Keddie, A. Maize stem borer colonization, establishment and crop damage levels in a maize-leucaena agroforestry system in Kenya. *Agric. Ecosyst. Environ.* **1999**, *76* (1), 1–15.

26. Zhu, Y.Y.; Chen, H.R.; Fan, J.H.; Wang, Y.Y.; Li, Y.; Chen, J.B.; Fan, J.X.; Yang, S.S.; Hu, L.P.; Leung, H.; Mew, T.W.; Teng, P.S.; Wang, Z.H.; Mundt, C.C. Genetic diversity and disease control in rice. *Nature* **2000**, *406* (6797), 718–722.

27. Pankhurst, C.E.; Doube, B.M.; Gupta, V.V.S.R., Eds. *Biological Indicators of Soil Health;* CAB International: Wallingford, UK, 1997.

28. Orgiazzi, A., Bardgett, R. D., Barrios, E., Behan-Pelletier, V., Briones, M. J. I., Chotte, J.-L., De Deyn, G. B., Eggleton, P., Fierer, N., Fraser, T., Hedlund, K., Jeffery, S., Johnson, N. C., Jones, A., Kandeler, E., Kaneko, N., Lavelle, P., Lemanceau, P., Miko, L., Montanarella, L., Moreira, F. M. S., Ramirez, K. S., Scheu, S., Singh, B. K., Six, J., van der Putten, W. H., and Wall, D. H. Eds. Global *Soil Biodiversity Atlas*; Publications Office of the European Union: Luxembourg, 2016.

29. Critchley, C.N.R.; Fowbert, J.A.; Sherwood, A.J.; Pywell, R.F. Vegetation development of sown grass margins in arable fields under a countrywide agri-environment scheme. *Biol. Conserv.* **2006**, *132* (1), 1–11.

30. Marshall, E.J.P.; West, T.M.; Kleijn, D. Impacts of an agrienvironment field margin prescription on the flora and fauna of arable farmland in different landscapes. *Agric. Ecosyst. Environ.* **2006**, *113* (1–4), 36–44.

31. Heink, U.; Kowarik, I. What are indicators? On the definition of indicators in ecology and environmental planning. *Ecol. Indic.* **2010**, *10* (3), 584–593.

32. Organisation for Economic Co-operation and Development (OECD). *Core Environmental Indicators: Development, Measurement, and Use*; OECD: Paris, France, 2003.

33. Rosskopf, E., Serrano-Pe´rez, P., Hong, J., Shrestha, U., del Carmen Rodrı´guez-Molina, M., Martin, K., Kokalis-Burelle, N., Shennan, C., Muramoto, J., and Butler, D. Anaerobic soil disinfestation and soilborne pest management. In *Organic Amendments and Soil Suppressiveness in Plant Disease Management*; Meghavansi, M. K., Varma, A., Eds.; Springer International Publishing: Switzerland, 2015; 277–305.

34. Burger, J. Bioindicators: Types, development, and use in ecological assessment and research. *Environ. Bioindic.* **2006**, *1* (1), 22–39.

35. BBC. 1986: Coal Mine Canaries Made Redundant. On This Day: 30th of December, available at http://news.bbc.co.uk/onthisday/hi/dates/stories/december/30/newsid_2547000/2547587.stm (accessed June 2019).

36. Richardson, R.E. The bottom fauna of the middle Illinois river, 1913–1925. *Ill. Nat. Hist. Surv. Bull.* **1928**, *17*, 387–475.

37. Kettlewell, H.B.D. How industrialisation can alter species. *Discovery* **1955**, *16* (12), 507–511.

38. Heck, W.W. The use of plants as indicators of air pollution. *Air Water Pollut.* **1966**, *10* (2), 99–111.

39. Burger, J. Bioindicators: A review of their use in the environmental literature 1970–2005. *Environ. Bioindic.* **2006**, *1* (2), 136–144.

40. United Nations Environment Programme (UNEP). Decisions Adopted by the Conference of the Parties to the Convention on Biological Diversity at Its Sixth Meeting, The Hague, 7–19 April 2002.

41. Paoletti, M.G. Using bioindicators based on biodiversity to assess landscape sustainability. *Agric. Ecosyst. Environ.* 1999, 74 (1–3), 1–18.

42. Buchs, W. Biodiversity and agri-environmental indicators—General scopes and skills with special reference to the habitat level. *Agric. Ecosyst. Environ.* **2003**, *98* (1–3), 35–78.

43. Innis, S.A.; Naiman, R.J.; Elliott, S.R. Indicators and assessment methods for measuring the ecological integrity of semi-aquatic terrestrial environments. *Hydrobiologia* 2000, 422–423, 111–131.

44. Girardin, P.; Bockstaller, C.; Van der Werf, H. Indicators: Tools to evaluate the environmental impacts of farming systems. *J. Sustainable Agric.* **1999**, *13* (4), 5–21.

45. Dudley, N.; Baldock, D.; Nasi, R.; Stolton, S. Measuring biodiversity and sustainable management in forests and agricultural landscapes. *Philos. Trans. R. Soc., B* **2005**, *360* (1454), 457–470.

46. Diame, L., Blatrix, R., Grechi, I., Rey, J. Y., Sane, C. A. B., Vayssieres, J. F., de Bon, H., and Diarra, K. Relations between the design and management of Senegalese orchards and ant diversity and community composition. *Agric. Ecosys. Environ.* **2015**, *212*, 94–105.

47. Rousseau, L., Fonte, S. J., Tellez, O., van der Hoek, R., and Lavelle, P. Soil macrofauna as indicators of soil quality and land use impacts in smallholder agroecosystems of western Nicaragua. *Ecol. Indic.* **2013**, *27*, 71–82.

48. Neher, D. Ecological sustainability in agricultural systems: Definition and measurement. *J. Sustainable Agric.* **1992**, *2*, 51–61. https://doi.org/10.1300/J064v02n03_05

49. Organisation for Economic Co-operation and Development (OECD). *Environmental Indicators for Agriculture: Methods and Results.* OECD Publishing: Paris, France, 2001; Vol. 3.

50. Organisation for Economic Co-operation Development (OECD). Agriculture and Biodiversity: Developing Indicators for Policy Analysis: Proceedings from an OECD Expert Meeting, Zurich, Switzerland. OECD Publishing: Paris, France, 2001.

51. The Commission of the European Communities. Evaluation of Agri-environment Programmes. State of Application of Regulation. (EEC) no. 2078/92. DGVI Commission Working Document VI/7655/98, 1998.

52. Mas, A.H.; Dietsch, T.V. Linking shade coffee certification to biodiversity conservation: Butterflies and birds in Chiapas, Mexico. *Ecol. Appl.* **2004**, *14* (3), 642–654.

53. The Millennium Assessment Board. The Millennium Ecosystem Assessment, 2005.

54. Niemeijer, D.; de Groot, R.S. A conceptual framework for selecting environmental indicator sets. *Ecol. Indic.* **2008**, *8* (1), 14–25.

55. Van Cauwenbergh, N.; Biala, K.; Bielders, C.; Brouckaert, V.; Franchois, L.; Cidad, V.G.; Hermy, M.; Mathijs, E.; Muys, B.; Reijnders, J.; Sauvenier, X.; Valckx, J.; Vanclooster, M.; Van der Veken, B.; Wauters, E.; Peeters, A. SAFE—A hierarchical framework for assessing the sustainability of agricultural systems. *Agric. Ecosyst. Environ.* 2007, 120 (2–4), 229–242.

56. Buchs, W.; Harenberg, A.; Zimmermann, J.; Wei, B. Biodiversity, the ultimate agri-environmental indicator? Potential and limits for the application of faunistic elements as gradual indicators in agroecosystems. *Agric. Ecosyst. Environ.* **2003**, *98* (1–3), 99–123.

57. Donn, S.; Griffiths, B.S.; Neilson, R.; Daniell, T.J. DNA extraction from soil nematodes for multi-sample community studies. *Appl. Soil Ecol.* **2008**, *38* (1), 20–26.

58. Ritz, K.; Black, H.I.J.; Campbell, C.D.; Harris, J.A.; Wood, C. Selecting biological indicators for monitoring soils: A framework for balancing scientific and technical opinion to assist policy development. *Ecol. Indic.* **2009**, *9* (6), 1212–1221.

59. Oppermann, R. Nature balance scheme for farms-evaluation of the ecological situation. *Agric. Ecosyst. Environ.* **2003**, *98* (1–3), 463.

60. Matzdorf, B.; Kaiser, T.; Rohner, M.-S. Developing biodiversity indicator to design efficient agri-environmental schemes for extensively used grassland. *Ecol. Indic.* **2008**, *8* (3), 256–269.

61. Osinski, E. Operationalisation of a landscape-oriented indicator. *Agric. Ecosyst. Environ.* **2003**, *98* (1–3), 371.

62. Gottschalk, T.K.; Dittrich, R.; Diekötter, T.; Sheridan, P.; Wolters, V.; Ekschmitt, K. Modelling land-use sustainability using farmland birds as indicators. *Ecol. Indic.* **2010**, *10* (1), 15–23.

63. Merckx, T.; Feber, R.E.; Riordan, P.; Townsend, M.C.; Bourn, N.A.D.; Parsons, M.S.; Macdonald, D.W. Optimizing the biodiversity gain from agri-environment schemes. *Agric. Ecosyst. Environ.* 2009, 130 (3–4), 177–182.

64. Rundlof, M.; Bengtsson, J.; Smith, H.G. Local and landscape effects of organic farming on butterfly species richness and abundance. *J. Appl. Ecol.* **2008**, *45* (3), 813–820.

65. Organisation for Economic Co-operation and Development (OECD). *Environmental Indicators for Agriculture: Volume 1, Concepts and Frameworks*; OECD Publisher: Paris, France, 1999.

66. Smeets, E.; Weterings, R. *Environmental Indicators: Typology and Overview*; European Environmental Agency: Copenhagen, Denmark, 1999.

67. Feest, A.; Aldred, T.D.; Jedamzik, K. Biodiversity quality: A paradigm for biodiversity. *Ecol. Indic.* **2010**, *10* (6), 1077–1082.

68. Feest, A.; van Swaay, C.; van Hinsberg, A. Nitrogen deposition and the reduction of butterfly biodiversity quality in the Netherlands. *Ecol. Indic.* **2014**, *39*, 115–119.

69. van Strien, A.J.; van Duuren, L.; Foppen, R.P.B.; Soldaat, L.L. A typology of indicators of biodiversity change as a tool to make better indicators. *Ecol. Indic.* **2009**, *9* (6), 1041–1048.

70. Butler, S.J.; Brooks, D.; Feber, R.E.; Storkey, J.; Vickery, J.A.; Norris, K. A cross-taxonomic index for quantifying the health of farmland biodiversity. *J. Appl. Ecol.* **2009**, *46* (6), 1154–1162.

71. Sieg, C.H.; Flather, C.H. Applicability of Montreal Process Criterion 1—Conservation of biological diversity—to rangeland sustainability. *Int. J. Sustainable Dev. World Ecol.* **2000**, *7* (2), 81.

72. Turnhout, E.; Hisschemöller, M.; Eijsackers, H. Ecological indicators: Between the two fires of science and policy. *Ecol. Indic.* **2007**, *7* (2), 215–228.

73. Bacon, C.; Méndez, V.E.; Gliessman, S.R.; Goodman D.; Fox, J. *Confronting the Coffee Crisis: Fair Trade, Sustainable Livelihoods and Ecosystems in Mexico and Central America*; MIT University Press: Cambridge, MA, 2008; 390pp.

74. Moguel, P.; Toledo, V.M. Biodiversity conservation in traditional coffee systems of Mexico. *Conserv. Biol.* **1999**, *13*, 11–21.

75. Perfecto, I.; Vandermeer, J.H.; Wright, A. *Nature's Matrix: Linking Agriculture, Conservation, and Food Sovereignty*; Earthscan: New York, 2010; 237pp.

76. Coral-Acosta, N.; Perez-Torres, J. Diversity of diurnal butterflies (Lepidoptera: Papilionoidea) associated with a shade coffee agroecosystem (Curiti, Santander). *Revista Colombiana De Entomologia* **2017**, *43*, 91–99.

77. Gliessman, S.R.; Putnam, H.; Cohen, R. Chapter 8: Agroecology and participatory knowledge production and exchange as a basis for food system change: The case of the Community Agroecology Network. In *Agroecological Practices for Sustainable Agriculture: Principles, Applications, and Making the Transition*; Wezel, A., Ed.; World Scientific Publishing: London, UK, 2017; 201–228.

78. IPES-Food. *Breaking Away from Industrial Food and Farming Systems: Seven Case Studies of Agroecological Transition*; International Panel of Experts in Sustainable Food Systems: Brussels, Belgium, 2018; 109pp.

79. Mendez, V.E.; Gliessman, S.R.; Gilbert, G.S. Tree biodiversity in farmer cooperatives of a shade coffee landscape in western El Salvador. *Agric. Ecosyst. Environ.* 2007, 119 (1–2), 145–159.

80. Bacon, C.M. Confronting the coffee crisis: Nicaraguan farmers' use of cooperative, fair trade, and agroecological networks to negotiate livelihoods and sustainability. PhD Dissertation, Department of Environmental Studies. University of California Santa Cruz, California, 2005.

81. Rizo-Patron, F.; Kumar, A.; Colton, M. B. M.; Springer, M.; Trama, F. A. Macroinvertebrate communities as bioindicators of water quality in conventional and organic irrigated rice fields in Guanacaste, Costa Rica. *Ecol. Indic.* 2013, *29*, 68–78.

82. Babini, M.S.; Bionda, C.D.; Salinas, Z.A.; Salas, N.E.; Martino, A. L. Reproductive endpoints of Rhinella arenarum (Anura, Bufonidae): Populations that persist in agroecosystems and their use for the environmental health assessment. *Ecotoxic. Environ. Safety*, **2018**, *154*, 294–301.

83. Atkinson, C.L.; Christian, A.D.; Spooner, D.E.; Vaughn, C.C. Long-lived organisms provide an integrative footprint of agricultural land use. *Ecol. Appl.* **2014**, *24*, 375–384.

84. Kiritani, K.; Ed. *A Comprehensive List of Organisms Associated with Paddy Ecosystems in Japan*, Revised Ed.; Nou-to shizen-no kenkyujo, Seibutsu-tayousei-nougyou shien senta: Fukuoka, Tokyo, Japan, 2010.

85. Kiritani, K., Ed. *Bioindicators for Paddy Ecosystems*; Nou-to shizen-no kenkyujo, Seibutsu-tayousei-nougyou shien senta: Fukuoka, Tokyo, Japan, 2009 (in Japanese).

86. Ito, K.; Mineta, T., Eds. *Plant Bioindicators for Paddy Ecosystems*; Nou-to shizen-no kenkyujo, Seibutsu-tayousei-nougyou shien senta: Fukuoka, Tokyo, Japan, 2009 (in Japanese).

87. Karr, J.R.; Fausch, K.D.; Angermeier, P.L.; Yant, P.R.; Schlosser, I.J. *Assessing Biological Integrity in Running Waters: A Method and Its Rationale. Special Publication 5*; Illinois Natural History Survey: Urbana, IL, 1986.

88. Kiritani, K. Integrated biodiversity management in paddy fields: Shift of paradigm from IPM toward IBM. *Integr. Pest Manage. Rev.* **2000**, *5* (3), 175–183.

89. Balmford, A.; Crane, P.; Dobson, A.; Green, R.E.; Mace, G.M. The 2010 challenge: Data availability, information needs and extraterrestrial insights. *Philos. Trans. R. Soc., B* **2005**, *360* (1454), 221–228.

90. Burger, J. Stakeholder involvement in indicator selection: Case studies and levels of participation. *Environ. Bioindic.* **2009**, *4* (2), 170–190.

91. Irmler, U. The spatial and temporal pattern of carabid beetles on arable fields in northern Germany (SchleswigHolstein) and their value as ecological indicators. *Agric. Ecosyst. Environ.* **2003**, *98* (1–3), 141–151.

92. Heyer, W.; Huelsbergen, K.J.; Wittmann, C.; Papaja, S.; Christen, O. Field related organisms as possible indicators for evaluation of land use intensity. *Agric. Ecosyst. Environ.* **2003**, *98* (1–3), 453–461.

93. Doring, T.F.; Hiller, A.; Wehke, S.; Schulte, G.; Broll, G. Biotic indicators of carabid species richness on organically and conventionally managed arable fields. *Agric. Ecosyst. Environ.* **2003**, *98* (1–3), 133.

94. Cavaliere, F.; Brandmayr, P.; Giglio, A. DNA damage in haemocytes of Harpalus (Pseudophonus) rufipes (De Geer, 1774) (Coleoptera, Carabidae) as an indicator of sublethal effects of exposure to herbicides. *Ecol. Indic.* **2019**, *98*, 88–91.

95. Pizzolotto, R.; Mazzei, A.; Bonacci, T.; Scalercio, S.; Iannotta, N.; Brandmayr, P. Ground beetles in Mediterranean olive agroecosystems: Their significance and functional role as bioindicators (Coleoptera, Carabidae). *PLoS ONE*, **2018**, *13*, e0194551.

96. Magagula, C.N.; Nzima, B.A. Interaction between habitat characteristics and insect diversity using ground beetles (coleoptera: carabidae) and ants (hymenoptera: formicidae) within a variety of agricultural habitats. *Appl. Ecol. Environ. Res.* **2015**, *13*, 863–876.

97. Oberg, S. Influence of landscape structure and farming practice on body condition and fecundity of wolf spiders. *Basic Appl. Ecol.* **2009**, *10* (7), 614–621.

98. Fernandez, D.E.; Cichon, L.I.; Sanchez, E.E.; Garrido, S.A.; Gittins, C. Effect of different cover crops on the presence of arthropods in an organic apple (*Malus domestica* Borkh) orchard. *J. Sustainable Agric.* **2008**, *32* (2), 197–211.

99. Birkhofer, K.; Fliessbach, A.; Wise, D.H.; Scheu, S. Generalist predators in organically and conventionally managed grass-clover fields: Implications for conservation biological control. *Ann. Appl. Biol.* **2008**, *153* (2), 271–280.

100. Clough, Y.; Holzschuh, A.; Gabriel, D.; Purtauf, T.; Kleijn, D.; Kruess, A.; Steffan-Dewenter, I.; Tscharntke, T. Alpha and beta diversity of arthropods and plants in organically and conventionally managed wheat fields. *J. Appl. Ecol.* **2007**, *44* (4), 804–812.

101. Isaia, M.; Bona, F.; Badino, G. Influence of landscape diversity and agricultural practices on spider assemblage in Italian vineyards of Langa Astigiana (northwest Italy). *Environ. Entomol.* **2006**, *35* (2), 297–307.

102. Schmidt, M.H.; Roschewitz, I.; Thies, C.; Tscharntke, T. Differential effects of landscape and management on diversity and density of ground-dwelling farmland spiders. *J. Appl. Ecol.* **2005**, *42* (2), 281–287.

103. Hough-Goldstein, J.A.; Vangessel, M.J.; Wilson, A.P. Manipulation of weed communities to enhance ground-dwelling arthropod populations in herbicide-resistant field corn. *Environ. Entomol.* **2004**, *33* (3), 577–586.

104. Marc, P.; Canard, A.; Ysnel, F. Spiders (Araneae) useful for pest limitation and bioindication. *Agric. Ecosyst. Environ.* **1999**, *74* (1–3), 229–273.

105. Mazzia, C.; Pasquet, A.; Caro, G.; Thenard, J.; Cornic, J.F.; Hedde, M.; Capowiez, Y. The impact of management strategies in apple orchards on the structural and functional diversity of epigeal spiders. *Ecotoxicology* **2015**, *24*, 616–625.

106. Kleijn, D.; van Langevelde, F. Interacting effects of landscape context and habitat quality on flower visiting insects in agricultural landscapes. *Basic Appl. Ecol.* **2006**, *7* (3), 201–214.

107. Schweiger, O.; Maelfait, J.P.; Van Wingerden, W.; Hendrickx, F.; Billeter, R.; Speelmans, M.; Augenstein, I.; Aukema, B.; Aviron, S.; Bailey, D.; Bukacek, R.; Burel, F.; Diekotter, T.; Dirksen, J.; Frenzel, M.; Herzog, F.; Liira, J.; Roubalova, M.; Bugter, R. Quantifying the impact of environmental factors on arthropod communities in agricultural landscapes across organizational levels and spatial scales. *J. Appl. Ecol.* **2005**, *42* (6), 1129–1139.

108. Thomas, J.A. Monitoring change in the abundance and distribution of insects using butterflies and other indicator groups. *Philos. Trans. R. Soc., B* **2005**, *360* (1454), 339–357.

109. Henle, K.; Dziock, F.; Foeckler, F.; Follner, K.; Husing, V.; Hettrich, A.; Rink, M.; Stab, S.; Scholz, M. Study design for assessing species environment relationships and developing indicator systems for ecological changes in floodplains—The approach of the RIVA project. *Int. Rev. Hydrobiol.* **2006**, *91* (4), 292–313.

110. Dormann, C.F.; Schweiger, O.; Augenstein, I.; Bailey, D.; Billeter, R.; de Blust, G.; DeFilippi, R.; Frenzel, M.; Hendrickx, F.; Herzog, F.; Klotz, S.; Liira, J.; Maelfait, J.P.; Schmidt, T.; Speelmans, M.; van Wingerden, W.; Zobel, M. Effects of landscape structure and land-use intensity on similarity of plant and animal communities. *Global Ecol. Biogeogr.* **2007**, *16* (6), 774–787.

111. Villa, M.; Santos, S.A.P.; Marrao, R.; Pinheiro, L.A.; Lopez-Saez, J.A.; Mexia, A.; Bento, A.; Pereira, J.A. Syrphids feed on multiple patches in heterogeneous agricultural landscapes during the autumn season, a period of food scarcity. *Agric. Ecosyst. Environ.* **2016**, *233*, 262–269.

112. Kevan, P.G. Pollinators as bioindicators of the state of the environment: Species, activity and diversity. *Agric. Ecosyst. Environ.* 1999, 74 (1–3), 373–393.

113. Pe'er, G.; Settele, J. The rare butterfly *Tomares nesimachus* (Lycaenidae) as a bioindicator for pollination services and ecosystem functioning in northern Israel. *Isr. J. Ecol. Evol.* 2008, 54 (1), 111–136.

114. Matos, M.C.B.; Sousa-Souto, L.; Almeida, R.S.; Teodoro, A.V. Contrasting patterns of species richness and composition of solitary wasps and bees (Insecta: Hymenoptera) according to land-use. *Biotropica* 2013, 45, 73–79.

115. Paoletti, M.G.; Hu, D.X.; Marc, P.; Huang, N.X.; Wu, W.L.; Han, C.R.; He, J.H.; Cai, L.W. Arthropods as bioindicators in agroecosystems of Jiang Han Plain, Qianjiang City, Hubei China. *Crit. Rev. Plant Sci.* 1999, 18 (3), 457–465.

116. Vitanovic, E.; Ivezic, M.; Kacic, S.; Katalinic, M.; Durbesic, P.; Barcic, J.I. Arthropod communities within the olive canopy as bioindicators of different management systems. *Spanish J. Agric. Res.* 2018, 16, e0301.

117. de Bruyn, L.A.L. Ants as bioindicators of soil function in rural environments. *Agric. Ecosyst. Environ.* 1999, 74 (1–3), 425–441.

118. Hoffmann, B.D. Using ants for rangeland monitoring: Global patterns in the responses of ant communities to grazing. *Ecol. Indic.* 2010, 10 (2), 105–111.

119. Groc, S.; Delabie, J.H.C.; Fernandez, F.; Petitclerc, F.; Corbara, B.; Leponce, M.; Cereghino, R.; Dejean, A. Litter-dwelling ants as bioindicators to gauge the sustainability of small arboreal monocultures embedded in the Amazonian rainforest. *Ecol. Indic.* 2017, 82, 43–49.

120. Magagula, C.N.; Nzima, B.A. Interaction between habitat characteristics and insect diversity using ground beetles (Coleoptera: Carabidae) and ants (Hymenoptera: Formicidae) within a variety of agricultural habitats. *Appl. Ecol. Environ. Res.* 2015, 13, 863–876.

121. Sanabria, C.; Lavelle, P.; Fonte, S.J. Ants as indicators of soil-based ecosystem services in agroecosystems of the Colombian Llanos. *Appl. Soil Ecol.* 2014, 84, 24–30.

122. Crepaldi, R.A.; Portilho, I.I.R.; Silvestre, R.; Mercante, F.M. Ants as bioindicators of soil quality in integrated crop-livestock system. *Ciencia Rural* 2014, 44, 781–787.

123. Springett, J.A.; Gray, R.A.J.; Bakker, L. Influence of agriculture on Enchytraeidae fauna of soils in the south-west of the north island of New Zealand. *Pedobiologia* 1996, 40 (5), 461–466.

124. Buckerfield, J.C.; Lee, K.E.; Davoren, C.W.; Hannay, J.N. Earthworms as indicators of sustainable production in dry-land cropping in southern Australia. *Soil Biol. Biochem.* 1997, 29 (3–4), 547–554.

125. Paoletti, M.G. The role of earthworms for assessment of sustainability and as bioindicators. *Agric. Ecosyst. Environ.* 1999, 74 (1–3), 137–155.

126. Suthar, S. Earthworm communities a bioindicator of arable land management practices: A case study in semiarid region of India. *Ecol. Indic.* 2009, 9 (3), 588–594.

127. Bartlett, M.D.; Briones, M.J.I.; Neilson, R.; Schmidt, O.; Spurgeon, D.; Creamer, R.E. A critical review of current methods in earthworm ecology: From individuals to populations. *Eur. J. Soil Biol.* 2010, 46 (2), 67–73.

128. Fusaro, S.; Gavinelli, F.; Lazzarini, F.; Paoletti, M. G. Soil biological quality index based on earthworms (QBS-e). A new way to use earthworms as bioindicators in agroecosystems. *Ecol. Indic.* 2018, 93, 1276–1292.

129. Campani, T.; Caliani, I.; Pozzuoli, C.; Romi, M.; Fossi, M.C.; Casini, S. Assessment of toxicological effects of raw and bioremediated olive mill waste in the earthworm Eisenia fetida: A biomarker approach for sustainable agriculture. *Appl. Soil Ecol.* 2017, 119, 18–25.

130. Bartz, M.L.C.; Pasini, A.; Brown, G.G. Earthworms as soil quality indicators in Brazilian no-tillage systems. *Appl. Soil Ecol.* 2013, 69, 39–48.

131. Pelosi, C.; Rombke, J. Enchytraeids as bioindicators of land use and management. *Appl. Soil Ecol.* 2018, 123, 775–779.

132. Filser, J.; Fromm, H.; Nagel, R.F.; Winter, K. Effects of previous intensive agricultural management on microorganisms and the biodiversity of soil fauna. *Plant Soil* 1995, 170 (1), 123–129.

133. Heisler, C. Collembola and Gamasina: Bioindicators for soil compaction. *Acta Zool. Fenn.* **1995**, *196*, 229–231.

134. Wardle, D.A.; Yeates, G.W.; Watson, R.N.; Nicholson, K.S. The detritus food-web and the diversity of soil fauna as indicators of disturbance regimes in agro-ecosystems. *Plant Soil* **1995**, *170* (1), 35–43.

135. Porazinska, D.L.; McSorley, R.; Duncan, L.W.; Graham, J.H.; Wheaton, T.A.; Parsons, L.R. Nematode community composition under various irrigation schemes in a citrus soil ecosystem. *J. Nematol.* **1998**, *30* (2), 170–178.

136. Porazinska, L.; McSorley, R.; Duncan, L.W.; Gallaher, R.N.; Wheaton, T.A.; Parsons, L.R. Relationships between soil chemical status, soil nematode community, and sustainability indices. *Nematropica* **1998**, *28* (2), 249–262.

137. van Straalen, N.M. Evaluation of bioindicator systems derived from soil arthropod communities. *Appl. Soil Ecol.* 1998, 9 (1–3), 429–437.

138. Enami, Y.; Shiraishi, H.; Nakamura, Y. Use of soil animals as bioindicators of various kinds of soil management in northern Japan. *JARQ* **1999**, *33* (2), 85–89.

139. Foissner, W. Soil protozoa as bioindicators: Pros and cons, methods, diversity, representative examples. *Agric. Eco- syst. Environ.* **1999**, *74* (1–3), 95–112.

140. Koehler, H.H. Predatory mites (Gamasina, Mesostigmata). *Agric. Ecosyst. Environ.* 1999, 74 (1–3), 395–410.

141. Paoletti, M.G.; Hassall, M. Woodlice (Isopoda: Oniscidea): Their potential for assessing sustainability and use as bioindicators. *Agric. Ecosyst. Environ.* **1999**, *74* (1–3), 157–165.

142. Yeates, G.W.; Bongers, T. Nematode diversity in agroecosystems. *Agric. Ecosyst. Environ.* **1999**, *74* (1–3), 113–135.

143. Lenz, R.; Eisenbeis, G. Short-term effects of different tillage in a sustainable farming system on nematode community structure. *Biol. Fertil. Soils* **2000**, *31* (3–4), 237–244.

144. Neher, D.A. Role of nematodes in soil health and their use as indicators. *J. Nematol.* **2001**, *33* (4), 161–168.

145. Li, Y.; Wu, J.; Chen, H.; Chen, J. Nematodes as bioindicator of soil health: Methods and applications. *Yingyong Shengtai Xuebao* **2005**, *16* (8), 1541–1546.

146. Briar, S.S.; Jagdale, G.B.; Cheng, Z.; Hoy, C.W.; Miller, S.A.; Grewal, P.S. Indicative value of soil nematode food web indices and trophic group abundance in differentiating habitats with a gradient of anthropogenic impact. *Environ. Bioindic.* **2007**, *2* (3), 146–160.

147. Gulvik, M.E. Mites (Acari) as indicators of soil biodiversity and land use monitoring: A review. *Pol. J. Ecol.* **2007**, *55*, 415–440.

148. Menta, C.; Leoni, A.; Bardini, M.; Gardi, C.; Gatti, F. Nematode and microarthropod communities: Comparative use of soil quality bioindicators in covered dump and natural soils. *Environ. Bioindic.* **2008**, *3* (1), 35–46.

149. Gergocs, V.; Hufnagel, L. Application of oribatid mites as indicators (review). *Appl. Ecol. Environ. Res.* **2009**, *7* (1), 79–98.

150. Aspetti, G.P.; Boccelli, R.; Ampollini, D.; Del Re, A.A.M.; Capri, E. Assessment of soil-quality index based on microarthropods in corn cultivation in northern Italy. *Ecol. Indic.* **2010**, *10* (2), 129–135.

151. Sánchez-Moreno, S.; Jiménez, L.; Alonso-Prados, J.L.; García-Baudín, J.M. Nematodes as indicators of fumigant effects on soil food webs in strawberry crops in southern Spain. *Ecol. Indic.* **2010**, *10* (2), 148–156.

152. Pothula, S.K.; Grewal, P.S.; Auge, R.M.; Saxton, A.M.; Bernard, E.C. Agricultural intensification and urbanization negatively impact soil nematode richness and abundance: A meta-analysis. *J. Nematol.* **2019**, *51*, UNSP e2019-11.

153. du Preez, G.C.; Daneel, M.S.; Wepener, V.; Fourie, H. Beneficial nematodes as bioindicators of ecosystem health in irrigated soils. *Appl. Soil Ecol.* **2018**, *132*, 155–168.

154. Ozden, O.; Hodgson, D.J. The impact of tillage and chemical management on beneficial arthropods in Mediterranean olive groves. *Isr. J. Ecol. Evol.* **2017**, *63*, 14–18.

155. Rudisser, J.; Tasser, E.; Peham, T.; Meyer, E.; Tappeiner, U. The dark side of biodiversity: Spatial application of the biological soil quality indicator (BSQ). *Ecol. Indic.* **2015**, *53*, 240–246.

156. Wang, K.H.; Radovich, T.; Pant, A.; Cheng, Z.Q. Integration of cover crops and vermicompost tea for soil and plant health management in a short-term vegetable cropping system. *Appl. Soil Ecol.* **2014**, *82*, 26–37.

157. Nesbitt, J.E.; Adl, S.M. Differences in soil quality indicators between organic and sustainably managed potato fields in Eastern Canada. *Ecol. Indic.* **2014**, *37*, 119–130.

158. Antonious, G.F. Impact of soil management and two botanical insecticides on urease and invertase activity. *J. Environ. Sci. Health, Part B* 2003, B38 (4), 479–488.

159. Benintende, S.M.; Benintende, M.C.; Sterren, M.A.; De Battista, J.J. Soil microbiological indicators of soil quality in four rice rotations systems. *Ecol. Indic.* **2008**, *8* (5), 704–708.

160. Lagomarsino, A.; Moscatelli, M.C.; Di Tizio, A.; Manci-nelli, R.; Grego, S.; Marinari, S. Soil biochemical indicators as a tool to assess the short-term impact of agricultural management on changes in organic C in a Mediterranean environment. *Ecol. Indic.* 2009, *9* (3), 518–527.

161. Geisen, S.; Mitchell, E.A.D.; Adl, S.; Bonkowski, M.; Dunthorn, M.; Ekelund, F.; Fernández, L.D.; Jousset, A.; Krashevska, V.; Singer, D.; Spiegel, F.W.; Walochnik, J.; Lara, E. Soil protists: A fertile frontier in soil biology research. *FEMS Microbiol. Rev.* **2018**, *42*, 293–323.

162. Berard, A.; Rimet, F.; Capowiez, Y.; Leboulanger, C. Procedures for determining the pesticide sensitivity of indigenous soil algae: A possible bioindicator of soil contamination? Arch. *Environ. Contam. Toxicol.* **2004**, *46* (1), 24–31.

163. Heger, T.J.; Straub, F.; Mitchell, E.A.D. Impact of farming practices on soil diatoms and testate amoebae: A pilot study in the DOK-trial at Therwil, Switzerland. *Eur. J. Soil Biol.* **2012**, *49*, 31–36.

164. Stenberg, B. Monitoring soil quality of arable land: Microbiological indicators. *Acta Agric. Scand., Sect. B* **1999**, *49* (1), 1–24.

165. Kaur, A.; Chaudhary, A.; Choudhary, R.; Kaushik, R. Phospholipid fatty acid—A bioindicator of environment monitoring and assessment in soil ecosystem. *Curr. Sci.* **2005**, *89* (7), 1103–1112.

166. Jordan, D.; Miles, R.J.; Hubbard, V.C.; Lorenz, T. Effect of management practices and cropping systems on earthworm abundance and microbial activity in Sanborn field: A 115-year-old agricultural field. *Pedobiologia* **2004**, *48* (2), 99.

167. Subbarao, K.V.; Kabir, Z.; Martin, F.N.; Koike, S.T. Management of soilborne diseases in strawberry using vegetable rotations. *Plant Dis.* **2007**, *91* (8), 964–972.

168. Smukler, S.M.; Jackson, L.E.; Murphree, L.; Yokota, R.; Koike, S.T.; Smith, R.F. Transition to large-scale organic vegetable production in the Salinas Valley, California. *Agric. Ecosyst. Environ.* **2008**, *126* (3–4), 168–188.

169. Oehl, F.; Laczko, E.; Bogenrieder, A.; Stahr, K.; Bosch, R.; van der Heijden, M.; Sieverding, E. Soil type and land use intensity determine the composition of arbuscular mycorrhizal fungal communities. *Soil Biol. Biochem.* **2010**, *42*, 724–738.

170. Bell, N.L.; Adam, K.H.; Jones, R.J.; Johnson, R.D.; Mtandavari, Y.F.; Burch, G.; Cave, V.; Cameron, C.; Maclean, P.; Popay, A.J.; Fleetwood, D. Detection of invertebrate suppressive soils, and identification of a possible biological control agent for meloidogyne nematodes using high resolution rhizosphere microbial community analysis. *Front. Plant Sci.* 2016, *7*, 1946.

171. Elmholt, S. Microbial activity, fungal abundance, and distribution of *Pénicillium* and *Fusarium* as bioindicators of a temporal development of organically cultivated soils. *Biol. Agric. Hort.* **1996**, *13* (2), 123–140.

172. Padoa-Schioppa, E.; Baietto, M.; Massa, R.; Bottoni, L. Bird communities as bioindicators: The focal species concept in agricultural landscapes. *Ecol. Indic.* **2006**, *6* (1), 83–93.

173. Bar, A.; Loffler, J. Ecological process indicators used for nature protection scenarios in agricultural landscapes of SW Norway. *Ecol. Indic.* **2007**, *7* (2), 396–411.

174. Robledano, F.; Esteve, M.A.; Farinós, P.; Carreño, M.F.; Martínez-Fernández, J. Terrestrial birds as indicators of agricultural-induced changes and associated loss in conservation value of Mediterranean wetlands. *Ecol. Indic.* **2010**, *10* (2), 274–286.

175. Morelli, F.; Girardello, M. Buntings (Emberizidae) as indicators of HNV of farmlands: A case of study in Central Italy. *Ethol. Ecol. Evol.* **2014**, *26*, 405–412.

176. Albrecht, H. Suitability of arable weeds as indicator organisms to evaluate species conservation effects of management in agricultural ecosystems. *Agric. Ecosyst. Environ.* **2003**, *98* (1–3), 201–211.

177. Sturz, A.; Matheson, B.; Arsenault, W.; Kimpinski, J.; Christie, B. Weeds as a source of plant growth promoting rhizobacteria in agricultural soils. *Can. J. Microbiol.* **2001**, *47*, 1013–1024.

178. Aavik, T.; Liira, J. Agrotolerant and high nature-value species—Plant biodiversity indicator groups in agroecosystems. *Ecol. Indic.* **2009**, *9* (5), 892–901.

179. Höft, A.; Müller, J.; Gerowitt, B. Vegetation indicators for grazing activities on grassland to be implemented in outcome-oriented agri-environmental payment schemes in north-east Germany. *Ecol. Indic.* **2010**, *10* (3), 719–726.

180. Fanfarillo, E.; Kasperski, A.; Giuliani, A.; Cicinelli, E.; Latini, M.; Abbate, G. Assessing naturalness of arable weed communities: A new index applied to a case study in central Italy. *Biol. Agric. Hort.* **2018**, *34*, 232–244.

181. Une, Y. Illustrated Poster of Rice and Living Organisms; Nou-to-shizen-no-kenkyujo: Fukuoka, Japan, 2004 (in Japanese).

18

Ecological Indicators: Eco-Exergy to Emergy Flow

Simone Bastianoni,
Luca Coscieme, and
Federico M. Pulselli

Introduction

Living systems can be viewed as dissipative structures, self-organizing systems that dissipate energy for the maintenance of organization.[1] They exist by the transforming available energy potentials, building new structures as a consequence of the process.[2] To study the ecosystem in this view, we need proper tools able to take into account the biological time[3] that has been necessary for the "creative" planning and construction of dissipative structures in the system, through the concentration of energies and matter.

The study of the behavior of a single species in a system provides knowledge of the system's parts. Similar to a jigsaw puzzle, we need to know the single drawing on each piece, as well as the boundaries/shapes of the pieces, in order to properly choose where each piece must be placed, so that we can reconstruct the entire drawing. However, in a very complex context, we are not capable of reconstructing the whole drawing if we do not have a reference, a picture of the whole system that can guide us, e.g., the drawing on the box of the puzzle. Holistic indicators, orientors, or goal functions give us this kind of information, guiding the disciplinary research in a conjunction of efforts, in order to have the best description of the system. In fact, extending the study from a "simple" description of the system in a given time to a description of the system evolving through various states, the holistic indicators allow us to understand if the system under study is globally following a path that will take the system to a "better" or to a "worse" state.[4]

As maintained by Müller and Leupelt,[5] goal functions and orientors have been developed as holistic measures of the global performance of ecosystems, of what we could call, in a broad sense, complexity. Fath, Patten, and Choi[6] showed that 10 extremal principles involving orientors (power, storage, empower, emergy, ascendency, dissipation, cycling, residence time, specific dissipation, and empower/exergy ratio) can be unified by ecological network notation.

In its evolution, a system changes properties and behaviors, like human beings or other organisms change their "reactions" to the surrounding "signals" during their evolution. Every different holistic indicator is appropriate to describe the system in a particular state or phase of its evolution or a particular characteristic of the same system. When studying a dynamic system in evolution, holistic indicators (and the joint use of them) seem to be more powerful than nonsystemic indicators. Only this approach seems to be truly useful to describe the system and its dynamics, i.e., from a "young system" state, through intermediate states (i.e., of reorganization after possible perturbations), toward a climax mature stage.

This holistic approach is relatively new, but it is very efficacious and promising for ecosystem investigation and management.

In this entry, we describe the combined use of two holistic indicators, eco-exergy and emergy, composing a ratio that is able to describe the system during its evolution. Moreover, it adds information to that obtained from the use of the two orientors separately. This kind of approach is in line with the book *A New Ecology: Systems Perspective* proposed by Sven Jørgensen and coauthors[4] and with what Fath, Patten, and Choi[6] maintained in their fundamental paper on the use of a plurality of "goal functions," highlighting their complementarity and interdependency.

A Description of the System Using Emergy

A class of indicators, based on the concept of emergy, are able to evaluate the convergence of matter and energy (several inputs) to a system. On the basis of a thermodynamic hierarchy of energy, and starting from solar energy input to the earth, emergy provides a measure of the environmental work necessary to generate an item or a flow (which could be, for example, an input to any system).

The concept of emergy derives from a reflection about the concept of energy quality.[7] The second principle of thermodynamics states that energy transformations imply an irreversible degradation of energy. In living systems and ecosystems, inputs of energy are transformed, and the energy quality changes. A portion of diluted sunlight is lost as heat, and a portion is concentrated into forms that are more able to do work, and/or more flexible to be used. According to Odum, many joules of low quality are needed for a few joules of high quality. In the case of a typical web of connections, like the food chain, "at each stage, energy is degraded as a necessary part of transforming a lower quality energy to a higher quality one in lesser quantity. The energy flows decrease as one goes up the food chain."[7] "A joule of sunlight, a joule of coal, a joule of human effort are of different quality and represent vastly different convergences of energy in their making."[8] Therefore, in many cases, it is not correct to use energy to describe the dynamics and behavior of systems. Solar energy can be considered the basis upon which energy transformations in the biosphere occur. Emergy, expressed in solar emergy joule (seJ or, according to a recent proposal, semj) is defined as the quantity of solar energy directly or indirectly necessary to produce a flow or a product. In the case of ecosystems, the emergy flow (empower) is considered, which is the quantity of solar energy directly or indirectly necessary to support the system and its level of organization.[9–11]

To compare all kinds of energy on the common basis of solar energy, solar transformity has been utilized, defined as the solar energy directly and indirectly required to generate 1 J of a product. It is a conversion factor that takes into account the position of one energy form in a sort of thermodynamic hierarchy in the biosphere. The solar transformity of the sunlight absorbed by the earth is 1 seJ/J by definition.[12]

To quantify the emergy of a product or system, all the inputs to the system or production process must be quantified and expressed in seJ by means of suitable transformities, which are used to convert different flows of energy into equivalent solar energy.[9] In case of matter inputs, the specific emergy (expressed in seJ/g or another unit) is used to convert mass into equivalent solar energy. Recently, the concept of unit emergy value has been used, independently of the unit in which the flows are expressed (energy or mass). The total emergy of a system (EmS) is given, approximately, by the sum of the energy content (E_i) of the ith input to the system multiplied by the corresponding transformity (Tr_i), while avoiding double counting any inputs (see Bastianoni[13] for a thorough analysis of this calculation method):

$$Em_S = \Sigma E_i \cdot Tr_i \qquad (1)$$

Every flow can be expressed by means of its solar equivalent, and a system of environmental accounting based on emergy can be implemented.[14]

Emergy represents the convergence of different kinds of energy to a system (E_i) times the quantity of solar energy that has been necessary to make available one unit (Tr_i,). It is not a state function because it depends on the kinds of energy and the process that is used to obtain a certain item (that can be a product or a given state of the system). Therefore, emergy enables us to identify, quantify, and weigh the inputs that feed an ecosystem. If natural selection has been given time to operate, the higher the emergy flux necessary to sustain a system, component, or a process, the higher the hierarchical level and the usefulness that can be expected for that entity (maximum empower principle[7]).

In a pristine natural system, self-organizing from young states, this maximum empower principle (derived from the maximum power principle[15]) is realized, and the continuous increase of emergy is an indication of a proper evolution toward a mature system (climax stage).[16] Therefore, empower has been proposed as an ecosystem health indicator.[17]

In "real-world" cases, where natural systems interact with human systems and dynamics, the increase of emergy (as a consequence of the increase in energetic inputs that reach the ecosystem) is not always "good" in the sense that it will support the evolution of the system toward a climax stage. In fact, a portion of the inputs that the ecosystem receives is not used to build structures in order to maintain the nonequilibrium state (e.g., nutrient overflow, pollutants). Emergy flow alone cannot be used as a "reference direction" indicator[18] if the system is not a pure pristine system.

A Description of the System Using Eco-Exergy

The usable energy input to a system is converted into genetic information, biomass storages, and a relation network.[7,19,20] Structural complexity and biodiversity influence the possible evolutions of the system toward another (more or less) stable state.[21]

Eco-exergy expresses the development of ecosystems by the increase in the work capacity,[4] considering the biomass stored in a system and its genetic information.[22–25]

According to Jørgensen,[26,27] we can distinguish between technological exergy and eco-exergy: the former uses the environment as a reference state and is able to measure the useful energy provided by a production process; the latter uses as a reference state the same ecosystem at thermodynamic equilibrium. Eco-exergy thus estimates the distance of an ecosystem from thermodynamic equilibrium and is given by the formula

$$Eco - Ex = \Sigma \beta_i \cdot c_i \tag{2}$$

where c_i is the concentration of the ith component of the ecosystem and β_i is the weighting factor that accounts for the genetic information that the component carries (for a list of β values.[28]

When the available inputs to the system are used to build up new biomass and/or complexity, the system is tending to its climax stage (maximum exergy principle[4]). Eco-exergy is a state-based descriptor of a system's structure (and functions, networks, interactions) based on usable energy and information. It has also been used as an ecosystem health indicator.[24,25,29]

Ratio of Eco-Exergy to Emergy Flow

Emergy and eco-exergy can be considered as complementary entities, with the former accounting for the amount of basic energy (solar) required to support a process or an ecosystem and the latter being the level of organization reached by a system.

The need to compare the emergy flow that supports an ecosystem with the consequent ecosystem reaction was already clear to Odum, who tried to assess the ecosystem response using the emergy-to-information ratio[7] as a measure of information hierarchy. Odum used the mass of DNA as information carrier and looked at the ratio of emergy to information measured in bits, considering that the emergy required per bit may give an indication on the usefulness of the information in an adapted system.[7,30] The emergy-to-information ratio, also used by Keitt,[31] has a problem that consists in the arbitrariness of the choice of the system's basic element: an atom or an individual of a species, a letter of an alphabet, or a gene as the basic "symbol." Bastianoni and Marchettini[30] have proposed joining emergy with eco-exergy in order to measure the ecosystem structure. The ratio between eco-exergy and emergy flow indicates the organization or structure of an ecosystem per unit of solar emergy flow required to produce or maintain it.[30] Bastianoni and Marchettini[30] first introduced this- relation as the ratio of emergy (flow) to eco-exergy. This choice was made in order to maintain coherence with the definition of transformity (i.e., the emergy that contributes to a production system divided by the energy content of a product). In fact, the ratio of emergy flow to eco-exergy represents an empower converging to a certain system divided by the eco-exergy of the whole system. Actually, the inverse seems more comprehensible, where the effect (eco-exergy) is at the numerator and the requirement is at the denominator, as in any efficiency indicator.

The role of information and structure is fundamental when we approach the study of complex systems, such as an ecosystem. The use of eco-exergy adds something to the classical exergy approach (for an overview, see Jørgensen and Mejer,[22] Jørgensen,[27] and Sciubba and Wall[32]), which does not take into account information content. For instance, the difference between a living organism and a dead one is not related to the classical exergetic content that is in fact the same, but is related to the capability of the living system to use the information content in its DNA.[33]

The ratio of eco-exergy to emergy flow has also been considered as an indicator of the efficiency[30] in transforming available inputs, evaluated in emergy terms, into the structure, information, and ecosystem organization, evaluated in eco-exergy terms. In fact, it represents the state of the system (as eco-exergy) per unit input (as empower). Strictly speaking, its unit is JyrseJ^{-1} (we maintain this representation even if β values may influence the pure thermodynamic nature of eco-exergy and its unit). Since the dimensions are those of time, it cannot be regarded as a real efficiency (which is dimensionless) but more as a proxy of efficiency. The higher its value, the higher the efficiency of the system; if the eco-exergy/empower ratio tends to increase (apart from oscillations due to normal biological cycles), it means that natural selection is making the system follow a thermodynamic path that will bring the system to a higher organizational level.[33,34] As an efficiency indicator, the ratio of eco-exergy to empower enlarges the viewpoint of a pure exergetic approach, where the exergy degraded and the eco-exergy stored for various ecosystems are compared: using emergy, there is a recognition of the fact that solar radiation is the driving force of all the energy (and exergy) flows on the biosphere, which is essential when important "indirect" inputs (of solar energy) are also present in a process, and must be identified, weighted, and finally, taken into account.

Fath, Patten, and Choi[6] have identified the ratio of eco-exergy to emergy flow as one of the possible orientors of an ecosystem: they link the emergy flow to the total system throughput and eco-exergy to the total system storage, therefore connecting the maximization of the ratio of eco-exergy to emergy flow with the maximization of residence time.

Applications in Environmental Investigation and Management

A Definition of Ecosystem Development

Ecological orientors and goal functions can indicate some aspects of the degree of naturalness of ecosystems; they provide a good basis for finding usable indicators for ecosystem health, ecological integrity, and sustainability;[5] they can also be used to evaluate the strength of human impact and an ecosystem's structural carrying capacity.

The ratio of eco-exergy to empower has often been applied in order to assess ecosystem health: in fact, ecosystems with different empower and eco-exergy can be compared with each other, also regarding their behavior and performance. In general, we can say that in natural systems, where selection has acted undisturbed for a long time, the ratio of eco-exergy to empower is higher and decreases with the progressive introduction of artificial inputs and stress factors that make the emergy flow higher and lower the eco-exergy content of the ecosystem. In the evolutionary process, close to the steady state (climax), the ratio of eco-exergy to empower tends to increase, which means that the system uses all the materials and energy available to reach a higher eco-exergy content. The same systems, once the climax has been reached, will remain in such a state for some time and can then grow/develop again only if further energy and/or materials are available. In the latter case, new sources of energy (or better emergy) can be used to build up new biomass and/or complexity of the ecosystem (stored eco-exergy). In terms of the ratio of eco-exergy to empower, when a system is relatively young and acquires new inputs, the ratio tends to be lower; when the system is developing toward the climax stage, the ratio tends to rise.[2,35]

This approach is helpful when different kinds of ecosystems are compared, natural and artificial. The former might have different quantities and qualities of available inputs, while the latter cannot be compared only on the basis of its "state" but also considering its requirements to develop and sustain the state itself. The results of a study on eight different aquatic ecosystems[30,35–37] demonstrated that the highest level of efficiency (in the exergy/empower sense) is obtained by a seminatural system within the lagoon of Venice, a farming basin developed over several centuries. Its efficiency is of the same order of magnitude as natural systems, but it is higher than systems with limited human input. Furthermore, the efficiency of the seminatural system is two orders of magnitude greater than artificial ecosystems by virtue of a higher level of organization and less need for external input.[36,38] Application to agroecosystems can be found in Bastianoni et al.[39] The ratios of eco- exergy to emergy and eco-exergy to empower have been also proposed to assess the self-organization efficiency of forest ecosystems.[40]

A Definition of Ecosystem Health and Pollution

A qualitative or quantitative change in the set of inputs can contribute to a change in a system's self-organization pattern and a system's different responses. Moreover, controlled human intervention can make a positive contribution to the system in terms of organization, information complexity, etc., (the eco-exergy of the system) that more than offset the environmental cost of the same intervention (the emergy flow corresponding to the human-induced inputs to the system).

If we consider the emergy flow to a system to vary between two equal and contiguous intervals, we will indicate the variation of emergy flow with ΔEM. Consequent changes in system organization can be measured by the variation of the exergy content of the system ΔEX. The quantity $\sigma = \Delta EX/\Delta EM$, with the dimensions of $J \cdot s \cdot seJ^{-1}$, represents the change of level of organization (exergy) of the system under study, when it is related to a change of the emergy flow. It is a quantity that is specific to the inputs that are subtracted or added. If σ is positive, the addition of emergy input gives rise to further organization (increasing eco-exergy), whereas a lowering of emergy has a negative effect on the system (decreasing eco-exergy). On the other hand, when σ is negative, a higher emergy flow implies a decrease in organization, whereas a lower quantity of one or more inputs causes increasing organization. In these two last cases, the inputs can generally be viewed as pollutants: in an evolutionary perspective, if they are removed, the system self-organizes; if they are added, the system is damaged. This provides a definition of pollution based on two holistic orientors representing system dynamics. A first-level observation of the behavior of σ (and of the system as a whole) gives information on the existence of pollutants; a deeper analysis can identify the intensity of pollution that is given by the sensibility of eco-exergy relative to a change in emergy flow.[41]

Holistic Interpretation of Ecosystem Services

Emergy and eco-exergy can be related to the concept of "ecosystem services." Ecosystem services are the benefits human populations derive, directly or indirectly, from ecosystem functions (e.g., food provision or waste assimilation).[42] This concept derives from a reconceptualization of ecosystem functioning from an anthropocentric viewpoint.[43]

In general, ecosystems utilize flows of energy and matter from the environment to maintain themselves as far as possible from thermodynamic equilibrium and to survive, grow, and/or develop. The degree of development and the efficacy with which these flows are used up and processed depend on the state/structure/organization of the ecosystem, which is a particular configuration of the abiotic–biotic system components, characterized by specific relationships between living organisms and nonliving surroundings. The outputs of an ecosystem are all flows of energy and matter moving from the system to the environment, as well as all goods and services useful for humans.

A relationship between the inflows of energy and matter supporting the ecosystem and the services it provides has been investigated by Pulselli, Coscieme, and Bastianoni,[21] who noted that in this input–output representation of ecosystems, the emergy flow (input) supporting an ecosystem and the value of the services (output) it provides are rather independent from each other, because the former depends on natural dynamics and the latter on the utility humans (decide to) draw from nature, which may vary from case to case. Despite this, at the global level, it has been calculated that nature contributes to humans not only more (as Costanza et al.[42] demonstrated) but also in a more efficient way than do all the world economic infrastructures. In fact, if we divide the world ecosystem service value by the emergy flow to the biosphere, we obtain the amount of money that is, on average, produced by 1 seJ of solar emergy feeding the global ecosystem. This ratio combines an amount of money that is not really circulating in the global economy and the flow of all renewable resources that feed the planet (sunlight, geothermal heat, rain, wind, etc.). It can be considered as a potential efficiency of the entire biosphere in providing a kind of economic wealth for humans (since at least a portion of it can be converted into real economic utility/benefit) based only on its natural functioning.[21]

The reciprocal of the above relation, i.e., the ratio of the global emergy flow to the total value of ecosystem services, is between 5.09×10^{11} and 1.51×10^{n} seJ/€ (depending on the minimum and maximum values calculated by Costanza et al.[42]). Both the maximum and minimum values are lower than values traditionally calculated for national economies, which in emergy theory are known as the emergy-to-money ratio (EMR). The EMR is given by the ratio of the emergy flow of a country (including both natural and commercial man-induced flows) to its GDP, and its order of magnitude is, in general, 10^{12} seJ/€ or more (for an overview of national values.[44] This means that the global ecosystem uses, on average, less emergy than a national economy per unit money provided to humans.

Jørgensen[26] proposed an approach to connect an ecosystem's structure and organization descriptor (eco-exergy) to a user-side measure (the value of ecosystem services), highlighting a relation between a biophysical and an economic evaluation of the environment. The calculation of ecosystem services through eco-exergy resulted in values higher than those proposed by Costanza et al.,[42] because eco-exergy represents the annual work capacity increase of an ecosystem that can be translated into the set of the possible services it can offer (not only the services that anthropic systems actually utilize) and can be compared with the actual flow of services utilized.

Ecosystem dynamics can be represented through a generic and complete input–state–output scheme. In this sense, we can imagine a kind of 3-D diagram, with the inflows of resources, measured in terms of solar emergy, on the x axis; the work capacity embodied in the system biota, expressed in terms of eco-exergy, on the y axis; and the useful services for humans, valuable in economic terms, on the z axis.[45] This multidimensional holistic approach makes it clear that inputs are used up, directly or indirectly, to produce services in output and/or to develop the system, and enables us to have an indication of changes in ecosystem dynamics, structure, and services. Within this framework, a thermodynamic/

socio-ecological evolutionary time path can be acknowledged: from young systems, to climax-stage systems, to socio-ecological integrated systems.

Conclusion

The ratio of eco-exergy to emergy flow is the combination of two thermodynamics-based orientors: emergy flow, which quantifies the amount of resources necessary for the system to survive, and eco-exergy, which represents the actual state of the ecosystem in terms of work capacity and distance from thermodynamic equilibrium. The joint use of these two entities adds information that is useful for the investigation of the behavior of the system during its evolution. It is a measure of the ability of a system to reach and maintain a given structure (as eco-exergy) per unit input (as emergy); it is therefore a measure of the efficiency of the system in transforming available resources into organization. The use of the ratio of eco-exergy to emergy flow is important when investigating the evolution of an ecosystem or the effects on ecosystem dynamics of human intervention and infrastructures. Two further applications of this entity have been described too, which can be useful in the field of environmental management: according to the change in exergy due to a change in emergy flow, we can identify potential pollutants and define the intensity of pollution; the combined use of eco-exergy and emergy can also help in assessing the role of ecosystem services in human well-being.

References

1. Prigogine, I. *From Being to Becoming: Time and Complexity in the Physical Sciences*; Freeman: New York, 1980.
2. Bastianoni, S.; Pulselli, F.M.; Rustici, M. Exergy versus emergy flow in ecosystems: Is there an order in maximization? Ecol. Indic. **2006**, *6*, 58–62.
3. Tiezzi, E. *The End of Time*; Wit Press: Southampton, U.K., 2003.
4. Jørgensen, S.E.; Fath, B.D.; Bastianoni, S.; Marques, J.C.; Müller, F.; Nielsen, S.N.; Patten, B.C.; Tiezzi, E.; Ulano- wicz, R.E. *A New Ecology: Systems Perspective*; Elsevier: Amsterdam, 2007.
5. Müller, F.; Leupelt, M. *Eco Targets, Goal Functions, and Orientors*; Springer-Verlag: New York, 1998.
6. Fath, B.D.; Patten, B.C.; Choi, J.S. Complementarity of ecological goal functions. J. Theor. Biol. **2001**, *208*, 493–506.
7. Odum, H.T. Self organization, transformity and information. Science **1988**, *242*, 1132–1139.
8. Odum, H.T. Emergy and biogeochemical cycles. In *Ecological Physical Chemistry*; Rossi, C., Tiezzi, E., Eds.; Elsevier: Amsterdam, 1991; 25–26.
9. Odum, H.T. *Environmental Accounting. Emergy and Environmental Decision Making*; John Wiley and Sons: New York, 1996.
10. Odum, H.T. *Emergy of Global Processes, Folio #2. Handbook of Emergy Evaluation*; Center for Environmental Policy, University of Florida: Gainesville, FL, 2000.
11. Odum, H.T.; Brown, M.T.; Brandt-Williams, S. *Introduction and Global Budget, Folio #1. Handbook of Emergy Evaluation*; Center for Environmental Policy, University of Florida: Gainesville, FL, 2000.
12. Brown, M.T.; Ulgiati, S. Emergy analysis and environmental accounting. In *Encyclopedia of Energy*; Cleveland, C., Ed.; Academic Press, Elsevier: Oxford, 2004; 329–359.
13. Bastianoni, S.; Morandi, F.; Flaminio, T.; Pulselli, R.M.; Tiezzi, E.B.P. Emergy and emergy algebra explained by means of ingenuous set theory. Ecol. Modell. **2011**, 222, 2903–2907.
14. Hau, J.L.; Bakshi, B.R. Promise and problems of emergy analysis. Ecol. Modell. **2004**, *178*, 215–225.
15. Lotka, A.J. Contribution to the energetics of evolution. Proc. National Acad. Sci. U. S. A. **1922**, 8, 147–151.

16. Campbell, D.E. Proposal for including what is valuable to ecosystems in environmental assessments. Environ. Sci. Technol. **2001**, *35*, 2867–2873.

17. Campbell, D.E. Using energy systems theory to define, measure, and interpret ecological integrity and ecosystem health. Ecosyst. Health **2000**, *6* (3), 181–204.

18. Samhouri, J.F.; Levin, P.S.; James, C.A.; Kershner, J.; Williams, G. Using existing scientific capacity to set targets for ecosystem-based management: A Puget Sound case study. Mar. Policy **2011**, *35*, 508–518.

19. Jørgensen, S.E.; Patten, B.C.; Straškraba, M. Ecosystems emerging: 4. Growth. Ecol. Modell. **2000**, *126*, 249–284.

20. Ulanowicz, R.E. *Ecology, the Ascendent Perspective*; Columbia University Press: New York, 1997.

21. Pulselli, F.M.; Coscieme, L.; Bastianoni, S. Ecosystem services as a counterpart of emergy flows to ecosystems. Ecol. Modell. **2011**, *222*, 2924–2928.

22. Jørgensen, S.E.; Mejer, H.F. A holistic approach to ecological modelling. Ecol. Modell. **1979**, 7, 169–189.

23. Jørgensen, S.E.; Mejer, H.F. Application of exergy in ecological models. In *Progress in Ecological Modelling*; Dubois, D., Ed.; Editions CEBEDOC: Liege, Belgium, 1981; 311–347.

24. Jørgensen, S.E. Application of holistic thermodynamic indicators. Ecol. Indic. **2006**, *6*, 24–29.

25. Jørgensen, S.E. Eco-exergy as an ecosystem health indicator. In *Encyclopedia of Ecology*; J0rgensen, S.E., Fath, B., Eds.; Elsevier: Amsterdam, 2008; 977–979.

26. Jørgensen, S.E. Ecosystem services, sustainability and thermodynamic indicators. Ecol. Complexity **2010**, 7, 311–313.

27. Jørgensen, S.E. Exergy. In *Encyclopedia of Ecology*;, Jørgensen, S.E., Fath B.D., Eds.; Elsevier: Oxford, 2008; 1498–1509.

28. Jørgensen, S.E.; Ladegaard, N.; Debeljak, M.; Marques, J.C. Calculations of exergy for organisms. Ecol. Modell. **2005**, *185* (2–4), 165–175.

29. Zaldivar, J.M.; Austoni, M.; Plus, M.; De Leo, G.A.; Giordani, G.; Viaroli, P. Ecosystem health assessment and bioeconomic analysis in coastal lagoon. In *Handbook of Ecological Indicator for Assessment of Ecosystem Heath*; Jørgensen, S.E., Costanza, R., Xu, F.L., Eds.; CRC Press: Boca Raton, FL, 2005; 163–184.

30. Bastianoni, S., Marchettini, N. Emergy/exergy ratio as a measure of the level of organization of systems. Ecol. Modell. **1997**, *99*, 33–40.

31. Keitt, T.H. *Hierarchical Organization of Energy and Information in a Tropical Rain Forest Ecosystem*; M.S. Thesis, University of Florida: Gainesville, FL, 1991.

32. Sciubba, E.; Wall, G. A brief commented history of exergy from the beginnings to 2004. Int. J. Thermodyn. **2007**, *10* (1), 1–26.

33. Tiezzi, E. *Steps Towards an Evolutionary Physics*; Wit Press: Southampton, U.K., 2006.

34. Pulselli, F.M.; Gaggi, C.; Bastianoni, S. Eco-Exergy to emergy flow ratio for the assessment of ecosystem health. In *Handbook of Ecological Indicators for Assessment of Ecosystem Health*, 2nd Ed.; Jørgensen, S.E., Xu, F.L., Costanza, R., Eds.; CRC Press: Boca Raton, FL, 2010; 113–124.

35. Bastianoni, S. Eco-Exergy to emergy flow ratio. In *Encyclopedia of Ecology*; Jørgensen, S.E., Fath, B., Eds.; Elsevier: Oxford, 2008; 979–983.

36. Bastianoni, S. Use of thermodynamic orientors to assess the efficiency of ecosystems: A case study in the lagoon of Venice. Sci. World J. **2002**, *2*, 255–260.

37. Bastianoni, S. Emergy, empower and the eco-exergy to empower ratio: A reconciliation of H.T. Odum with Prigogine? Int. J. Ecodyn. **2006**, *1* (3), 226–235.

38. Bastianoni, S.; Marchettini, N.; Pulselli, F.M.; Rosini, M. The joint use of exergy and emergy as indicators of ecosystems performances. In *Handbook of Ecological Indicators for Assessment of Ecosystem Health*; Jørgensen, S.E., Costanza, R., Xu, F.L., Eds.; CRC Press: Boca Raton, FL, 2005; 239–248.

39. Bastianoni, S.; Nielsen, S.L.; Marchettini N.; J0rgensen, S.E. Use of thermodynamic functions for expressing some relevant aspects of sustainability. Int. J. Energy Res. **2005**, *29*, 53–64.

40. Lu, H.F.; Wang, Z.H.; Campbell, D.E.; Ren, H.; Wang, J. Emergy and eco-exergy evaluation of four forest restoration modes in southeast China. Ecol. Eng. **2011**, *37*, 277–285.

41. Bastianoni, S. A definition of 'pollution' based on thermodynamic goal functions. Ecol. Modell. **1998**, *113*, *2* (1–3), 163–166.

42. Costanza, R.; d'Arge, R.; De Groot, R.; Farber, S.; Grasso, M.; Limburg, K.; Naeem, S.; O'Neill, R.V.; Paruelo, J.; Raskin, R.G.; Sutton, P.; van den Belt, M. The value of the world's ecosystem services and natural capital. Nature **1997**, *387*, 253–260.

43. De Groot, R.S.; Wilson, M.A.; Boumans, R.M.J. A typology for the classification, description and valuation of ecosystem functions, goods and services. Ecol. Econ. **2002**, *41*, 393–408.

44. Sweeney, S.; Cohen, M.J.; King, D.; Brown, M.T. Creation of a global emergy database for standardized national emergy synthesis. In *Emergy Synthesis 4: Theory and Application of Emergy Methodology*; Brown, M., Ed.; University of Florida: Gainesville, FL, 2007; 23.1–23.15.

45. Pulselli, F.M.; Coscieme, L.; Jørgensen, S.E.; Bastianoni, S. Thermodynamics-based orientors for holistic interpretation of ecosystem services. In *Acts of the XXIV Congress of the Italian Chemical Society*; University of Salento: Lecce, Italy, 11–16 September 2011; 101.

19

Ecological Indicators: Ecosystem Health

Felix Müller,
Benjamin Burkhard,
Marion Kandziora,
Claus Schimming,
and Wilhelm
Windhorst

Introduction

Environmental management has to operate in an extremely complex framework, which can be characterized by a multitude of different components and an even higher number of interrelations between these parts.[1] Ecological as well as human and societal influences have to be taken into account, and the dominant role of indirect effects has to be realized.[2] Consequently, there is a very high demand for holistic management concepts, which approach the management object from an ecosystem-based starting point. The biggest challenge of such concepts arises from the enormous complexity of human–environmental systems.[3,4] To cope with this challenge, indicators can be suitable tools. They enable quantitative statements in spite of the complex environment, but this potential is attained due to simplification, aggregation, modeling, and abstraction. Therefore, indicators often are correlated with a high uncertainty, i.e., whether the indicated object is really well represented by the indicator.[5]

Ecosystem health is an environmental management concept that directly meets these problems.[6] It follows the important demand to manage the environment from a system-based viewpoint, acknowledging inherent complexities. Therefore, the selection of suitable indicators plays a major role for quantitative applications of the health concept.[7] In the following text, some focal approaches on how these challenging demands are met are documented. The leading questions are the following:

- What are the basic features and requirements for ecosystem-based indicators, and how are they related to the concept of ecosystem health?
- What are the focal ideas of ecosystem health, and how can they be translated into ecological indication concepts?

- What are the problems and experiences of health indication in different ecosystem types?
- How do recent indicator approaches cope with the challenging demands for ecosystem-based indication?

The entry starts with a short statement on the general requirements for ecological indicators. Thereafter, the actual indicandum—ecosystem health—is sketched, and a literature survey of established ecosystem health indicators is presented. In the final sections, indicator applications and aggregations are discussed, and demands for future development are formulated based on a comparison of identified requirements and the state of the art.

Basic Features of Ecological Indicators

Environmental management should be based upon qualitative or quantitative key variables that can be used to demonstrate the demand for management actions and the outcomes of the respective activities. Such indicators provide aggregated information on certain complex phenomena,[8–10] which often are not directly accessible.[11–13] Indicators are developed on the basis of specific management purposes; they often include an integrating, synoptical value, and they should be capable of showing the differences between an existing situation and an aspired-to target state.[14,15] Thus, indicators are signals for attracting attention on changes in complex humanenvironmental systems. Heink and Kowarik[16] propose the following indicator definition: "An indicator in ecology and environmental planning is a component or a measure of environmentally relevant phenomena used to depict or evaluate environmental conditions or changes or to set environmental goals."

Being applied for the management of human–environmental systems, indicators have to satisfy very different and challenging demands: On the one hand, scientific correctness is a major requirement, and on the other, transparency and public utility in the decision making processes are significant demands (see Tables 1 and 2). Therefore, indicator applications should be based on satisfying scientific hypotheses and relevant cause–effect relations, while they also have to translate the high complexity of ecosystems in a scientifically sound way to meet the needs of politicians and decision makers for common acceptance. Furthermore, the indicator should comprise an optimal sensitivity for the related disturbance, and it should be characterized by a clear representation of the indicandum by the indicator—in this case, it should represent the challenging properties of ecosystem health.

Basic Components of Ecosystem Health

The pioneering ecologist Aldo Leopold's writings about land sickness[17] created the basic ideas for the ecosystem health concept. In the following decades, definitions of ecosystem health have been constantly evolving toward an increasing integration of human and societal contexts in order to understand what is

TABLE 1   Scientific Demands on Good Indicators

Good indicator sets should provide scientific correctness basing upon the following:
- A clear representation of the indicandum by the indicator.
- Clear proof of relevant cause–effect relations.
- An optimal sensitivity of the representation.
- Information for adequate spatio-temporal scales.
- A very high transparency of the derivation strategy.
- A high degree of validity and representativeness of the available data sources.
- A high degree of comparability in and with indicator sets.
- An optimal degree of aggregation.
- Good fulfillment of statistical requirements concerning verification, reproduction, representativity, and validity.

Source: Wiggering an Müller.[9]

TABLE 2 Applied Demands on Good Indicators

Good indicator sets should provide practical applicability basing upon the following:

- Information and estimations of the normative loadings.
- High political relevance concerning the decision process.
- High comprehensibility and public transparency.
- Direct relations to management actions.
- An orientation toward environmental targets.
- A high utility for early warning purposes.
- A satisfying measurability.
- A high degree of data availability.
- Information on long-term trends of development.

Source: Wiggering an Müller.[9]

considered to be a healthy ecosystem.[18,19] The concept has gained special popularity in the United States and in Canada, where ecosystem health has been integrated in legislation. Today, ecosystem health is part of various international political programs, like the Rio convention on sustainable development.[20] Here, it has been demanded in principle 7 that "states shall cooperate in a spirit of global partnership to conserve, protect and restore the health and integrity of the Earth's ecosystem." Ecosystem health does not only take into account ecological components but also requires a linkage with social, economic, and cultural dimensions (see Table 3).

De Kruijf and Van Vuuren[25] analyzed that "the present definitions of ecosystem health contain several of the following elements:

- Healthy ecosystems are free from ecosystem distress syndrome …;
- Healthy ecosystems are resilient … they recover from natural perturbations and disturbances;
- Healthy ecosystems are self-sustaining and can be perpetuated without subsidies or drawing down natural capital;
- Healthy ecosystems do not impair adjacent ecosystems …;
- Healthy ecosystems are free from risk factors;
- Healthy ecosystems are economically viable;
- Healthy ecosystems sustain healthy human communities."

Summarizing the different approaches, a working definition of ecosystem health can be given as follows: Ecosystem health refers to the ability of an ecosystem to maintain its structure and function over time under external stress, safeguarding a sustainable provision of ecosystem goods and services contributing to human well-being.

The facts mentioned above have to be reflected while defining appropriate sets of indicators that can be applied for environmental management.[16] Suitable ecosystem health indicator sets have to consider ecological structures as well as ecological functions on different spatial and temporal scales. As shown by the Millennium Ecosystem Assessment,[26] most ecosystems on our planet have already been

TABLE 3 Different Approaches Defining Ecosystem Health

Haskell et al.[21]: An ecological system is healthy and free from "distress syndrome" if it is stable and sustainable—that is, if it is active and maintains its organization and autonomy over time and is resilient to stress.'

Karr[22]: Ecosystem health is related to "the condition in which a system realizes its inherent potential, maintains a stable condition, preserves its capacity for self-repair when perturbed, and needs minimal external support for management."

Rapport et al.[23]: Ecosystem health refers to a "condition where the parts and functions of an ecosystem are sustained over time and where the system's capacity for self-repair is maintained, such that goals for uses, values, and services of the ecosystem are met."

Xu and Mage[24]: Ecosystem health refers to "the system's ability to realize its functions desired by society and to maintain its structure needed both by its functions and by society over a long time."

degraded under the pressure of increasing human demands. If ecosystem health shall be achieved on a long-term perspective, preventative and restorative environmental management strategies are needed. When looking at different typical ecosystem health indicators, like species diversity or water quality, many of the ecosystems on our planet can be considered unhealthy.[18] As a consequence, many ecosystem functions needed for the provision of ecosystem services have been altered. Ecosystem health refers to systems that are manipulated to satisfy human needs.[17] Therefore, ecosystem health provides a suitable conceptual framework describing the linkages between ecosystem functions, services, and human well-being.

The explicit integration of societal components makes it different from other ecosystem management concepts, for example, ecological integrity. Ecological integrity refers to the functioning of ecosystems based on self-organized processes, while ecosystem health also includes resilience and sustainability with regard to the provision of ecosystem services. Therefore, different ecological concepts like homeostasis, diversity, complexity, emergent properties, or hierarchy principles are closely related to the health concept.[19]

Basic Requirements for the Indication of Ecosystem Health

A focal problem of these concepts is the complexity of ecosystems that arises from the high number of components, relations, and interactions. Hence, for environmental practice and decision making, this complexity has to be reduced. Ecosystem theories provide an applicable basis for such a reduction. Some of the respective theoretical fundamentals of the ecosystem health concept are listed in Table 4.

These requirements are summarized in the "V-O-R model,"[21] describing the ecosystem vigor, organization, and resilience. Vigor is indicated by activity, metabolism, or primary productivity, while organization represents the diversity and number of interactions between the system components. Resilience is understood as a system's capacity to maintain structure and function in the presence of (external) stress. When resilience is exceeded, the system can shift to an alternate state. By including this approach, ecosystem health is closely related to the concepts of stress ecology, where vigor, system organization, resilience, and the absence of ecosystem distress are the main factors for a system's condition.

This model is correlated with a very high demand for comprehensive data sets and long-time series to determine resilience. Additionally, linkages of environmental and social–economic attributes and attributes representing structures as well as functions and organization have to be included. All these demands can hardly be fulfilled. Therefore, quantification deficits have to be expected as one of the main problems of the ecosystem health approach.

Utilized Indicators of Ecosystem Health

In the following paragraphs, a short literature survey of health indicators is presented, whereby different approaches have been distinguished: In the beginning, community-based indicators are listed, followed by aggregated theoretical indicators and indicators based on ecosystem analysis. Finally, a link will be developed toward the indication of ecosystem services.

TABLE 4   Axioms of Ecosystem Health

Dynamism: Nature is a set of processes, more than a composition of structures.
Relatedness: Nature is a network of interactions.
Hierarchy: Nature is built up by complex hierarchies of spatio-temporal scales.
Creativity: Nature consists of self-organizing systems.
Different fragilities: Nature includes various sets of different resiliences.

Source: Wiggering an Müller.[9]

Species- and Community-Based Indicators and Indices

Biodiversity loss is one of the characteristic signs for ecosystems under stress[18] and thus is a major issue in environmental management.[26] Consequently, there are many initiatives and concepts to describe and assess biodiversity.[27] Biodiversity indicators and indices are based on the abundance, absence, or composition of selected species or communities. They vary from single-species indicators to complex composite indicators. Suitable indicator species have to be selected in order to be representative for certain phenomena or sensitive to particular environmental changes.[28] Therefore, the appearance and dominance of certain communities can be associated with states of ecosystem health.[7]

The parameters used to quantify respective indicators can be derived from direct measurements and observations of selected species' abundance. Species- and community- based indicators can be linked to numerous international and national policy instruments, for which biodiversity indicators need to be derived (e.g., Bern Convention 1979; Bonn Convention 1979; Convention on Biological Diversity CBD 1992; the Millennium Development Goals to be achieved by 2015). Most policy and decision makers rely on indices that aggregate biodiversity data across large numbers of species,[29] but also, a limited number of key taxa are frequently used to indicate ecosystem health.[30] Some of these biodiversity indicator concepts—mainly related to the indicator collections of Marquez et al.[31] and Joergensen et al.[6]—are presented in the following.

Species Richness

The most established way to indicate biodiversity is based on species counts and composite indices. One advantage is that the number of species in a certain area is a measure that is easily established and understood by a broad range of people. The Shannon–Wiener index and the Simpson index are the most commonly applied indicators.[29] The Shannon-Wiener index H' originates in information theory and integrates species' number and evenness:

$$H' = -\sum p_i \log 2 p_i \qquad (1)$$

p_i is the proportion of individuals found in species i. The values of this index can vary between 0 and 5. H' has a maximum value if the individuals of all species occur with the same density. The Simpson index refers to the number of species present and the relative abundance of each species.[32] Species richness provides important information on ecosystem conditions. However, its application in ecosystem health assessments can be misleading, for example, concerning nonnative (exotic or "invasive") species. Their abundance will increase the value of standard biodiversity indicators, but their increased dominance in biotic communities can be a typical sign of ecosystem stress.[18]

Indicator Taxa

Indicator taxa (or bioindicators, indicator species) are species or higher taxonomic groups whose properties can be used as proxies for assessments of ecosystem health.[33,34] Respective species have to be selected in order to react on ecosystem alterations by changing their abundance, density, conditions, or activities. Therefore, indicator species have to be selected objectively and must represent clear indicator-indicandum relationships. Pollinators have been suggested as useful bioindicators for ecosystem health by Kevan[35] as they are crucial for the functioning of almost all terrestrial ecosystems. Moreover, pollinators are needed for the provision of manifold ecosystem services. Further indicators based on the abundance of selected species have been suggested by Jørgensen et al.[6,36] and Burkhard et al.[7]

- Saprobic Classification: The saprobe index gives information about the degree of water pollution.[37] The different saprogenic stages are related to certain indicator organisms like bacteria, fungi, algae, amoeba, mussels, worms, insect larvae, or fishes. The stages range from polysaprobic (very highly polluted), α-mesosaprobic (highly polluted), β-mesosaprobic (medium polluted), to oligosaprobic (rather clean and clear water).

- Bellan's Pollution Index: Aquatic species like *Platynereis dumerilii, Theosthema oerstedi, Cirratulus cirratus,* and *Dodecaria concharum* are used as indicators for water pollution, whereas species like *Syllis gracillis* and *Typosyllis prolifera* indicate clear water conditions.[38] Bellan's pollution index equation is

$$IP = \sum \text{dominance of pollution indicator species/clear water indicator} \qquad (2)$$

Index values higher than 1 indicate a pollution-based disturbance in the community.

- AZTI Marine Biotic Index: AZTI Marine Biotic Index (AMBI) distinguishes the soft bottom macrofauna into five groups in accordance with their sensitivity to increasing stres[39]:
  - I. Species very sensitive to organic enrichment and eutrophication, present only under unpolluted conditions.
  - II. Species indifferent to organic enrichment, occurring in low densities only and with no significant variations over time.
  - III. Species tolerant to excess organic matter enrichment, usually supported by organic enrichment, that can also be found under normal conditions.
  - IV. Second-order opportunist species.
  - V. First-order opportunist species (deposit feeders).

The coefficient is calculated as follows:

$$AMBI = \left\{ (0 \times \%I) + (1.5 \times \%II) + (3 \times \%III) + (4.5 \times \%IV) + (6 \times \%V) \right\} / 100 \qquad (3)$$

The AMBI values vary among the following: normal (0.0–1.2), slightly polluted (1.2–3.2), moderately polluted (3.2–5.0), highly polluted (5.0–6.0), or very highly polluted (6.0–7.0).

- BENTIX

BENTIX is based on the AMBI but uses three groups only[40]:
  - I. Species generally sensitive to disturbances.
  - II. Species tolerant to stress or disturbance. Populations may respond to organic enrichment or other pollution sources.
  - III. First-order opportunistic species (pioneer, colonizers, or species that are tolerant to hypoxia).

The indicator values are calculated as follows:

$$Benix = \left\{ (6 \times \% I) + 2(\% II + \% III) \right\} / 100 \qquad (4)$$

The results represent different states of aquatic ecosystems: normal (4.5–6.0), slightly polluted (3.5–4.5), moderately polluted (2.5–3.5), highly polluted (2.0–2.5), or very highly polluted (Bentix = 0).

- Macrofauna Monitoring Index: Twelve indicator species are included in the macrofauna monitoring index. Each indicator species is assigned a score, based on the ratio of its abundance. The actual index value is the average score of species that are present in the sample.[41]
- Umbrella, Flagship, and Keystone Species: Umbrella species have high demands for their habitat conditions with regard to habitat size and quality. When protecting these species, many other species will be supported automatically. Flagship (or charismatic) species are organisms whose necessity for protection can be easily communicated. Keystone species provide an extraordinary importance for the maintenance of ecosystem structures and functions as well as for other species in the same ecosystem. Therefore, the identification and protection of keystone species can be crucial for the management of ecosystem health.[42]

Examples for the utilization of these indicator types are the Species Trend Index,[43] Red Lists,[44] or the Living Planet Index (LPI), which has been developed for land, freshwater, and marine vertebrate species. The average population trends over time are documented in the LPI. The actual calculations are based on a data set of more than 2500 species and 8000 population time series over the past 30 years. Three indices are calculated: 1) terrestrial species population index; 2) freshwater species population index, and 3) marine species population index. Each of these indices is set to a baseline of 100 in 1970, and all are given an equal weighting.[45,46]

Ratios between Different Classes of Organisms or Elements

The increase or decrease of one species in relation to others provides information about changes in ecosystems, for example, Nygard's algal index[36] or the diatoms/nondiatoms ratio.[47]

Indicators Based on Ecological Strategies

Different ecological strategies are altered by human activities or during different stages of natural development. Hence, indicators for the distinct behavior of different taxonomic groups under environmental stress situations were developed, e.g., the nematodes/copepods index, the polychaetes/ amphipods ratio, and the index of r/k strategists, which considers different taxa: Most communities in ecosystems in rather late developmental stages show dominance of k- selected or conservative species with large body sizes and long life spans. R-selected or opportunistic species have shorter life spans and are often numerically dominant. After a significant disturbance and during the following reorganization, the opportunistic species can become dominant in biomass as well as in number, whereas the conservative species are usually less favored.[36,48] Another strategy-related indicator is the trophic infaunal index, which refers to organisms' different feeding strategies (distinction of macrobenthos species into suspension feeders, interface feeders, surface deposit feeders, and subsurface deposit feeders).[49]

Additionally, there are several attempts aiming at harmonizing existing biodiversity indicator initiatives. Two examples are sketched here.

- Streamlining European 2010 Biodiversity Indicators: The Streamlining European Biodiversity Indicators (SEBI) were established in 2005 under the umbrella of the Convention on Biological Diversity (CBD). It is a process to select a set of biodiversity indicators to monitor progress toward the 2010 target of halting biodiversity loss and help achieve progress toward the target.[50,51] The SEBI is a regionally coordinated program that has been initiated in Europe as collaboration between the European Environment Agency and other European and United Nations institutions. The SEBI proposes a list of 26 indicators within the 7 CBD focal areas: status and trends of the components of biological diversity, threats to biodiversity, ecosystem integrity and ecosystem goods and services, sustainable use, status of access and benefit sharing, status of resource transfers, and public opinion.[50,51]
- Group on Earth Observations Biodiversity Observation Network: The Group on Earth Observations Biodiversity Observation Network (GEO BON) is a global partnership helping to collect, manage, analyze, and report biodiversity data.[52] It is a voluntary partnership of 73 national governments and 46 participating organizations and was launched in 2002. The GEO BON aims at providing a framework to coordinate data and observations within the Global Earth Observation System of Systems (GEOSS). Biodiversity has been named as one of nine GEOSS priority societal benefit areas. The GEO BON will integrate key ecosystem functional parameters into a Terrestrial Ecosystem Function Index (TEFI). The TEFI will integrate data of measurements of the energy, carbon, and nutrient balance.[52]

Aggregated Theory-Based Indicators and Indices

In contrast to the biotic approaches, which mainly can be used as structural indicators, the health component "vigor" is included in ecosystem theory based-indicators and indicator sets. The following four aggregations stem from thermodynamics, network, and information theories. Their basic target is a holistic aggregation of ecosystem properties into one guiding variable.

- Exergy and Exergy Indices: Exergy is that energy fraction that can be transformed into useful mechanical work. In ecological terms, it can, for example, be measured by the total biomass of the system. Eco-exergy is a refinement of the exergy concept in which biomass is weighted by the genetic complexity of the species observed.[4,36] Further holistic indicators are the exergy index and the specific exergy.[4] In ecosystems, the captured exergy is used to build up biomass and structures during successions.[48] Therefore, more complex systems also have more built-in exergy than simpler ones. Both exergy and specific exergy have been used as indicators for ecosystem health.[6] Relations between the exergy values and other ecosystem health characteristics like diversity, structure, or resilience can be found. For example, a very eutrophic ecosystem has a very high exergy due to the high biomass, but the specific exergy is low as the biomass is dominated by algae with low β values. The combination of exergy index and specific exergy provides a satisfactory structural and holistic description of ecosystem health.
- Entropy: Entropy production is one result of any metabolic activity. It can be measured by the system's respiration or the total system's export. As life is a very effective producer of entropy, this indicator has been proposed as an ecological orientor to represent maturity as well as ecosystem stress.[53,54]
- Emergy: Emergy (embodied energy) accounts for the differences between distinct biomass fractions in ecosystems basing upon the energy that has been used to build up the respective structure.[55] Conversion values, called transformities, have been derived to allow the calculation of emergy values for many ecological entities as well as socio-ecological products.[56]
- Ascendency: Ascendency is a holistic indicator that is based upon the energy flows in ecological systems and the information associated with the particular network configuration.[57,58] It represents the total system throughput and the flow diversity as a result of the food web structure. The respective network configuration is indicated by the average mutual information. Ascendency is measured by the total system throughput times the average mutual information, providing helpful information on an ecosystem's energy flow schemes and efficiencies.

Ecosystem Analytical Indicators and Indices

While most of the approaches mentioned before are aiming at one focal dimension and one value to characterize the state of an ecosystem, the following indicators have been constructed as multidimensional approaches. They try to represent ecosystem structures (biotic and abiotic structures), functions (water, matter, energy flows), and (in some cases) their relevance for human systems.

- Integrity Indicators and Orientors: Several ecosystem assessments are based on the concept of ecosystem integrity, which is closely related to ecosystem health.[18,59] The focal difference can be found in the origins of the concepts: While integrity was related to wilderness as a target function, health has been referring to ecosystems under human pressures from the beginning.[60] Meanwhile, the core conceptions have become rather similar. Therefore, one approach of integrity indication will be included in the following paragraphs.

  Taking into account the focal ideas of the sustainable development concept, "meet the needs of future generations "means "keep available ecosystem services on a long-term, intergenerational and broad scale, intragenerational level." From a synoptic viewpoint, all ecosystem services are strongly dependent on the performance of the system's regulation capacity (see the section on "Ecosystem Service Indicators"). Taking into account that the integral of regulating ecosystem services represents self-organized processes,[61] it becomes clear that the respective benefits are dependent on the degrees and potentials of self-organization. To maintain these services, the ability for future self-organizing processes has to be preserved.[53,62] Under this viewpoint, Barkmann et al.[63] have defined ecological integrity as a "management target for the preservation against nonspecific ecological risks, that are general disturbances of the self-organising capacity of ecological systems. Thus, the goal should be a support and preservation of those processes and structures which are essential prerequisites of the ecological ability for self-organisation".

In ecosystem theory, many different approaches (see Joergensen[4]) are highly compatible with the theory of self-organization. The consequences have been condensed within the orientor approach,[64–66] a system-based theory about ecosystem development, which is founded on the ideas of nonequilibrium thermodynamics,[54,62] and network development on the one hand and succession theory on the other.[67,68] The basic idea is that throughout the undisturbed complexifying development of ecosystems, certain characteristics are increasing steadily, developing toward an attractor state, which is restricted by the specific site conditions. For instance, the food web will become more and more complex; heterogeneity, species richness, and connectedness will be rising; and many other attributes will follow a similar long-term trajectory.

Many of these orientors cannot be easily measured or modeled under usual circumstances. Therefore, the selected indicators have to be represented by variables that are accessible by "traditional" methods of ecosystem quantification. Furthermore, it has to be reflected that the number of indicators should be reduced as far as possible, providing a small set consisting of the most important items that can be calculated or measured in many local instances. The focal subsystems that should be taken into account to represent ecosystem organization are ecosystem structures with the biotic and abiotic diversity and functions, represented by the energy, water, and matter balances (for a detailed justification, see Müller[61]).

On the basis of these features, a general indicator set to describe the ecosystem or landscape state in terrestrial environments has been derived. It is shown in Table 5. The basic hypothesis concerning this set is that a holistic representation of the degree of and the capacity for complexifying ecological processes on the basis of an accessible number of indicators can be fulfilled by these variables. They also represent the basic trends of ecosystem development; thus, they show the developmental stage of an ecosystem or a landscape. As a whole, this variable set represents the degree of self-organization in the investigated system. For quantifications, see Müller[69] or Müller and Burkhard.[70]

- The holistic ecosystem health indicator: An expansion toward an integration of human items is provided by the holistic ecosystem heath indicator (HEHI) system. It was developed in 1999 in Costa Rica as an integrative indicator that might be an appropriate tool for assessing and

TABLE 5 Proposed Indicators to Represent the Organizational State of Ecosystems and Landscapes

Orientor Group	Indicator	Potential Key Variable
Biotic structure	Biodiversity	Number of species
Abiotic structures	Biotope heterogeneity	Index of heterogeneity
Energy balance	Exergy capture	Gross or net primary production
	Entropy production	Entropy production[71] Entropy production[72] Output by evapotranspiration and respiration
	Metabolic efficiency	Respiration per biomass
Water balance	Biotic water flows	Transpiration per evapotranspiration
Matter balance	Nutrient loss	Nitrate leaching
	Storage capacity	Intrabiotic nitrogen
	Soil organic carbon	Soil organic matter

Source: Müller.[69]

TABLE 6   Elements of the Holistic Ecosystem Health Indicator Set

Ecological Elements	Social Elements	Interactive Elements
Soil quality	Income	Land use and distribution
Riparian zone	Access to Services	Watershed protection
Water quality	Job stability	Land degradation
Biomass	Gender roles	Citizen involvement
Land use	Demographics	Implementation of legislation
Primary production	Community	Environmental awareness
Regeneration	Strength	
Biodiversity		
Erosion		

Source: Aguilar.[73]

evaluating health of managed ecosystems.[73] The HEHI follows a hierarchical structure starting with three main branches: ecological, social, and interactive. The interactive branch includes measures relating to land use and management decisions that characterize the interactions between the human communities and the ecosystem. Furthermore, each branch is subdivided into categories or criteria (see Table 6). Each category is given a target or a benchmark, which is based on references available in scientific literature, policies, etc.

Ecosystem Service Indicators

As the explicit target figure of ecosystem health indicators has more and more been moved toward human wellbeing, the respective indicanda—ecosystem goods and services—are mentioned here. Their implementation seems to be very significant as a criterion of success in ecosystem health management. Ecosystem services are the benefits people obtain from natural structures and functions. Since ecosystems are dynamic and complex units, the assessment of their services is strongly facilitated by the categorization into functional groups, which are exemplarily listed in Table 7, referring to the following:

- Provisioning services: products obtained from ecosystems, e.g., food, water.
- Regulating services: benefits from regulating ecosystem processes, e.g., flood regulation, disease regulation.
- Cultural services: nonmaterial benefits, e.g., recreation, spiritual benefits, information.

Ecosystem structures and functions can be indicated by ecological integrity, as described above. Regulating ecosystem services is strongly related to ecosystem functions, and some regulating services are even overlapping with ecological integrity processes (e.g., processes related to nutrient or water regulation.[74] Thus, clear definitions of ecological integrity variables and regulating ecosystem service indicators are mandatory. Most ecosystem functions are difficult to quantify under natural conditions, but the application of ecological models can help. Perhaps the best data are available for provisioning ecosystem services. Normally, production and trade quantities and their market prices are used. Cultural ecosystem services again are rather difficult to quantify due to each individual's subjective and situation-dependent appreciation of related values.[26]

Ecosystem services are not a linear chain from means to ends because ecosystems as well as societal systems are complex, dynamic, and adaptive.[3] There exist multiscale relationships between services and benefits. When ecosystems are stressed and degraded, their service provision is affected too, which in turn has impacts on human activities and health.[75] The fact that there is a high correlation between decline in ecosystem health and service provision leads to the suggestion that ecosystem services are good integrative and aggregate measures, showing the consequences of the respective ecosystem health conditions.[76]

TABLE 7   List of Ecosystem Services

Regulating Ecosystem Services	
Local climate regulation	Effects on temperature, wind, radiation, precipitation
Global climate regulation	For example, carbon sequestration, greenhouse gas emission
Flood protection	Extreme flood event dampening
Groundwater recharge	Runoff, flooding, aquifer recharge
Air quality regulation	Removal of toxic and other elements from the atmosphere
Erosion regulation	Soil retention and prevention of landslides
Nutrient regulation	(Re)cycling of, e.g., N, P, or other elements
Water purification	Removal of impurities from fresh water
Pollination	For example, by wind and bees
Provisioning Ecosystem Services	
Crops	Edible plants
Livestock	Edible animals
Fodder	Animal fodder
Capture fisheries	Fish accessible for fishermen
Aquaculture	Terrestrial or marine aquaculture
Wild foods	For example, berries, mushrooms, hunting, fishing
Timber	Trees or plants for construction
Wood fuel	Trees or plants for heating, cooking
Energy (biomass)	Trees or plants for energy generation
Biochemicals and medicine	Production of biochemicals, medicines
Freshwater	For example. for drinking, irrigation
Cultural Ecosystem Services (selection)	
Recreation and aesthetic values	Landscape and visual qualities
Intrinsic value of biodiversity	Value of nature and species themselves

Source: Belcher and Boehm,[87] Schönthaler et al.[111] and Reuter et al.[112]

Health Indicators in Different Ecosystem Types

To illuminate the wide field of health indicators, in the following sections, some utilizations of the concept in different ecosystem types are presented.

Agroecosystem Health

An agroecosystem can be defined as "a socio-ecological system, managed primarily for the purpose of producing food, fiber and other agricultural products, comprising domesticated plants and animals, biotic and abiotic elements of the underlying soils, drainage networks, and natural vegetation and wildlife."[77–79] The health status of agro ecosystems has been described by Rapport,[80] and Rapport et al.[81] In this entry, a valuable overview about various approaches to indicate agroecosystem health on various spatial scales is provided. For example, Zhang et al.[82] merge geographical information systems (GIS)-based land use analysis data with the use of pesticides and their pathways through the environment. This approach allows the assessment of ecosystem health as a ratio between the amounts of pesticides applied in one grid to the maximum dose applied in the study area. Kaffka et al.[83] suggest that the capacity to retain nutrients like N and P might be useful to evaluate the ecosystem health of a catchment area, while Mitchell et al.[84] consider the content of soil organic matter (SOM) of agricultural areas as a focal indicator to assess ecosystem health in the foreground.

Hopkins[85] indicates ecosystem health via the number and size of wildlife patches in an area dominated by intensive agriculture.

In comparison to the aforementioned authors, Altieri and Nicholls[86] base their suggestions to achieve healthy agroecosystems on the avoidance capacity of diseases—indicated by optimal recycling of nutrients, closed energy flows, water and soil conservation, and biological pest regulation. Belcher and Boehm[87] use yield, soil N and P, soil water, SOM, soil erosion, and CO_2 emissions as major attributes in their sustainable agroecosystem model. However, the assessments of agroecosystem health regularly focus on resources that have to be classified as internal to the system.[88] Hence, the options to assess the health of an agroecosystem in relation to its ability to adapt to variations in its changing socio-economic and ecological context are rarely realized according to Waltner-Toews[79] or Ikerd.[89]

Forest Ecosystem Health

In the United States, 20 years ago, a sound definition of forest health had already been derived to sustain healthy conditions of ecosystem development and productivity in a long-term perspective.[90] The term "forest health" increasingly found evidence in mandates concerning environmental management and protection, mostly supported by the idea that the conventional measures for describing forest states (e.g., crown conditions, tree growth, loss of nutrients, soil potentials, biodiversity) can also be used to indicate forest ecosystem health. Regarding forests, only recently in Europe, Ecosystem Health has been advertised in the context of deposition of air pollutants[91] and the consequent change of chemical states, particularly the degree of eutrophication and acidification which relate to the holistic aspects of productivity and biogeochemical cycling. Information on nutrient recycling, imbalance with the inputs and outputs and on energy use are the essentials of the Ecosystem Assessment Health concept and the adequate indicators proposed by Jørgensen.[6] This kind of concept, basically productivity related and holistic, does greatly comply with ecosystem theories and opens on for indicating Forest Ecosystem Health combing both, utilitarian and ecosystem perspectives[92] respecting the conditions under which selforganization of forest ecosystems can take place.[50]

Whereas atmospheric deposition of acidifying air pollutants and eutrophic nitrogen is identified as one of the major environmental problems, an ecosystem process-orientated indicator has been demanded for quantifying marginal loads for damaging structures and for interference with ecosystem functions.[93–95] Since the respective concept of critical loads is a stoichiometric approach and a function of the load quantity on chemical effects on ecosystems, intensity criteria are needed to provide adequate threshold values.[96–98] The conduction of the concept depends on combined balances of mass and charge provided by mineral elements, nitrogen, and free acidity completing the ion composition of internal transfers and the matter–flux relationship with the abiotic environment. Imbalance between input and recycling respecting mass and ionic charge indicates the efficiency of nutrient use while the quantities can be related with the intensities of effective concentrations and free acidity. In this regard soil chemistry is a function of the extensity of production on the intensities of nutrient availability, free acidity and effectiveness of toxic concentrations under influence of ecosystem self-organization.

The performance of cycling and imbalance between input rates and the degree of recycling provide useful information on the Ecosystem Health aspect. Moreover, as imbalance of mineral element cycling and related losses from the ecosystem are irreversible processes for terrestrial ecosystems, decreasing alkalinity in combination with increasing free acidity are directing to maturity of ecosystems. Regarding abiotic structures, maturity emerges by development of soil structures, which is related to loss of potentials. Based on these principles, Ulrich[99] suggested the indication of stages of forest ecosystems, for instance, by the structural properties of soil constituents providing acid neutralization and buffer capacity.

Aquatic Ecosystem Health

Aquatic ecosystems (wetlands, marine and estuarine zones, lakes, groundwater, lagoons, and rivers) consist of complex structures and fulfill important functions for the provision of numerous ecosystem services.[26] Some of these aquatic ecosystems are heavily endangered, e.g., wetlands were turned into agricultural land with dramatic consequences due to the loss of the buffer capacity for pesticides, nutrients, and floods as well as the loss of habitat functions. Marine and estuarine ecosystems are heavily influenced by humans due to population growth and the associated consequences such as pollution, growing demands for resources, eutrophication, overfishing, and habitat modification (e.g., mangrove clearing). Estuarine and marine ecosystems are interdependent as estuarine areas are the nursery grounds for many species, which are then provided as successful functioning marine commercial stocks. Probably the most important function of lakes is the freshwater storage, which provides ecosystem services for society and economy, but these systems also are heavily endangered.

Aquatic ecosystems are focal areas of ecosystem health assessments. Therefore, many of the previously mentioned indicator concepts mainly provide information on aquatic ecosystem health (see Utilized indicators of ecosystem health or Joergensen et al.[6]). Besides these long indicator listings, Boesch and Paul[100] highlight the following traditional indicators: contaminant levels, material input (e.g., nutrients, sediments), water quality (e.g., dissolved oxygen), fish catch, extent of certain habitats (e.g., wetlands), community structure, toxicity biomarkers, and indicators of human pathogens).

Urban Ecosystem Health

A very special aspect is provided by the health concept in urban systems, i.e., because the human factor plays a dominant role in the relevant literature.[101–105] For instance, Su et al.[101] state that "an urban ecosystem consists of residents and their environment in certain time and space scales, in which, ecologically-speaking, consumers are the dominant component lacking producers and decomposers."

Therefore, urban ecosystem health must be assessed by very comprehensive, integrative indicator sets, which also include variables of human health. Consequently, Hancock[104] has determined six basic elements for healthy cities:

1. Population health and distribution.
2. Societal well-being.
3. Government, management, and social equity.
4. Human habitat quality and convenience.
5. Natural environment quality.
6. Impact of the urban ecosystem on the larger-scale natural ecosystem.

A similar approach can be found in the indicator sets of Su et al.,[101] who generally distinguish human and environmental subsystems (see Table 8).

TABLE 8 Some Indicanda for Urban Health

Human Subsystem	Environmental Subsystem
Public health	Provisioning services
Health expenditure	Environmental quality
Nutrition	Atmospheric quality
Budget and finance situation	Water quality
Urban infrastructure	Forest coverage
Human housing conditions	Farmland area
Education	Emergy density
Employment	Carrying capacity

Source: Su et al.[101]

Indicator Application and Aggregation in Management

Management toward ecosystem health is directed to improve human well-being. However, in order to environmentally manage, alternatives to achieve higher levels of human well-being and ecosystem health or to stabilize the present level in a changing world have to be identified. Hence, indicators have to facilitate the comparability of states in space and time, in order to allow informed decisions based on assessments of the present state and the state achievable by management options in respect to the power and competences in the hands of the manager. Concerning ecosystem health, at least, decision makers acting on the following levels have to be equipped with indicators:

1. Site management, e.g., field, forest, lake
2. Unit management, e.g., farm, forest district, catchment, nature sanctuary
3. Public management, e.g., community, county, nation
4. Public, e.g., citizens, interest groups (nongovernmental organizations)

All levels are interrelated. For example, 1) different kinds of pest management on the site level change not only the local ecosystem but also the food quality and availability. 2) The fodder quality available on the farm level depends on the productivity of the fields and impacts the economic efficiency of the farm. 3) The socio-economic state on the community level depends, on the one hand, on the economic viability of the hosted unit (item 2), but the provision of public services likewise constrains the range of activities of the units. 4) Adaptation to changing global constraints of public (item 3), unit (item 2), and site (item 1) management largely depends on the public awareness and level of satisfaction, e.g., human well-being realized at present and achievable in the future.

To account for all levels requires a nested approach in indication of ecosystem health plus mirroring the mutual synergistic and antagonistic interactions between the different levels. The need for such an approach has been articulated by Walter-Toewe and Wall[88] and Rapport and Singh.[106] However, a broadly accepted scheme meeting these requirements is not in place yet.[107] The present state of the art is largely influenced by the concept of ecosystem services and human well-being presented by the Millennium Assessment[26] and the TEEB Study.[108] With respect to "management," the further development of indicators should be constrained by the range of management options in space and time available to decision makers on the indicated levels (1–4) and be limited to parameters relatable to indicators of human well-being.

The need for indicator aggregation evolves out of the definition of ecosystem health. According to Waltner-Toews and Wall,[88] a nested approach is required in this context also. Another constraining aspect to be considered is societal interests, which are embedded in the definition as well. In addition, in case that the state of ecosystem health is considered to be poor, new or changed management activities have to be initiated and monitored by indicators. Thus, the goal of indicator aggregation is to transfer the information about the ecosystem state to those spatiotemporal scales on which management is possible and performed. The respective levels range from site management via communities, counties, states, nations up to the international institutions. Hence, a satisfying overlap between the spatial extent of the ecosystems at stake and the respective management unit is required. Systems theory and hierarchy theory provide a theoretical background to facilitate indicator aggregation for such nested systems.

To practically deal with this means to accept generalizations, including losses of information, as never can all interactions causing emergent properties relevant for human well-being be known. Hence, bottom-up aggregation is normally severely restricted. Furthermore, the focus on ecosystems enforces the integration of components, which are measured with parameters that cannot be aggregated on the base of concise units, e.g., the mortality rate of a specific species in forests of a watershed is already challenging to determine but analytically impossible to fuse with the number of pathogens endangering the fish population in a lake of the same catchment. A feasible approach to deal with this challenge is to work with relative indicator values, i.e., to indicate relative changes of the selected variables. A suitable option to facilitate spatiotemporal comparisons is to define a particular ecosystem state as reference state and to study the relative alterations from this reference state. Examples harnessing this approach have been presented by Windhorst et al.[109,110] A suitable method to facilitate aggregation and to deal with spatiotemporal interactions is to use simulation models (Belcher and Boehm,[87] Schönthaler et al.[111] and Reuter et al.[112]).

In any case, all ecosystem health indicators and respective aggregation approaches should be connectable to the different components of human well-being. The Millennium Assessment's[26] categories were "security, basic material for good life, health, good social relations and freedom of choice and action." They can be indicated for the management units at stake. Hence, multiple interactions to be considered take place and further are conceivable for future situations, creating a fuzzy environment, obstructing the development of generally applicable aggregation procedures. However, progress in identifying suitable aggregation procedures can be achieved by answering the following questions for each management unit at stake:

- Can the aggregated indicator of ecosystem health be addressed and modified by at least one management action?
- Are changed values of the aggregated indicator indicative for different ecosystem states of the management unit at stake?
- Do changed values of the aggregated indicator indicate betterment of at least one category of human well-being?

Conclusions

The previous paragraphs have shown that there is an enormous variety of indicators proposed to represent ecosystem health. Many of these indicators have been used to assess ecosystem health in different ecosystem types, and in many of those cases, the indicators have been useful tools for environmental management. On the other hand, these applications illuminate some general problems of health indicators that should be solved in the future:

- The health approach is very challenging, i.e., due to its comprehensive character. Therefore, health status can hardly be represented by one variable alone. Instead, comprehensive indicator sets are necessary to include the basic elements of vigor, organization, and resilience.
- Indicators or sets selected have often failed in fulfilling the comprehensive criteria of ecosystem health assessments because they are mostly specific for one particular environmental problem to be solved[6] and do not adequately represent ecosystem complexity. Thus, there are difficulties in satisfying the original ecosystem health idea advertised.
- Consequently, the meaningfulness of solely structural indicators is limited, as they do not reflect processes of ecosystem functions. Therefore, many of the listed indicators should be understood as elements of indicator sets, not as single indicators of ecosystem health.
- As the health concept has been outlined in strict contact with human systems, the linkage between man and environment should be included, at least in indicator selection. That linkage up to now can hardly be found in the literature.
- Due to the metaphoric character of the "health" approach, it has been very successful in some areas, while in other nations, it has not been applicable due to critical viewpoints on the concept's theoretical or even philosophical character.
- Furthermore, the indication of ecosystem health also covers the health status of the human population, as well as socio-economic and cultural dimensions in relation to the vigor, organization, and resilience of ecosystems.[18] Hence, parameters to indicate security, basic material for good life, health, and good social relations are indispensable to analyze overall ecosystem health, which is in line with the assessment of Rapport et al.[75] An integrative approach to bundle a suite of indicators and to attach meaningful values is the concept of ecosystem distress syndromes,[113]

which can be seen as a forerunner of the environmental degradation syndromes elaborated by the German Advisory Council on Global Change (WBGU). The three major syndromes have been named by WBGU[105]: 1) utilization, which includes the overexploitation of marginal land; 2) development which includes the destruction of ecosystems as a result of large-scale projects; and 3) sink, comprising environmental degradation resulting from large-scale diffusion of long-lived substances.

Coming back to our initial questions, we can summarize the following:

- The basic features and requirements of the ecosystem health approach demand for comprehensive and integrative indicator sets, which is a big scientific and practical challenge.
- Therefore, several proposals exist, and the health concept is used as a reference in several cases, but very often, the interdisciplinary demands of the approach are not fulfilled.
- Applications can be found mainly in aquatic ecosystems, mostly being quite distant from the involvement of human factors.
- Good chances for future development can be seen by enhancing the integration with integrity and ecosystem services.

References

1. Wilson, G.A.; Bryant, R.L. *Environmental Management: New Directions for the 21st Century;* Taylor and Francis Group: London, 2002.
2. Fath, B.D.; Patten, B.C. Review of the foundations of network environ analysis. Ecosystems **1999**, 2, 167–179.
3. Allen, C.R.; Holling, C.S. *Discontinuities in Ecosystems and Other Complex Systems*; Columbia University Press, New York: 2008.
4. Joergensen, S.E.; Fath, B.; Bastianoni, S.; Marquez, J.; Müller, F.; Nielsen, S.N.;Patten, B.; Tiezzi, E.; Ulanowicz, R. *A New Ecology—The Systems Perspective;* Elsevier Publishers: Amsterdam, 2007.
5. Bossel, H. Indicators for sustainable development: Theory, method, applications. A report to the Balaton Group. International Institute for Sustainable Development: Winnipeg, 1999.
6. Jørgensen, S.E.; Xu, F.L.; Costanza, R., Eds. *Handbook of Ecological Indicators for Assessment of Ecosystem Health*, 2nd Ed. (Applied Ecology and Environmental Management); CRC Press: London, 2010.
7. Burkhard, B.; Müller, F.; Lill, A. Ecosystem health indicators: Overview. In *Ecological Indicators. Vol. [2] of Encyclopedia of Ecology*, 5 vols; Jørgensen, S.E., Fath, B.D., Eds.; Elsevier: Oxford, 2008; 1132–1138.
8. Dale, V.H.; Beyeler, S.C. Challenges in the development and use of ecological indicators. Ecol. Indic. **2001**, 1, 3–10.
9. Wiggering, H.; Müller, F., Eds. *Umweltziele und Indikatoren;* Springer-Verlag: Berlin, Heidelberg, New York, 2003.
10. Müller, F.; Lenz, R. Ecological indicators: Theoretical fundamentals of consistent applications in environmental management. Ecol. Indic. **2006**, 6, 1–5.
11. Walz, R. Development of environmental indicator systems: Experiences from Germany. Environ. Manage. **2000**, 26, 613–623.
12. Turnhout, E.; Hisschemoller, M.; Eijsackers, H. Ecological indicators: Between the two fires of science and policy. Ecol. Indic. **2007**, 7, 215–228.
13. Niemeijer, D.; De Groot, R.S. A conceptual framework for selecting environmental indicators sets. Ecol. Indic. **2008**, 8, 14–25.
14. Girardin, P.; Bockstaller, C.; van der Werf, H.M.G. Indicators: Tools to evaluate the environmental impacts of farming systems. J. Sustainable Agric. **1999**, 13, 5–21.
15. EEA. Environmental Indicators: Typology and Use in Reporting; EEA: Copenhagen, 2003; 20 pp.

16. Heink, U.; Kowarik, I. What are indicators? On the definition of indicators in ecology and environmental planning. Ecol. Indic. **2010**, *10*, 584–593.

17. Leopold, A. Wilderness as a land laboratory. Living Wilderness **1941**, *6*, 3.

18. Rapport D.J. Regaining healthy ecosystems: The supreme challenge of our age. In *Managing for Healthy Ecosystems*; Rapport, D.J., Ed.; Lewis Publisher: Boca Raton, 2003; 5–10.

19. Costanza, R.; Norton, B.G.; Haskell, B.D., Eds. *Ecosystem Health: New Goals for Environmental Management*; Island Press: Washington, DC, 1992; 279 pp.

20. UNCED (United Nations Conference for Environment and Development). *Agenda 21*; United Nations: New York, 1992.

21. Haskell B.D.; Norton, B.G.; Costanza, R. Introduction: What is ecosystem health and why should we worry about it? In *Ecosystem Health: New Goals for Environmental Management*; Costanza, R., Norton, B.G., Haskell B.D., Eds.; Island Press: Washington, DC, 1992; 3–20.

22. Karr, J.D. Measuring biological integrity: Lessons from streams. In *Ecological Integrity and the Management of Ecosystems*; Woodley, S., Kay, J., Francis, G., Eds.; CRC Press: Boca Raton, FL, 1993; 83–104.

23. Rapport, J.D.; Rolston, E.D.; Qualset, O.C.; Damania, B.A.; Lasley, L.W. *Managing for Healthy Ecosystems*; CRC Press: Boca Raton, London, New York, Washington, DC, 2002.

24. Xu, W.; Mage, J.A. A review of concepts and criteria for assessing agroecosystem health including a preliminary case study from southern Ontario. Agric., Ecosyst. Environ. **2001**, *83*, 215–233.

25. De Kruijf, H.A.M.; Van Vuuren, D.P. Following sustainable development in relation to the North–South dialogue: Ecosystem health and sustainability indicators. Ecotoxicol. Environ. Safety **1989**, *40*, 414.

26. MA (Millennium Ecosystem Assessment). *Ecosystems and Human Wellbeing*: Synthesis. Island Press, World Resources Institute: Washington, DC, 2005.

27. Scholes, R.J.; Mace, G.M.; Turner, W.; Geller, G.N.; Jürgens, N.; Larigauderie, A.; Muchoney, D.; Walther, B.A.; Mooney, H.A. Toward a global biodiversity observing system. Science **2008**, *321*, 1044–1045.

28. van Strien, A.J.; van Duuren, L.; Foppen, R.P.B.; Soldaat, L.L. A typology of indicators of biodiversity change as a tool to make better indicators. Ecol. Indic. **2009**, *9*, 1041–1048.

29. Lamb, E.G.; Bayne, E.; Holloway, G.; Schieck, J.; Boutin, S.; Herbers, J.; Haughland, D.L. Indices for monitoring biodiversity change: Are some more effective than others? Ecol. Indic. **2009**, *9*, 432–444.

30. Rossi, J.P. Extrapolation and biodiversity indicators: Handle with caution! Ecol. Indic. **2010**, doi:10.1016/ j.ecolind.2010.09.002.

31. Marquez, J.C.; Salas, F.; Patricio, J.; Teixera, H.; Neto, J.M. *Ecological Indicators for Coastal and Estuarine Environmental Assessment: A User Guide*; WIT Press: Southampton, 2009.

32. Simpson, E.H. Measurement of diversity. Nature **1949**, *163*, 688.

33. Hilty, J.; Merenlender, A. Faunal indicator taxa selection for monitoring ecosystem health. Biol. Conserv. **2000**, *92*, 185–197.

34. Pinto, R.; Patricio, J.; Baeta, A.; Fath, B.; Neto, J.M.; Marques, J.C. Review and evaluation of estuarine biotic indices to assess benthic condition. Ecol. Indic. **2009**, *9*, 1–25

35. Kevan, P.C. Pollinators as bioindicators of the state of the environment: Species, activity and diversity. Agric., Eco-syst. Environ. **1999**, *74*, 373–393.

36. Jørgensen, S.E. Introduction. In *Handbook of Ecological Indicators for Assessment of Ecosystem Health*, 2nd Ed. (Applied Ecology and Environmental Management); Jørgensen, S.E., Xu, F.L., Costanza, R., Eds.: CRC Press: London, 2010.Jørgensen, S.E.; Xu, F.L.; Salas, F.; Marques, J.C. *Application of Indicators for the Assessment of Ecosystem Health. Handbook of Ecological Indicators for the Assessment of Ecosystem Health*; CRC Press, Boca, Raton: 2005.

37. Kolkwitz, R.; Marsson, M. Grundsätze für die biologische Beurteilung des Wassers nach seiner Flora und Fauna. Mitt. aus d. kgl. Prüfungsanstalt für Wasserversorgung und Abwässerbeseitigung **1902**, *1*, 33–72.

38. Bellan, G. Pollution Indices. In *Encyclopedia of Ecology*, 5 vols; Jørgensen, S.E., Fath, B.D., Eds.; Elsevier: Oxford, 2008; 2861–2868.
39. Borja, A.; Franco, J.; Perez, V. A marine biotic index to establish the ecological quality of soft-bottom benthos within European estuarine and coastal environments. Mar. Pollut. Bull. **2000**, *40* (12), 1100–1114.
40. Simboura, N.; Zenetos, A. Benthic indicators to use ecological quality classification of Mediterranean soft bottom marine ecosystems, including a new biotic index. Mediterr. Mar. Sci. **2002**, *3* (2), 77–111.
41. Roberts, R.D.; Gregory, M.G.; Foster, B.A. Developing an efficient macrofauna monitoring index from an impact study—A dredge spoil example. Mar. Pollut. Bull. **1998**, *36* (3), 231–235.
42. Simberloff, D. Flagships, umbrellas, and keystones: Is single-species management passé in the landscape era? –Biol. Conserv. **1998**, *83* (3), 247–257.
43. Cocciufa, C.; Petriccione, B.; Framstad, E.; Bredemeier, M. Biodiversity Assessment in LTER sites. An EC Report (Deliverable 3.R2.D1) from ALTER Net (A Long Term Biodiversity, Ecosystem and Awareness Research Network); 2007.
44. Brito, D.; Ambal, R.G.; Brooks, T.; De Silva, N.; Foster, M.; Hao, W.; Hilton-Taylor, C.; Paglia, A.; Rodríguez, J.P.; Rodríguez, J.V. How similar are national red lists and the IUCN Red List? Biol. Conserv. **2010**, *143*, 1154–1158.
45. UNEP-WWF. *The Living Planet Report*; 2004.
46. Loh, J. *Living Planet Report 2000*. Gland, World Wide Fund for Nature; 2000.
47. Lenhart, H.J. Effects of river nutrient load reduction on the eutrophication of the North Sea, simulated with the ecosystem model ERSEM. Senckenbergiana maritima **2001**, *31* (2), 299–311.
48. Burkhard, B.; Fath, B.D.; Müller, F. Adapting the adaptive cycle: Hypotheses on the development of ecosystem properties and services. Ecol. Modell. **2011**, doi:10.1016/ j.ecolmodel.2011.05.016.
49. Word, J.Q. Classification of benthic invertebrates into infaunal trophic index feeding groups. Coastal Water Research Project Report 1979–1980; Los Angeles, 1979; 103–121.
50. EEA. Halting the Loss of Biodiversity by 2010: Proposal for a First Set of Indicators to Monitor Progress in Europe, EEA Technical Report No 11/2007; 2007.
51. EEA. *Progress towards the European 2010 Biodiversity Target*, EEA Technical Report No 5/2009; 2009.
52. GEO BON (Group on Earth Observations Biodiversity Observation Network). Detailed Implementation Plan. GEO BON. Version 1.0, May 22, 2010.
53. Kay, J.J. On the nature of ecological integrity: Some closing comment. In *Ecological Integrity and the Management of Ecosystems*; Woodley, S., Kay, J., Francis, G., Eds.; St. Lucie Press: Delray, FL, 1993; 210–212.
54. Schneider, E.D.; Kay, J.J. Life as a manifestation of the second law of thermodynamics. Math. Comput. Modell. **1994**, *19*, 25–48.
55. Odum, H.T. Environmental Accounting: Emergy and Environmental Policy Making; John Wiley and Sons: New York, 1996; 370 pp.
56. Brown, M.T.; Ulgiati, S. Updated evaluation of exergy and emergy driving the geobiosphere: A review and refinement of the emergy baseline. Ecol. Modell. **2010**, *221*, 2501–2508.
57. Ulanowicz, R.E. *Growth and Development: Ecosystems Phenomenology*; Springer-Verlag: NY, 1986; 203 pp.
58. Ulanowicz, R.E. *Ecology the Ascendent Perspective*; Columbia University Press: NY, 1997; 201 pp.
59. Callicott, J.B. A review of some problems with the concept of ecosystem health. Ecosyst. Health **1995**, *1* (2), 101–112.
60. Westra, L. The ethics of ecological integrity and ecosystem health: The interface. In *Managing for Healthy Ecosystems*; Rapport, D.J., Ed.; Boca Raton, London, New York, Washington, DC, 2003; 31–40.

61. Müller, F. Ecosystem indicators for the integrated management of landscape health and integrity. In *Ecological Indicators for Assessment of Ecosystem Health*; Joergensen, S.E., Costanza, R., Xu, F.L., Eds.; Taylor and Francis: Boca Raton, FL, 2004; 277–304.

62. Kay, J.J. Ecosystems as self-organised holarchic open systems: Narratives and the second law of thermodynamics. In *Handbook of Ecosystem Theories and Management*; Joergensen, S.E., Müller, F., Eds.; Lewis Publishers: Boca Raton, 2000; 135–160.

63. Barkmann, J.; Baumann, R.; Meyer, U.; Müller, F.; Windhorst, W. Ökologische Integrität: Risikovorsorge im Nachhaltigen Landschaftsmanagement. Gaia **2001**, *10* (2), 97–108.

64. Bossel, H. Sustainability: Application of systems theoretical aspects to societal development. In *Handbook of Ecosystem Theories and Management*; Jørgensen, S.E., Müller, F., Eds.; CRC Press, Boca Raton: 2000; 519–536.

65. Müller, F.; Leupelt, M. *Eco Targets, Goal Functions, and Orientators*; Springer: Berlin, 1998; 619 pp.

66. Müller, F.; Jørgensen, S.E. Ecological orientors: A path to environmental applications of ecosystem theories. In *Handbook of Ecosystem Theories and Management*; Jørgensen, S.E., Müller, F., Eds.; CRC Publishers: New York, 2000; 561–576.

67. Odum, E.P. The strategy of ecosystem development. Science **1969**, *164*, 262–270.

68. Odum, E.P. *Fundamentals of Ecology*; WB Saunders WB: Philadelphia, 1971.

69. Müller, F. Indicating ecosystem and landscape organization. Ecol. Indic. **2005**, J (4), 280–294

70. Müller, F.; Burkhard. B. Ecosystem indicators for the integrated management of landscape health and integrity. In *Handbook of Ecological Indicators for Assessment of Ecosystem Health*, 2nd Ed.; Jorgensen, S.E., Xu, L., Costanza, R., Eds.; Taylor and Francis, 2010; 391–423.

71. Aoki, I. Entropy and exergy in the development of living systems: A case study of lake ecosystems. J. Phys. Soc. Jpn. **1998**, *67*, 2132–2139

72. Svirezhev, Y.M.; Steinborn, W. Exergy of solar radiation: Thermodynamic approach. Ecol. Modell. **2001**, *145*, 101–110.

73. Aguilar, B.J. Applications of ecosystem health for the sustainability of managed systems in Cost Rica. Ecosyst. Health **1999**, *5*, 1–13.

74. Burkhard, B.; Kroll, F.; Nedkov, S.; Müller, F. Mapping supply, demand and budgets of ecosystem services. Ecol. Indic. **2011**, doi:10.1016/j.ecolind.2011.06.019.

75. Rapport, D.J.; Costanza, R.; McMichael, A.J. Assessing ecosystem health. TREE **1998**, *13* (10), 397–402.

76. Rapport, D.J. Ecosystem services and management options as blanket indicators of ecosystem health. J. Aquat. Ecosyst. Health **1995**, *4*, 97–105.

77. Gallopin, G. The potential of agroecosystem health as a guiding concept for agricultural research. Ecosyst. Health *1*, 129–140.

78. Waltner-Toews, D. Ecosystem health: A Framework for implementing sustainability in agriculture. BioScience *46* (9), 686–689.

79. Waltner-Toews, D. Agro-ecosystem health: Concept and principles. In *Agro-Ecosystem Health*, Proceedings of a seminar held in Wageningen, Sep 26, 1996; van Bruchem, J., Ed.; NRLO-rapport nr. 97/31; 1997 9–22.

80. Rapport, D.J. Evaluating ecosystem health. J. Aquat. Eco- syst. Health **1992**, *1*, 15–24.

81. Rapport, D.J.; Lalsley, W.L.; Rolston, E.R.; Nielsen, N.O.; Qualset, C.Q.; Damania, A.B. *Managing for Healthy Ecosystems*, Proceedings of the 1st Ecosystem Management Congress, 1999; CRC Press LLC, 2003; 1510 pp.

82. Zhang, M.; Smallwood, K.S.; Anderson, E. Relating indicators of ecosystem health and ecological integrity to assess risks to sustainable agriculture and native biota—A case study of Yolo County, California. In *Managing for Healthy Ecosystems*, Proceedings of the 1st Ecosystem Management Congress, 1999; Rapport, D.J., et al., Eds.; CRC Press LLC, 2003; 757–768.

83. Kaffka, S.R.; Dhawan, A.; Kirby, D.W. Irrigation, agricultural drainage, and nutrient loading in the Upper Klamath Basin. In *Managing for Healthy Ecosystems*, Proceedings of the 1st Ecosystem Management Congress, 1999; Rapport, D.J., et al., Eds.: CRC Press LLC, 2003; 1011–1026.

84. Mitchell, J.P.; Lanini, W.T.; Temple, S.R.; Brostrom, P.N.; Herrrero, E.V.; Miyao, E.M.; Prather, S.; Hembree, K.J. Reduced-disturbance agroecosystems in California. In *Managing for Healthy Ecosystems*, Proceedings of the 1st Ecosystem Management Congress, 1999; Rapport, D.J., et al., Eds.; CRC Press LLC, 2003; 993–997.

85. Hopkins, D.H. Fallow LAND Patches and ecosystem health in California's Central Valley agro-ecosystem. In *Managing for Healthy Ecosystems*, Proceedings of the 1st Ecosystem Management Congress, 1999; CRC Press LLC, 2003; 981–991.

86. Altieri, M.A.; Nicholls, C.I. Ecologically based pest management: A key pathway to achieving agroecosystem health. In *Managing for Healthy Ecosystems*, Proceedings of the 1st Ecosystem Management Congress, 1999; Rapport, D.J., et al., Eds.; CRC Press LLC, 2003; 999–1010.

87. Belcher, K.W.; Boehm, M. Evaluating agroecosystem sustainability using an integrated model. In *Managing for Healthy Ecosystems*, Proceedings of the 1st Ecosystem Management Congress, 1999; Rapport, D.J., et al., Eds.; CRC Press LLC, 1209–1226.

88. Waltner-Toews, D.; Wall, E. Emergent perplexity of postnormal questions for community and agroecosystem health. Soc. Sci. Med. **1997**, *45* (11), 1741–1749.

89. Ikerd, J. Assessing the health of agroecosystems: A socioeconomic perspective. In Proceedings of the 1st International Ecosystem Health and Medicine Symposium, University of Missouri; 1994; 9 pp.

90. Kolb, T.E.; Wagner, M.R.; Covington, W.W. Utilitarian and ecosystem perspectives. Concepts of forest health. J. For. **1994**, *92* (2), 10–15.

91. Percy, K.E.; Ferretti, M. Air pollution and forest health: Towards new monitoring concepts. Environ. Pollut. **2004**, 130, 113–126.

92. USDA Forest Service. Healthy forests for America's future: A strategic plan, USDA Forest Service MP-1513; 1993; 58 pp.

93. Spranger, T.; Lorenz, U.; Gregor, H.D., Eds. *Manual on Methodologies and Criteria for Modelling and Mapping Critical Loads and Levels and Air Pollution effects, Risks and Trends*; Federal Environmental Agency (Umweltbundesamt): Berlin, 2004.

94. Gauger, T.; Haenel, H.D.; Rösemann, C.; Dämmgen, U.; Bleeker, A.; Erisman, J.H.; Vermeulen, A.T.; Schaap, M.; Timmermanns, R.M.A.; Builtjes, P.J.H.; Duyzer, J.H. *National Implementation of the UNECE Convention on Long-range Transboundary Air Pollution (Effects)—Part 1: Deposition Loads: Methods, Modelling and Mapping Results, Trends*, Environmental Research of the Federal Ministry of the Environment, Natural Conservation and Nuclear Safety Research Report 204 63 252, No. 38/2008, UBA-FB 001189E; Federal Environmental Agency (Umweltbundesamt): Dessau, 2010a.

95. Gauger, T.; Haenel, H.D.; Rösemann, C.; Nagel, H.D.; Becker, R.; Kraft, P.; Schlutow, A.; Schütze, G.; Weigelt- Kirchner, R.; Anshelm, F. Nationale Umsetzung UNECE- Luftreinhaltekonvention (Wirkungen) Teil 2: Wirkungen und Risiokoabschätzungen Critical Loads, Biodiversität, Dynamische Modellierung, Critical Levels Überschreitungen, Materialkorrosion, Environmental Research of the Federal Ministry of the Environment, Natural Conservation and Nuclear Safety Research Report 204 63 252l No. 39/2008, UBA-FB 001189; Federal Environmental Agency (Umweltbundesamt): Dessau, 2010b.

96. UNECE Convention on Long-range Transboundary Air Pollution. Manual on methodologies for mapping critical loads/levels and geographical areas where they are exceeded, http://www.rivm.nl/thema/images/mapman-2004_tcm61-48383.pdf.

97. Warfinge, P.; Sverdrup, H. Calculating critical loads of acid deposition with profile—A steady soil chemistry model. Water, Air Soil Pollut. **1992**, *63*, 119–143.

98. Warfinge, P.; Sverdrup, H. Critical loads of acidity to Swedish forest soils, methods, data and results. Dep. Chem. Eng., Rep. Ecol. Environ. Eng. 5; Lund University: Lund, Sweden, 1995; 104 pp.

99. Ulrich, B. Stability, elasticity, and resilience of terrestrial ecosystems with respect to matter balance. In *Potentials and Limitations of Ecosystems Analysis*; Schulze, E.D., Zwölfer, H., Eds.; Springer: Berlin, 1987; 435 pp.

100. Boesch, D.F.; Paul, J.F. An overview of coastal environmental health indicators. Hum. Ecol. Risk Assess. 2001, *7* (5), 1–9.

101. Su, M.; Fath, B.D.; Yang, Z. Urban ecosystem health assessment: A review. Sci. Total Environ. **2010**, *408*, 2425–2434.

102. Bell, M.L.; Cifuntes, L.A.; Davis, D.L.; Cushing, E.; Gusman Telles, A.; Gouveia, N. Environmental health indicators and a case study of air pollution in Latin American Cities. Environ. Res. **2011**, 111, 57–66.

103. Alberti, M. Maintaining ecological integrity and sustaining ecosystem function in urban areas. Curr. Opin. Environ. Sustainability **2010**, 2, 178–184.

104. Hancock, T. Health and sustainability in the urban environment. Environ. Impact Assess. Rev. **1996**, *16*, 259–277.

105. Tzoulas, K.; Korpela, K.; Venn, S.; Ylipelkonen, V.; Kazmierczak, A.; Niemela, J.; James, P. Promoting ecosystem and human health in urban areas using green infrastructure: A literature review. Landscape Urban Plann. **2007**, *81*, 167–178.

106. Rapport, D.J.; Singh, A. An EcoHealth-based framework for state of environment reporting. – Ecol. Indic. **2005**, *6*, 409–428.

107. Li, B.; Xie, H.L.; Guo, H.H.; Hou, Y. Study on the assessment method of agroecosystem health based on the pressure-state-response mode. In Proceedings of the IEEE International Conference on Industrial Engineering and Engineering Management, Hong Kong, Dec 8–11, 2009; 2458–2462.

108. Kumar, P., Ed. *The Economics of Ecosystems and Biodiversity: Ecological and Economic Foundations*; Earthscan, 2010; 456 pp.

109. Windhorst, W; Müller, F.; Wiggering, H. Umweltziele und Indikatoren für den Ökosystemschutz. In *Umweltziele und Indikatoren*; Müller, F., Wiggering, H., Eds.; Springer Verlag, 2003; 345–373.

110. Windhorst, W.; Colijn, F.; Kabuta, S.; Laane, R.P.; Lenhart, H. Defining a good ecological status of coastal waters—A case study for the Elbe plume. In *Managing European Coasts;* Vermaat, J.E., Bouwer, L., Turner, K., Salomons, W., Eds.; Springer Verlag, 2005; 59–74.

111. Schönthaler, K.; Meyer, U.; Pokorny, D.; Reichenbach, M.; Schuller, D.; Windhorst, W. *Ökosystemare Umweltbeobachtung—Vom Konzept zur Umsetzung*; Erich-Schmidt-Verlag: Berlin, 2004; 370 pp.

112. Reuter, H.; Middelhoff, U.; Schmidt, G.; Windhorst, W.; Schröder, W.; Breckling, B. Up-scaling the environmental effects of genetically modified plants—Assessing potential impact on nature conservation areas in Northern Germany. In *Multiple Scales in Ecology*; Schröder, B., Reuter, H., Reineking, B., Eds.; Peter Lang Internationaler Verlag der Wissenschaften: Frankfurt am Main. Band 13 edition. Theorie in der Ökologie, 2007; 95–109.

113. Rapport, D.J.; Regier, H.A.; Hutchinson, T.C. Ecosystem behavior under stress. Am. Nat. **1985**, *125* (5), 617–640.

20

Sustainable Fisheries: Models and Management

Fabian
Zimmermann
and Katja Enberg

Introduction: Population Dynamics and Fisheries

Population dynamics models are a key component of fisheries sciences to describe the changes in populations over time and their responses to fishing. The need for modeling approaches originates from the difficulties to observe fish directly. Together with the large socioeconomic relevance of fisheries, this has put fisheries models at the forefront of modeling biological systems. The focus has traditionally been on describing changes in population biomass through the growth and decay of a population's biomass. Historically, models have been divided into biomass models that lump entire populations into one biomass and models that are structured by age or size, allowing for more specific dynamics such as growth of body size, maturation, reproduction, recruitment and mortality. In this section, we will contrast biomass models with their structured counterparts, present models of growth, stock-recruitment and mortality, and outline current applications in a fisheries context.

Biomass, Age and Size Structure

In the 1950s, progress in industrial fishing triggered an increasing need to analytically describe the dynamics of fish populations in response to fishing, serving as basis to estimate the productivity of populations and maximize the yield of fisheries. Some of the first models to address these questions were the surplus production models, notably the Schaefer model by Milner B. Schaefer. These types of models describe a population as undifferentiated biomass that grows in response to the population size. Specifically, the Schaefer model (Schaefer 1954) assumes that changes in biomass are governed by population growth rate r, the carrying capacity K and the fisheries catch C:

$$\frac{dB}{dt} = rB\left(1 - \frac{B}{K}\right) - C \tag{1}$$

C can be defined in different ways, most typically it is introduced as a product of B and a fishing rate (which in turn is usually fishing effort times a catchability coefficient). The core of the equation is, however, the logistic growth of the population, which levels out to zero at $B = K$ (and at $B = 0$) and reaches a maximum of $\frac{rK}{4}$ at $B=\frac{K}{2}$. Consequently, in this model, B remains in perpetuity at $\frac{K}{2}$ if $C=\frac{rK}{4}$, which therefore maximizes the sustainable yield of a fishery. This idea builds the foundation of the concept of a maximum sustainable yield (MSY) that to date dominates global fisheries policies (Hey 2012).

There are various modifications of the Schaefer model (Haddon 2010), such as, for instance, the Fox model:

$$\frac{dB}{dt} = rB\left(1 - \frac{\log(B)}{\log(K)}\right) \tag{2}$$

which accounts for the fact that productivity in many fish species is assumed to be not a symmetric parabola but exhibit a maximum at $B < \frac{K}{2}$. An attempt to integrate different functional forms into one generalized surplus production model is the Pella and Tomlinson model:

$$\frac{dB}{dt} = \frac{r}{s}B\left(1 - \left(\frac{B}{K}\right)^s\right) \tag{3}$$

Here the parameter s defines the shape of the relationship between biomass and productivity, corresponding to the Schaefer model $s = 1$ and displaying a left- or right-skewed parabola when $s < 1$ or $s > 1$, respectively.

Surplus production models can be considered as the first analytical models that have been applied systematically in fisheries science, both to assess the state of fish stocks as well as to determine reference points for management or explore conceptual questions. The biggest advantage of this class of models is their simplicity that requires comparatively little information or input data to generate (somewhat) meaningful predictions. It is for this reason why biomass-based models are still widely in use today, mostly in data-poor fisheries where the lack of information prevents the use of more sophisticated models. However, the simplicity is also the biggest caveat since omitting largely the biology and ecology of a population tends to oversimplify true dynamics. The demographic composition of a population and traits linked to demographics such as growth, maturity and fecundity are important factors of population dynamics. Ignoring these will, therefore, often result in biased or even completely wrong predictions. In the following sections, we will therefore focus on structured models of fisheries dynamics and their different components.

Growth

Besides the recruitment of new individuals to the population, increases in population biomass are caused by increasing body size of individual fish, making individual growth one of two major factors of a population's production. Fish typically grow from millimeter-sized eggs to adults that can reach up to several meters in some species while undergoing dramatic changes in their ecology and morphology. By constraining the available food and potential predation, the size of a fish is a major determinant of its ecological niche and habitat choice, while food acquisition and environmental conditions drive changes in size. Growth is a trait that is shaped by the entire life history and ecology of a species (Enberg et al. 2012), such that, typically, short-lived species tend to grow faster than long-lived ones. Furthermore, the size of fish commonly determines their fecundity and reproductive potential, and thus, the size composition of a population may be important for its productivity (Hixon Johnson, and Sogard 2014). Understanding growth is, therefore, essential to model population dynamics.

Based on the model design and their application, growth models in fish can be separated into two groups: process-based growth models and statistical growth models (Enberg, Dunlop, and Jørgensen 2008).

The former type of models defines growth through the underlying biology of an organism and the environment it experiences, shaping the processes that govern growth. Typically, process-based growth models use insights from bioenergetics to predict growth as a product of energy acquisition and transformation, accounting for metabolic processes, behavior, food intake, temperature or life-history trade-offs, as for instance between growth and reproductive investment. Process-based growth models aim at a fundamental understanding of growth, providing an approximation to the mechanisms that determine growth and enabling them to predict more accurately growth and how it may change in response to environmental and anthropogenic disturbances. In contrast, the second group of growth models focuses on a statistical description of observed sizes with little to none of the underlying biology of growth included. These models contain often fewer parameters and assume a direct relationship between age or size of an organism with growth. Examples are the logistic growth model

$$G_t = K\left(1 - \frac{W_t}{W_\infty}\right) \tag{4}$$

or the Gompertz growth model

$$G_t = K\left(\log(W_\infty) - \log(W_t)\right) \tag{5}$$

that assume that the growth rate G_t is a linear function of absolute or log-transformed weight W at time t subject to growth parameter K and an asymptotic maximum weight W_∞. While ignoring the biology of growth processes, statistical growth models can be applied to size-at-age data that is readily available for many commercially harvested fish populations. Process-based growth models, on the other hand, require more knowledge of the growth processes and data to fully parameterize the model. Particularly the necessary data on bioenergetic processes is often missing, turning the larger flexibility and predictive power of such models into a disadvantage in a practical context. Consequently, statistical growth models tend to be more common in many areas of fisheries science such as stock assessment, as they are sufficient to describe the observed size compositions of a population.

There is, however, an overlap between the two groups of growth models. Statistical growth models can be extended with additional physical or biological parameters, approximating better the actual drivers behind growth. On the other hand, process-based growth models are often used in purely statistical approaches. A good example for such applications is the von Bertalanffy growth model (VBGM), which stems from a model that includes anabolic and catabolic processes but is commonly fitted to size data like a statistical model.

Ludwig von Bertalanffy originally proposed a general growth model describing the change in length as a differential equation of length at time t, the maximum length L_∞, and a growth parameter r:

$$\frac{dL}{dt} = r\left(L_\infty - L_t\right). \tag{6}$$

Mechanistically, this is founded on the differences in how anabolic and catabolic processes scale with body weight W:

$$\frac{dW}{dt} = aW_t^{m_1} - cW_t^{m_2} \tag{7}$$

where a and c are the proportionality coefficients for anabolism and catabolism, respectively, and m_1 and m_2 the corresponding scaling exponents. Assuming that the acquisition and thus anabolic processes are proportional to body mass by $m_1 = \frac{2}{3}$ and catabolic processes (metabolism and maintenance) by $m_1 = 1$, this results with increasing body size in a higher proportion of available resources spent on catabolic processes, leaving less for growth.

The VBGM has later been introduced by Beverton and Holt into fisheries, where it has become and remained the dominating model for fish growth, mainly because it provides typically good fits to length or weight data of most fish species for both individuals and population averages. The standard form of the model is used to calculate the length or weight at time t and results in an asymptotic shape, i.e., size approaches a maximum while growth increments decrease over time:

$$L_t = L_\infty \left(1 - e^{-k(t-t_0)}\right) \tag{8}$$

$$W_t = aW_\infty \left(1 - e^{-k(t-t_0)}\right)^b \tag{9}$$

Length or weight are, therefore, determined by the growth coefficient k and an asymptotic (maximum) length L_∞ or weight W_∞. The exponent b is derived from the age–length relationship $L_t = aW_t^b$ (and in most applications simplified to a cubic relationship), and t_0 is an (hypothetical, negative) age when size is zero. The latter is included to avoid that length or weight equal to zero at hatching ($t = 0$).

The VBGM assumes discrete time steps that correspond in most applications to years. However, this may be inadequate from a biological perspective since temperate and boreal species typically do not grow equally throughout the year but show distinct intra-annual growth patterns that align with the seasonal food availability. To account for such dynamics, the standard VGBM can be modified by introducing cyclical growth patterns:

$$L_t = L_\infty \left(1 - e^{-\left(s_1 sin\left(\frac{2\pi(t-s_2)}{s_3}\right) + k(t-t_0)\right)}\right), \tag{10}$$

with s_1 and s_2 determining the shape of the oscillation and s_3 the frequency by subdividing the time step, e.g., in the most common case of annual time steps into months ($s_3 = 12$) or weeks ($s_3 = 52$).

Such modifications also reveal the main limitation of the standard VBGM: lacking functional flexibility and very simplified or erroneous biology. Specifically, the underlying scaling of anabolic and catabolic processes with size has been empirically shown as very similar, with values ranging between 0.7 and 0.8 for both exponents in most fish species. Furthermore, the VBGM neglects crucial insights from life-history theory and, therefore, captures juvenile growth less accurately than adult growth. The reason for this is that the model does not account for maturation and reproduction. Reproductive investment is very energy-demanding and, thus, requires organisms to allocate a major share of the acquired energy to it. Since resources for basic maintenance processes can only be reduced to a very limited extent, reproductive investment mainly comes at the expense of growth. Consequently, there is a trade-off between growth and reproduction, resulting in different growth trajectories before and after maturation. The VBGM does not incorporate these dynamics and, therefore, tends to underestimate the growth rates of juvenile fish, which typically approximates linear growth. The following two models attempt to address these limitations by allowing for more functional flexibility or specifically incorporating life-history considerations.

Generalized models serve the purpose of aggregating different models into one equation that can take different functional forms depending on the parameter values. The model developed by Schnute and Richards generalizes several of the growth models used in fisheries, including the VBGM and logistic model as well as models previously proposed by Gompertz, Chapman, Richards, or Schnute:

$$L_t = L_\infty \left(1 - ae^{-bt^c}\right)^{-\frac{1}{d}} \tag{11}$$

Setting parameters c and d equal to one, for instance, reduces the model to the VBGM. A generalized model such as the one from Schnute and Richards allows therefore for a better representation of observed data and biology of a population. Nevertheless, this presents a statistical approach rather than a processed-based one, possibly explaining data well but without biological understanding. Additionally, the large number of parameters may make the fitting process challenging, particularly if no prior (mechanistic) knowledge is included.

A different approach has been taken by biphasic growth models that build explicitly on life-history theory and incorporate the trade-off between growth and reproduction. These build on bioenergetics and the assumption that the change in somatic weight W depends on the acquired energy E_t and the energy invested in gonadal weight G_{t+1}:

$$W_{t+1} = W_t + E_t - G_{t+1} \tag{12}$$

Provided that weight is a cubic function of length, $W = L^3$, this can be used to model the growth in length. The model by Roff (1983) is one application of this idea, assuming a linear growth for juvenile fish, i.e., when age a is lower than the maturation age a_{mat}, whereas growth will depend on the gonado-somatic index (GSI) from the onset of maturation:

$$L_{t+1} = \begin{cases} L_t + l_0 \\ \left. L_t + l_0 \middle/ (1+R_{t+1})^{\frac{1}{3}} \right. \end{cases} \quad \text{for} \begin{cases} a < a_{mat} \\ a \geq a_{mat} \end{cases} \tag{13}$$

GSI is gonad mass divided by somatic mass, which means the larger the investment in R_{t+1}, the more will growth in length be reduced.

A very similar approach has been taken by the modified model based on Quince et al. (2008) in which the growth is also shaped by the reproductive investment R_{t+1} from of the GSI (Boukal et al 2014):

$$L_{t+1} = {}^{(1-\alpha)\beta}\sqrt{\frac{L_t^{(1-\beta)\alpha} + (1-\beta)cb^{-(1-\beta)}}{1+q^{-1}(1-\beta)R_{t+1}}} \tag{14}$$

For fish below a_{mat}, R_{t+1} equals zero, whereas from a_{mat} onward $R_{t+1} > 0$. The modified Quince et al. model resolves some limitations of Roff's model and incorporates a larger functional flexibility. This includes a less constrained length–weight relationship than Roff's model by assuming $W_t = aL_t^\alpha$, a conversion factor between somatic and gonadic investment q, and by not enforcing strictly linear growth for juvenile individuals. Both models, however, incorporate the same key feature that results in biphasic growth trajectories, separating quasi-linear growth prior to maturation from a decreased growth after the onset of maturation that depends on the reproductive investment. These models illustrate how our understanding of life-history processes can be included in growth models to represent better how growth can vary between different life stages and achieve better fits to empirical data.

Besides the trade-off between growth and reproduction, there are other trade-offs that may directly or indirectly affect growth. A major driver is survival and, therefore, everything that affects mortality, most notably predation and fishing. For instance, very size-selective mortality can increase survival for fish that invest more into growth (instead into basic maintenance processes such as the immune system) to grow faster through the size range of increased mortality. Similarly, behavioral adaptations may result in decreased or increased growth. For instance, passive behavior such as hiding can be used to reduce predation risks at the expense of reduced foraging, reducing the resources available for growth and reproduction. In contrast, growth may be increased through more active foraging and a bolder behavior; however, this may also expose fish to higher predation and reduce the probability of survival (Claireaux, Jørgensen, and Enberg 2018).

Essential life-history traits such as growth, reproductive investment, survival, and behavior are to a large part determined by an individual's inherited genotype. This means that traits are shaped by evolution and subject to evolutionary change that depends on the reproductive success of a specific life history within a given environment. Consequently, how resources are acquired and allocated into growth or reproduction is less of an individual decision than the result of an inherited life-history strategy. The trade-offs between growth, reproductive investment, and survival are key to this process and determine the success of a specific strategy under the current environmental conditions in reproducing and thus inheriting the same strategy to the next generation. Because mortality is a major driver of natural selection, changes in the degree or selectivity of mortality may affect the selection landscape and result in evolutionary adaptations in growth or traits that influence growth. This means growth trajectories within a population are not stationary over time but may change in response to environmental change and anthropogenic perturbation such as fishing or increasing sea water temperatures.

Although the growth of an organism is fundamentally shaped by life-history evolution, most observed changes in growth and thus size-at-age occur in the short term as a result of phenotypic plasticity. The most important driver is environmental variability, specifically physical and ecological conditions that influence metabolism and food availability. The latter is particularly relevant, since it determines directly the available energy that can be acquired by an organism and invested into processes such as growth. Food availability per capita is the combined result of food production through the food web and the competition for the available food sources. On a seasonal or annual time scale, growth of an organism can therefore be determined by bottom-up effects through variation in the ecosystem productivity, e.g., through varying nutrient inflow or temperatures, as well as through the abundance of its own population and other competing species. Feedbacks occur between the environmental variability and density dependence, for instance, when high food availability leads to increasing population abundance(s) and thus to competition in the future. Although the causes of environmental variability are often difficult to determine and parametrize in models, density-dependent growth has been empirically established (Zimmermann, Ricard, and Heino 2018) and may affect the sustainability of fisheries (van Gemert and Andersen 2018). A simple approach to implement density dependence in growth models is the use of an asymptotic length $L_{\infty, t}$ that decreases as a function of a density dependence coefficient d and the population biomass B_t in each year: $L_{\infty, t} = L_{\infty} - dB_t$ (Lorenzen and Enberg 2002). This example illustrates that not only insights from bioenergetics, physiology, and life-history theory are important to modeling growth but also insights from population ecology.

Recruitment

Recruitment is a key component of population dynamics and contributes together with body growth to the increase in biomass within a population. Because of the enormous reproductive potential of most fish species, recruitment tends to be the most important factor for the overall productivity of a population and the major driver of fluctuation in population size. Recruitment as such is the combined result of the total number of eggs produced by the mature part of a population and the survival throughout the early life stages from egg to juveniles, which explains the large variation in number of recruits observed in most fish. Typically, fish produce very large numbers of eggs per individuals, reaching millions per spawning event. At the same time, the early life stages are very vulnerable to unsuitable physical conditions, predation or insufficient food, causing large inter-annual variation in survival. Consequently, recruitment can result in favorable years in very large cohorts that sustain a population for many years during which recruitment may be average or fail completely. This variability in recruitment, however, poses also a challenge for any attempt to model and predict recruitment. Nevertheless, because recruitment is fundamental for population dynamics and thus fisheries, various recruitment models have been established. Most of these models rely on the basic assumption that recruitment must be related to the mature part of the population and is subject to some form of density-dependent reduction. The two models that remain most widely used until today are the stock-recruitment models developed by Ricker (1954) and Beverton and Holt (1957).

The Beverton–Holt model assumes a stock–recruitment relationship that increases with increasing biomass of mature fish B at time t, however, with decreasing number of recruits per spawning individual and thus approaching an asymptotic maximum of recruitment:

$$R_t = \frac{\alpha B_t}{\beta + B_t} \tag{15}$$

with α representing the asymptotic maximum for a given B_t and β the population biomass where $\alpha/2$ is reached, defining the steepness of the curve. Biologically, α stands for the maximum spawning output that linearly increases with population biomass, whereas β defines the density-dependent regulation in recruitment and therefore the productivity of a population at specific biomass levels. The underlying mechanism is the density-dependent survival of early life stages, which is assumed to decrease with increasing amount of eggs spawned by a larger population biomass due to intra-cohort competition for resources, particularly food (Van Poorten, Korman, and Walters 2018), and stronger predation pressure.

The Ricker model takes a similar approach as the Beverton–Holt model, except that it assumes an overcompensatory effect of increasing population biomass on recruitment. This implies that the Ricker stock-recruitment curve reaches a peak recruitment after which the realized number of recruits decreases again, instead of simply approaching an asymptotic maximum. The typical equation to describe this relationship is denoted as:

$$R_t = \alpha B_t e^{-\beta B_t} \tag{16}$$

Here, α defines the recruitment at a low biomass of the spawning population and scales the total number of recruits, whereas β determines the density-dependent decrease in recruits per spawning biomass. As in the Beverton–Holt model, α represents the reproductive output of the mature population and β the density-dependent mortality experienced by early life stages after spawning. The key difference between the two models is that the recruits per spawning biomass in the Ricker model do not remain at a value larger than zero but approach zero, suggesting that a population biomass above a certain level has such detrimental effects on recruitment that it overcompensates the marginal increase in spawning output. Such effects can occur through substantial negative inter-cohort interactions through cannibalism or competition (Ricard, Zimmermann, and Heino 2016), i.e., older cohorts that deplete the resources of following cohorts or prey directly on them, or other negative feedbacks, for instance when growth at early life stages is density-dependent while predation is size-dependent. This may add up to a substantially increased mortality when cohort density decreases growth rates.

It is noteworthy that the Ricker curve can take an almost identical shape as the Beverton–Holt curve for an observed range of population biomass and number of recruits, making the Ricker model more flexible in representing populations with different recruitment patterns. A step further in this direction is taken by generalized recruitment models that allow for a large functional flexibility with other models as special cases. One example for such an approach is a model suggested by Deriso and later modified by Schnute:

$$R_t = \alpha B_t (1 - \beta \gamma B_t)^{\frac{1}{\gamma}} \tag{17}$$

Here α and β take equivalent roles as in the Beverton–Holt or Ricker models, while parameter γ determines the form of the recruitment curve. For instance, when γ goes to 0, the Deriso model corresponds to the Ricker model, and $\gamma = -1$ transforms it into a Beverton–Holt-type model. A generalized model of this kind avoids the need for prior assumptions on the type of relationship, enabling better fits to data or to test effects of gradual changes in the functional form. The downside of such an approach is, however, that it takes mainly a statistical and not a process-orientated approach to incorporate biological knowledge. Furthermore, recruitment data typically turns out to be very noisy, which may make it very difficult to find reasonable parameter estimates for a model with a high degree of functional freedom.

Empirically, the biomass of the spawning population is in most cases an insufficient predictor of recruitment. As a consequence, models such as Beverton–Holt or Ricker typically fit poorly to data.

Recruitment data is in general very noisy and shows a lot of variation. Because survival at early life stages tends to be much more important than the total reproductive output of the mature individuals, other factors such as environmental conditions, food availability and predation that directly or indirectly affect mortality of eggs, larvae and juveniles are key drivers of recruitment (Zimmermann, Claireaux, and Enberg 2019). Especially large-scale atmospheric and oceanographic processes with cyclical patterns have been identified as important forcing factors. A simple approach to model such cyclical patterns in recruitment is to include an autoregressive term in a stock-recruitment model:

$$R = \alpha B_t e^{-\beta B_t} + u_t \tag{18}$$

representing a Ricker model that includes a $AR(1)$ process defined as $u_t = \varphi u_{t-1}$ with φ as autocorrelation coefficient. This allows for capturing temporal autocorrelation in recruitment time series and simulating inter-annual cyclical patterns.

A different approach to extend a Ricker model is to include explicitly an additional variable that is related with recruitment:

$$R = \alpha B_t e^{-\beta B_t} e^{-\gamma X_t} \tag{19}$$

Here X is the second variable besides mature biomass B and γ is the corresponding coefficient. Examples for X could include any environmental or ecological factor that is expected to affect the recruitment of a specific population, such as annual sea surface temperature, zooplankton indices or the biomass of a predator. This approach can be further extended with additional, equally defined terms.

Classic stock-recruitment models assume that the relationship between mature population biomass and recruitment remains stationary over time. However, this assumption may often not hold because the reproductive potential of a population as well as the mechanisms of density regulation can change over time due to external factors, both anthropogenic and natural. Regime shifts can, for instance, occur when changes in the environment or anthropogenic impacts alter the ecosystem productivity with implications on the reproductive success of a population and thus the relationship between the mature population and recruitment (Vert-pre et al. 2013). To incorporate two different regimes, a Ricker model can be modified:

$$R_t = \alpha_i B_t e^{-\beta_i B_t} \tag{20}$$

α and β depend now on $i = 1, 2$ which can be defined as two different periods in a time series.

The number of regimes can be extended further if required. This comes, however, with an equally increasing number of parameters as caveat. Considering that time series of recruitment data usually only cover a few decades, the number of parameters can become easily disproportional compared to the number of data points. A solution to this problem is to introduce a time-invariant parameter that captures gradual changes in the relationship between population and recruitment over time (Perälä and Kuparinen 2015). For instance, if we assume that the productivity of the mature biomass varies with time, time-variant α_t can be introduced:

$$R_t = \alpha_t B_t e^{-\beta B_t} \tag{21}$$

where parameter α follows a random walk process $\alpha_t = \alpha_{t-1} + \sigma_t$, with σ assumed to be normally distributed. This enables the productivity term in the stock-recruitment relationship to vary gradually over time, representing changes in the population's reproductive output for instance through changes in the population's demographic structure due to fishing or because of increasing temperatures, as well as subsequent evolutionary adaptation (Enberg et al 2010).

Mortality and Population Dynamics

Whereas individual growth and recruitment represent the increase in biomass of a population, mortality constitutes the loss term in the equation. Survival is therefore a crucial component in fisheries models, making them very sensitive to the underlying assumptions and specifications of mortality. Mortality is typically given in rates that define the survival of individual fish over time. Two main sources of mortality are distinguished: mortality from natural causes, most notably predation and diseases, commonly denominated as natural mortality M, and mortality from anthropogenic harvesting, commonly termed fishing mortality F. These instantaneous rates of mortality can be summed to total mortality $Z = F + M$, which translates into proportional survival after a given time step as e^{-Z} and, reciprocally, relative mortality as $1 - e^{-Z}$. In an age-structured population, the abundance reduces in one time step t by:

$$N_t = N_{t-1}e^{-Z} \tag{22}$$

or alternatively an abundance at a given age a of:

$$N_a = N_0 e^{-Za}, \tag{23}$$

with N_0 as initial abundance of a cohort, typically corresponding to recruitment R_t.

Mortality is, however, in most fish species dependent on size and, thus, age of an individual. This applies both for F and M. While other sources of mortality such as diseases, parasite or starvation can be relevant as well, M is to a large degree shaped by predation. Vulnerability to predation is influenced by various traits, including the individual behavior, but most importantly body size because most predators have a size window for prey based on limitations in perception and handling. For a given stock, smaller and younger fishes tend to be much more vulnerable to predation than bigger and older ones, which means that mortality decreases substantially with increasing size and age.

F is subject to similar size selectivity, indirectly because parts of populations such as pre-recruits often do not share the same habitat as the ones targeted by fisheries and directly through size-selective fishing gear. In contrast to M, F increases in most cases with age and size: while small, young fishes are usually excluded from fisheries and experience very low fishing, the F for the targeted age and size classes can be substantial and outweigh M. These dynamics can be captured by using age-specific mortality rates F_a and M_a, leading to

$$N_{a,t} = N_{a,t-1}e^{-(F_a+M_a)} \tag{24}$$

This model can be further extended by allowing F and M also to vary in time:

$$N_{a,t} = N_{a,t-1}e^{-(F_{a,t}+M_{a,t})} \tag{25}$$

The two-dimensional mortality matrices here are age- and time-specific, which is common for stock assessment models where annual F_a's are estimated.

Stock Assessment

Managing fish stocks depends on knowing the state of the stock and its trajectory. Because fish are typically numerous, mobile and difficult to observe, this can be much more difficult than for classic terrestrial resources such as forests. Finding methods to assess a fish stock and use this information as the basis for regulative decisions is therefore at the core of sustainable fisheries management. This so-called stock assessment process relies commonly on observations, notably from commercial catches or standardized scientific sampling, and mathematical models to transform these often nonrepresentative, biased or snapshot-like observations into information on changes in the stock. The type and complexity of the model used is determined by the quantity, quality and structure of the available data.

The main distinction follows our previous differentiation of model complexity: biomass or age-/length-structured models. The latter require consistent information on the age or size of individual fish in the catch or scientific samples, which implies comparatively large and costly efforts in data collection over many years. Consequently, only a minority of global fish stocks are assessed in such a framework, mostly in commercially valuable fisheries in industrialized countries. All other fish stocks are classified as data-limited and are assessed with a wide range of methods (Chrysafi and Kuparinen 2015). Typically, these produce time series of absolute or relative biomass based on combining data on fisheries catch and effort with life-history information. However, to inform on the key parameters of population growth r and K (see Eq. 1), data of stock sizes at very high and low levels are necessary to give good indications on maximum growth rate and carrying capacity, respectively, which typically requires relatively long time series that ought to include unfished and overfished periods, what is rarely available. Alternatively, life-history parameters can act as predictors of stock productivity and resilience and therefore indicate adequate catch levels. This comes with the benefit that information gaps even in cases with very little or no data can be bridged by tapping into existing knowledge on ecology and life-history theory to incorporate it as priors into Bayesian assessment models (Kindsvater et al. 2018).

Age- or length-structured models can be divided into two major categories: statistical methods and cohort dynamics models. Both are based on the population dynamics detailed before in Eqs. 10 and 24 and, most importantly, the catch equation that relates total catch to age- or size-specific abundance and mortality:

$$C_{a,t} = \frac{F_a}{F_a + M_a} N_{a,t} \left(1 - e^{-(F_a + M_a)}\right) \tag{26}$$

In stock assessment, this equation provides an approach to determine the unknown variables F_a and N_a from catch data. Because time series of catch exist in essentially all commercial fisheries where assessments are conducted, the catch equation represents the core of almost all assessment models.

The main difference between statistical methods and cohort dynamics is the way these variables and other parameters are derived. Cohort dynamics, most commonly known as virtual population analysis (VPA) or more advanced derivatives such as extended survivor analysis (XSA), do not estimate their parameters but are instead based on an iterative process where the size of each cohort is calculated backwards in time and the result reseeded until the values of $N_{a,t}$ and $F_{a,t}$ approximate the observed $C_{a,t}$ sufficiently well. For this approach to work, other parameters, especially M_a, need to be given as fixed values. Later versions such as XSA were mostly developed to allow for the inclusion of catch indices from commercial or scientific surveys. VPA-type assessment models were particularly widespread in the assessments of European fisheries but have been increasingly replaced by statistical models in recent years.

Statistical catch-at-age models estimate their parameters directly by fitting Eq. 26 and other underlying equations to data by optimizing a given objective function. This offers greater flexibility and actual statistical evaluation compared to cohort dynamics models, yet comes with risks of over-parametrization and over-fitting when using ill-defined models. This can easily be illustrated with a model that assumes all $N_{a,t}$ and $F_{a,t}$ as independent parameters, which would result with, e.g., 10 age classes and a time series of only 20 years in 200 parameters alone, twice as many as data point. The problem is typically circumvented by linking parameters such as $F_{a,t}$ among each other through, e.g., a selectivity-at-age function or restricting changes over time. Together with increasing computational power and efficient optimization algorithms, this has elevated the utility of these models considerably and therefore also the number of stocks assessed with such, even though limitations remain (Subbey 2018).

Two of the most common frameworks for statistical assessment models are Stock Synthesis (Methot and Wetzel 2013) and, in recent years, State-space Assessment Model (SAM, Nielsen and Berg 2014). The strength of Stock Synthesis and reason for its popularity is mainly its flexible adjustment to a range of data situations, from relatively data-limited stocks to such with many data series and process knowledge available. SAM on the other hand is a state-space model framework which allows, thus, for a separation of observation and process error and the estimation of the development in F_a over time. Benefits include that time-varying selectivity can be directly estimated and uncertainty incorporated, representing possible solutions for two major problems in stock assessment. Both Stock Synthesis and SAM utilize comparable automatic differentiation algorithms that can estimate a large number of parameters efficiently based on maximum likelihood (or, in case of stock synthesis, also Bayesian inference).

Despite technical advances in assessment methods and increasing data availability, major challenges persist in stock assessment. Generally, there's a widespread lack of process knowledge and true mechanistic understanding that leads to biased or erroneous estimates (Maunder and Piner 2015). This includes insufficient knowledge on key biological processes, notably natural mortality, which are often described and parametrized in highly simplified ways that build a stark contrast to the increasing wealth of data and complexity when it comes to commercial and scientific catch indices. Natural mortality is in the large majority of assessed stocks kept constant over all age classes and over time. Similarly, maturity at age and growth are often described with time-invariant equations, despite existing knowledge that all these traits and the underlying processes can show large variation over time with significant impacts on stock dynamics. Ultimately, this is also reflected in the fact that most stock assessments maintain an isolated perspective on one specific stock and ignore its ecological context (Skern-Mauritzen et al. 2016), even though this excludes important information on drivers of dynamics and stock productivity. Incorporating ecosystem dynamics, for instance, by integrating them into natural mortality and growth in assessment models and their forecasts may therefore provide an important step forward in improving the prediction power and subsequently also the management of fisheries. Currently, the downside is the required data and effort that adds to an already work-intensive assessment process. One potential solution to this problem could be reduced assessment frequencies to free up capacities (Zimmermann and Enberg 2017) and focus on collecting the necessary ecosystem data to improve the quality of assessments.

Advice and Management

A key goal of modeling fish stocks and assessing them is to inform management strategies and shape regulative fisheries policies. The main outputs of the assessment process are therefore indicators of a stock's current state to evaluate them against reference points, forming the basis of the management advice. Reference points are important targets or thresholds in the state of a stock or the catch levels, which are typically linked to desirable objectives (notably, the stock size or fishing mortality that produces MSY) or unwanted situation (e.g., the stock size below which the risk of collapse exceeds a predefined limit). These thresholds are estimated from the available information, i.e., the specific approach depends again on data availability. If a complete analytical assessment exists, the estimated parameters (such as stock–recruitment relationship, length-at-age, natural mortality, fisheries selectivity-at-age, uncertainty) can be applied in stochastic simulation models of the population dynamics to calculate the desired reference points. In data-limited situations, there is often the need for different approaches due to the lack of a full assessment model. These utilize either catch and effort data or qualitative measurements of stock status (Froese et al. 2017) or are derived from biological information (Brooks, Powers, and Cortés 2009). It is noteworthy that in data-limited stocks, methods that rely on time series of catches tend to not perform better than simpler methods with lower data requirements (Carruthers et al. 2014).

Management advice to the regulatory agents is provided as management strategies and their outcomes, often with clear recommendations accompanied by a decision table that lists different options and how they perform in respect to reference points or policy objectives. Risk assessment is an important part of management advice, typically focusing on estimated probabilities of falling below limit thresholds of stock size or yield. Ideally, this translates into risk management that implements a precautionary approach to mitigate harmful impacts on fish stocks or fisheries (Hilborn et al. 2001). To develop and test management strategies, the most common framework today are management strategy evaluations (MSEs) that apply a suite of management strategies in simulation models to evaluate their long-term performance. These can include a broad range of (generic) management strategies or specifically developed ones. Main challenges hereby are the representation of uncertainty and the comprehensive transfer of results to managers and decision makers (Punt et al. 2016). The dimensions of a MSE increase with each management strategy or reference point considered, easily resulting in very large and complex outputs. It is therefore crucial that the results of MSE are summarized and presented in the best possible way, otherwise they may fail to convey the necessary information to managers and lead to suboptimal decision-making.

Ultimately, the key to successful fisheries management is the clear definition of objectives. A common reason for management failure is ill-specified or conflicting objectives (Hilborn 2007). Many objectives in fisheries management do not align or are even mutually exclusive. Classic trade-offs among objectives include food production and conservation or economic benefits and employment in the fisheries sector. It is, therefore, crucial that stakeholders such as policy makers, managers, fishermen, scientists and the public agree on the objectives that fisheries ought to achieve, to facilitate the design of suitable management strategies and their evaluation. Nevertheless, the success also hinges on the entire process of assessing stocks and, therefore, the adequate use of models to represent and predict stock dynamics sufficiently well. The major challenge of fisheries science remains therefore to collect the necessary information, use data efficiently and maximize the knowledge gain through suitable model choices.

The ecosystem approach to fisheries management has increasingly been recognized as a next step towards a more holistic and sustainable use of marine resources. Fish stocks are part of an ecosystem that is shaped by its environment, species composition and anthropogenic pressures. Climatic and oceanographic processes and trophic interactions among fish stocks and their prey and predators are important sources of variation in population dynamics that need to be accounted for if these dynamics ought to be accurately explained or predicted. Interactions between stocks are, for instance, important for management strategies because they not only affect recruitment and natural mortality but also result in trade-offs in reference points such as MSY (Voss et al. 2014) and rebuilding strategies of overfished stocks (Zimmermann and Yamazaki 2017). Furthermore, adaptations to selective harvesting (Heino et al. 2013), density effects (van Gemert and Andersen 2018) or climate change (Britten, Dowd, and Worm 2016) can affect the productivity of fish stocks and, thus, shift reference points. Ignoring such dynamics may therefore result in flawed reference points that increase the risk stock collapse and overfishing or underfishing, that is managing a fish stock suboptimally in respect to the biological or economic objectives. Finding robust and beneficial ways of increasingly integrating ecological information into assessment and management is therefore a major goal in our quest towards sustainable fisheries.

Currently, ecological interactions and the variation caused by them are still rarely included in stock assessment and management (Skern-Mauritzen et al. 2016). A major obstacle in the process towards ecosystem-based management has been that higher model complexity results in the curse of dimensionality and increasing uncertainty (Collie et al. 2016). Models that were meant to build the basis for ecosystem-based management by replicating entire food webs or ecosystems have turned out too complex, lack uncertainty quantification and are difficult to parametrize on a year-to-year basis, particularly for the many fisheries where data or knowledge are limited. For tactical fisheries management, current ecosystem models are therefore too data hungry and, most importantly, show poor estimation and prediction capabilities (Planque 2016). This is problematic because for management, the short-term prediction skills are crucial (Dietze et al. 2018) and there is no gain from incorporating ecosystem dynamics into management without sufficient prediction power (Basson 1999). A way forward is, therefore, an approach that focuses on integrating specific ecological or environmental interactions that can be effectively monitored, have high predictive power and significantly impact the dynamics of a population. Future research will subsequently focus on how such ecological interactions can in a systematic way be identified and incorporated into fisheries management processes.

References

Basson, M. 1999. "The importance of environmental factors in the design of management procedures." *ICES Journal of Marine Science* 56 (6):933–942. doi:10.1006/jmsc.1999.0541.

Beverton, Raymond J.H., and Sidney J. Holt. 1957. "On the dynamics of exploited fish populations." In Fisheries and Food, Fisheries Investigations *Series 2*, Vol. 19. London: UK Ministry of Agriculture.

Boukal, D., Dieckmann, U., Enberg, K., Heino, M., Jørgensen, C. (2014) Life-history implications of the allometric scaling of growth. *Journal of Theoretical Biology* **359**: 199–207.

Britten, Gregory L., Michael Dowd, and Boris Worm. 2016. "Changing recruitment capacity in global fish stocks." *Proceedings of the National Academy of Sciences* 113 (1):134–139.

Brooks, Elizabeth N., Joseph E. Powers, and Enric Cortés. 2009. "Analytical reference points for age-structured models: Application to data-poor fisheries." *ICES Journal of Marine Science* 67 (1):165–175. doi:10.1093/icesjms/fsp225.

Carruthers, Thomas R., André E. Punt, Carl J. Walters, Alec MacCall, Murdoch K. McAllister, Edward J. Dick, and Jason Cope. 2014. "Evaluating methods for setting catch limits in data-limited fisheries." *Fisheries Research* 153:48–68. doi:10.1016/j.fishres.2013.12.014.

Chrysafi, Anna, and Anna Kuparinen. 2015. "Assessing abundance of populations with limited data: Lessons learned from data-poor fisheries stock assessment." *Environmental Reviews* 24 (1):25–38. doi:10.1139/er-2015-0044.

Claireaux, M., Jørgensen, C., Enberg, K. (2018) Evolutionary effects of fishing gear on foraging behavior and life-history traits. *Ecology and Evolution* 8, 10711–10721.

Collie, Jeremy S., Louis W. Botsford, Alan Hastings, Isaac C. Kaplan, John L. Largier, Patricia A. Livingston, Éva Plagányi, Kenneth A. Rose, Brian K. Wells, and Francisco E. Werner. 2016. "Ecosystem models for fisheries management: Finding the sweet spot." *Fish and Fisheries* 17:101–125.

Dietze, M.C., Fox, A., Beck-Johnson, L.M., et al. (2018) Iterative near-term ecological forecasting: Needs, opportunities, and challenges. *Proceedings of the National Academy of Sciences* 115, 1424–1432.

Enberg, K., E.S. Dunlop, and C. Jørgensen. 2008. Fish growth. In Encyclopedia of Ecology, Edited by S.E. Jørgensen and B.D. Fath. Elsevier, Oxford. Pp 1564–1572.

Enberg, K., Jørgensen, C., Mangel, M. (2010) Fishing-induced evolution and changing reproductive ecology of fish: the evolution of steepness. *Canadian Journal of Fisheries and Aquatic Sciences* 67: 1708–1719.

Enberg, K., C. Jørgensen, E.S. Dunlop, Ø. Varpe, D.S. Boukal, L. Baulier, S. Eliassen, and M. Heino. 2012. "Fishing induced-evolution of growth: Concepts, mechanisms and the empirical evidence." *Marine Ecology* 33 (1):1–25. doi:10.1111/j.1439–0485.2011.00460.x.

Froese, Rainer, Nazli Demirel, Gianpaolo Coro, Kristin M. Kleisner, and Henning Winker. 2017. "Estimating fisheries reference points from catch and resilience." *Fish and Fisheries* 18 (3):506–526. doi:10.1111/faf.12190.

Haddon, Malcolm. 2010. *Modelling and Quantitative Methods in Fisheries*. Second ed. Boca Raton, FL: CRC Press.

Heino, Mikko, Loïc Baulier, David S. Boukal, Bruno Ernande, Fiona D Johnston, Fabian M Mollet, Heidi Pardoe, Nina O Therkildsen, Silva Uusi-Heikkilä, and Anssi Vainikka. 2013. "Can fisheries-induced evolution shift reference points for fisheries management?" *ICES Journal of Marine Science: Journal du Conseil* 70 (4):707–721.

Hey, Ellen. 2012. "The persistence of a concept: Maximum sustainable yield." *The International Journal of Marine and Coastal Law* 27 (4):763–771.

Hilborn, R. 2007. "Defining success in fisheries and conflicts in objectives." *Marine Policy* 31 (2):153–158.

Hilborn, R., J.J. Maguire, A.M. Parma, and A.A. Rosenberg. 2001. "The precautionary approach and risk management: Can they increase the probability of successes in fishery management?" *Canadian Journal of Fisheries and Aquatic Sciences* 58 (1):99–107.

Hixon, Mark A., Darren W. Johnson, and Susan M. Sogard. 2014. "BOFFFFs: On the importance of conserving old-growth age structure in fishery populations." *ICES Journal of Marine Science* 71 (8):2171–2185. doi:10.1093/icesjms/fst200.

Kindsvater, Holly K., Nicholas K. Dulvy, Cat Horswill, Maria-José Juan-Jordá, Marc Mangel, and Jason Matthiopoulos. 2018. "Overcoming the data crisis in biodiversity conservation." *Trends in Ecology & Evolution*. doi:10.1016/j.tree.2018.06.004.

Lorenzen, K., and K. Enberg. 2002. "Density-dependent growth as a key mechanism in the regulation of fish populations: Evidence from among-population comparisons." *Proceedings of the Royal Society B: Biological Sciences* 269 (1486):49–54.

Maunder, Mark N., and Kevin R. Piner. 2015. "Contemporary fisheries stock assessment: Many issues still remain." *ICES Journal of Marine Science* 72 (1):7–18. doi:10.1093/icesjms/fsu015.

Methot, Richard D., and Chantell R. Wetzel. 2013. "Stock synthesis: A biological and statistical framework for fish stock assessment and fishery management." *Fisheries Research* 142:86–99. doi:10.1016/j.fishres.2012.10.012.

Nielsen, Anders, and Casper W. Berg. 2014. "Estimation of time-varying selectivity in stock assessments using state-space models." *Fisheries Research* 158: 96–101.

Perälä, Tommi, and Anna Kuparinen. 2015. "Detecting regime shifts in fish stock dynamics." *Canadian Journal of Fisheries and Aquatic Sciences* 72 (11):1619–1628.

Planque, Benjamin. 2016. "Projecting the future state of marine ecosystems, 'la grande illusion'?" *ICES Journal of Marine Science: Journal du Conseil* 73 (2):204–208.

Punt, André E., Doug S. Butterworth, Carryn L. de Moor, José A. A. De Oliveira, and Malcolm Haddon. 2016. "Management strategy evaluation: Best practices." *Fish and Fisheries* 17 (2):303–334. doi:10.1111/faf.12104.

Quince, C., P.A. Abrams, B.J. Shuter, and N.P. Lester. 2008. "Biphasic growth in fish I: Theoretical foundations." *Journal of Theoretical Biology* 254 (2):197–206.

Ricard, D., F. Zimmermann, and M. Heino. 2016. "Are negative intra-specific interactions important for recruitment dynamics? A case study of Atlantic fish stocks." *Marine Ecology Progress Series* 547:211–217. doi:10.3354/meps11625.

Ricker, William E. 1954. "Stock and recruitment." *Journal of the Fisheries Board of Canada* 11 (5):559–623.

Roff, Derek A. 1983. "An allocation model of growth and reproduction in fish." *Canadian Journal of Fisheries and Aquatic Sciences* 40 (9):1395–1404.

Schaefer, M.B. 1954. "Some aspects of the dynamics of populations important to the management of the commercial marine fisheries." *Inter-American Tropical Tuna Commission Bulletin* 1 (2):27–56.

Skern-Mauritzen, Mette, Geir Ottersen, Nils Olav Handegard, Geir Huse, Gjert E. Dingsør, Nils C. Stenseth, and Olav S. Kjesbu. 2016. "Ecosystem processes are rarely included in tactical fisheries management." *Fish and Fisheries* 17 (1):165–175.

Subbey, Sam. 2018. "Parameter estimation in stock assessment modelling: Caveats with gradient-based algorithms." *ICES Journal of Marine Science.* doi: 10.1093/icesjms/fsy044.

van Gemert, Rob, and Ken H. Andersen. 2018. "Challenges to fisheries advice and management due to stock recovery." *ICES Journal of Marine Science.* doi:10.1093/icesjms/fsy084.

Van Poorten, B., Korman, J., Walters, C. (2018) Revisiting Beverton–Holt recruitment in the presence of variation in food availability. *Reviews in Fish Biology and Fisheries* **28**: 607–624.

Vert-pre, Katyana A., Ricardo O. Amoroso, Olaf P. Jensen, and Ray Hilborn. 2013. "Frequency and intensity of productivity regime shifts in marine fish stocks." *Proceedings of the National Academy of Sciences of the United States of America* 110 (5):1779–1784. doi:10.1073/pnas.1214879110.

Voss, Rudi, Martin Quaas, Joern O. Schmidt, and Julia Hoffmann. 2014. "Regional trade-offs from multi-species maximum sustainable yield (MMSY) management options." *Marine Ecology Progress Series* 498: 1–12.

Zimmermann, Fabian, Marion Claireaux, and Katja Enberg. 2019. "Common trends in recruitment dynamics of north-east Atlantic fish stocks and their links to environment, ecology and management." *Fish and Fisheries.* doi:10.1111/faf.12360.

Zimmermann, Fabian, and Katja Enberg. 2017. "Can less be more? Effects of reduced frequency of surveys and stock assessments." *ICES Journal of Marine Science* 74 (1):56–68. doi:10.1093/icesjms/fsw134.

Zimmermann, Fabian, Daniel Ricard, and Mikko Heino. 2018. "Density regulation in Northeast Atlantic fish populations: Density dependence is stronger in recruitment than in somatic growth." *Journal of Animal Ecology* 87 (3):672–681. doi:10.1111/1365-2656.12800.

Zimmermann, Fabian, and Satoshi Yamazaki. 2017. "Exploring conflicting management objectives in rebuilding of multi-stock fisheries." *Ocean & Coastal Management* 138:124–137. doi: 10.1016/j.ocecoaman.2017.01.014.

V

ENT: Environmental Management Using Environmental Technologies

21

Bioremediation: Contaminated Soil Restoration

Sven Erik Jørgensen

Introduction

Contaminated soil is an increasing problem in industrialized countries, and the high cost of remediation has driven the interest in the direction of ecological engineering applications of bioremediation technologies. They apply biological processes, mainly microorganisms or plants. It is often possible to solve the pollution problem satisfactorily by this ecotechnologically based methodology without the hazard and expense involved in removing polluted materials for treatment elsewhere for the use of traditional environmental technological methods.

Ecotechnological bioremediation may be applied for both organic waste and heavy metals, although the methods applied in practice may differ and they are therefore treated below in two different sections, on organic compounds and heavy metals. The success of any bioremediation technology depends on a number of factors including site characteristics, environmental factors such as temperature, pH, redox potential, concentrations of nutrients, the contaminant, the presence of microorganisms, and bioavailability. It is therefore necessary to look into these factors to comprehend the applicability of these methods.

Bioavailability of Toxic Organic Compounds

Bioavailability is a crucial factor for the application of bioremediation. It is defined as the amount of contaminant present that can be readily taken up by organisms. The bioavailability controls the biodegradation rates for organic contaminants because microbial cells must expend energy to induce the catabolic processes used in biodegradation. If the contaminant concentration is too low, induction will not occur. Soil microbial populations are typically slow-growing organisms and often exposed to nutrient-poor environments.[1] Bioavailability also determines the toxicity of both organic and inorganic contaminants to organisms other than those applied for bioremediation. There is therefore an increased need for bioremediation when bioavailability is high, which fortunately makes bioremediation more attractive. Three cases can be envisioned that would result from different bioavailabilities of contaminants:[2]

1. Biodegradation will not occur because the concentration of the bioavailable contaminant is insufficient and/or the biodegradation rate for the contaminant is too low to justify the energy expenditure to induce biodegradation.
2. Microbial cells may degrade the contaminant at low bioavailable concentrations and/or low biodegradation rates, but in a resting or maintenance stage rather than in a growing stage.
3. At a sufficient bioavailability and biodegradation rate, there is enough bioavailable contaminant to induce biodegradation in a growing stage. That will allow for optimal rates of remediation.

The biodegradability of organic contaminants is highly dependent on the physical and chemical structure[3] of the contaminant and the soil. The section on "Biodegradation" will discuss this topic. A coarse but still applicable rule for a very first estimation of the biodegradability of organic compounds is given in the entry entitled "Biodegradation." Moreover, the software EEP (Estimation of Ecotoxicological Parameters) is able to give some first estimation of biodegradability.

The bioavailability of heavy metals is also a significant factor for the applicability of bioremediation. Heavy metals are of course not degraded but removed, mainly by plant uptake. The uptake by organisms of heavy metals, which determines the overall removal efficiency, is entirely controlled by the bioavailable amount of heavy metals. An ecological model presented in the section on *Uptake of Heavy Metals by Plants* will illustrate the strong dependence of the bioavailability of heavy metals.[4]

Bioavailability is influenced by a number of factors:

1. Low water solubility
2. Sorption on the solid phase of soil
3. Physical makeup of the soil (pore size distribution)
4. Microbial adaptations

Low water solubility can limit availability of the substrate to bacterial cells and hence constrain biodegradation.[5,6] Microbial cells are 70%–90% water, and the food they utilize comes from the water surrounding the cells. Plants take up water to cover the evapotranspiration needed for the maintenance of their life functions. Therefore, uptake and transport are only feasible for water-soluble material. If first-order biodegradation kinetics is presumed, the biodegradation rate becomes proportional to the concentration in the water phase. It means for components with low water solubility that they are biodegraded very slowly. There are clear relationships between the water solubility of an organic compound and the chemical structure that can be utilized to estimate the water solubility.[3] EEP (a software containing many equations to estimate ecotoxicological parameters) and other estimation equations utilize these relationships to make estimation of the water solubility and of K_{ow}. Side reactions may change the water solubility. This is of particular interest for heavy metal ions, which can increase the solubility by the formation of complexes either with organic or with inorganic compounds. The formation of complexes with humic acid and fluvic acid plays a major role for the solubility of metal ions in soil water. Hydrocarbons that are frequently found as soil contaminant have a low water solubility: 2–6 µg/L for pentacyclic aromatic hydrocarbons and *n*-alkanes of chain length 18–30. The solubility decreases with increasing molecular weight.[3]

The state of the contaminant in combination with the water solubility is also of importance. There is evidence that liquid-phase hydrocarbons are more bioavailable than solid-phase hydrocarbons.[7] In practical terms, this means that the maximum growth rate occurs in different solubility ranges for liquid-phase (0.01–1 mg/L) and solid-phase (1–10 mg/L) components. Degradation can be described by a Michaelis–Menten expression. Water solubility increases with increasing temperature and usually an Arrhenius expression can be applied with the temperature coefficient 1.06 or 1.07.

Many authors have found that surfactants increase mineralization rate due to increased dissolution. Also, surfactants may provide an additional carbon source, which is preferentially utilized by the bacteria. There may, how-ever, also be a negative effect by surfactants due to their toxicity to the bacterial population.

Sorption on the solid phase of soil may be a limiting factor for biodegradation of microorganisms and uptake by plants. There are several reports that suggest that organic chemicals are not mineralized while associated with solid phases.[8,9] Experiments by Robinson et al.[10] show that sorbed-phase substrate was not degraded and that longterm biodegradation was limited by the slowly desorbing fraction of substrate. These results suggest that rate-limited mass transfer processes (primary desorption) may significantly affect the rate at which a compound is degraded in the presence of a solid phase.

The model presented in the section on "Uptake of Heavy Metals by Plants" uses the fraction soluble in the soil water of heavy metal ions to determine the uptake. The sorption is dependent of pH, redox potential, and humus, clay, and sand fractions in the soil. The relationship between these factors and the sorption is included in the model. If the sorption of organic compounds to soil is not known, the soil–water partition coefficient, K_{oc}, can be estimated from the octanol–water partition coefficient by the following equations:

$$\log K_{oc} = -0.006 + 0.937 \log K_{ow} \tag{1}$$

$$\log K_{oc} = -0.35 + 0.99 \log K_{ow} \tag{2}$$

In the case that the carbon fraction of organic carbon in soil is f, the distribution coefficient, K_D, for the ratio of the concentration in soil and in water can be found as $K_D = K_{oc} f$.

It has been suggested that there are different stages of sorption processes and that newly sorbed material is more labile and therefore more bioavailable than aged sorbed material. Numerous experiments have demonstrated that aging affects bioavailability in soil due to changes in the soil structure, resulting in slower desorption processes. The sorption can frequently be described by either Freundlich or Langmuir adsorption isotherm, expressed respectively by the following equations:

$$a = kc^b \tag{3}$$

$$a = k'c/(c + b') \tag{4}$$

where a is the concentration in soil, c is the concentration in water, and k, k', b, and b' are constants. Equation 3, corresponding to Freundlich adsorption isotherm, is a straight line with slope b in a log–log diagram, since $\log a = \log k + b \log c$.[11] The Langmuir adsorption isotherm is an expression similar to the Michaelis–Menten equation. If $1/a$ is plotted versus $1/c$,[11] we obtain a straight line, Lineweaver-Burk's plot, as $1/a = 1/k' + b'/k'c$. When $1/a = 0$, $1/c = -1/b'$ and when $1/c = 0$, $1/a = 1/k'$. This plot can be applied to assess the expression of the type used in Michaelis–Menten's equation and in Langmuir's adsorption isotherm; it is observed that b is often close to 1 and c is for most environmental problems small. This implies that the two adsorption isotherms get close to $a/c = k$, and k becomes a distribution coefficient. k for 100% organic carbon is usually denoted K_{oc} (see above).

The sorption determines the uptake of organic contaminants by plants as it is expressed in the following equation:[11]

$$\text{BCF} = f_{\text{lipid}} K_{ow}{}^b / h f K_{ow}{}^a, \tag{5}$$

where BCF, the biological concentration factor, expresses the ratio between the concentration in soil and in the plant (or the microorganisms); f_{lipid} is the lipid fraction in the plant; f is, as shown above, the fraction of organic carbon in the soil; and a, b, and h are constants. The denominator expresses the fraction of the organic matter that is dissolved in the soil water. h is therefore the constant determined by Equations 1 and 2. If we use Equation 1, Equation 5 may be reformulated to the following equation:

$$BCF = 1.01 f_{lipid} K_{ow}^{0.063} / f \qquad (6)$$

As it is seen, BCF becomes almost independent of K_{ow} and mainly dependent on the ratio between f_{lipid} and f. A high BCF means that the concentration in the plants (eventual microorganisms) is high and a significant amount of the toxic compound is removed; the lipid fraction in the plants has to be high and the carbon content of the soil has to be low.

Physical makeup of the soil (pore size distribution) is of importance for bioavailability. Bacteria may be excluded from the microporous domain since most bacteria range from 0.5 to 2 μm. If such an exclusion occurs, biodegradation cannot take place in the microporous domain. The degradation rate is therefore limited by the diffusion of solute from the microporous to the macroporous domain. This is obviously of particular importance for organic contaminants with a high molecular weight. In a field situation, it is difficult to separate the effects of sorption and micropore exclusion, as some residues are protected from biodegradation by both mechanisms.

Microbial adaptations: Microorganisms have developed several strategies to increase the bioavailability of organic contaminant. One strategy is the development of increased cell affinity for hydrophobic surfaces. It allows the microorganisms to attach to the hydrophobic substrate and directly adsorb it. A second strategy is the production and release of surface active agents or biosurfactants.[12,13]

The biological adaptation is a current change of the properties of the microbial population by a selection of the microorganisms that are best fitted to survive and grow under the prevailing conditions. They are determined by the properties of the environment including the concentrations and characteristics of the contaminants. A biological adaptation is widely used to prepare a microorganism population for bioremediation. It is often possible, although not general, that a 10 times faster decomposition can be achieved by the use of adapted microorganisms.[2]

Biodegradation

See the entry entitled "Biodegradation," where a general presentation of this process included methods for estimation of the biodegradation rate from the chemical structure of the toxic organic compounds.

The usual applied procedure to follow for the utilization of microbiological biodegradation to reduce the concentration of a toxic organic matter in contaminated soil has six steps:

1. Spatial mapping of quantitative distribution: the contaminant is developed by analytical chemistry intensively. Analyses of pore water are often applied to evaluate the extent of environmental risks.
2. Laboratory test/treatability studies to verify the applicability of bioremediation.
3. Calculation (often by development of a model) to assess the feasibility of the method in situ.
4. Production of an adapted strain of the microorganisms in sufficient amount.
5. Implementation in situ. If the groundwater table is high, it is usually lowered. Injection pits are introduced into the soil, and air is blown into the soil to reinforce the decomposition of organic matter. In case of chlorinated compounds, a mixture of methane and air may be applied.
6. The results are followed by use of a wide spectrum of analytical methods including radioactive tracers, detection of intermediary metabolites, and respiration rate.

Uptake of Heavy Metals by Plants

Plants are contaminated by heavy metals originating from deposition of heavy metals (waste sites), air pollution, the application of sludge from municipal wastewater plant as a soil conditioner, and the use of fertilizers. The uptake of heavy metals from municipal sludge by plants has previously been modeled.[4] The model is based on a mass balance for cadmium in a typical Danish soil (see Figure 1).

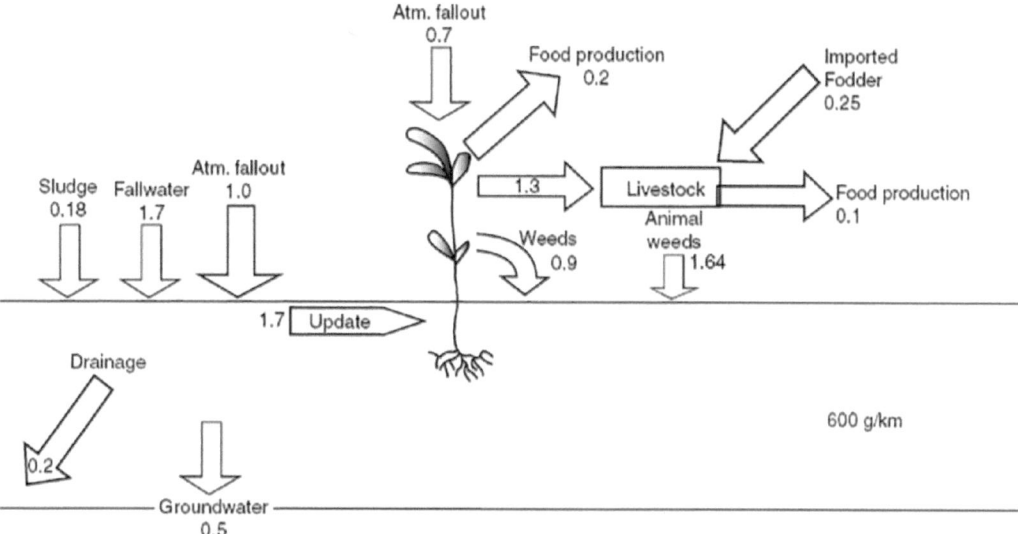

FIGURE 1 Cadmium balance of 1 ha of an average Danish agriculture land. All rates are expressed as grams of Cd per hectare per day.

The model can briefly be described as follows: Depending on the soil composition, it is possible to find for various heavy metal ions a distribution coefficient, i.e., the fraction of the heavy metal that is dissolved in the soil water relative to the total amount. The distribution coefficient was found by examination of the dissolved heavy metals relative to the total amount for several different types of soil. Correlation between pH, the concentration of humic substances, clay, and sand in the soil on the one hand, and the distribution coefficient on the other, was also determined. The uptake of heavy metals was considered a first-order reaction of the dissolved heavy metal. It is, how-ever, also possible to test acid volatile sulfide and organic carbon to describe the metal binding capacity of sediment in constructed wetlands. This will give approximately the same ratio "bound" to "bioavailable" heavy metals as the above-mentioned correlation. The basic idea is the same, namely, to find easily measurable soil proper-ties that determine the metal binding capacity, which is crucial for the uptake of heavy metals by plants.

In addition to the uptake from soil water, the model presented below considers the following:

1. Direct uptake from atmospheric fallout onto the plants.
2. Other sources of contamination such as fertilizers and the long-term release of heavy metal bound to the soil and the non-harvested parts of the plants.

Published data on lead and cadmium contamination in agriculture are used to calibrate and validate the model that is intended to be used for the following:

1. Risk assessment for the use of fertilizers and sludge that contain heavy metals as contaminants.
2. A risk involved in the use of plants harvested from a waste site.
3. Determining the possibilities of removal of heavy metals by plants that have a particular ability to take up heavy metals. This last intended application of the model makes it useful for determina-tion of the result of application of bioremediation.

Figure 2 shows a conceptual diagram of the Cd version of the model. The STELLA software was applied. As can be seen, it has four state variables: Cd-bound, Cd-soil, Cd- detritus, and Cd-plant. An attempt was made to use one or two state variables for cadmium in the soil, but to get acceptable accordance between data and model output, three state variables were needed. This can be explained by the presence of several soil components that bind the heavy metal differently.

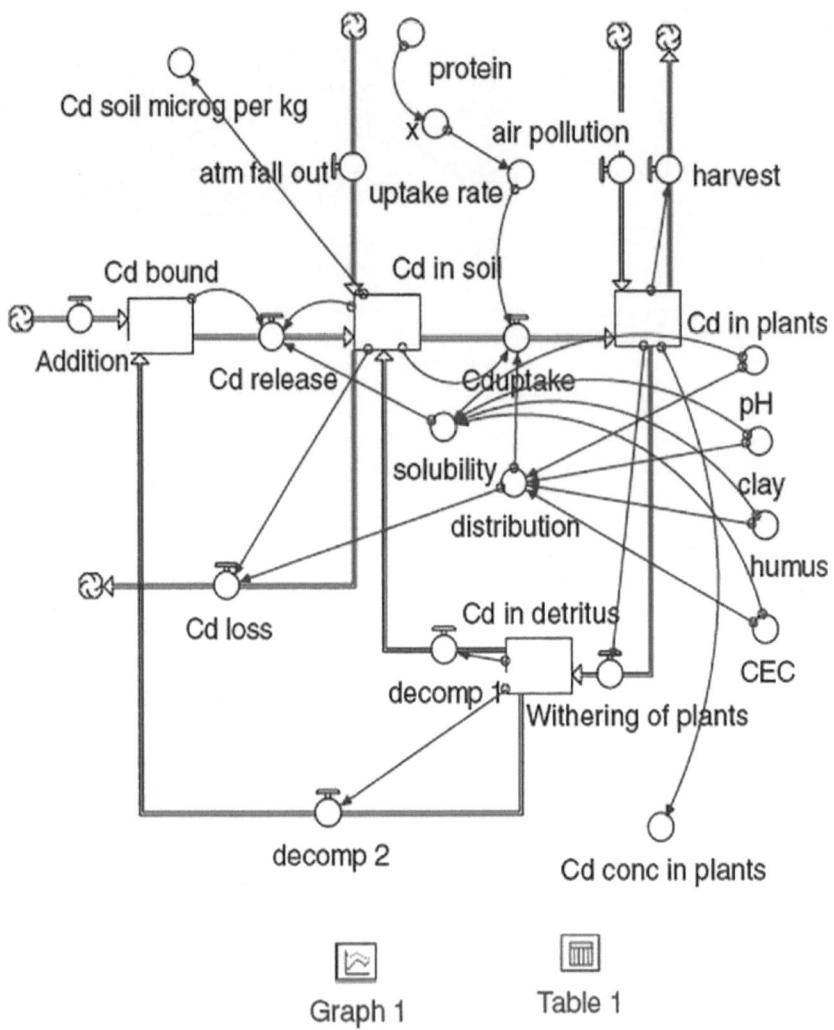

FIGURE 2 Conceptual diagram of the model. Boxes show state variables, double-line arrows denote flows, circles show functions, and single-line arrows show feedback mechanisms.

The loss covers transfer to the soil and groundwater below the root zone. It is expressed as a first-order reaction with a rate coefficient dependent on the distribution coefficient that is found from the soil composition and pH, according to the correlation found by Jørgensen.[14] The transfer from Cd-bound to Cd-soil indicates the slow release of cadmium due to a slow decomposition of the more or less refractory material to which cadmium is bound. The cadmium uptake by plants is expressed as a first-order reaction, where the rate is dependent on the distribution coefficient, as only dissolved cadmium can be taken up. It is furthermore dependent on the plant species. As will be seen, the uptake is a step function that, here (grass), is 0.0005 during the growing season, and, of course, zero after the harvest and until the next growing season starts. Cd-waste covers the transfer of plant residues to detritus after harvest. It is therefore a pulse function, which here is 60% of the plant biomass, as the remaining 40% has been harvested. Cd-detritus covers a wide range of biodegradable matter and the mineralization and is therefore accounted for in the model by use of two mineralization processes: one to Cd-soil and one to Cd-bound. The first one is rapid and is given a higher rate for the first 180 days as the addition of municipal sludge in this case is at day 0. The second one is at about the same rate, but as the cadmium is transferred to the Cd-bound, the slow release rate is considered by the very slow transfer from Cd-bound to Cd-soil. Similar models can be erected for other heavy metals, but the distribution coefficient is of course different for the other heavy metals.[14]

Plants with a high protein content will generally take up heavy metals with a higher efficiency. Several applications of plants with a high uptake efficiency have been reported in the literature; see Tongbin et al.,[15] where a very effective uptake of arsenic is reported, and Feng et al.,[16] where simultaneous removal of arsenic and antimony is reported. The use of algae for removal of heavy metals has also been successfully tested.[17]

Models are used increasingly to solve the problems of contaminated soil (see UNEP-EITC and Copenhagen University[18]). Spatial models have been developed to consider the distribution of the contaminants in eco-spatial time scale. These models are very useful in setting up and optimizing time-bound action plans of bioremediating the contaminated soil.

References

1. Roszak, D.B.; Colwell, R. Survival strategies of bacteria in natural environment. Microbiol. Rev. **1987**, *51*, 365–379.
2. Maier, R.M. Biovailability and its importance for bioremediation. In *Bioremediation*; Valdes, J.J., Ed.; Kluwer: Dordrecht, the Netherlands, 2000; 58–79.
3. Jørgensen, S.E.; Halling-Sørensen, B.; Mahler, H. *Handbook of Estimation Methods in Ecotoxicology and Environmental Chemistry*; Taylor and Francis Publ.: Boca Raton, FL, 1997; 230 pp.
4. Jørgensen, S.E.; Fath, B. *Fundamentals of Ecological Modelling*, 4th Ed.; Elsevier: Amsterdam, 2011; 398 pp.
5. Fogel et al. EPA Report 560/5-82-015 Washington, DC, 1981.
6. Zhang, Y.; Miller, R.M. Enhanced octadecane dispersion and biodegration by *Pseudomonas rhamnolipid* surfactant. Appl. Environ. Microbiol. **1992**, *58*, 3276–3282.
7. Miller, R.M. Surfactant-enhanced bioavailability of slightly soluble organic compounds. In *Bioremediation: Science and Applications*; Skipper, H., Turco, R., Eds.; Specila Publications. Soil Science Society of America: Madison, WI, 1995; 33–54.
8. Greer, L.E.; Shelton, D.R. Surfactant-enhanced bioavailability of slightly soluble organic compounds. In *Bioremediation*; Skipper, H., Turco, R., Eds.; Soil Science Society of America: Madison, WI, 1992; 33–54.
9. Miller, M.E.; Alexander, M. Kinetics of bacterial degradation of benzylamine in montmorillonic suspension. Environ. Sci. Technol. **1991**, *25*, 240–245.
10. Robinson K.G. et al. Availability of sorbed toluene in solids for biodegradation by acclimatized bacteria. Water Res. **1990**, *24*, 345–350.
11. Jørgensen, S.E. *Principles of Pollution Abatement*; Elsevier: Amsterdam, 2000; 520 pp.
12. Rosenberg, E. Microbial surfactants. Crit. Rev. Biotechnol. **1986**, *3*, 109–132.
13. Fiechter, A. Biosurfactant: moving towards industrial application. Trends Biotechnol. **1992**, *10*, 208–217.
14. Jørgensen, S.E. Do heavy metals prevent the agricultural use of municipal sludge? Water Res. **1975**, *9*, 163–170.
15. Tongbin, C. et al. Arsenic hyperaccumulator *Pteris vittata* L. and its arsenic accumulation. Chin. Sci. Bull. **2002**, *47*, 902–905.
16. Feng R. et al. Simultaneous hyper-accumulation of arsenic and antimony in Cretan brake fern: Evidence of plant uptake and sub-cellular distributions. Microchem. J. **2011**, *97*, 38–43.
17. Mitsch, W.J.; Jørgensen, S.E. *Ecological Engineering and Ecosystem Restoration*; John Wiley: New York, 2004; 410 pp.
18. UNEP-IETC and Copenhagen University. *Handbook of Phytotechnology for Water Quality Improvement and Wetland Management through Modelling Applications*, 2005.

22

Biotechnology: Pest Management

Maurizio G. Paoletti

Benefits of Genetic Engineering in Pest Control

Since 1987, many crops have been genetically modified for features such as resistance to insects, resistance to pathogens (including viruses) and herbicides, and for improved features such as longer-lasting ripening, higher nutritional status, protein content, seedless fruit, and sweetness. Up to 34 new genetically engineered crops have been approved to enter into the market.

In 1998, 27.8 million ha of engineered crops were planted in countries such as the United States, Argentina, Canada, and Australia. The United States alone contains 74% of the modified crop land-planted. Globally, 19.8% of this area has been planted with herbicide-tolerant crops, 7.7% with insect-resistant crops, and 0.3% with insect and HRCs. Five crops—soybean, corn, cotton, canola, and potato—cover the largest acreage of engineered crops.[1,2]

Disease Resistance in Crops

The crops currently on the market that have been engineered for resistance to plant pathogens are listed in Table 1. Disease-resistant engineered crops have some potential advantages because few current pesticides can control bacterial and viral diseases of crops. In addition, these engineered plants help reduce problems from pesticides.

The large-scale cultivation of plants expressing viral and bacterial genes might lead to adverse ecological consequences. The most significant risk is the potential for gene transfer of disease resistance from cultivated crops to weed relatives. For example, it has been postulated that a virus-resistant squash could transfer its newly acquired virus-resistant genes to wild squash (*Cucurbita pepo*), which is native to the southern United States. If the virus- resistant genes spread, newly disease-resistant weed squash could become a hardier, more abundant weed. Moreover, because the United States is the origin for squash, changes in the genetic make-up of wild squash could conceivably lessen its value to squash breeders.

TABLE 1 Plants Genetically-Engineered for Virus Resistance That Have Been Approved for Field Tests in the United States from 1987 to July 1995

Crop	Disease(s)	Research Organization
Alfalfa	Alfalfa mosaic virus,	Pioneer Hi-Bred
	Tobacco mosaic virus (TMV),	
	Cucumber mosaic virus (CMV)	
Barley	Barley yellow dwarf virus (BYDV)	USDA
Beets	Beet necrotic yellow vein virus	Betaseed
Cantelope and/or	CMV, papaya ringspot virus (PRV)	Upjohn
squash	Zucchini yellow mosaic virus (ZYMV),	
	Watermelon mosaic virus II (WMVII)	
	CMV	Harris Moran Seed
	ZYMV	Michigan State University
	ZYMV	Rogers NK Seed
	Soybean mosaic virus (SMV)	Cornell University
	SMV, CMV	New York State Experiment Station
Corn	Maize dwarf mosaic virus (MDMV)	Pioneer Hi-Bred
	Maize chlorotic mottle virus (MCMV),	
	Maize chlorotic dwarf virus (MCDV)	
	MDMV	Northup King
	MDMV	DeKalb
	MDMV	Rogers NK Seed
Cucumbers	CMV	New York State Experiment Station
Lettuce	Tomato spotted wilt virus (TSWV)	Upjohn
Papayas	PRV	University of Hawaii
Peanuts	TSWV	Agracetus
Plum Trees	PRV, plum pox virus	USDA
Potatoes	Potato leaf roll virus (PLRV),	Monsanto
	Potato virus X (PVX),	
	Potato virus Y (PVY)	
	PLRV, PVY, late blight of potatoes	Frito-Lay
	PLRV	Calgene
	PLRV, PRY	University of Idaho
Potatoes	PLRV, PVY	USDA
	PYV	Oregon State University
Soybeans	SMV	Pioneer Hi-Bred
Tobacco	ALMV, tobacco etch virus (TEV),	
	Tobacco vein mottling virus	
	TEV, PVY	University of Florida
	TEV, PVY	North Carolina State University
	TMV	Oklahoma State University
	TEV	USDA
Tomatoes	TMV, tomato mosaic virus (TMV)	Monsanto
	CMV, tomato yellow leafcurl virus	
	TMV, ToMV	Upjohn
	ToMV	Rogers NK Seed
	CMV	PetoSeed
	CMV	Asgrow
	CMV	Harris Moran Seeds
	CMV	New York State Experiment Station
	CMV	USDA

Source: Krimsky and Wrubel[4] and McCullum et al.[5]

Some plant pathologists have also suggested that development of virus-resistant crops could allow viruses to infect new hosts through transencapsidation. This may be especially important for certain viruses, e.g., luteoviruses, where possible heterologus encapsidation of other viral RNAs with the expressed coat protein is known to occur naturally. With other viruses, such as the PRV that infects papaya, the risk of heteroencapsidation is thought to be minimal because the papaya crop itself is infected by very few viruses.

Virus-resistant crops may also lead to the creation of new viruses through an exchange of genetic material or recombination between RNA virus genomes. Recombination between RNA virus genomes requires infection of the same host cell with two or more viruses. Several authors have pointed out that recombination could also occur in genetically engineered plants expressing viral sequences of infection with a single virus, and that large-scale cultivation of such crops could lead to increased possibilities of combinations. It has recently been shown that RNA transcribed from a transgene can recombine with an infecting virus to produce highly virulent new viruses.

A strategy for reduced risk would include: 1) identification of potential hazards; 2) determination of frequency of recombination between homologous, but nonidentical sequences in crops and weeds; and 3) determination of whether or not such recombinants can have selective advantage.

Assessment of Transgenic Virus-Resistant Potatoes in Mexico

An in-depth assessment of potential socioeconomic implications related to the introduction of some genetically modified varieties of virus-resistant potatoes (PVY, PVX, PIRV) in Mexico underscores the importance of this technology. This type of genetic modification could prove especially beneficial to large-scale farmers, but only marginally beneficial to small-scale farmers, because most small farmers use red potato varieties that are not considered suitable for transformation. In addition, 77% of the seeds that small farmers use come from informal sources, not from the seed providers that could sell the new resistant varieties.

The mycoplasma and virus diseases in Mexico are not currently controlled with pesticides, and rank second and third in economic damages. The major pest, the fungus *Phytophtora infestans,* ranks first in economic damages and requires, in some cases, up to 30 fungicide applications. Thus, the interesting new genetically altered varieties of potatoes are of little benefit to crop production for small farmers.

HRCS

Several engineered crops that include herbicide resistance are commercially available; 13 other key crops in the world are ready for field trials (Table 2). In addition, some crops (e.g., corn) are being engineered to contain both herbicide (glyphosate) and biotic insecticide resistance (BT α-endotoxin).

Herbicides adopted for herbicide-resistant crops employ lower doses when compared with atrazine, 2,4-D, and alachlor. However, the resistance of the crop to the target herbicide would, in practice, suggest to the farmer to apply dosages higher than recommended. In addition, costs for this new technology of HRCs are about 2-times higher in corn than the recommended herbicide use and cultivation weed control program.

Integrated pest management (IPM) could benefit from some HRCs, if alternative non-chemical methods can be applied first to control weeds and the target herbicide could be used later, only when and where the economic threshold of weeds is surpassed. Generally, though, the use of herbicide resistant crops will lead to increased use of herbicides and environmental and economic problems. Most HRCs were developed for Western agriculture. For example, in Northern African countries, most crops, such as sorghum, wheat, and canola (oilseed rape), have wild weed relatives, thereby increasing the risk that genes from the herbicide- resistant crop varieties could be transferred to wild weed relatives.

The risk of herbicide-resistant genes from a transgenic crop variety being transferred to weed relatives has been demonstrated for canola (oilseed rape) and sugar beet.

TABLE 2 Herbicide-Resistant Crops (HRCs) Approved for Field Tests in the
United States from 1987 to July 1995

Crop	Herbicide	Research Organization
Alfalfa	Glyphosate	Northrup King
Barley	Glufosinate/Bialaphos	USDA
Canola (oilseed rape)	Glufosinate/Bialaphos	University of Idaho
	Glyphosate	Hoechst-Roussel/AgrEvo
		InterMountain Canola
		Monsanto
Corn	Glufosinate/Bialaphos	Hoechst-Roussel/AgrEvo
		ICI
		UpJohn
		Cargill
		DeKalb
		Holdens
		Pioneer Hi-Bred
		Asgrow
		Great Lakes Hybrids
		Ciba-Geigy
		Genetic Enterprises
	Glyphosate	Monsanto
		DeKalb
	Sulfonylurea	Pioneer Hi-Bred
		Du Pont
	Imidazolinone	American Cyanamid
Cotton	Glyphosate	Monsanto
		Dairyland Seeds
		Northrup King
	Bromoxynil	Calgene
		Monsanto
		Rhone Poulenc
	Sulfonylurea	Du Pont
		Delta and Pine Land
	Imidazolinone	Phytogen
Peanuts	Glufosinate/Bialaphos	University of Florida
Potatoes	Bromoxynil	University of Idaho
		USDA
	2,4-D	USDA
	Glyphosate	Monsanto
	Imidazolinone	American Cyanamid
Rice	Glufosinate/Bialaphos	Louisiana State University
Soybeans	Glyphosate	Monsanto
	Glyphosate	UpJohn
		Pioneer Hi-Bred
		Northrup King
		Agri-Pro
	Glufosinate/Bialaphos	UpJohn
	Sulfonylurea	Hoechst/AgrEvo
		Du Pont

(Continued)

TABLE 2 (*Continued*) Herbicide-Resistant Crops (HRCs) Approved for Field Tests in the United States from 1987 to July 1995

Crop	Herbicide	Research Organization
Sugar Beets	Glufosinate/Bialaphos	Hoechst-Roussel
	Glyphosate	American Crystal Sugar
Tobacco	Sulfonylurea	American Cyanmid
Tomatoes	Glyphosate	Monsanto
	Glufosinate/Bialaphos	Canners Seed
Wheat	Glufosinate/Bialaphos	AgrEvo

Source: Krimsky and Wrubel[4] McCullum et al.,[5] and Agribusiness[8]

Repeated use of herbicides in the same area creates problems of weed herbicide resistance. For instance, if glyphosate is used with HRCs crops on about 70 million ha, this might accelerate pressure on weeds to evolve herbicide resistant biotypes. Sulfonylureas and imidazolinones in HRCs are particularly prone to rapid evolution of resistant weeds. Extensive adoption of HRCs will increase the hectarage and surface treated, thereby exacerbating the resistance problems and environmental pollution problems.

Bromoxynil has been targeted in herbicide resistant cotton by Calgene and Monsanto (Table 2). This herbicide has been used on winter cereals, cotton, corn, sugarbeets, and onions to control broad leaf weeds. Drift of bromoxynil has been observed to damage nearby grapes, cherries, alfalfa, and roses. In addition, legumious plants can be sensitive to this herbicide, and potatoes can be damaged by it. Herbicide residues above the accepted standards have been detected in soil and groundwater, and as drift fallout. Rodents demonstrate some mutagenic responses to bro- moxynil. Beneficial *Stafilinid* beetles show reduced survival and egg production, even at recommended dosages of bromoxynil. Crustaceans (*Daphnia magna*) have also been severely affected by this herbicide.

Toxicity of Herbicides and HRCs

Toxic effects of herbicides to humans and animals also have been reported. For example, the Basta surfactant (sodium polyoxyethylene alklether sulfate) has been shown to have strong vasodialatative effects in humans and cardio- stimulative effects in rats. Treated mice embryos exhibited specific morphological defects.

Most HRCs have been engineered for glyphosate resistance. Although adverse effects of herbicide-resistant soybeans have not been observed when fed to animals such as cows, chickens, and catfish, genotoxic effects have been demonstrated on other non-target organisms. Earthworms have been shown to be severely injured by the glyphosate herbicide at 2.5–10.1/ha. For example, *Allolobophora caliginosa,* the most common earthworm in European, North American, and New Zealand fields, is killed by this herbicide. In addition, aquatic organisms, including fish, can be severely injured or killed when exposed to glyphosate. The beneficial nematode, *Steinerema feltiae*, a useful biological control organism, is reduced by 19%–30% by the use of glyphosate.

There are also unknown health risks associated with the use of low doses of herbicides. Due to the common research focus on cancer risk, little research has been focused on neurological, immunological, developmental, and reproductive effects of herbicide exposures. Much of this problem is due to the fact that scientists may lack the methodologies and/or the diagnostic tests necessary to properly evaluate the risks caused by exposure to many toxic chemicals, including herbicides.

While industry often stresses the desirable characteristics of their HRCS, environmental and agricultural groups, and other scientists, have indicated the risks. For example, research has shown that the application of glyphosate can increase the level of plant estrogens in the bean, *Vicia faba*. Feeding experiments have shown that cows fed transgenic glyphosate-resistant soybeans had a statistically significant difference in daily milk-fat production as compared to control groups. Some scientists are concerned that the increased milk-fat production by cows fed these transgenic soybeans may be a direct consequence of higher estrogen levels in these transgenic soybeans.

Economic Impacts of HRCs

Some analysts project that switching to bromoxynil for broadleaf weed control in cotton could result in savings of 37 million dollars each year. Furthermore, recent problems with use of glyphosate-resistant cotton in the Mississippi Delta region—crop losses resulting in up to $500,000 of this year's cotton crop— suggest that this technology needs to be further developed before some farmers will reap economic benefits. In addition, a recent study of herbicide- resistant corn suggests that the costs of weed control might be about two times more expensive than normal herbicide and cultivation weed control in corn.

While some scientists suggest that use of HRCs will cause a shift to fewer broad spectrum herbicides, most scientists conclude that the use of HRCs will actually increase herbicide use.

BT for Insect Control

More than 40 BT crystal protein genes have been sequenced, and 14 distinct genes have been identified and classified into six major groups based on amino acids and insecticidal activity. Many crop plants have been engineered with the BT α-endotoxin, including alfalfa, corn, cotton, potatoes, rice, tomatoes, and tobacco (Table 3). The amount of toxic protein expressed in the modified plant is 0.01%–0.02% of the total soluble proteins.

TABLE 3   Transgenic Insect Resistant Crops Containing BT δ-Endotoxins. Approved Field Tests in United States from 1987 to July 1995

Crop	Research Organization
Alfalfa	Mycogen
Apples	Dry Creek
	University of California
Corn	Asgrow
	Cargill
	Ciba-Geigy
	Dow
	Genetic Enterprises
	Holdens
	Hunt-Wesson
	Monsanto
	Mycogen
	NC+Hybrids
	Nortrup King
	Pioneer Hi-Bred
	Rogers NK Seed
Cotton	Calgene
	Delta and Pineland
	Jacob Hartz
	Monsanto
	Mycogen
	Northrup King

(Continued)

TABLE 3 (*Continued*) Transgenic Insect Resistant Crops Containing BT δ-Endotoxins. Approved Field Tests in United States from 1987 to July 1995

Crop	Research Organization
Cranberry	University of Wisconsin
Eggplant	Rutgers University
Poplar	University of Wisconsin
Potatoes	USDA
	Calgene
	Frito-Lay
	Michigan State University
	Monsanto
	Montana State University
	New Mexico State University
	University of Idaho
Rice	Louisiana State University
Spruce	University of Wisconsin
Tobacco	Auburn University
	Calgene
	Ciba-Geigy
	EPA
	Mycogen
	North Carolina State University
	Roham and Haas
Tomatoes	Campbell
	EPA
	Monsanto
	Ohio State University
	PetoSeeds
	Rogers NK Seeds
Walnuts	University of California, Davis
	USDA

Source: Krimsky and Wrubel[4] and Agribusiness.[7]

Some trials with corn demonstrate a high level of efficacy in controlling corn borers. Corn engineered with BT endotoxin has the potential to reduce corn borer damage by 5%–15% over 28 million ha in the US, with a potential economic benefit of $50 million annually. Some suggest that corn engineered with BT toxin will increase yields by 7% over similar varieties. However, it is too early to tell if all these benefits will be realized consistently. Potential negative environmental effects also exist because the pollen of engineered plants contains BT, which is toxic to bees, beneficial predators, and endangered butterflies like the Karka Blue and Monarch Butterflies.

Cotton was the first crop plant engineered with the BT α-endotoxin. Caterpillar pests, including the cotton boll- worm and budworm, cost U.S. farmers about $171 million/ yr as measured in yield losses and insecticide costs. Benedict et al.[3] predict that the widespread use of BT cotton could reduce insecticide use and thereby reduce costs by as much as 50% to 90%, saving farmers $86 to $186 million/yr.

The development of insect resistance to transgenic crop varieties is one highly possible risk associated with the use of BT D-endotoxin in genetically engineered crop varieties. Resistance to BT has already been demonstrated in the cotton budworm and bollworm. If BT- engineered plants become resistant, a key insecticide that has been utilized successfully in IPM programs could be lost. Therefore, proper resistance management strategies with use of this new technology are imperative. Another potential risk is that the BT a-endotoxin could be harmful to non-target organisms. For example, it is not clear what potential effect the BT D-endotoxin residues that are incorporated into soils will have against an array of non-target useful invertebrates living in the rural landscape. It has also been demonstrated that predators, such as the lacewing larvae (*Crysoperla carnea*) that feed on corn borers (*Ostrinia nubilalis*), grown on engineered BT corn have consistently higher mortality rates when compared to specimens fed with non-engineered corn borers. In addition, the treated larvae need three more days to reach adulthood than lacewings fed on prey from non-BT corn.

Discussion

Both pesticides and biotechnology have definite advantages in reducing crop losses to pests. At present, pesticides are used more widely than biotechnology, and thus are playing a greater role in protecting world food supplies. In terms of environmental and public health impacts, pesticides probably have a greater negative impact at present because of this more widespread use.

Genetically engineered crops for resistance to insect pests and plant pathogens could, in most cases, be environmentally beneficial, because these more resistant crops could allow a reduction in the use of hazardous insecticides and fungicides in crop production. In time, there may also be economic benefits to farmers who use genetically engineered crops; this will depend, however, on the prices charged by the biotechnology firms for these modified, transgenic crops.

There are, however, some environmental problems associated with the use of genetically engineered crops in agriculture. For example, adding BT to crops like corn for insect control can result in any of the following negative environmental consequences: 1) development of resistance to BT by pests species in corn and other crops; 2) health risks from exposure to the BT toxin to humans in their food and to livestock in feed; 3) the toxicity of the pollen from the BT-treated corn to honey bees, beneficial natural enemies, and endangered species of insects that feed on the modified corn plants or come into contact with the drifting pollen; engineered plant residues incorporated into soil can produce undesirable effects on soil micros and mesofauna.

A major environmental and economic concern associated with genetically engineered crops is the development of HRCs. Although in rare instances HRCs may result in a beneficial reduction of toxic herbicide use, it is more likely that the use of HRCs will increase herbicide use and environmental pollution. In addition, farmers will suffer because of the high costs of employing HRCs—in some instances, weed control with HRCs may increase weed control costs for the farmer threefold.

More than 40% of the research by biotechnology firms is focused on the development of HRCs. This is not surprising, because most of the biotechnology firms are also chemical companies who stand to profit if herbicide resistance in crops result in greater pesticide sales. Theoretically, the acceptance and use of engineered plants in sustainable and integrated agriculture should consistently reduce current use of pesticides, but this is not the current trend. In addition, most products and new technologies are designed for Western agriculture systems, not for poor or developing countries. For instance, if terminator genes enter into the seed market, there will be no possibility of traditional and small farmers using their plants to produce their seeds. Thus, genetic engineering could promote improvements for the environment; however, the current products—especially the herbicide-resistant plants and the BT-resistant crops—do have serious environmental impacts, similar to the consequences of pesticide use.

References

1. James, C. Global Review of Commercialized Transgenic Crops: 1998. In *ISAAA Briefs*; Cornell University: Ithaca, New York, 1998; 8–1998.
2. Moff, A.S. Toting up the early harvest of transgenic plants. Science **1998**, *282*, 2176–2178.

3. Benedict, J.H.; Ring, D.R.; Sachs, E.S.; Altman, D.W.; DeSpain, R.R.; Stone, T.B.; Sims, J.R. Influence of Transgenic BT Cottons on Tobacco Budworm and Bollworm Behavior Survival, and Plant Injury. In *Proceedings Beltwide Cotton Council*; Herber, J., Richter, D.A., Eds.; National Cotton Council: Memphis, Tennessee, 1992; 891–895.

4. Krimsky, S.; Wrubel, R.P. *Agricultural Biotechnology and the Environment*; University of Illinois Press: Urbana, Illinois, 1996.

5. McCullum, C.; Pimentel, D.; Paoletti, M.G. Genetic Engineering in Agriculture and the Environment: Risks and Benefits. In *Biotechnology and Safety Assessment*; Thomas, J.A., Ed.; Taylor and Francis: Washington, D.C., 1998; 177–217.

6. Agribusiness. The Gene Exchange, Fall 1997; http://www.uc- susa.org/Gene/F97.agribusiness.html (accessed July 5, 2001).

7. Agribusiness. The Gene Exchange, Winter 1996; http://www. ucsusa.org/Gene/W96.agribusiness. html (accessed July 5, 2001).

23

Plant Pathogens (Fungi): Biological Control

Timothy Paulitz

Introduction

In the classical definition of biological control, certain fungi, termed biocontrol agents (BCAs), can reduce the amount of inoculum or disease-producing activity of a plant pathogen, usually another fungus.[1] The net result is a reduction of plant disease and crop loss. This section will cover mechanisms of how these fungi antagonize the pathogen, what part of the pathogen life cycle can be targeted, how the BCAs can be applied, and examples of commercially available products. Within the past 20 years, there has been a tremendous increase in interest and research on the subject, spurred by a search for more environmentally benign methods of disease control. But fungal BCAs have limitations that have restricted the number of products that are currently on the market. Table 1 shows some products used against soilborne pathogens on the market as of January 1999.

Mechanisms of Biological Control by Fungi

The strategy behind managing pathogens is to target or interrupt part of the pathogen life cycle.[1,2] Like any microbe, pathogens start from inoculum in the environment, which can be spores, mycelia, or other dormant survival structures. These germinate on the plant surface, penetrate and infect the plant, and reproduce and sporulate on the plant to produce new inoculum. Many pathogens can also grow saprophytically on dead organic matter and plant debris. In biological control, the pathogen can be targeted in three ways.[1,2] First, the inoculum of the pathogen can be reduced or destroyed. This is most effective for soilborne pathogens, where the inoculum is dormant in the soil and the monocyclic disease is determined by the initial inoculum present in the field. BCAs can also interfere with inoculum formation by pathogens growing saprophytically on organic matter and plant debris. However, this strategy is not very effective for foliar polycyclic diseases, where inoculum comes from outside the field and the initial inoculum has little effect on the final outcome of the disease. Another strategy is one of protection, where a population of the BCA is established on the infection site of the plant before the pathogen attacks, thus preventing the pathogen's entry. These infection sites can be on seeds, bulbs, roots, leaves, fruit, flowers, or wounds. Finally, nonpathogenic or avirulent fungi can stimulate the plant to a higher level of resistance to a later-attacking pathogen, a concept termed induced resistance.

TABLE 1 Some Commercial Biocontrol Products for Use against Soilborne Crop Diseases

Biocontrol Fungus	Trade Name	Target Pathogen/Disease	Crop	Manufacturer
Ampelomyces quisqualis M-10	AQ10 Biofungicide	Powdery mildew	Cucurbits, grapes, ornamentals, strawberries, tomatoes	Ecogen Inc., Langhorne, Pennsylvania
Candida oleophila I-182	Aspire	Botrytis, Penicillium	Citrus, pome fruit	Ecogen Inc., Langhorne, Pennsylvania
Fusarium oxysporum (nonpathogenic)	Biofox C	Fusarium oxysporum	Basil, carnation, cyclamen, tomato	S.I.A.P.A., Galliera, Bologna, Italy
Trichoderma harzianum and *T. polysporum*	Binab T	Wilt and root rot pathogens, wood decay pathogens	Fruit, flowers, ornamentals, turf, vegetables	Bio-innovation, Algaras, Sweden
Coniothyrium minitans	Contans	Sclerotinia sclerotiorum and S. minor	Canola, sunflower, peanut, soybean, lettuce, bean, tomato	Prophyta Biologischer Pflanzenschutz, Malchow/Poel, Germany
Fusarium Oxysporum (nonpathogenic)	Fusaclean	Fusarium oxysporum	Basil, carnation, cyclamen, gerbera, tomato	Natural Plant Protection, Nogueres, France
Pythium oligandrum	Polygandron	Pythium ultimum	Sugar beet	Plant Protection Institute, Bratislavsk, Slovak Republic
Trichoderma harzianum and *T. viride*	Promote	Pythium, Rhizoctonia, Fusarium	Greenhouse, nursery transplants, seedlings	JH Biotech, Ventura, California
Trichoderma harzianum	RootShield, Bio-Trek T-22G, Planter Box	Pythium, Rhizoctonia, Fusarium, Sclerotinia homeocarpa	Trees, shrubs, transplants, ornamentals, cabbage, tomato, cucumber, bean, corn, cotton, potato, soybean, turf	Bioworks, Geneva, New York
Phlebia gigantea	Rotstop	Heterobasidium annosum	Trees	Kemira Agro Oy, Helsinki, Finland
Gliocladium virens GL-21	SoilGard (formerly GlioGard)	Damping-off and root pathogens, Pythium, Rhizoctonia	Ornamentals and food crops grown in greenhouses, nurseries, homes, interiorscapes	Thermo Triology, Columbia, Maryland
Trichoderma harzianum	Trichodex	Botrytis cinerea, Colletotrichum, Monilinia laxa, Plasmopara viticola, Rhizopus stolonifer, Sclerotinia sclerotiorum	Cucumber, grape, nectarine, soybean, strawberry, sunflower, tomato	Makhteshim Chemical Works, Beer Sheva, Israel
Trichoderma Harzianum and *T. viride*	Trichopel, Trichoject	Armillaria, Botryosphaeria, Fusarium, Nectria, Phytophthora, Pythium, Rhizoctonia		Agrimm Technologies, Christchurch, New Zealand

Source: Information provided by the U.S. Department of Agriculture, Agriculture Research Service, Beltsville, Maryland, and was compiled by D. Fravel (http://www.barc.usda.gov/psi/bpdl/bodlpood/bioprod.htm).

The most direct way a fungus can attack a fungal pathogen is by mycoparasitism, where the BCA uses the pathogen as a source of food.[3] The hyphae of the mycoparasite contact, penetrate, and colonize the hyphae, spores, or survival structures of the host fungus. Many of these mycoparasites produce enzymes that degrade the cell walls of the fungal host, including β-1–3 glucanase and chitinase. Most mycoparasites are necrotrophic and eventually kill their fungal host. Much of this research has focused on reducing the inoculum of soilborne pathogens. Classic examples include *Trichoderma* and *Gliocladium* spp. parasitizing *Rhizoctonia solani* and *Pythium* spp., which cause seed, seedling, and root rots.[4–6] *Pythium* spp. such as *Pythium oligandrum* and *P. nunn* parasitize pathogenic species of *Pythium*. *Coniothyrium minitans* and *Sporodesmium sclerotivorum* parasitize sclerotia of *Sclerotinia* spp., such as the white mold pathogen *S. sclerotiorum*, which attacks hundreds of plant species (57). *Ampelomyces quisqualis* parasitizes cleistothecia of powdery mildews. Major limitations to this strategy are that mycoparasites are slow acting and large amounts of mycoparasite inoculum must be added to the soil to ensure it will encounter the propagules of the pathogen. However, a promising strategy demonstrated with *S. sclerotivorum* is to render a soil suppressive to the pathogen by an inoculative release at a lower inoculum density, and allowing the mycoparasite to build up over successive seasons, using the pathogen as a food source. This is similar to the classic predator–prey relationship found in insect biocontrol.

Some fungi can produce antibiotic compounds that are toxic to other microbes, including plant pathogens. *Trichoderma* spp. produce volatile and nonvolatile antifungal compounds, including peptiabols, pyrones, and terpenoid antibiotics.[4] *Gliocladium virens* produces glioviren and gliotoxin that inhibit *R. solani* and *Pythium ultimum*. This mechanism is most effective when the BCA can grow to high populations and has an energy source to produce the antibiotic. An example would be *Trichoderma* or *Gliocladium* spp. applied to seeds or where a food base is added to the inoculum.[4,5]

Plant pathogens require carbon, nitrogen, iron, and other nutrients to grow. Many spores have an exogenous requirement for these nutrients, supplied by the plant rhizosphere or phyllosphere, in order to germinate. BCAs can compete with the pathogen for these limiting nutrients. For example, nonpathogenic species of *Fusarium oxysporum* can compete with pathogenic forma speciales for these limiting nutrients, resulting in control of wilt diseases. Competition by yeasts or hyphal fungi may protect flowers and foliage against nectotrophic pathogens such as *Botrytis cinerea* by colonizing the senescent tissue or nutrient-rich flower petals.[8,9] This mechanism, although difficult to prove experimentally, is probably one of the primary ways BCAs can protect a plant surface through preemptive exclusion of the pathogen.

Fungal biocontrol agents can also affect the pathogen by acting indirectly on the plant to make it more resistant to pathogen attack. Nonpathogenic microbes can induce a systemic resistance in plants (79). When the plant recognizes the inducing BCA, a signal is transduced systemically to the entire plant, bringing the defenses to a "high state of alert," so that a subsequent challenge by a pathogen is reduced. Nonpathogenic isolates of *F. oxysporum* induce a defense reaction against pathogenic isolates of *F. oxysporum*. This mechanism has several advantages. Once induced, the resistance is systemic, the entire plant becomes more resistant, and high populations of the BCA do not need to be maintained. It can also protect parts of the plant that cannot be protected directly by the BCA, including new growth of shoots and roots. However, more research is needed to investigate the applicability of this technology under greenhouse and field conditions.

Application of Fungal Biocontrol Agents

How are fungal BCAs applied? Most are applied in an inundative strategy in large amounts to build up the population of the BCA high enough to overwhelm and have an effect on the pathogen.[1,5] Most are also targeted toward soilborne pathogens. However, one limitation of this strategy is the large amount of inoculum that must be applied and the high cost of production of spores, conidia, biomass, or chlamydospores.[5,9] Another problem is the erratic performance of many biocontrol agents under field conditions, due to unfavorable environmental conditions for the BCA, and the problem of establishing the BCA in a niche already occupied by competing microflora. Therefore most of the commercially available products have targeted applications that avoid these problems. For example, the greenhouse and nursery markets are prime targets because of the controlled environmental conditions and the high economic value of the crops. Another method is to use a protective strategy and apply the BCA directly to the infection court when it is small. For example, high populations of *Trichoderma* or *Gliocladium* conidia can be coated onto seeds to protect against damping-off pathogens such as *R. solani* and *P. ultimum*. Transplant cuttings and bulbs can be treated with liquid suspensions of products before planting in the greenhouse or field. Products such as formulations of *G. virens* can be mixed directly into the soil or soilless mixes in the greenhouse. Granules of *Trichoderma* spp. can be added to seed furrows, mixed with seeds in a planter box, or added to a commercial seed slurry.[4]

Since many pathogens gain access to plants through wounds, biocontrol agents can also be applied to transplant or pruning wounds. A classic example, and one of the first commercially used fungal biocontrol agents, is the application of *Phlebia* (= *Peniophora*) *gigantea* to cut pine stumps to prevent the stumps from being colonized by the pathogen *Heterobasidium annosum*. The pathogen can spread from these stumps to the entire plantation via the root system.

Postharvest pathogens are weak pathogens that require wounds on fruit to gain access. Yeast-like organisms such as *Candida* spp. can be applied to fruit during processing to exclude rot pathogens such as *Penicillium* spp. from colonizing wounds.[8] However, these applications require more stringent testing for registration, since they are applied directly to a food product. Competition is preferable to antibiosis for this application, since antifungal compounds would also have to be tested for animal and human toxicity.

Roots are a difficult infection court to protect, since the susceptible tips are constantly growing and moving through space encountering new inoculum. One strategy is to treat the entire rooting medium in the greenhouse or nursery. Another approach is to use a fungus that can colonize the root system from a seed or furrow treatment and protect the expanding root surface. This characteristic, called rhizosphere competence, has been demonstrated in some strains of *Trichoderma* spp.

Foliar applications of fungi are the least common, although this is the most common method of fungicide application. One example is *Pseudozyma* (= *Sporothrix*) *flocculosa*, a yeast-like fungus which is being developed for control of powdery mildews on greenhouse roses and cucumbers in Canada. *Trichoderma* and *Gliocladium* spp. can be applied to foliage and flowers and can prevent infection by necrotrophic fungi such as *B. cinerea* and *S. sclerotiorum*.

Future of Biological Control by Fungi

In conclusion, the full potential of controlling plant diseases with fungi still has not been realized. Only a small number of products are on the market, but this is a vast improvement compared to only five years ago. There are still many economic constraints in terms of the cost of development and registration of products and the low cost production of organisms in liquid fermentation or solid on substrates.[5,9] Like chemicals, the risks of fungal BCAs need to be addressed, including the displacement of nontarget microbes, allergenicity to humans and other animals, and toxigenicity and pathogenicity to nontarget organisms.[10] However, there are no existing chemical controls for many diseases, because of deregistration of pesticides, pathogen resistance to pesticides, and environmental concerns. These diseases may be the niches for fungal biocontrol agents. It is unlikely that biological control will succeed alone, but it needs to be integrated with other disease management strategies, including cultural control and genetic disease resistance.

References

1. Cook, R.J.; Baker, K.F. *The Nature and Practice of Biological Control of Plant Pathogens*; American Phytopathologi-cal Society Press: St. Paul, MN, 1983; 539.
2. *Principles and Practice of Managing Soilborne Plant Pathogens*; Hall, R., Ed.; American Phytopathological Society Press: St. Paul, MN, 1996; 442.

3. *Fungi in Biological Control Systems*; Burge, M.N., Ed.; Manchester University Press: Manchester, 1988; 269.
4. *Trichoderma and Gliocladium*; Harman, G.E., Kubicek, C.R., Eds.; Taylor and Francis: London, 1998; 1 and 2, 278–393.
5. *Pest Management: Biologically Based Technologies*; Vaughn, J.L., Lumsden, R.D., Eds.; American Chemical Society: Washington, DC, 1993; 435.
6. *Biological Control of Plant Diseases. Progress and Challenges for the Future*; Papavizas, G.C., Cook, R.J., Tjamos, E.C., Eds.; Plenum Press: New York, 1992; 462.
7. Whipps, J.M. Biological control of soil-borne plant pathogens. Advances in Botanical Res. **1997**, *26*, 1–134.
8. *Plant-Microbe Interactions and Biological Control*; Kuykendallm, L.D., Boland, G.L., Eds.; Marcel Dekker, Inc.: New York, 1998; 442.
9. *Integrated Pest and Disease Management in Greenhouse Crops;* Lodovica Gullino, M., van Lenteren, J.C., Elad, Y., Albajes, R., Eds.; Kluwer Academic Publishers: Dordrecht, the Netherlands, 1999; 545.
10. *Biological Control: Benefits and Risks*; Hokkanen, H.M.T., Lynch, J.M., Eds.; Cambridge University Press: Cambridge, 1995; 304.

24

Plant Pathogens (Viruses): Biological Control

Hei-Ti Hsu

Economic Loss

Damage to crop plants due to virus and viroid infections is difficult to assess. The actual figures for global crop loss are not available. Plant disease losses are estimated at $60 billion annually. Losses due to virus and viroids have been considered second to those caused by fungi. Unlike diseases caused by fungi, bacteria, and nematodes, where control measures using chemical, biological, and integrated pest management approaches have been effective, diseases caused by viruses or viroids are far more difficult to manage.

Economic crop loss resulting from virus and viroid disease is due to the reduced growth and vigor of infected plants which, in turn, causes a reduction in yield. In some instances, a virus infection may kill a plant. Apart from yield reduction, the quality and market value of commercial end products may be affected. There are also costs of attempting to maintain crop health such as vector control, production of pathogen-free propagation materials, and quarantine and eradication programs. In addition, resources are being diverted to research, extension, and education as well as toward breeding for resistance to virus or viroid infection.

World Impact

No single country is exempt from crop losses. Production of food, fiber, and horticultural crops are seriously affected worldwide by virus or viroid infection of plants.[1] This is even more so in developing countries that depend on one or a few major crops; for example, *Cassava mosaic virus* in cassava plants in Kenya, *Citrus tristeza virus* in citrus trees in Africa and South America, and *Cacao swollen shoot virus* in cacao trees in Ghana. Recently, *Papaya ringspot virus* (PRV) infection has affected every region where papaya plants are grown. The virus induces a lethal disease in papaya. The widespread aphid-transmitted PRV has changed the way papayas are grown in many parts of the world. Normally, papayas are produced annually for a number of years over the life of the papaya plant. For proper management of the disease due to PRV infection, papaya has now become an annual crop in which healthy seedlings are planted each year. Even so, productivity is still below the average yield obtained before PRV became a problem. Viroids infect a limited number of crops when compared with viruses. However, they can cause severe problems in specific crops, for instance, cadang-cadang disease of coconuts, potato spindle tuber disease, and chrysanthemum stunt disease.

Control Measures

No direct chemical control means are available to combat virus infections in plants. Control of viral diseases is achieved primarily by sanitary practices that involve reducing sources of inoculum from outside, preventing spreading within the crop, and limiting the population of insects, mites, nematodes, and fungi that may serve as vectors for many plant viruses.[1] Virus disease testing programs are now common in many parts of the world where the economic importance of growing virus-free plants is recognized. Although seeds and seedlings certified as virus-free are more expensive than those that have not been tested for certain viruses, testing provides assurance of virus-free production materials. Early detection of virus in a field and removal of the infected plants minimizes spread of the virus.

Plants may be protected from development of severe disease symptoms by first introducing a mild strain of virus into a healthy plant. A plant systemically infected with a mild strain of virus is protected from infection by a severe strain of the same virus. This phenomenon in called "cross protection" and has been observed for many plant viruses.[2] It is also observed to occur between viroids or plant virus satellites. In practice, cross protection is of great interest since it has been utilized to protect plants against severe virus strains (*Citrus tristeza virus, Papaya ringspot virus, Zucchini yellow mosaic virus, Tomato mosaic virus,* etc.), in the field.

Another approach toward controlling plant virus diseases is to develop resistant or tolerant plants.[3] Historically, long-term manipulation of crop plants through breeding has produced many valuable commercial varieties resistant to plant viruses. Breeding plants resistant to vectors may also offer control of the virus they transmit. Conventional breeding of crossing and back crossing commercial varieties with plants bearing virus resistance traits takes years to develop. In order for a new variety to be commercially acceptable, undesirable traits from the resistant parent breeding line must be selected out. The process is labor intensive and time consuming. Advances in science have allowed new technology to precisely manipulate resistance genes at the molecular level.[4] Biotechnology represents the fastest growing area of biological research. The application of biotechnology in breeding for resistance to virus infection is a major area of research. Successful control of viral disease through resistance breeding will undoubtedly reduce the use of synthetic pesticides for vector control.[5]

Introducing virus resistance and vector resistance into a cultivar by gene transfer technology (genetic engineering) has been successful in combating plant viruses.[6] The technology has several major advantages over conventional cross breeding. It is a relatively fast procedure. Desirable genes can be introduced without disturbing the balanced genome of target plants. Furthermore, there is no restriction on the source of the transgenes allowing the use of genes from other plant species or even from outside the plant kingdom (Table 1).[1,7]

Several approaches for producing transgenic virus- resistant plants have been explored. Among these, plants expressing virus coat protein genes, parts of other viral genes, or virus satellite ribonucleic acids (RNAs) have been shown to offer the best control.[2,8,9] Plants expressing antisense viral RNAs, ribozymes, pathogen-related proteins, or virus-specific antibody genes may also confer resistance to virus infection. Control of virus vectors by introducing insect toxins such as trypsin inhibitor, lectin, and *Bacillus thuringiensis* (Bt) toxin genes into plants would undoubtedly contribute toward achieving the goal of controlling plant virus diseases.

Prospects

Use of resistant cultivars is considered the best approach to combat virus infection in plants. Biotechnology, no doubt, will play a significant role in the economic growth of many countries. Molecular breeding, however, will not replace but complement the efforts of conventional cross breeding. Much attention has been given to engineering resistance to plant viruses. Recently, genetic engineering of crop plants has been closely scrutinized and criticized due to increasing public concerns regarding human health and environmental impact. Careful assessment of the benefits and potential risks involving the release of

TABLE 1   Genes That Contribute or May Contribute Toward Control of Virus Diseases in Plants

Virus-Derived Gene Sequences

Coat proteins
Replicases
Movement proteins
Polyprotein proteases
Sense RNAs
Antisense RNAs

Plant Host-Derived Transgenes

Pathogen-related proteins
Anti-viral proteins
Proteinase inhibitors
Natural resistance genes
Lectins

Other Transgenes and Sequences

Satellite RNAs
Virus-specific antibodies
Interferon-induced mammalian oligoadenylate synthetase
Insect toxins
Anti-viral ribozymes (catalytic RNA)

Source: Khetarpal et al.[1]

genetically modified plants into the environment and their consumption is necessary before these crops become widely accepted by the public.[10,11]

References

1. Khetarpal, R.K., Koganezawa, H., Hadidi, A., Eds. *Plant Virus Disease Control*; APS Press: St. Paul, MN, 1998, 1–684.
2. Beachy, R.N. Coat-protein-mediated resistance to tobacco mosaic virus: discovery mechanisms and exploitation. Phil. Trans. R. Soc. Lond. B. **1999**, *354*, 659–664.
3. Salomon, R. The evolutionary advantage of breeding for tolerance over resistance against viral plant disease. Israel J. Plant Sci. **1999**, *47*, 135–139.
4. Kawchuk, L.M.; Prufer, D. Molecular strategies for engineering resistance to potato viruses. Can. J. Plant Pathol. **1999**, *21*, 231–247.
5. Barker, I.; Henry, C.M.; Thomas, M.R.; Stratford, R. Potential Benefits of the Transgenic Control of Plant Viruses in the United Kingdom. In *Plant Virology Protocols: From Virus Isolation to Transgenic Resistance;* Foster, G.D., Taylor, S.C., Eds.; Humana Press, Inc.: Totowa, NJ, 1998; 81, 557–566.
6. Dempsey, D.A.; Silva, H.; Klessig, D.F. Engineering disease and pest resistance in plants. Trends Microbiol. **1998**, *6*, 54–61.
7. Gutierrez-Campos, R.; Torres-Acosta, J.A.; Saucedo-Arias, L.J.; Gomez-Lim, M.A. The use of cysteine proteinase inhibitors to engineer resistance against potyviruses in transgenic tobacco plants. Nat. Biotechnol. **1999**, *17*, 1223–1226.
8. Maiti, I.B.; Von Lanken, C.; Hong, Y.; Dey, N.; Hunt, A.G. Expression of multiple virus-derived resistance determinants in transgenic plants does not lead to additive resistance properties. J. Plant Biochem. Biotech. **1999**, 8, 67–73.

9. Prins, M.; Goldbach, R. RNA-mediated virus resistance in transgenic plants. Arch. Virol. **1996**, *141*, 2259–2276.

10. Hammond, J.; Lecoq, H.; Raccah, B. Epidemiological risks from mixed virus infections and transgenic plants expressing viral genes. Adv. Virus Res. **1999**, *54*, 189–314.

11. Kaniewski, W.K.; Thomas, P.E. Field testing for virus resistance and agronomic performance in transgenic plants. Mol. Biotechnol. **1999**, *12*, 101–115.

25

Stored-Product Pests: Biological Control

Lise Stengård
Hansen

Introduction

Stored grain and other durables of plant origin, raw and processed, can become infested by insect and mite pests. As the environment in storage facilities in general is conducive to rapid pest development, this often results in substantial quantitative and qualitative losses. Many stored-product pests are internal feeders developing inside whole-grain cereals and legumes in storage. Another group of pests are external feeders developing on broken kernels, flour, etc., and are mainly found in processing facilities, stores of processed cereals and bulk grain, and packaged foodstuffs (Table 1).

TABLE 1 Examples of Pests and Natural Enemies in Stored Products

Pest Type	Commodity	Examples of Pest Species	Examples of Natural Enemies
Internal feeders or primary pests	Whole-grain cereals	*Sitophilus spp.* (C) *Rhyzopertha dominica* (C) *Sitotroga cerealella* (L) *Prostephanus truncatus* (C)	*Lariophagus distinguendus* (l) *Theocolax elegans* (l) *Anisopteromalus calandrae* (l) *Teretrius nigrescens* (p)
	Whole legumes	*Bruchus* spp. (C) *Callosobruchus* spp. (C)	*Dinarmus* spp. (l)
External feeders or secondary pests	Broken kernels, flour, milled rice, dried fruit, spices, nuts	*Tribolium* spp. (C) *Cryptolestes* spp. (C) *Oryzaephilus* spp. (C)	*Trichogramma spp.* (e) *Holepyris silvanidis* (l) *Cephalonomia* spp. (l) *Xylocoris flavipes* (p)
		Ephestia kuehniella (L) *Cadra cautella* (L) *Plodia interpunctella* (L)	*Trichogramma spp.* (e) *Venturia canescens* (l) *Habrobracon hebetor* (l) *Xylocoris flavipes* (p)
		Storage mites (Acarina)	*Cheyletus eruditus* (p)

Source: Schöller,[2] Subramanyam and Hagstrum.[3,4]
C, Coleoptera; e, egg parasitoid; L, Lepidoptera; l, larval parasitoid; p, predator.

Beneficial insects are often found in storage facilities and may prevent or delay pest development. The internal feeders are attacked by a range of parasitoids specialized in detecting infested kernels and placing their progeny on the larvae within the kernel. The external feeders are more freely exposed to the activities of natural enemies and are attacked by both predators and parasitoids (Table 1).

Although the concept is not new, biological control of stored-product pests is still not widely used. The following description gives some examples of the recent development within biological control of stored-product pests using beneficial insects and mites.

For general reviews on stored-product pest species and biological control of stored-product pests.[1–4]

Factors that Promote Successful Biocontrol in Storage

Storage facilities are ideally suited for development and reproduction of pests. However, these factors are also conducive to successful biocontrol: 1) Climatic conditions are relatively stable and, at least during part of the year, favorable for insect development; 2) the storage structure provides protection from temperature extremes and precipitation as well as a physical barrier to dispersal; and 3) in many storage situations, time is not a limiting factor and thus often sufficient for natural enemies to establish and exert their control. These factors can be manipulated to be more favorable to the natural enemies, e.g., grain temperature can be reduced by aeration to a level that offers a greater advantage to the activities of the beneficial insect than to the pest.[5]

"Classical" Biological Control

In "classical" biological control, a natural enemy is imported and released for establishment to control an introduced (exotic) pest. An example of this among stored-product pests is the larger grain borer, *Prostephanus truncatus,* which was accidentally introduced from Central America into Africa, where it has become a serious pest of many products, e.g., stored maize and cassava chips. A predator, *Teretrius nigrescens,* was introduced from Central America and, after large scale releases, established successfully in many locations in Africa. *T. nigrescens* has been credited with reducing losses of stored maize to *P. truncatus.* However, despite the presence of the predator, *P. truncatus* densities often reach outbreak levels in many countries. An analysis in Holst and Meikle[6] led to the conclusion that *T. nigrescens* alone is unable to exert control to an acceptable level inside a store due to the predator's intraspecific density-dependence and low growth rate compared with its prey. In this case, biocontrol must be supplemented with other integrated control measures.[7]

Biocontrol of Internally Feeding Pests

The larvae of internally feeding insect pests develop within whole-grain cereals and legumes. These species are important primary pests in stores of bulk grain. Several species of parasitoids are specialized to live on these pests. The adult parasitoids enter the bulk of grain and find infested kernels, probably by means of acoustic or olfactory cues. They then drill into the kernel, paralyze the host, and deposit an egg from which the parasitoid larva emerges. After consuming the host, the parasitoid exits the kernel as an adult. The ability of the parasitoid to locate its host within grain varies among species: *Anisopteromalus calandrae* is primarily active at the grain surface whereas *Theocolax elegans* and *Lariophagus distinguendus* are able to find infested kernels down to a depth of 2.2 and 4 m, respectively.[8,9] Great differences in life history parameters occur among different parasitoid strains.[8] These factors as well as grain characteristics such as kernel size and grain variety all affect the ability of the natural enemy to exert effective control of the pest.

A constraint against biocontrol in stored products is a reluctance to increase the amount of insects by releasing natural enemies. However, these minute parasitoids (<2 mm) can easily be removed by grain cleaning procedures. In a study using *T. elegans* against *Rhyzopertha dominica* in wheat, the number of insect fragments in the resulting flour was reduced by 89%.[10]

Biocontrol of Externally Feeding Pests

Both beetles and moths are represented among external pests that feed on broken kernels and debris as well as flour. They are important pests in grain stores, in cereal processing facilities such as flourmills, and in warehouses storing cereal products.

Many species of parasitoids, particularly Trichogram-matidae, attack the egg stage of external pests. The impact of these species is increased by their host feeding behavior, which can account for half of the mortality of host eggs.[11] These egg parasitoids are able to parasitize a wide range of species, but in nature they generally show affinity to a specific habitat; it is thus important to select strains that are adapted to the stored-product environment. *Trichogramma* species have been successfully released against pyralid moth pests in experimental peanut storages in the U.S.A.[12] and in wholesale stores and industrial bakeries in Germany.[13] These egg parasitoids do not establish within the premises and must be released on a regular basis.

Larvae of moth pests are attacked by both ectoparasitoids, e.g., *Habrobracon hebetor* and endoparasitoids, e.g., *Venturia canescens*. Both species are cosmopolitan, often occurring together in flourmills and both species show potential for biocontrol of moth pests.[12] However, *H. hebetor* may affect populations of *V. canescens* negatively by feeding on hosts parasitized by this species.

The predatory bug *Xylocoris flavipes* is a generalist living on eggs and larvae of a wide range of beetles and moths. Almost 30 species of stored-product pests have been reported as prey of *X. flavipes*.[12] In residues in empty maize stores in the U.S.A., many of these species occur together in the same store. A single introduction of *X. flavipes* led to population reductions of 70%-100% of externally feeding beetle pests, whereas internal feeders and late instar moth larvae were less affected. It was suggested that releases of predatory bugs combined with parasitoids for the moths and internal feeders might eliminate or greatly reduce residual pests before the next storage season.[12]

This strategy of introducing predators in empty stores was widely practiced to control storage mites (Acarina) in the Czech Republic. The predatory mite *Cheyletus eruditus* led to reductions in storage mite populations of 88%, compared with the 18% reduction obtained with a pesticide treatment.[14]

Conclusions

The above-mentioned examples show the great potential of biological control of stored-product pests. This research field benefits from extensive international collaboration and applicability, as the pest species as well as their natural enemies have a cosmopolitan distribution as a result of international grain trade. A great amount of faunistic surveys and laboratory research on the natural enemies has been carried out, but very few field trials have been conducted. Application of natural enemies to control pests in bulk grain as well as in empty grain stores prior to introduction of newly harvested grain is considered to hold potential in the near future. However, widespread use depends on crucial experience to be obtained from field trails. The next step is ensuring reliable supplies of natural enemies, designing introduction strategies, and establishing quality control during mass rearing and shipment of beneficials. From these starting points, other pest species and other storage situations can be covered. These activities can contribute to satisfying public demands of more environmentally friendly pest control methods and food production that is focused on consumer safety.

References

1. *Ecology and Management of Food-Industry Pests*; Gorham, J.R., Ed.; Food and Drug Administration Technical Bulletin 4; Association of Official Analytical Chemists: Arlington, 1991.
2. Schöller, M. Biologische Bekämpfung vorratschä-dlicher Arthropoden mit Räubern und Parasitoiden—Samelbericht und Bibliografie (Biological control of arthropod pests in stored product protection with predators and parasitoids— review and bibliography). Mitt. Biol. Bundesanst. Land-Forstwirtsch. **1998**, *342*, 85–189.

3. *Integrated Management of Insects in Stored Products;* Subramanyam, B., Hagstrum, D., Eds.; Marcel Dekker, Inc.: New York, 1996.

4. *Alternatives to Pesticides in Stored-Product IPM;* Subramanyam, B., Hagstrum, D., Eds.; Kluwer Academic Publishers: Boston, 2000.

5. Flinn, P.W. Temperature effects on efficacy of *Choetospila elegans* (Hymenoptera: Pteromalidae) to suppress *Rhyzopertha dominica* (Coleoptera: Bostrichidae) in stored wheat. J. Econ. Entomol. **1998,** *91,* 320–323.

6. Holst, N.; Meikle, W.G. *Teretrius nigrescens* against larger grain borer *Prostephanus truncates* in African maize stores: biological control at work? J. Appl. Ecol. **2003,** *40,* 307–319.

7. Markham, R.; Borgemeister, C.; Meikle, W.G. Can biological control resolve the larger grain borer crisis? In *Stored Product Protection,* Proceedings of the 6th International Working Conference on Stored-product Protection, Canberra, Australia, Apr 17–23, 1994; Highly, E., Wright, E.J., Banks, H.J., Champ, B.R., Eds.; CAB: Wallingford, U.K., 1994; 1087–1097.

8. Steidle, J.L.M.; Schöller,M. Fecundity and ability of the parasitoid *Lariophagus distinguendus* (Hymenoptera: Pteromalidae) to find larvae of the granary weevil *Sitophilus granarius* (Coleoptera: Cucurlionidae) in bulk grain. J. Stored Prod. Res. **2002,** *38,* 43–53.

9. Press, J.W. Comparative penetration efficacy in wheat between the weevil parasitoids *Anisopteromalus calandrae* and *Choetospila elegans.* J. Entomol. Sci. **1992,** *27,* 154–157.

10. Flinn, P.W.; Hagstrum, D.W. Augmentative releases of parasitoid wasps in stored wheat reduces insect fragments in flour. J. Stored Prod. Res. **2001,** *37,* 179–186.

11. Hansen, L.S.; Jensen, K.-M.V. Effect of temperature on parasitism and host-feeding of *Trichogramma turkestanica* (Hymenoptera: Trichogrammatidae) on *Ephestia kuehniella* (Lepidoptera: Pyralidae). J. Econ. Entomol. **2002,** *95,* 50–56.

12. Brower, J.H.; Smith, L.; Vail, P.V.; Flinn, P.W. Biological control. In *Integrated Management of Insects in Stored Products;* Subramanyam, B., Hagstrum, D., Eds.; Marcel Dekker, Inc.: New York, 1996; 223–286.

13. Schöller, M.; Flinn, P.W. Parasitoids and predators. In *Alternatives to Pesticides in Stored-Product IPM;* Subramanyam, B., Hagstrum, D., Eds.; Kluwer Academic Publishers: Boston, 2000; 229–271.

14. Zdarkova, E. Personal experience with biological control of stored food mites. IOBC=WPRS Bull. **1998,** *21* (3), 89–93.

26

Weeds (Insects and Mites): Biological Control

Peter Harris

Introduction

The usual targets for classical biocontrol are introduced weeds of uncultivated land that lack enemies and have spread to dominate large areas. Such weeds cause huge habitat and forage losses, reduce diversity of the native flora and fauna, and displace rare species.[1] The concern about establishing their natural enemies is that desirable plants might also be attacked. Testing to ensure that this does not happen costs about two scientist years per agent ($900,000). About a third of the agents fail to establish, a third remain scarce, and nearly half of those that become abundant do not achieve control. On the other hand, costs of biocontrol increase little with infestation size and the same agent may achieve control in climates as different as Australia and Canada. Thus, the cost per ha can be small and biocontrol is less damaging to nontarget plants than most alternatives. Release approval for the agents in North America rests with Animal and Plant Health Inspections Service-United States Drug Administration and the Canadian Food Inspection Agency who consult advisory groups.

History and Impact of Classical Weed Biocontrol

Classical weed biocontrol was the serendipitous result of introducing the mealybug (*Dactylopus ceylonicus* thinking it was *D. coccus*), to India in 1795.[2] The purpose was to start a dye industry on the impenetrable stands of the South American prickly pear cactus, *Opuntia vulgaris*. The result was rapid cactus kill and little dye production. After 68 years and several eradication attempts the government distributed the insect for *O. vulgaris* control.

Hawaii took the ball in 1902 by importing 23 insect species to control the introduced shrub *Lantana camara*.[2] Their intention was to establish a complex to attack all parts of the plant. Only 12 species arrived alive in sufficient numbers for release and eight established. Success was claimed for a fly that destroyed 86% of the seed, but none of the criteria for success were demonstrated: that the weed is reduced to a low density in several locations, it remains at a low density, or it returns to a high density when protected from the agent.[3]

The mealy bug *D. ceylonicus* practically eliminated *O. vulgaris* in Australia but did not affect two more abundant cacti, which by 1925 had infested 60 million acres. Concerns that introduced insects might attack cultivated plants were met with the "no-choice" test in which only species that starve on representative crop plants were released. They tested 49 insect species, released 24, of which 12 established. Control was attributed to just two: the mealy bug *D. opuntia* and the cladode-boring moth *Cactoblastis cactorum*, which, two years after increase, caused a crash of cactus from 5000 to 11 plants/acre.[4]

Australia then tackled St. John's wort *(Hypericum perforatum)*.[1] Eight insect species were released, four established and, up to seven years after release, the leaf-feeding beetle *Chrysolina quadrigemina* increased to control the weed. The four established insects were then released in California, where 40,500 ha of open range that were infested in northern California in 1929 had increased to almost 2 million ha in western North America by 1940. Increases of *C. quadrigemina* returned most of the infested area to a native bunch grass community with fair to good productivity and an increased diversity of plant species, and the weed stabilized at about 1% of its former density.[5] The two beetle species were released in Canada in 1952. Up to 13 years later, *C. quadrigemina* increased to control the weed on summer-dry sites and *C. hyperici* on summer moist sites, but both failed above 900 m elevation.[6] This program terminated chemical control of the weed in British Columbia and removed it from the Ontario noxious weed list.

Both *Chrysolina* spp. developed on Canadian *Hypericum* spp. in no-choice tests, but in the field they have not been attacked.[6] The plant genus contains photosensitizing compounds lethal to sun exposed larvae. In nature the larvae feed at dawn and dusk on a winter mat of foliage, which is absent in the Canadian species. Two nontarget species are attacked, but are poor hosts: the introduced evergreen ornamental *H. calycinum,* which has hard mature foliage, and in California the native *H. concinnum*, which has diffuse foliage (presumably shading is reduced and the need for climbing by inactive larvae is increased).

It is more predictive to use the no-choice tests to show that the larval host range is restricted to a taxonomic group of plants rather than that individual crop plants are unacceptable. Some reviewers felt that tests of crop plants were still necessary for public confidence, but gradually these tests have been eliminated. The laboratory larval host range limit exaggerates the field host range, since congeners not attacked in nature usually support development in the test. The larva role is to stay on the plant on which it hatched and eat, which may involve distinguishing the host from intermingled vegetation, but not host finding. Host finding is done by the adult female using habitat and plant cues.[7] The larval host range limit was a satisfactory measure of risk as long as native congeners were not an issue. This changed in 1997 when the attack of native *Cirsium* spp. by the seed-head weevil *Rhinocyllus conicus* was deplored.[8] The 1967 petition for release of the weevil against the introduced thistle *Carduus nutans* stated that its host range included *Cirsium*, but this caused no concern. Today the view is mixed. Some deplore the weevil, but to many a reduction of native thistles is not a concern and the control of most stands of *C. nutans* with increased forage production, makes it an eminent success.

Three Current Issues

Host-Specificity Tests

Most introduced weeds in North America have native congeners. A total of 23 of 27 insect species established in Canada developed on congeners in laboratory no-choice tests, but only one species in Canada and three in the United States have attacked the North American relatives of the weed in the field. Exclusion of species with larvae that develop on native congeners in the laboratory will practically end weed biocontrol. A scientifically sound alternative is to base release approval on the adult host range. The larval no-choice test does not distinguish between species with adults that oviposit on the weed's relatives in the field from those that exclude them. However, adult tests in large field cages or open releases allow assessment of this choice. Thus, although the beetle *Altica carduorum* develops on all *Cirsium* spp., the adult only "sees" *C. arvense* in large arenas. This combined with a low suitability of other *Cirsium* spp. for the immature stages means there is selection pressure against host plant shifts and accounts for the field specificity of the beetle in Eurasia.[9] However, the development of new host screening procedures are currently at an impasse because some reviewers still feel larval ability to complete development in the laboratory is an unacceptable risk even if the adult does not recognize the plant as a host.

Costs and Agent Success

Past cost increases of weed biocontrol have been partly compensated for by using fewer agents and by improving establishment success. The first Hawaiian and first two Australian projects tried to release 55 insect species of which 44% established and three reduced weed density. The current world establishment rate is 65% with the release of 4.1 species against each successfully controlled weed.[10] The idea that all parts of the plant need to be attacked is not supported by the fact that 81% of the successes result from single species. Where several agents contributed, they tend to do so in different habitats such as *C. quadrigemina* and *C. hyperici*.

New host tests are likely to increase costs. An obvious solution is to eliminate the approximately 80% of agents released that contribute little. Present practice is to be more discriminating with expensive agents than those that are cheap simply because testing was done by another country (Table 1). However, establishing any agent may have ecological effects. Thus, it is prudent to reduce the number released to the necessary minimum. No agent trait associated with effectiveness has been found, although establishment is broadly correlated with a high rate of increase, long adult life, the number of generations a year, and small size. However, effectiveness may be plant related. For example, root feeders have been more effective on tansy ragwort, *Senecio jacobaea*, and leafy spurge, *Euphorbia esula*, than defoliators and seed feeders, which have been effective on other plants. If this were determined in preliminary studies, host specificity tests could be restricted to insects in the relevant feeding guild. This combined with studies of the insect's ecological requirements could halve the number of failures.

Legislation

The enabling legislation used for classical weed biocontrol is the Federal Plant Pest Act of 1957 in the United States and the Plant Protection Act of 1990 in Canada. The purpose of both is to prevent pest establishment and spread, so release of nonpests is allowed by reverse logic. However, with no explicit mention of biocontrol, other acts, with various purposes, come into play. For example, the U.S. Endangered Species Act of 1973 requires federal agencies to ensure their actions are not likely to jeopardize the continued existence of endangered or threatened species. The ability of larvae to develop on an endangered plant may indicate jeopardy, but the issue is complicated. The act can save habitat from human destruction, but it cannot stop invading weeds, such as leafy spurge, which are displacing the

TABLE 1 Success of Weed Biocontrol Agents Released in Canada (1981–1990)

	Sponsor of Pre-Release Studies	
	Canada	United States
No. released	18	6
% Established	73	33
Species contributing to control	8	1
Canadian costs/agent	2 SY[a]	0.04 SY
Canadian costs/success	6.6 SY	0.24 SY

[a] SY = scientist year, currently about $450,000.

threatened northern prairie skink in Canada and the endangered western prairie fringed orchid in both countries. Biocontrol is the best hope for their survival; however, there are rare Florida spurges that will support larval development of leafy spurge agents. It is not possible under the act to weigh the benefits against the risks, although the risks are small if the agent has an obligatory winter diapause that prevents survival in Florida. Nevertheless, the approval of agents for the major North American leafy spurge problem is currently stalled. The Australian Biocontrol Act of 1984 solves the issue by instructing reviewers to approve the release of agents when the expected benefits outweigh the risks. A further need is to judge risk on a holistic assessment of adult and immature needs and habits. It would also be helpful to make weed biocontrol more open to public input and have procedures for resolving conflicts of interest.

References

1. Pimentel, D.L.; Lach, L.; Zuniga, R.; Morrison, D. Environmental and Economic Costs Associated with Non-Indigenous Species in the United States. Available at http://www.news. cornell.edu/releases/Jan99/species_costs.htm (accessed Feb 25, 1999).
2. Goeden, R.D. Biological Control of Weeds. In *Introduced Parasites and Predators of Arthropod Pests and Weeds: A World Review;* Clausen, C.P., Ed.; Part II, Agricultural Handbook 480, USDA: Washington, DC, 1978; 357–414.
3. Smith, H.S.; DeBach, P.; The measurement of the effect of entomophagous insects on population densities of their hosts. J. Econ. Entomol. **1942**, *3*, 845–849.
4. Crawley, M.J.; The success and failures of weed biocontrol using insects. Biocontrol. News Inf. **1989**, *10*, 2213–2223.
5. Huffaker, C.B.; Kennet, C.E.; A ten-year study of vegetation changes associated with biological control of klamath weed. J. Range Manage. **1959**, *12*, 69–82.
6. Harris, P. Status of Introduced and Main Indigenous Organisms on Weeds Targeted for Biocontrol in Canada. Available at http://res.agr.ca/leth/weedbio.htm (accessed Dec 20, 1999).
7. Fox, C.W.; Lalonde, R.G.; Host confusion and the evolution of insects' diet breadths. Oikos **1993**, *67*, 577–581.
8. Louda, S.M.; Kindal, D.; Connor, J.; Simberloff, D. Ecological effects of an insect introduced for the biological control of weeds. Science **1997**, *277*, 1088–1090.
9. Harris, P. Evolution of classical weed biocontrol: meeting survival challenges. Bull. Ent. Soc. Can. **1998**, *30*, 134–143.
10. Julien, M.J. Biological control of weeds worldwide: trends, rates of success and the future. Biocontrol. News Inf. **1989**, *10*, 299–306.

VI

NEC: Natural Elements and Chemicals Found in Nature

IV

27

Antagonistic Plants

Philip Oduor-Owino

Introduction

Plant parasitic nematodes cause significant crop losses in Africa and other parts of the world. Infected plants suffer from water deficiency and low yields, and have necrotic and/or galled roots. Control of nematodes has been mainly through the use of chemicals and host resistance. However, the existence of physiological races in the pathogen's population has complicated efforts to breed for resistant cultivars. Chemical control is effective but difficult to sustain for long-term benefits. The high cost of nematicides and their toxic effects also make them less attractive. Some nematicides such as Nemagon and Fumazon have now been banned from the world market and this has placed severe constraints on strategies for nematode control. Interest in developing alternative control measures that are safe and economically attractive has now intensified world-wide. The use of antagonistic plants is viewed as a viable nematode management option.[1–3]

Nature of Antagonistic Plants

Antagonistic plants are defined as plants that produce chemicals in their roots that are toxic and/ or repellant to phytonematodes in the soil ecosytem.[4] These plants include *Tagetes erecta* L; *Tagetes patula* L; *Datura stramonium* L; *Ricinus communis* L; and *Asparagus officinalis*. Fresh roots of asparagus produce asparaguric acid glycoside that is toxic, even when diluted, to most plant parasitic nematodes. Root exudates from *Tagetes, Datura,* and *Ricinus* spp. induces premature hatching of nematode eggs, blocks the processes of mitosis and meoisis, and reduces galling intensity on roots of susceptible plants. This has been attributed, in part, to the toxic effects of the alkaloids terthienyl, hyosine, and ricinine present in *Tagetes, Datura,* and *Ricinis* spp., respectively. These compounds may also disrupt female taxis to roots or male taxis to female.[4] Other plants with antagonistic properties include some crucifers and citrus. Root diffusates from crucifers reduce the pathogenicity of nematodes on potato, while a compound in citrus roots is toxic to *Tylechulus semipenetrans*.[1]

Antagonistic Plants in Cultural Pest Control

Antagonistic plants may have a great nematode-control potential in agriculture if properly utilized in crop rotation and intercropping systems.[2] For example, intercropping food crops with nematicidal plants is now a nematode management strategy in Tanzania, India, and Zimbabwe and has also been

TABLE 1 Effect of Tomato Intercropping with *Datura stramonium*, *Ricinus communis*, and *Tagetes minuta*, and Soil Treatment with Captafol and Aldicarb, on Gall Index, Number of Juveniles, Tomato Growth and Fungal Parasitism of *Meloidogyne javanica* Eggs by the Fungus *Paecilomyces lilacinus*, 50 Days after Inoculation

Soil Treatment[a]	Egg Parasitism (%)	No. of Juveniles/ 300 cm³ soil	Shoot Dry Weight (g)	Shoot Height (cm)	Gall[b] Index (0–4)
Ne "only" untreated	1.0de	670a	1.5f	26.6f	4.0a
Soil "only" untreated	0.0e	0d	3.5b	43.1ab	0.0e
F + Ne	23.2c	660a	1.6f	30.1e	3.0ab
F + Cap + Ne	1.3de	635a	1.9f	31.2e	3.9a
F + Ald + Ne	27.6b	12de	4.6a	45.3a	1.4d
F + Tag + Ne	29.8ab	161c	2.4e	38.4b	2.1c
F + Dat + Ne	28.3ab	173c	2.5de	36.4cd	2.2c
F + Ric + Ne	30.9a	210c	2.9cd	35.4cde	2.4c
Tag + Ne	2.1de	209c	2.8d	33.4cde	2.9b
Dat + Ne	3.0d	183c	3.0cd	36.6c	2.8b
Ric + Ne	2.8de	204c	3.2bc	37.7bc	3.0ab
Ald + Ne	1.2de	14d	4.5a	46.1a	3.8a
Cap + Ne	0.0e	460b	1.9f	36.1d	4.0a

Source: Oduor-Owino.[7]

Note: Numbers are means of 10 replications. Means followed by different letters within a column are significantly different ($P = 0.05$) according to Duncan's Multiple Range Test.

[a] Ne, nematode; F, *P. lilacinus*; Cap, Captafol; Ald, Aldicarb; Tag, *T. minuta*; Ric, *R. communis*.

[b] Gall index was based on a 0–4 rating scale, where 0 = no galls and 4 = 76–100% of the root system galled.

recommended for Pakistan.[5] Field trials with *T. minuta*, *D. stramonium*, and *R. communis* are promising.[2,5,6] These plants reduce galling intensity and enhance tomato performance significantly. In India, a rotation of *D. stramonium*, maize, tomato, and pepper reduced the population of root-knot nematodes by 30% but the level of nutrient depletion by the antagonistic plant was 15%.[4] Integration of these plants with the biological agent, *Paecilomyces lilacinus* Thom (Sam), gave better results in Kenya.[3,7,8] Tomato plants grown in soils planted with the various antagonistic plants in combination with *P. lilacinus* develop significantly heavy shoots and roots and relatively fever root galls compared to controls (Table 1). Cases where antagonistic plants are used in crop rotation or intercropping systems are now increasing.[4] For instance, in Indonesia, *Tagetes* sp. *Crotalaria usaramoensis*, corn, and sweet potato (*Ipomea batatas*) are used to reduce *Meloidogyne* spp. density in the soil. For cereal-based cropping systems, the following crop sequences for root-knot nematode control are recommended in the Philippines: rice-mung bean (*Phaseolus* aureus)-corn-cabbage-rice, rice-tobacco (*Nicotiana tobacum*)-rice and rice-tobacco and *Tagetes* spp. There is also considerably less galling by *Meloidogyne* spp. on potato (*Solanum tuberosum*) roots intercropped with onion (*Allium* sp.), corn, and marigold compared with galling found on potatoes alone. Although antagonistic plants are gaining popularity as pest management tools, their benefits and risks must be understood thoroughly before one can exploit their potential in pest control.

Benefits and Risks of Antagonistic Plants

Benefits and risks associated with the utilization of antagonistic plants in agriculture are varied. Phytonematoxic plants such as *R. communis*, *D. stramonium*, *Tagetes* spp. *Crotalaria* spp. *A. najus,* and *Datura metel* L. are traditionally gaining popularity due to their medicinal significance.[9] The flowers of *D. metel* are used against asthma, while *Crotalaria* spp. is a nitrogen-fixing legume. Castor oil from *R. communis* is used for making soaps and waxes: rinolecic acid from castor seeds is a valuable laxative.[9] Despite these attributes, antagonistic plants may pose a serious threat to food production if not well utilized. They may compete with economically important crops for space and nutrients.

In addition they are slow in action, an attribute that makes them less attractive for use in a commercial setting. Because of this scenario, it is important that scientific disciplines work together in order to develop a viable pest control system. What is good for the nematologist may not be good for either the agronomist or economist.

Future Concerns

There is increasing internal awareness of the value of natural plants and their products in the development of new drugs and formulation of materials that can be used for pest control. Since some of the antagonistic plants can also be used to treat human ailments,[9] they may attract intensive scientific evaluation, recognition, and funding. However, more work should be done to reexamine the future of antagonistic plants in nematode control and in the drug industry. Efficacy of these plants against nematodes and their utilization in the pharmaceutical industry will depend highly on the concentrations of the active ingredients in their tissues.[10] It will also depend on whether they can stimulate activity of most biocontrol fungi and plant growth consistently. They have so far enhanced tomato growth in the greenhouse and in the field significantly[2,7] (Tables 2 and 3), but more trials are needed in order to understand the relationship between antagonistic plants, natural enemies, and crop performance.

TABLE 2 Effect of Soil Treatment with Aldicarb, *Tagetes minuta*, *Datura metel*, and *Datura stramonium* on Root-Knot Nematodes in Tomatoes (Greenhouse Test)

Soil Treatment[a]	Shoot Height (cm)	Shoot Dry Weight (g)	Gall Index (0–5)	Galls (no.g⁻¹) root Weight	Nematodes, No. (300 mL)⁻¹ Soil
Soil + Ne[b], untreated (control)	42.4c[c]	1.3e	4.4a	510.0a	564.1a
Soil only, untreated	97.8a	4.9b	0.0c	0.0d	0.0e
Soil + Ne + Aldicarb	116.3a	6.1a	1.0c	23.3d	18.4d
Soil + Ne + *D. metel*	73.6d	3.7c	2.1b	77.4c	176.3c
Soil + Ne + *T. minuta*	65.4b	2.6d	2.0b	134.9b	310.4b
Soil + Ne + *D. stramonium*	73.4b	3.1c	2.2b	88.4c	170.4c
Soil + Ne + *R. communis*	69.0b	3.3c	2.4b	90.0c	173.0c

Source: Oduor-Owino, P. Effects of Aldicarb and Selected Medicinal Plants of Kenya on Tomato Growth and Root-Knot Severity, unpublished data, 1992.
[a] Autoclaved soil used.
[b] Ne, nematode eggs added to soil.
[c] Means followed by the same letter within each column are not significantly different at the 5% level (Duncan's Multiple Range Test).

TABLE 3 Effect of Soil Treatment with Aldicarb, *Tagetes minuta*, *Datura metel*, and *Datura stramonium* on Infection of Tomato by Root-Knot Nematodes (Field Test)

Treatment	Shoot Height (cm)	Shoot Dry Weight (g)	Fruit Yield (g)	Galls (No. g⁻¹ Root Weight)	Nematodes, No. (300 mL)⁻¹ Soil
Untreated (control)	80.3d[a]	40.5d	380.3e	69.1a	150.4a
Aldicarb	187.3a	135.1a	3800.4a	4.50d	6.4d
D. metel	157.1b	89.3b	2590.1b	6.4c	17.3bc
T. minuta	107.1c	45.1d	761.1c	11.4b	21.1b
D. stramonium	150.1b	69.4c	2030.4b	9.6b	18.4c

Source: Oduor-Owino.[2]
[a] Means followed by the same letter within each column are not significantly different at the 5% level (Duncan's Multiple Range Test).

See also *Biological Control of Nematodes*, pages 6163; *Risks of Biological Control*, pages 720–722; *Toxins in Plants*, pages 840–842; *Pest-Host Plant Relationships*, pages 593–594; *Intercropping for Pest Management*, pages 423–425.

References

1. Caswell, E.P.; Tan, C.S.; De Frank, J.; Apt, W.J. The influence of root exudates of *Chloris gayana* and *Tagetes patula on Rotylenchulus reniformis*. Revue de Nematologie **1991**, *14* (2), 581–587.
2. Oduor-Owino, P. Effects of Aldicarb, *Datura stramonium, Datura metel* and *Tagetes minuta* on the pathogenicity of root-knot nematodes in Kenya crop protection. Organic Soil Amendment **1993**, *12* (4), 315–317.
3. Oduor-Owino, P.; Sikora, R.A.; Waudo, S.W.; Schuster, R.P. Effects of aldicarb and mixed cropping with *Datura stramonium, Ricinus communis* and *Tagetes minuta* on the biological control and integrated management of *Meloidogyne javanica*. Nematologica **1996**, *42*, 127–130.
4. Yeates, G.W. How plants affect nematodes. Advances in Ecological Research **1987**, *17* (2), 61–137.
5. Oduor-Owino, P.; Waudo, S.W. Effects of antagonistic plants and chicken manure on the biological control and fungal parasitism of root-knot nematode eggs in naturally infested field soil. Pakistan Journal of Nematology **1995**, *13* (2), 109–117.
6. Oduor-Owino, P.; Waudo, S.W. Comparative efficacy of nematicides and nematicidal plants on root-knot nematodes. Trop. Agric. **1994**, *71* (4), 272–274.
7. Oduor-Owino, P. *Fungal Parasitism of Root-knot Nematode Eggs and Effects of Organic Matter, Selected Agrochemicals and Intercropping on the Biological Control of Meloidogyne javanica on Tomato*; Ph.D. Thesis, Kenyatta University: Nairobi, Kenya, 1996; 132.
8. Oduor-Owino, P.; Sikora, R.A.; Waudo, S.W.; Schuster, R.P. Tomato growth and fungal parasitism of *Meloidogyne javanica* eggs as affected by nematicides, time of harvest and intercropping. East African Agricultural and Forestry Journal **1995**, *61* (1), 23–30.
9. Oduor-Owino, P.; Waudo, S.W. Medicinal plants of Kenya: effects on *Meloigogyne incognita* and the growth of okra. Afro-Asian Journal of Nematology **1992**, *2* (1), 64–66.
10. Oduor-Owino, P. Effects of marigold leaf extract and captafol on fungal parasitism of root-knot nematode eggskenyan isolates. Nematologia Mediteranea **1992**, *20*, 211–213.
11. Oduor-Owino, P.; Waudo, S.W.; Sikora, R.A. Biological control of *Meloidogyne javanica* in Kenya: effect of plant residues, benomyl, and decomposition products of mustard *(Brassica campestris)*. Nematologica **1993**, *39*, 127–134.

28

Arthropod Host-Plant Resistant Crops

Gerald E. Wilde

Introduction

Because of the many advantages plant resistance offers, virtually every cultivated crop has been evaluated for this trait and one or more resistant sources have been identified. The challenge has been to incorporate these resistant sources into agronomically adapted and consumer acceptable, high-yielding cultivars. In addition to traditional breeding methods, the use of modern breeding techniques and genetic transformation of crops has opened the door to other ways of identifying, incorporating, and employing pest-resistance genes to effectively and economically manage arthropod pest populations. The use of resistant cultivars contributes significant economic and social benefits and sustainable agricultural systems to the world's farmers. The positive effects of resistant cultivars have been demonstrated repeatedly in crops as diverse as wheat, alfalfa, grape, sorghum, maize, rice, apple, and cotton.

Percentage of Crops That Have Some Degree of Pest Resistance

Plant resistance has been employed to a greater or lesser degree in practically all of the major food, feed, and fiber crops. Table 1 lists a number of major crops grown in the world and the number of pests for which resistance has been employed to at least some extent in the field. Hectarage planted to resistant cultivars varies for each pest and crop and over time as new varieties and hybrids (both susceptible and resistant) are grown and, in some instances, as new pest biotypes (pest populations that are capable of damaging previously resistant sources) develop. For example, most of the modern rice varieties and hybrids grown in China, India, and other countries are resistant to one or more major pests. Resistant American grape rootstocks have been used extensively over the world to control *Phylloxera vittifolae* (Fitch). A large percentage of the alfalfa planted in the United States is comprised of varieties resistant to aphid species. Sorghum hybrids with resistance to the greenbug have occupied up to 80% of the hectarage in the United States. Significant hectarages of wheat and barley in the United States, Canada, and North and South Africa have resistance to at least one pest. Most commercial soybean

TABLE 1 List of Some Major Crops Grown in the World and Number of Arthropod Pests for Which Resistant Cultivars Have Been Used in the Field by Growers for Pest Management

Crop	No. of Pests
Alfalfa	6
Apple	1
Asparagus	1
Barley	3
Bean	1
Cassava	2
Chickpea	0
Cotton	6
Grape	1
Lettuce	1
Maize	10
Millet	1
Oat	1
Pea	1
Peanut	4
Potato	1
Raspberry	1
Rice	14
Rye	1
Sorghum	6
Soybean	1
St. Augustine grass	1
Sugar beet	1
Sugarcane	3
Sunflower	1
Sweet clover	1
Sweet potato	1
Wheat	7

varieties are resistant to the potato leafhopper. Several cotton varieties carrying genes for resistance to jassids (*Empoasca* sp) are grown widely in Africa, India, and the Philippines. In the United States, more than 65% of commercial maize hybrids have some resistance to corn leaf aphid, >90% have some resistance to first generation European corn borer, and >75% have some resistance to second generation corn borer.

However, many more resistance genes have been identified in all crops than have been used in modern commercial varieties and hybrids, because incorporating them into high yielding cultivars acceptable to growers has been difficult. Recently, transgenic crops have been utilized to combat major insect pests. Hybrids or varieties with insect-resistance genes have been developed in cotton, maize, and potato. An estimated 6.7 million hectares of transgenic corn resistant to the European corn borer, 2.5 million hectares of transgenic cotton resistant to several pests, and 20,000 hectares of transgenic potato resistant to Colorado potato beetle were grown in the world in 1998. The hectares planted to transgenic crops are likely to increase as additional countries register these products and this technology is used on additional crops. For example, specific biotechnology applications are being field tested for rice and wheat, which together occupy 400 million hectares globally.

Effect of Plant Resistance on Pest Populations

The growing of pest-resistant cultivars can be used as a major control tactic or adjunct to other measures. Historically, the use of resistant cultivars combined with other tactics has resulted in a reduction of many pest species to subeconomic levels. Even small increases in resistance enhance the effectiveness of cultural, biological, and insecticidal controls. The extent to which growing resistant plants affects pest populations is dependent upon the level of resistance expressed, the mechanisms of resistance involved, and the number of hectares grown. The growing of resistant wheat on 50% of the hectarage in Kansas has been shown to reduce Hessian fly populations to extremely low levels. Resistance in wheat to wheat curl mite (ca. 25% of the hectarage) was effective in limiting the spread of wheat streak mosaic virus, which the mite transmits. The incorporation of leaf and stem pubescence into most commercial soybean varieties has resulted in population suppression of the potato leafhopper over the past 60 years. As the hectarage of sorghum resistant to the greenbug increased to >50%, the area of sorghum treated with insecticide was reduced by 50%. Tenfold reductions in pest populations have been observed where insect-resistant rice cultivars have been grown widely.

Economic and Social Benefits

Assessing the economic benefits of plant resistance is difficult in the context of integrated pest management programs and is likely to be underestimated frequently and substantially. Even determining the obvious advantages (yield benefits and reduced production costs) may be difficult over a large area where pest populations vary from locality to locality and year to year. Other environmental benefits, such as cleaner water and food, reduced risks to farmers, more flexibility in planting and cropping systems, reduced disease transmission, and reduced secondary pest outbreaks, also are difficult to quantify. Nevertheless, some specific estimates are available. In the United States alone the estimated valued of using arthropod-resistant alfalfa, barley, corn, sorghum, and wheat cultivars is more than $1.4 billion each year. The net economic benefit of greenbug resistance in U.S. sorghum production is estimated at close to $400 million annually. The global economic value of arthropod-resistant wheat has been estimated at $250 million annually. The value of resistance to aphids in alfalfa in the major alfalfa-producing states of the United States is estimated at more than $100 million annually. Breeding for pest resistance in rice has been estimated to be responsible for one-third of recent yield increases and $1 billion of additional annual income to rice producers. The net return of insect-resistant Bt maize in the United States and Canada has been estimated in some studies at $42.00–$67.30 per hectare, but other studies have indicated less of an economic return. The average net economic return of insect-resistant Bt cotton in 1997 was $133 per hectare.

Bibliography

1. Antle, J.M.; Pingali, P.L. Pesticides, productivity and farmer health: A philippine case study. Am. J. Agric. **1994**, *76*, 418–430.
2. *Global Plant Genetic Resources for Insect-Resistant Crops;* Clement, S.L., Quisinberry, S.S., Eds.; CRC Press: New York, 295.
3. Harvey, T.L.;Martin,T.J.; Seifers, D.L. Importanceof plant resistance to insect and mite vectors in controlling virus diseases of plants: resistance to the wheat curl mite (Acari: Eriophyidae). J. Agric. Entomol. **1994**, *11*, 271–277.
4. Hyde, J.; Martin, M.A.; Preckel, P.V.; Edwards, L.R. The economics of Bt corn: valuing protection from the European corn borer. Rev. Agric. Econ. **1999**, *21*, 442–454.
5. James, C. *Global Review of Commercialized Transgenic Crops;* ISAAA Briefs No. 8, ISAA: Ithaca, NY, 1998, 43.

6. In *Insect Resistant Maize: Recent Advances and Utilization*, Proceedings of an International Symposium, International.
7. Maize and Wheat Improvement Center (CIMMYT), Nov 27–Dec 3, 1994; Mihm, J.A., Ed.; CIMMYT: Mexico, D.F., 1997; 302.
8. Painter, R.H. *Insect Resistance in Crop Plants;* University of Kansas Press: Lawrence, KS, 1968, 520.
9. Smith, C.M. *Plant Resistance to Insects. A Fundamental Approach;* John Wiley and Sons: New York, 1989, 286.
10. Smith, C.M.; Quisinberry, S.S. Value and use of plant resistance to insects in integrated pest management. J. Agric. Entomol. **1994**, *11*, 189–190.
11. van Emden, H.F. Host-Plant Resistance to Insect Pests. In *Techniques for Reducing Pesticides: Environmental and Economic Benefits;* Pimentel, D., Ed.; John Wiley: Chichester, England, 1997, 124–132.
12. In *Economic, Environmental, and Social Benefits of Resistance in Field Crops*, Proceedings, Thomas Say Publications in Entomology, Wiseman, B.R., Webster, J.A., Eds.; Entomological Society of America: Lanham, MD, 1999, 189.
13. Contribution No. 00–252-B of the Kansas Agricultural Experiment Station.

29

Biomass

Alberto Traverso
and David Tucker

Introduction

This work is organized into seven main sections. The first section provides the reader with a general overview on biomass, including definition, environmental benefits, energetic properties, and a short list of biomass types that can be used as energy sources. The second section illustrates the mechanical processes to produce standardized solid biomass fuels. The third section describes one of the major technologies for converting biomass into energy, combustion. The fourth section analyzes pyrolysis and gasification as promising techniques for efficient exploitation of biomass, still in a precommercial phase. The fifth section is concerned with biochemical processes for producing biogas and biofuels for transportation. The sixth section outlines the major benefits from biomass exploitation for energy purposes. The seventh section reports about the potential carbon dioxide (CO_2) emission reduction due to extensive use of biomass as a renewable energy resource. The eighth section concludes this entry.

Generalities about Biomass

In general, biomass is whatever substance produced or by-produced by biological processes. Commonly, biomass refers to the organic matter derived from plants and generated through photosynthesis. Biomass provides not only food but also construction materials, textiles, paper, medicines, and energy. In particular, biomass can be regarded as solar energy stored in the chemical bonds of the organic material or as a reduced state of carbon. CO_2 from the atmosphere and water absorbed by the plant roots are combined in the photosynthetic process to produce carbohydrates (or sugars) that form the biomass.

TABLE 1   Typical Types of Biomass for Energy Use

Supply Sector	Type	Example
Forestry	Dedicated forestry	Short rotation plantations (e.g., willow, poplar, eucalyptus)
	Forestry by-products	Wood blocks, wood chips from thinnings
Agriculture	Dry lignocellulosic energy crops	Herbaceous crops (e.g., miscanthus, reed canary-grass, giant reed)
	Oil, sugar and starch energy crops	Oil seeds for methylesters (e.g., rape seed, sunflower)
		Sugar crops for ethanol (e.g., sugar cane, sweet sorghum)
		Starch crops for ethanol (e.g., maize, wheat)
	Agricultural residues	Straw, prunings from vineyards and fruit trees
	Livestock waste	Wet and dry manure
Industry	Industrial residues	Industrial waste wood, sawdust from sawmills
		Fibrous vegetable waste from paper industries
Waste	Dry lignocellulosic	Residues from parks and gardens (e.g., prunings, grass)
	Contaminated waste	Demolition wood
		Organic fraction of municipal solid waste
		Biodegradable landfilled waste, landfill gas
		Sewage sludge

Source: Adapted from European Biomass Industry Association[1] and DOE Biomass Research and Development Initiative.[2]

The solar energy that drives photosynthesis is stored in the chemical bonds of the biomass structural components. During biomass combustion, oxygen from the atmosphere combines with the carbon and hydrogen in biomass to produce CO_2 and water. The process is therefore cyclic because the carbon dioxide is then available to produce new biomass. This is also the reason why bioenergy is potentially considered as carbon- neutral, although some non-recoverable CO_2 emissions occur due to the use of fossil fuels during the production and transport of biofuels.

Biomass resources can be classified according to the supply sector, as shown in Table 1.

The chemical composition of plant biomass varies among species. Yet, in general terms, plants are made of approximately 25% lignin and 75% carbohydrates or sugars. The carbohydrate fraction consists of many sugar molecules linked together in long chains or polymers. Two categories are distinguished: cellulose and hemi- cellulose. The lignin fraction consists of non-sugar-type molecules that act as a glue holding together the cellulose fibers.

Energy Content of Biomass

Bioenergy is energy of biological and renewable origin, normally derived from purpose-grown energy crops or by-products of agriculture and forestry. Examples of bioenergy resources are wood, straw, bagasse, and organic waste. The term *bioenergy* encompasses the overall technical means through which biomass is produced, converted, and used. Figure 1 summarizes the variety of processes for energy production from biomass.

The calorific value of a fuel is usually expressed as higher heating value (HHV) and/or lower heating value (LHV). The difference results from the vaporization of water formed from the combustion of hydrogen in the material and the original moisture.

The most important property of biomass feedstocks with regard to combustion—and to the other thermochemical processes—is the moisture content, which influences the energy content of the fuel. Wood, just after falling, has a typical 55% water content and an LHV of approximately 7.1 MJ/kg; logwood after 2–3 years of air-drying may present a 20% water content and an LHV of 14.4 MJ/kg; pellets show a quite constant humidity content of about 8% with an LHV equal to 17 MJ/kg.

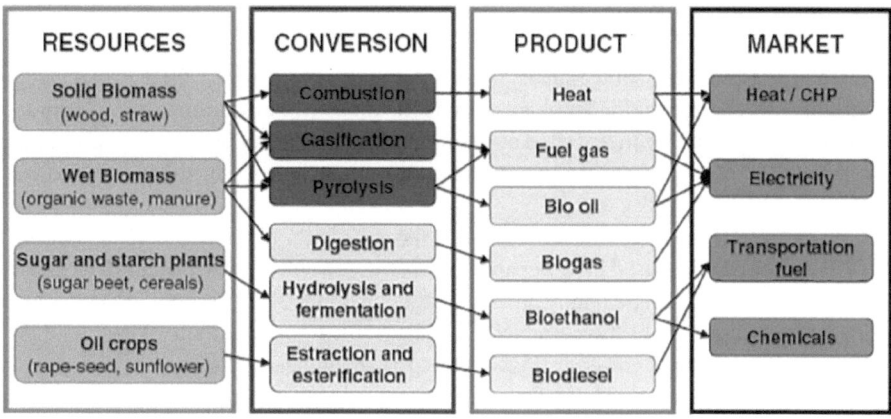

FIGURE 1 Processes to convert biomass into useful energy, i.e., bioenergy.
Source: European Biomass Industry Association,[1] Overend et al.[3] and Risoe National Laboratory.[6]

Mechanical Processes for Energy Densification

Some practical problems are associated with the use of biomass material (sawdust, wood chips, or agricultural residues) as fuel. Those problems are mainly related to the high bulk volume, which results in high transportation costs and requires large storage capacities, and to the high moisture content, which can result in biological degradation as well as in freezing and blocking the in-plant transportation systems. In addition, variations in moisture content make difficult the optimal plant operation and process control. All those problems may be overcome by standardization and densification. The former consists in processing the original biomass in order to obtain fuels with standard size and heating properties, while the latter consists in compressing the material, which needs to be available in the sawdust size, to give it more uniform properties.

Table 2 reports the main features of pellets, briquettes, and chips.

TABLE 2 Comparison of Different Solid Wood Fuels

	Pellets	Briquettes	Chips
Appearance			
Raw material	Dry and ground wood or agricultural residues	Dry and ground wood or agricultural residues. Raw material can be more coarse than for pelleting, due to the larger dimensions of final product	Dry wood logs
Shape	Cylindrical (generally Ø 6 to 12 mm, with a length 4 to 5 times the Ø).	Cylindrical (generally Ø 80 to 90 mm) or parallelepiped ($150 \times 70 \times 60$ mm)	Irregularly parallelepiped ($70 \times 30 \times 3$ mm)

Biomass Combustion

The burning of wood and other solid biomass is the oldest energy technology used by man. Combustion is a well- established commercial technology with applications in most industrialized and developing countries, and development is concentrated on resolving environmental problems, improving the overall performance with multifuel operation, and increasing the efficiency of the power and heat cycles (combined heat and power, CHP).

The devices used for direct combustion of solid biomass fuels range from small domestic stoves (1 to 10 kW) to the large boilers used in power and CHP plants (>5 MW). Intermediate devices cover small boilers (10 to 50 kW) used in single family houses heating, medium-sized boilers (50 to 150 kW) used for multifamily house or building heating, and large boilers (150 kW to more than 1 MW) used for district heating. Cofiring in fossil-fired power stations enables the advantages of large-sized plants (>100 MWe) that are not applicable for dedicated biomass combustion due to limited local biomass availability.

To achieve complete burnout and high efficiencies in small-scale combustion, downdraft boilers with inverse flow have been introduced, applying the two-stage combustion principle. An operation at very low load should be avoided as it can lead to high emissions. Hence, it is recommended to couple log wood boilers to a heat storage tank. Since wood pellets are well suited for automatic heating at small heat outputs, as needed for buildings nowadays, pellet furnaces are an interesting application with increasing propagation. Thanks to the well-defined fuel at low water content, pellet furnaces can easily achieve high combustion quality. They are applied both as stoves and as boilers and find increased acceptance in urban areas, due to the high efficiency of modern pellet stoves now used for home heating. While a conventional fireplace is less than 10% efficient at delivering heat to a house, an average modern pellet stove achieves 80%–90% efficiency. Technology development has led to the application of strongly improved heating systems, which are automated and have catalytic gas cleaning equipment. Such systems significantly reduce the emissions from fireplaces and older systems while at the same time significantly improving the efficiency.

Understoker furnaces are mostly used for wood chips and similar fuel with relatively low ash content, while grate furnaces can also be applied for high ash and water content. Special types of furnaces have been developed for straw, a very low density material that is usually stored in bales. Beside conventional grate furnaces operated with whole bales, cigar burners and other specific furnaces are in operation. Stationary or bubbling fluidized bed (SFB) as well as circulating fluidized bed (CFB) boilers are applied for large-scale applications and often used for waste wood or mixtures of wood and industrial wastes, e.g., from the pulp and paper industry.

Co-Combustion

Bioenergy production might be hampered by limitations in the supply and/or fuel quality. In those cases, cofiring of several types of biomass or of biomass with coal ensures flexibility in operation, both technically and economically. Several concepts have been developed:

- Co-combustion or direct cofiring. The biomass is directly fed to the boiler furnace, if needed after physical preprocessing of the biomass such as drying, grinding, torrefaction, or metal removal is applied. This typically takes place in SFB or CFB combustors. Such technologies can be applied to a wide range of fuels, even for very wet fuels like bark or sludge. Multifuel fluidized bed boilers achieve efficiencies of more than 90%, while flue gas emissions are lower than for conventional grate combustion due to lower combustion temperatures.
- Indirect cofiring. Biomass is first gasified and the fuel gas is then cofired in the main boiler. Sometimes, the gas has to be cooled and cleaned, which is more challenging and implies higher operation costs.
- Parallel combustion. The biomass is burnt in a separate boiler for steam generation. The steam is used in a power plant together with the main fuel.

Problems in Biomass Combustion

Biomass has a number of characteristics that makes it more difficult to handle and combust than fossil fuels. The low energy density, the high water content, and the tendency to "bridge" in tanks or pipes are the main problems in handling and transport of the biomass, while the difficulty in using biomass as fuel relates to its content of inorganic constituents. Some types of biomass used contain significant amounts of chlorine, sulfur, and potassium. The salts, KCl and K_2SO_4, are quite volatile, and the release of these components may lead to heavy deposition on heat transfer surfaces, resulting in reduced heat transfer and enhanced corrosion rates. Severe deposits may interfere with operation and cause unscheduled shutdowns.

In order to minimize these problems, various fuel pretreatment processes have been considered, including washing the biomass with hot water or using a combination of pyrolysis and char treatment.

Thermochemical Conversion of Biomass

Pyrolysis and gasification are the two most typical thermochemical processes because they convert the original bioenergy feedstock into more convenient energy carriers such as producer gas, oil, methanol, and char,[3] instead of producing useful energy directly.

Pyrolysis

Pyrolysis is a process for thermal conversion of solid fuels, like biomass or wastes, in the complete absence of oxidizing agent (air/oxygen) or with such limited supply that gasification does not occur to any appreciable extent. Commercial applications are focused on either the production of charcoal or the production of a liquid product, the bio-oil, and pyro-gas. Charcoal is a very ancient product. Traditional processes (partial combustion of wood covered by a layer of earth) were very inefficient and polluting, but modern processes presently used in industry such as rotary kiln carbonization have been optimized to maximize efficiency and minimize environmental impact. Bio-oil production (or wood liquefaction) is potentially very interesting as a substitute for fuel oil and as a feedstock for production of synthetic gasoline or diesel fuel. Pyro-gas has higher energy density than gasification gas (syngas) because it has been created without oxygen (and nitrogen, if air is employed); hence, it does not contain the gaseous products of partial combustion.

Pyrolysis takes place at temperatures in the range 400–800°C, and during this process, most of the cellulose and hemicellulose and part of the lignin will disintegrate to form smaller and lighter molecules, which are gases at the pyrolysis temperature. As these gases cool, some of the vapors condense to form a liquid, which is the bio-oil and tar. The remaining part of the biomass, mainly parts of the lignin, is left as a solid, i.e., the charcoal. It is possible to influence the product mix through a control of heating rate, residence time, pressure, and maximum reaction temperature, so that either gases, condensable vapors, or the solid charcoal is promoted.

Gasification

Modern gasification technology has been developed since the 18th century (first historical hints about gasification date from the Chinese Han Dynasty, between 206 BC and AD 220), but it is still at a development phase.[4,5] Gasification is a conversion process that involves partial oxidation at elevated temperature. It is intermediate between combustion and pyrolysis: in fact, oxygen (or air) is present but it is not enough for complete combustion. This process can start from carbonaceous feedstock such as biomass or coal and convert them into a gaseous energy carrier. The overall gasification process may be split into two main stages: the first is the pyrolysis stage, i.e., where oxygen is not present but temperature is high; typical pyrolysis reactions take place here; the second stage is the partial combustion, where

TABLE 3 Qualitative Comparison of Technology for Energy Conversion of Biomass

Process	Technology	Economics	Environment	Market Potential	Present Deployment
Combustion—heat	+++	€	+++	+++	+++
Combustion—electricity	++(+)	€€	++(+)	+++	++
Gasification	+(+)	€€€	+(++)	+++	(+)
Pyrolysis	(+)	€€€	(+++)	+++	(+)

Source: Adapted from European Biomass Industry Association[1] and Risoe National Laboratory.[6]
Note: +, low; +++, high; €, cheap; €€€, expensive.

oxygen is present and reacts with the pyrolysed biomass to release heat necessary for the process. In the latter stage, the actual gasification reactions take place, which consist of almost complete charcoal conversion into lighter gaseous products (e.g., carbon monoxide and hydrogen), through the chemical oxidizing action of oxygen, steam, and carbon dioxide: such gases are injected into the reactor near the partial combustion zone (normally, steam and carbon dioxide are mutually exclusive). Gasification reactions require temperature in excess of 800°C to minimize tar and maximize gas production. The gasification output gas, called "producer gas," is composed of hydrogen (18%–20%), carbon monoxide (18%–20%), carbon dioxide (8%–10%), methane (2%–3%), trace amounts of higher hydrocarbons like ethane and ethene, water, nitrogen (if air is used as oxidant agent), and various contaminants such as small char particles, ash, tars, and oils. The incondensable part of producer gas is called "syngas," and it represents the useful product of gasification. If air is used, syngas has a high heating value in the order of 4–7 MJ/m^3, which is exploitable for boiler, engine, and turbine operation, but, due to its low energy density, it is not suitable for pipeline transportation. If pure oxygen is used, the syngas high heating value almost doubles (approximately 10–18 MJ/m^3 high heating value). Such a syngas is suitable for limited pipeline distribution as well as for conversion to liquid fuels (e.g., methanol and gasoline). However, the most common technology is air gasification because it avoids the costs and the hazards of oxygen production and usage. With air gasification, the syngas efficiency, describing the energy content of the cold gas stream in relation to that of the input biomass stream, is on the order of 55%–85%, and typically 70%.

Comparison of Thermal Conversion Methods of Biomass

Table 3 reports a general overview on specific features of the conversion technologies analyzed here, showing the related advantages and drawbacks.

Biochemical Conversion of Biomass

Biochemical conversion of biomass refers to processes that decompose the original biomass into useful products. Commonly, the energy product is either in the liquid or in the gaseous form; hence, it is called "biofuel" or "biogas," respectively. Biofuels are very promising for the transportation sector, while biogas is used for electricity and heat production. Normally, biofuels are obtained from dedicated crops (e.g., biodiesel from seed oil), while biogas production results from concerns over environmental issues such as elimination of pollution, treatment of waste, and control of landfill greenhouse gas (GHG) emissions.

Biogas from Anaerobic Digestion

Biogas is produced most commonly by anaerobic digestion of biomass. Anaerobic digestion refers to the bacterial breakdown of organic materials in the absence of oxygen. This biochemical process produces a gas called biogas, principally composed of methane (30%–60% in volume) and carbon dioxide. Such a biogas can be converted to energy in the following ways:

- Biogas converted by conventional boilers for heating purposes at the production plant (house heating, district heating, industrial purposes).
- Biogas for CHP generation.
- Biogas and natural gas combinations and integration in the natural gas grid.
- Biogas upgraded and used as vehicle fuel in the transportation sector.
- Biogas utilization for hydrogen production and fuel cells.

An important production of biogas comes from landfills. Anaerobic digestion in landfills is brought about by the microbial decomposition of the organic matter in refuse. Landfill gas is, on average, 55% methane and 45% carbon dioxide. With waste generation increasing at a faster rate than economic growth, it makes sense to recover the energy from that stream, through thermal or fermentation processes.

Biofuels for Transport

A wide range of chemical processes may be employed to produce liquid fuels from biomass. Such fuels can find a very high level of acceptance by the market, thanks to the relatively easy adaptation to existing technologies (i.e., gasoline and diesel engines). The main potential biofuels are outlined below.

- Biodiesel is a methyl-ester produced from vegetable or animal oil to be used as alternative to conventional petroleum-derived diesel fuel. Compared to pure vegetable or animal oil, which can be used in adapted diesel engines as well, biodiesel presents lower viscosity and slightly HHV.
- Pure vegetable oil is produced from oil plants through pressing, extraction, or comparable procedures, crude or refined but chemically unmodified. Usually, it is compatible with existing diesel engines only if blended with conventional diesel fuel, at rates not higher than 5%–10% in volume. Higher rates may lead to emission and engine durability problems.
- Bioethanol is ethanol produced from biomass and/or the biodegradable fraction of waste. Bioethanol can be produced from any biological feedstock that contains appreciable amounts of sugar or other matter that can be converted into sugar, such as starch or cellulose. Also, ligno-cellulosic materials (wood and straw) can be used, but their processing into bioethanol is more expensive. Application to modified spark ignition engines is possible.
- Bio-ETBE (ethyl-tertio-butyl-ether) is ETBE produced on the basis of bioethanol. Bio-ETBE may be effectively used for enhancing the octane number of gasoline (blends with petrol gasoline).
- Biomethanol is methanol produced from biomass. Methanol can be produced from gasification syngas (a mixture of carbon monoxide and hydrogen) or wood dry distillation (old method with low methanol yields). Virtually all syngas for conventional methanol production is produced by steam reforming of natural gas into syngas. In the case of biomethanol, a biomass is gasified first to produce a syngas from which the biomethanol is produced. Application to spark ignition engines and fuel cells is possible. Compared to ethanol, methanol presents more serious handling issues, because it is corrosive and poisonous for human beings.
- Bio-MTBE (methyl-tertio-butyl-ether) is a fuel produced on the basis of biomethanol. It is suitable for blends with petrol gasoline.
- Biodimethylether (DME) is dimethylether produced from biomass. Bio-DME can be formed from syngas by means of oxygenate synthesis. It has emerged only recently as an automotive fuel option. Storage capabilities are similar to those of LPG. Application to spark ignition engines is possible.

Benefits from Biomass Energy

There is quite a wide consensus that, over the coming decades, modern biofuels will provide a substantial source of alternative energy. Nowadays, biomass already provides approximately 11%–14% of the world's primary energy consumption (data vary according to sources).

TABLE 4 Benefits in Reduction of GHG Emissions

+	Avoided mining of fossil resources
−	Emission from biomass production
+	Avoided fossil fuel transport (from producer to user)
−	Emission from biomass fuel transport (from producer to user)
+	Avoided fossil fuel utilization

Source: Risoe National Laboratory.[6]
Note: +, positive; −, neutral.

There are significant differences between industrialized and developing countries; in particular, in many developing countries, bioenergy is the main energy source, even if it is used in very low efficient applications (e.g., cooking stoves have an efficiency of about 5%–15%). Furthermore, inefficient biomass utilization is often associated with the increasing scarcity of hand-gathered wood, nutrient depletion, and the problems of deforestation and desertification.

One of the key drivers to bioenergy deployment is its positive environmental benefit regarding the global balance of GHG emissions. This is not a trivial matter, because biomass production and use are not entirely GHG neutral. In general terms, the GHG emission reduction as a result of employing biomass for energy is as reported in Table 4.

Since the energy cost associated with collection and transport of biomass is a significant portion, bio-energy is a decentralized energy option whose implementation presents positive impacts on rural development by creating business and employment opportunities. Jobs are created all along the bioenergy chain, from biomass production or procurement, to its transport, conversion, distribution, and marketing.

Bioenergy is a key factor for the transition to a more sustainable development.

Potential for CO$_2$ Emission Reduction

When biomass is used for energy production, the carbon contained in it is ultimately transformed into CO$_2$. In fact, such a biomass-derived CO$_2$ does not contribute to global warming, as it equals the CO$_2$ absorbed by the biomass during its growth; the relatively short time of such carbon cycle makes the biomass a carbon-neutral energy resource.

Abundant resources and favorable policies[7] enable bio-power to expand in Northern Europe (mostly cogeneration from wood residues), in the United States, and in countries producing sugar cane bagasse (e.g., Brazil).

In the short term, cofiring remains the most cost-effective use of biomass for power generation, along with small-scale, off-grid use. In the mid-long term, gasification plants and biorefineries for biofuel production could expand significantly (mainly ethanol, lignocellulosic ethanol, biodiesel). International Energy Agency projections suggest that the biomass share in electricity production may increase from the current 1.3% to some 3%–5% by 2050, depending on assumptions.[8] This is a small contribution compared to the estimated total biomass potential, but biomass are also used for heat generation and to produce fuels for transport.

Today, biomass supplies some 50 EJ/yr (1 EJ = 10^{18} joules [J] = 10^{15} kilojoules [kJ] = 24 million tons of oil equivalent [Mtoe]) globally, which represents 10% of global annual primary energy consumption (Figure 2). This is mostly traditional biomass used for cooking and heating.

Based on this diverse range of feedstocks, the technical potential for biomass is estimated in the literature to be possibly as high as 1500 EJ/yr by 2050, although most biomass supply scenarios that take into account sustainability constraints indicate an annual potential of between 200 and 500 EJ/yr (excluding aquatic biomass). Forestry and agricultural residues and other organic wastes (including municipal solid waste) would provide between 50 and 150 EJ/yr, while the remainder would come from energy crops, surplus forest growth, and increased agricultural productivity.

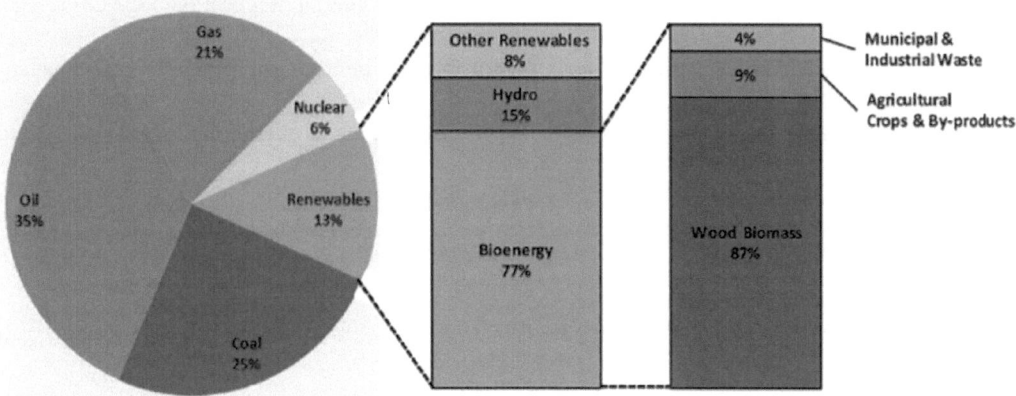

FIGURE 2   Share of bioenergy in the world primary energy mix. (*Source:* International Energy Agency.[9])

Projected world primary energy demand by 2050 is expected to be in the range of 600 to 1000 EJ, compared to about 500 EJ in 2008.[9] Scenarios looking at the penetration of different low-carbon energy sources indicate that future demand for bioenergy could be up to 250 EJ/yr. This projected demand falls well within the sustainable supply potential estimate, so it is reasonable to assume that biomass could sustainably contribute between a quarter and a third of the future global energy mix. Whatever is actually realized will depend on the cost competitiveness of bioenergy and on future policy frameworks, such as GHG emission reduction targets.

Given the CO_2-neutral nature of biomass, and assuming that biomass will primarily substitute fossil fuels, the potential for reduction in CO_2 emission in 2050 can then be estimated as the same figure (i.e., around 20%–30% of anthropogenic CO_2), compared to a business-as-usual scenario. Definitely, biomass will play a determinant role towards a CO_2-free development.

Conclusions

Biomass refers to a very wide range of substances produced by biological processes. In the energy field, special focus has been and will be placed on vegetable biomass, such as wood and agricultural by-products, because of the energy potential as well as economic and environmental benefits. Size and humidity standardization of biomass is a necessary step to make it suitable for effective domestic and industrial exploitation: chips, briquettes, and pellets are modern examples of standard solid fuels.

Biomass can be converted into energy in three pathways: combustion, thermochemical processing, and biochemical processing. The combustion of solid biomass for the production of heat or electricity and heat is the most viable technology, while pyrolysis and gasification still face economic and reliability issues. Among biochemical processes, anaerobic digestion is often used to reduce the environmental impact of hazardous waste and landfills. Biochemical processes are also concerned with the conversion of biomass into useful fuels for transportation, such as biodiesel, bioethanol, and biomethanol. All of them can effectively contribute to the transition to a more sustainable transportation system at zero GHG emissions.

Biomass represents a viable option for green energy resources of the 21st century.

References

1. European Biomass Industry Association, available at http:// www.eubia.org/.
2. DOE Biomass Research and Development Initiative, available at http://www.bioproducts-bioenergy.gov/.

3. Overend, R.P.; Milne, T.A.; Mudge, L.K. *Fundamentals of Thermochemical Biomass Conversion;* Elsevier Applied Science Publishers Ltd.: New York, 1985.
4. Bridgewater, A.V. The technical and economic feasibility of biomass gasification for power generation. Fuel J. **1995**, *74* (6–8), 557–564.
5. Franco, A.; Giannini, N. Perspectives for the use of biomass as fuel in combined cycle power plants. Int. J. Thermal Sci. **2005**, *44*, (2), 163–177.
6. Risoe National Laboratory, Denmark, available at http:// www.risoe.dk/.
7. Biomass for Power Generation and CHP, International Energy Agency (IEA), Energy Technology Essentials, ETE03, Jan 2007.
8. World Energy Outlook; International Energy Agency (IEA), 2002.
9. Bioenergy—A Sustainable and Reliable Energy Source, International Energy Agency (IEA), available at http://www.ieabioenergy.com, 2009.

30

Nematodes:
Biological Control

Simon Gowen

Introduction

Many pathogens and predators of nematodes are known but few have the necessary characteristics of specificity, mobility, or speed of colonization to have a significant influence on a pest population. Attempts at their commercial exploitation as field treatments have not been successful largely because of their inconsistency. Understanding the subtleties associated with the deployment of biocontrol agents will require considerable research effort. Additionally, the recommended rates of application and the formulation on suitable carriers and nutrient sources poses a problem in practicability and in the interpretation of the biological processes involved.

Contemporary research has shown that natural control does exist and that in certain crop/nematode/pathogen situations nematode populations will decline as they are attacked by components of the soil microflora. Soils where this occurs are known as suppressive, but well-documented examples of naturally occurring suppressiveness to particular nematode pests are uncommon.

During the life of a crop the population densities of many of the serious nematode pests can increase by 1000fold. Economic damage may result from initial population densities of one nematode per gram of soil. To be effective therefore a biocontrol agent must have an impact on the numbers of nematodes that would invade a host and not simply eliminate the surplus individuals that may never locate or invade a root. This being so, those pathogens and predators that are relatively unspecific (trapping, ingesting, or parasitizing all types of free-living nematodes in soil) may be considered less promising than those that parasitize specific pests.

Significant progress has been made in the recognition and deployment of such microorganisms parasitic on some of the species of sedentary nematodes such as the root-knot nematodes, *Meloidogyne* spp., and some of the cyst nematodes, *Heterodera* spp., and *Globodera* spp.

Root-knot and cyst nematodes produce eggs either in clusters on roots or contained within or attached to the cuticle of the female nematode. Biocontrol agents that prevent these nematodes from reproducing may have more impact from an epidemiological point of view than those that kill the free-living individuals in the soil.

Biocontrol Agents Specific to Certain Nematode Pests

Verticillium chlamydosporium is a facultative, soil-dwelling fungus that parasitizes eggs in egg masses exposed on the root surface. Under the right conditions, such fungi will have a significant effect on nematode populations. The efficacy of *V. chlamydosporium* is partly dependent on its root colonizing ability; this can vary according to the plant host. Skill is required in selecting crops that support and/or increase the root colonization by the fungus but are also less favored hosts of root-knot nematodes. *V. chlamydosporium* may be less effective when it is deployed with plants that are highly susceptible and large galls are produced in response to the nematode infection. In such cases, many egg masses may not be exposed on the root surface and so escape infection.

Paecilomyces lilacinus is another fungus commonly found infecting the eggs of sedentary nematodes such as the root-knot and the cyst nematodes, and, like *V. chlamydosporium* being relatively easy to produce on defined growth media, has good potential for commercial development.

Pasteuria penetrans, an obligate bacterial parasite of root-knot nematodes begins its life cycle on free-living juveniles in the soil. Spores attach to the juveniles as they move in search of host roots. Parasitic development begins after the nematode enters a root and continues in synchrony with that of its host. The nematode eventually is overcome by its parasite; it fails to produce eggs; and its body, filled with the spores of the bacterium, eventually ruptures releasing spores into the soil.

The efficiency of *P. penetrans* as a biocontrol agent of root-knot nematodes depends on the concentrations of spores in the soil, the chances of contact with the juvenile stage, and the specificity of the particular *P. penetrans* population. Commercial success will depend therefore on finding techniques for mass-producing the bacterium and on developing populations with a broad spectrum of pathogenicity.

Other *Pastearia* species parasitic on some sedentary *(Heterodera)* and migratory *(Pratylenchus)* species have been described.

Biological control agents such as *V. chlamydosporium*, *P. lilacinus*, and *P. penetrans* could provide an adequate replacement for nematicides in some cropping systems but the lack of immediate effects, such as are provided by nematicide or fumigant treatments, is a disadvantage. Protection is normally needed in the early stages of plant growth such as in nursery beds. In this situation, integration with other practices such as nematicides, rotation, solarization, and mulches is necessary.

There are several reports of the successful deployment of these biocontrol agents. Small field plots treated once with *P. penetrans* spores (produced by an in vivo system) caused a decline in numbers of root-knot nematodes and increases in yield over a series of crop cycles using root-knot nematode susceptible crops. In other locations, where treatments with *P. penetrans* were combined with *V. chlamydosporium*, organic manures and grass mulches showed similar declines in nematode populations. These two organisms acted against root-knot nematodes in a complementary fashion. As part of this strategy, root systems containing spore-filled cadavers were deliberately left to disintegrate in the soil after each crop. No field treatments were effective after only one crop indicating that some crop loss must be expected during the development of suppressiveness. *P. penetrans* was also effective when used in combination with a nematicide in permanent beds within a plastic polytunnel. Better control of root-knot nematodes was achieved if the biocontrol agent was combined with other control strategies. With such treatments, beneficial effects may develop over one crop cycle.

The chlamydospores of *V. chlamydosporium* do not have the persistence of the spores of *P. penetrans*, which can remain viable for many years.

Nonspecific Biocontrol Agents

There is a long history of interest in the fungi that trap nematodes in soil such as species of *Arthrobotrys*. These are commonly found in all soils but despite much research effort the problems of the unreliability of soil applications have not been solved and none have become established as successful commercial products.

There are several rhizosphere colonists that have potential for alleviating nematode damage. The precise mechanisms are not clear. Some produce toxins but others may affect root exudation and thus indirectly the attractiveness of roots to nematodes. Experiments have shown that strains of *Pseudomonas fluorescens* can reduce root invasion by different plant parasites but as with the trapping fungi, poor consistency hinders successful development of these microorganisms as commercial products.

Future Prospects

Recently, the nematicidal (and insecticidal) effects of the toxins produced by the bacteria associated with entomopathogenic nematodes (*Photorhabdus* spp., *Xenorhabdus* spp., and *Pseudomonas oryzihabitans*) have been demonstrated.

Success in the commercial development of biocontrol agents does appear promising with those microorganisms that can be formulated as a standard product with proven reliability; others may have a future as single treatment introductions in the more intensively managed protected cropping systems but commercialization may be difficult.

Research is still needed to develop reliable methods of production, formulation, and application. The challenge is to provide a sufficient duration of protection. Such treatments will need to be part of a package of control measures.

Bibliography

1. Aalten, P.M.; Vitour, D.; Blanvillain, D.; Gowen, S.R.; Sutra, L. Effect of rhizosphere fluorescent *Pseudomonas* strains on plant parasitic nematodes *Radopholus similis* and *meloidogyne* spp. Letters Appl. Microbiol. **1998**, *27*, 357–361.
2. Bourne, M.; Kerry, B.R.; De Leij, F.A.A.M. The importance of the host plant on the interaction between root-knot nematodes (*Meloidogyne* spp.) and the nematophagous fungus, *Verticillium chlamydosporium* Goddard. Biocon. SciTech. **1996**, *6*, 539–548.
3. Crump, D.J. A method for assessing the natural control of cyst nematode populations. Nematologica **1987**, *33*, 232–243.
4. Gowen, S.R.; Bala, G.; Madulu, J.; Mwageni, W.; Trivino,C.T. In *Field Evaluation of Pasteuria penetrans for the Management of Root-Knot Nematodes*, The 1998 Brighton Conference on Pests and Diseases, **1998**, *3*, 755–760.
5. Kerry, B.R.; Jaffee, B.A. Fungi as Biocontrol Agents for Plant Parasitic Nematodes. In *The Mycota IV Environmental and Microbial Relationships*; Wicklow, D. T., Soderstrom, B.E., Eds.; Springer-Verlag: Berlin, Heidelberg, 1997; 204–218.
6. Samaliev, H.Y.; Andreoglou, F.I.; Elawad, S.A.; Hague, N.G.M.; Gowen, S.R. The Nematicidal Effects of the Bacteria *Pseudomonas oryzihabitans* and *Xenorhabdus Nematophilus* on the Root-Knot Nematode, *Meloidogyne javanica*. Nematology, *in press*.
7. Stirling, G.R. Biological Control of Plant Parasitic Nematodes. CAB International: Wallingford, U.K., 1991; 282.
8. Tzortzakakis, E.A.; Channer, A.G. de R.; Gowen, S.R.; Ahmed, R. Studies on the potential use of *Pasteuria penetrans* as a biocontrol agent of root-knot nematodes (*Meloidogyne spp.*). Plant Pathol. **1997**, *46*, 44–55.
9. Tzortzakakis, E.A.; Gowen, S.R. Evaluation of *Pasteuria penetrans* alone and in combination with oxamyl, plant re sistance and solarization for control of *Meloidogyne* spp. on vegetables grown in greenhouses in Crete. Crop Prot. **1994**, *13*, 455–462.
10. Weibelzahl-Fulton, E.; Dickson, D.W.; Whitty, E.B. Suppression of *Meloidogyne incognita* and *M. javanica* by *Pasteuria penetrans* in field soil. J. Nematol. **1996**, *28*, 43–49.

PRO: Basic Environmental Processes

VII

PRO: Basic
Environmental
Processes

31

Agroforestry: Water Use Efficiency

James R. Brandle,
Laurie Hodges,
and Xinhua Zhou

Introduction

Agroforestry is the intentional integration of trees and shrubs into agricultural systems. Windbreaks, riparian forest buffers, alley-cropping, silvopastoral grazing systems, and forest farming are the primary agroforestry practices found in temperate regions of North America.[1] Placing trees and shrubs on the landscape changes the surface energy balance, influences the surrounding microclimate, and has the potential to alter water use and productivity of adjacent crops.[2,3]

In agricultural systems, water is often the major factor limiting growth. When water availability is limited as a result of limited supply or high cost, its efficient use becomes critical to successful production systems. For example, proper irrigation at the appropriate stage of crop development minimizes pumping costs and increases yield; reducing soil tillage conserves soil water and may enhance yield, and reducing surface runoff or trapping snow improves soil water storage for future crop use. These water conservation efforts contribute to the efficient use of available water and are determined primarily by management practices. In contrast, Tanner and Sinclair[4] distinguish between the efficient use of water and water use efficiency (WUE). WUE is primarily a function of physiological responses of plants to environmental conditions. This review focuses on WUE defined as the amount of biomass (or grain) produced per unit of land area for each unit of water consumed.[4]

Soil water may be consumed by evaporation from the soil surface or by the transport of water through the plant and subsequent evaporation from the leaf surface. The rate of water consumption is determined by the microclimate of the crop. Because agroforestry practices alter the microclimate of adjacent fields, they affect WUE of plants growing in those fields.

Discussion

Windbreaks, riparian forest buffers or alley-cropping systems are the practices most likely to be integrated into crop production systems. In all three practices, trees and shrubs tend to be arranged in narrow barriers adjacent to the crop field. Microclimate responses downwind of any of these types of barriers are similar and the following discussion applies to all three types of barriers. As wind approaches these barriers, it is diverted up and over the barrier creating two zones of protection, a larger zone to the lee of the barrier (the side away from the wind) and a smaller zone on the windward side of the barrier. In these zones, wind speed is reduced and turbulence and eddy structure in the vicinity of the barrier are altered. As a result of these changes, the transfer coefficients for heat and mass between the crop and the atmosphere are altered; the gradients of temperature, humidity, and carbon dioxide concentration above the soil and canopy are changed;[5] and the plant processes of transpiration and photosynthesis are altered.[6]

McNaughton[5] defined two regions within the leeward zone of protection: the *quiet zone,* extending from the top of the barrier down to a point in the field located approximately 8H leeward (H is the height of the barrier) and a *wake zone,* lying beyond the quiet zone and extending from approximately 8H to a distance of 20H to 25H from the barrier. Within the quiet zone where turbulence is reduced, we expect conditions to be such that the canopy is "uncoupled" from the atmospheric conditions above the sheltered zone, while in the wake zone where turbulence is increased, we expect the canopy to become more strongly "coupled" to the atmosphere above. In both locations we would expect the rates of photosynthesis and transpiration to be altered and WUE to change.

The magnitude of change in wind speed, as well as the extent of microclimate modifications within the quiet and wake zones, are largely determined by the structure of the windbreak or barrier and the underlying meteorological conditions. Structure refers to the amounts of solid material and open space and their arrangement within the barrier. Dense barriers, for example, multiple rows of conifers, generally result in greater wind speed reduction but more turbulence. More porous barriers, for example, single rows of deciduous species, result in less wind speed reduction but also less turbulence. The downwind extent of the protected area is generally greater for more porous barriers. As a result, narrow, less dense barriers (40%–60% density) are typically used to protect crop fields.

The overall influence of wind protection on plant water relations is complex and linked to temperature, humidity, wind speed, and other meteorological conditions found in the protected zone, the amount of available soil water, crop size, and stage of development.[2,3,7] Until recently, the major effect of wind protection and its influence on crop growth and yield were assumed to be due primarily to soil water conservation and reduced water stress of sheltered plants.[8,9] There is little question that the evaporation rate from bare soil is reduced in the protected zone.[3] However, the effect of reduced wind speed on transpiration, evaporation from the plant canopy, and overall plant water status is less clear.[2,3,7]

According to Grace,[9] transpiration rates may increase, decrease, or remain unaffected by wind protection depending on wind speed, atmospheric resistance, and saturation vapor pressure deficit. Cleugh[3] suggests that as stomatal resistance increases, evaporation from the canopy may actually be increased with a reduction in wind speed. When stomatal resistance is high and water is limited, stomatal resistance controls the rate of evaporation from the leaf surface, not the amount of turbulence. Under these conditions a decrease in wind speed and turbulent mixing may increase the potential for evaporation from the leaf surface.[3]

Evaporation from the leaf surface consists of two phases, an energy driven phase and a diffusion driven phase. Movement of water through the plant and out the stomata is driven by the water potential gradient within the plant. This gradient is influenced by the plant's energy balance. On the lee side of the buffer, reduced wind speed and turbulent mixing lead to increases in leaf temperature and transpiration to meet the increased energy load on the plant. If adequate water is available, it is moved through the plant to the leaf surface and the potential for evaporation from the leaf surface is increased. If water is limited, the stomata partially or completely close, transpiration is reduced, and evaporation from the leaf surface declines.

In contrast, movement of water vapor across the leaf boundary layer is controlled by the vapor pressure gradient and the thickness of the leaf boundary layer. As windspeed decreases, the thickness of this boundary layer increases, the vapor pressure gradient decreases, and the rate of evaporation from the leaf surface decreases. The relative magnitude of the two processes determines whether or not transpiration and subsequent evaporation from the canopy are increased, decreased, or remain unchanged.[7,9,10]

While these theoretical considerations are important in understanding the process, several studies[11–13] have demonstrated a good correlation between wind protection, conservation of soil water, and enhanced crop yield. Even so, the effect of wind protection on WUE is neither constant throughout the growing period[7] nor is it consistent over varying meteorological conditions.

Agroforestry practices impact the water relations of the crop by affecting the loss of water through damaged leaves. On soils subject to wind erosion, windbreaks or other agroforestry buffers provide significant reductions in the amount of wind blown soil and subsequent abrasion of plant parts and cuticular damage.[9,14] Loss of cuticular integrity or direct tearing of the leaves[15] reduces the ability of the plant to control water loss.

Agroforestry buffers have a direct effect on the distribution of precipitation, both rain and snow. In the case of snow, a porous barrier will result in a more uniform distribution of snow across the field, providing additional soil water for the crop.[16] In the case of rain, the barrier has minimal influence on the distribution of precipitation across the field; however, in the area immediately adjacent to the barrier a rain shadow may occur on the leeward side. On the windward side, the barrier may lead to slightly higher levels of measured precipitation at or near the base of the trees due to increased stem flow or dripping from the canopy.

Trees and shrubs used in agroforestry practices also consume a portion of the available water. In the area immediately adjacent to the barrier, competition for water between the crop and the barrier has a negative impact on yield. These same areas are also subject to some degree of shading depending on the orientation of the barrier. These changes in radiation load influence the energy balance and thus the growth and development of the crop and the utilization of water.[2]

Summary

In summary, agroforestry practices such as windbreaks, riparian forest buffers and alley-cropping systems generally improve both the efficient use of water by the agricultural system and the WUE of the individual crop. In the case of efficient water use, the evidence is clear. In the case of crop WUE, the evidence leaves some unanswered questions. How do we account for the varied crop yield responses reported in the literature? In many cases yields are increased but no clear relationship to crop water budget is shown. In other cases crop yield response is minimal. Under what meteorological conditions are the effects of agroforestry practices most valuable to water balance questions? Final crop yield is a integration of the environmental conditions over the entire growing season. Many different combinations of environmental conditions may result in similar plant responses. How do we address the numerous combinations of plant stress and plant growth to determine "a response" to wind protection? To answer many of these questions it will be necessary to intensify the numerical modeling methods developed by Wilson[17] and Wang and Takle.[18] With a better model to describe the turbulence fields and the transport of water, heat, and carbon dioxide as influenced by agroforestry practices, it should be possible to assess the numerous combinations of environmental factors influencing crop growth in these systems.

References

1. Lassoie, J.P.; Buck, L.E. Development of agroforestry as an integrated land use management strategy. In *North American Agroforestry: An Integrated Science and Practice*; Garrett, H.E., Rietveld, W.J., Fisher, R.F., Eds.; American Society of Agronomy, Inc.: Madison, WI, 2000; 1–29.

2. Brandle, J.R.; Hodges, L.; Wight, B. Windbreak practices. In *North American Agroforestry: An Integrated Science and Practice;* Garrett, H.E., Rietveld, W.J., Fisher, R.F., Eds.; American Society of Agronomy, Inc.: Madison, WI, 2000; 79–118.

3. Cleugh, H.A. Effects of windbreaks on airflow, microclimates and crop yields. Agrofor. Syst. **1998**, *41,* 55–84.

4. Tanner, C.B.; Sinclair, T.R. Efficient water use in crop production: research or re-search? In *Limitations to Efficient Water Use in Crop Production*; Taylor, H.M., Jordan, W.R., Sinclair, T.R., Eds.; American Society of Agronomy, Inc.: Madison, WI, 1983; 1–27.

5. McNaughton, K.G. Effects of windbreaks on turbulent transport and microclimate. Agric. Ecosyst. Environ. **1988**, *22/23*, 17–39.
6. Grace, J. Some effects of wind on plants. In *Plants and Their Atmospheric Environment*; Grace, J., Ford, E.D., Jarvis, P.G., Eds.; Blackwell Scientific Publications: Oxford, 1981; 31–56.
7. Nuberg, I.K. Effect of shelter on temperate crops: a review to define research for Australian conditions. Agrofor. Syst. **1998**, *41*, 3–34.
8. Caborn, J.M. *Shelterbelts and microclimate,* Forestry Commission Bulletin No. 29; Her Majesty's Stationery Office: Edinburgh, 1957; 135 pp.
9. Grace, J. Plant response to wind. Agric. Ecosyst. Environ. **1988**, *22/23*, 71–88.
10. Thornley, J.H.M.; Johnson, I.R. *Plant and Crop Modeling: A Mathematical Approach to Plant and Crop Physiology*; Clarendon Press: New York, 1990; 669 pp.
11. Song, Z.M.; Wei, L. The correlation between windbreak influenced climate and crop yield. In *Agroforestry Systems in China*; Zhu, Z.H., Cai, M.T., Wang, S.J., Jiang, Y.X., Eds.; International Development Research Centre (IDRC, Canada), Regional Office for Southeast and East Asia, published jointly with the Chinese Academy of Forestry: Singapore, 1991; 21–115.
12. Wu, Y.Y.; Dalmacio, R.V. Energy balance, water use and wheat yield in a Paulownia-wheat intercropped field. In *Agroforestry Systems in China*; Zhu, Z.H., Cai, M.T., Wang, S.J., Jiang, Y.X., Eds.; International Development Research Centre (IDRC, Canada), Regional Office for Southeast and East Asia, published jointly with the Chinese Academy of Forestry: Singapore, 1991; 54–65.
13. Huxley, P.A.; Pinney, A.; Akunda, E.; Muraya, P. A tree/crop interface orientation experiment with a *Grevillea robusta* hedgerow and maize. Agrofor. Syst. **1994**, 26, 23–45.
14. Kort, J. Benefits of windbreaks to field and forage crops. Agric. Ecosyst. Environ. **1988**, *22/23*, 165–190.
15. Miller, J.M.; Böhm, M.; Cleugh, H.A. *Direct Mechanical Effects of Wind on Selected Crops: A Review,* Technical Report Number 67; CSIRO Center for Environmental Mechanics: Canberra, Australia, 1995; 68 pp.
16. Scholten, H. Snow distribution on crop fields. Agric. Ecosyst. Environ. **1988**, *22/23*, 363–380.
17. Wilson, J.D. Numerical studies of flow through a windbreak. J. Wind Eng. Ind. Aerodyn. **1985**, *21*, 119–154.
18. Wang, H.; Takle, E.S. A numerical simulation of boundary-layer flows near shelterbelts. Boundary-Layer Meteorol. **1995**, *75*, 141–173.

32

Bacterial Pest Control

David N. Ferro

Introduction

Although many genera of bacteria are found to be associated with insects—such as *Clostridium, Strategus, Pseudomonas, Proteus, Diplococcus, Serratia, Bacillus,* and *Enterobacter*—only *Bacillus* and *Serratia* represent agents that cause suppression of insect populations, i.e., that perform as biological control agents. Bacteria are the most widely used microbial agents for controlling insect pests. Some species of *Bacillus* and *Serratia* kill by replicating within the host, while strains of *Bacillus thuringiensis* produce protein toxins that kill soon after being ingested. Bacteria that replicate within their hosts and that persist in the environment by maintaining an infection cycle are biological control agents in the traditional sense. However, products of *B. thuringiensis* that produce toxins that kill insect pests and are applied the way an insecticide would be applied are often not considered to be biological control agents. Bacteria in the genera *Photorhabdus* and *Xe-norhabdus* are symbiotic with nematodes in the families Heterorhabditidae and Steinernematidiae, respectively. The nematodes serve as vectors that mechanically penetrate into the insect hemocoel and deposit the bacteria. The bacteria then replicate and kill the host quickly by causing septicemia. The only commercially available bacterial products are from strains of *B. popillae, B. thuringiensis*, and *Serratia entomophila*.

Paenibacillus (Formerly Bacillus) Popilliae (Dutky)

Milky disease was first observed in Japanese beetle larvae (grubs) in New Jersey in 1933. *P. popilliae* is an obligate pathogen of larvae in the family Scarabaeidae, as it is only found associated with its host or in the soil surrounding its host. *P. popilliae* and *Paenibacillus lentimorbus* (Dutky) both cause milky disease of scarab beetles; however, most discussions of milky disease refer to strains of *P. popilliae*. *P. popilliae* produces a crystal or parasporal body, which allows it to survive for many years in the soil in the absence of its host. Although there are dozens of strains of *P. popilliae* that infect scarab hosts, only *P. popilliae* has been used commercially as a biological control agent of the Japanese beetle, *Popillia japonica* (Newman), a major pest of turf.

The term "milky disease" describes the advanced stages of infection in scarab larvae where the host is turned a milky white by the build-up of *Bacillus* spores in the hemolymph. The infection process begins with the scarab larvae ingesting spores while feeding on roots and organic matter in the soil. The spores then undergo germination and outgrowth in the cells of the lumen of the alimentary canal. The vegetative rods penetrate the epithelial cells of the midgut, and then move into the hemolymph where they multiply and sporulate. Death often occurs a month or more after ingestion. It is unclear what the role of the proteinaceous parasporal body is in the infection process.

Culture and Control

Many attempts have been made to rear *P. popillae* on an artificial diet. Even though spores and vegetative rods from field-collected larvae can be plated on agar media, the inability of the milky disease bacteria to grow and sporulate on standard microbiological media has made it extremely costly to produce for commercial purposes. Products, to date, are made from milky larvae, primarily from naturally infected larvae collected from the field.

The spores are formulated on talc and contain 10^8 spores/g of powder. The powder is applied at about 20 kg/ha using a fertilizer spreader or by punching holes in the soil and adding bacteria. Infection can occur in all three larval stages. For optimal replication to occur, soil temperatures need to exceed 20°C. Large overwintered larvae usually pupate before soil temperatures are high enough in late spring. For this reason, applications are targeted against small larvae late in the summer when the small larvae are actively feeding near the soil surface. Control seems to be greatest when larval densities exceed $300/m^2$; however, economic losses in turf occur at densities above $100/m^2$. Unless a more virulent strain is found or a more cost-effective way to produce spores is developed, the use of this bacterium is likely to be restricted to lawns and playing fields that can tolerate higher densities of larvae.

Serratia Entomophila (Grimont et al.)

Amber disease of the New Zealand grass grub *Costelytra zealandica* (White) is a chronic infection of the larval gut caused by *S. entomophila*. This disease was first observed in New Zealand in 1981. Following ingestion of bacterial cells while feeding on grass roots, the bacteria adhere to the foregut and multiply in the region of the cardiac valve; the larvae cease feeding after 2–5 days and become amber colored due to clearance of the gut. Death does not occur until 1–3 mo after ingestion. As the disease progresses, the larvae become shrunken due to a general degradation of the fat cells. Invasion of the hemocoel does not occur until late stages of the disease, when general septicemia is accompanied by death of the insect.

Culture and Control

S. entomophila is produced in large fermentors as nonsporeforming bacteria to be applied as a live microbial pesticide. Recently, the Industrial Processing Division of DSIR, New Zealand produced 4×10^{10} bacteria/mL, and field trials have shown that $>4 \times 10^{13}$/ha are needed for control. The problem with using live bacteria (vs. spores) is the difficulty of maintaining viability on the shelf and in the field prior to ingestion. Currently, refrigerated product can be kept for only 3 mo.

Grass grub larvae live in the soil as pests of low-value grasslands. Because *S. entomophila* is applied as live bacteria rather than as spores, it is more vulnerable to UV light and desiccation. For this reason, it is important to place the formulated material 2–5 cm below the soil surface using a subsurface applicator, such as a modified seed drill. This approach allows for 90% survival of the bacteria. Bacteria applied in this way quickly start an epizootic, which then spreads through the grass grub population.

Bacillus Thuringiensis (Berliner)

B. thuringiensis is a spore-forming bacterium that produces a parasporal crystal (protein delta-endotoxin). After the susceptible insect larva ingests the endotoxin, in the absence or presence of the spore, the crystal is solubilized and activated by alkaline (pH 10.5) gut proteases. The toxic subunits bind to receptor sites on the midgut epithelium within minutes of ingestion. This is quickly followed by lysis of these cells, causing a cessation of feeding within 10–15 min of ingestion. Although the spores pass into the hemocoel through pores in the epithelium of the midgut, it is the starvation in conjunction with infection that kills the insect. The toxins from these bacteria are formulated in much the same way as a synthetic toxin, and do not cause an epizootic.

There are several subspecies (= strains) of *B. thuringiensis* based on the serotype of flagellar antigens, and these subspecies produce different endotoxins, or at least different amounts of endotoxins that are relatively host specific. For example, *B.t. israelensis* is effective against Nematocera (Diptera) larvae such as mosquito larvae, *B.t. kurstaki* against Lepidoptera, *B.t. aizawai* against Lepidoptera, and *B.t. tenebrionis* against Chrysomelidae (Coleoptera). Notation for the gene that encodes for the toxin is in lowercase; for example, Cry3A gene regulates the production of the Cry3A toxin. Table 1 includes a list of some of the subspecies and toxins they produce. Because these bacteria are so host-specific, they can be quickly incorporated into a pest management program in which biological control agents are an integral component.

Culture and Control

B. thuringiensis can be produced in large quantities using commercial fermentors. Formulations can be applied to foliage or other larval substrates in the same manner as most insecticides. However, several operative factors affect the effectiveness of these bacterial agents.

B.t.s are most effective against early instars (Table 2). Their effectiveness is very dependent upon ambient temperatures; the protein endotoxin is not very persistent; thorough coverage of foliage is necessary; and they are host-specific. This host specificity allows for control of the target pest without killing other insect biological control agents; however, in many cropping systems, there is a complex of insect pests

TABLE 1 *B. thuringiensis* Subspecies and Crystal Protein Toxins

Crystal protein	B.t. Subspecies			
	B.t aizawai	*B.t. kurstaki*	*B.t. tenebrionis*	*B.t. israelensis*
CryIAa	*	*		
CryIAb	*	*		
CryIAc		*		
CryIC	*			
Cry ID	*			
Cry2A		*		
Cry2B		*		
Cry3A			*	*
Cry4A				*
Cry4B				*
Cry4C				*
Cry4D				*
CytA				*

TABLE 2   Stage-Specific Larval Mortality for the Colorado Potato Beetle
Fed Foliage Treated with *B. thuringiensis san diego* (= *tenebrionis*)

Larval stage	LC50 (mg/l)	Larval weight (mg)	95% CI Lower	95% CI Upper
Early 1st instar	2.03	1.0	1.46	2.60
Late 1st instar	3.92	2.3	2.02	6.27
Early 2nd instar	4.35	4.0	3.30	5.56
Late 2nd instar	14.45	7.8	10.75	19.50
Early 3rd instar	14.86	15.6	9.95	20.48

and often these need to be controlled at the same time, which may require using the *B.t.* product with a synthetic insecticide, if natural controls fail. Novel ways have been developed to deliver the toxin for ingestion by the pest.

One of the genes that control the production of the toxin has been inserted into *Pseudomonas fluorescens*. After the fermentation has been completed, the broth is chemically treated and heated to kill the bacteria. During this process, the protein toxin becomes encapsulated by the bacterial cell wall. The encapsulation process appears to protect the toxin from degradation in the field, making it more persistent. Several genes have also been inserted into plants that express the toxin in its tissues. In the case of potatoes, the transgenic plants are highly resistant to the Colorado potato beetle, which has considerably reduced the insecticide load on potatoes.

Potential Biological Control AGENTS

Bacillus sphaericus (Neide) has been shown to be toxic only to larvae of culicid Diptera mosquitoes. This bacterium can be easily produced via fermentation. Insecticidal activity is due to crystalline toxins associated with the cell wall. The toxin is released by digestion after the host insect has consumed the bacteria. *B. alvei* and *B. brevis* are infectious for larvae of several mosquito species. There is no evidence that these species are significant biocontrol agents. The success of these bacteria in the field is likely to be dependent on selection of strains that are more virulent and that can persist in a range of aquatic environments.

Bibliography

1. Crickmore, N.; Zeigler, D.R.; Feitelson, J.; Schnepf, E.; Van Rie, J.; Lereclus, D.; Baum, J.; Dean, D.H. Revision of the nomenclature for the *Bacillus thuringiensis* pesticidal crystal proteins. Am. Soc. Micro. **1998**, *62*, 807–813.
2. Glare, T.R.; Jackson, R.A. *Use of Pathogens in Scarab Pest Management*; Intercept: Andover, England, 1992; 43–61, 179–198.
3. Jackson, T.A.; Huger, A.M.; Glare, T.R. Pathology of amber disease in the New Zealand grass grub *Costelytra zealandica* (Coleoptera: Scarabaeidae). J. Invert. Pathol. **1993**, *61*, 123–130.
4. Tanada, Y.; Kaya, H.K. *Insect Pathology*; Academic Press: New York, 1993; 83–146.
5. Van Driesche, R.G.; Bellows, T.S., Jr. *Biological Control*; Chapman and Hall: New York, 1996.

33

Bioaccumulation

Tomaz Langenbach

Molecular Properties for Bioaccumulation

Bioaccumulation occurs only with molecules with low degradability and correlates with their grade of lipophilicity.[1] Organic substances with main bonds of aliphatic and aromatic C—C, C—H, and C—Cl (or other halogens) are predominantly nonpolar molecules (Lipophilic) with low water solubility and high stability. They are less susceptible to chemical reactions of hydrolysis, oxidation, and enzymatic attack.[2] On the other hand, bonds with different functional groups with O, P, N, and other elements turn molecules more polar, soluble, and degraded more easily. Bioaccumulation can occur with molecules between 100 and 600 units of molecular weight with the maximum of 350.[3] Probably this is related to membrane permeability capacity.

A common feature of bioaccumulation is the molecular stability of lipophilic organic substances and the nondegradability of heavy metals. The severity of heavy metals is due to many factors.

1. Metals with Hg, Cd, Zn, Cu, and Pb are the most toxic and most studied types followed by metals containing Ni, Al, As, Cr, and other elements.[4] Bioaccumulation can also occur also with essential metals such as Fe, Zn, Cu, Mo, Na, and Ca.
2. Speciation is the anions or other components that constitute the heavy metal molecules. This is important in defining solubility that, for example, is high for sulfate and low for sulphide.[5] Heavy metals bound to organic molecules such as methyl, ethyl or other aliphatic or aryl groups increase penetration capacity through membranes and consequently, have a poisoning effect.
3. The sensivity to the toxic effects of heavy metals and other xenobionts is dependent on the biological material being a microorganism, plant, animal, or type of tissue.

Bioaccumulation and the Environment

The pollution sources can be released by discharge of substances with uneven distribution in air, water, and soil. The movement of these substances up to bioaccumulation can occur by different routes mainly mediated by the food chain. This process can involve water, suspended particles, sediments, food, soil, and air particles (Figure 1). An important part of these substances can be concentrated in nonliving components. From these sources, persistent organic pollutants (POP) or heavy metals can be released to biota.[6] The final distribution presented by the mass balance in the environment with a group of

FIGURE 1   Bioaccumulation in the environment. The black points represent the pollutant molecules.

organochlorines and polyaromatics, shows that most are found in soil or sediment, whereas for highly volatile substances, most remain in the atmosphere. Less than 0.7% of the total remains in vegetation and no more than 2×10^{-3}% can be found in the aquatic biota.[7] The relationship between biota and environment shows that concentrations of bioaccumulated heavy metals in organisms are always higher than in water, but are usually lower than in sediments.[4]

Water Environment

After pollution reaches water bodies, different processes can occur to incorporate it into nonliving components as sediments and biota represented by microorganisms, plants, crustaceans, fishes, etc. The route of pollutant uptake in the biota if from waterborne, adsorption, filtration, or by food chain is an important factor in bioaccumulation.[8] Along the food chain the step-wise increase of concentration from lower to higher trophic levels, called biomagnification, can reach the bioconcentration factor (BCF) up to 100,000. In this process terrestrial animals as well as birds can be heavily contaminated by eating polluted fish.

In the global marine environment the apparent final fate of persistent organic pollutants (POP) is in the flora, fauna, and sediment of the abyss.[9] The main transport of POP follows the downstream movement of the organic flow in the water and in the long term these chemicals are incorporated in the sediment that function as final sink. It was observed that the bioaccumulation in the deep water fishes are up to 10 times higher than in surface water fishes.[9]

In the flora and fauna some heavy metals can bioaccumulate up to threshold values and others maintain a correlation with the concentration in the environment.[4] In aquatic plants, fish, and other metazoarians the distribution of the substances are quite distinct between tissues.[10] Lipophilic substances are preferentially found in adipose tissues with high lipid content.

In the Soil

Soil is polluted in large areas by pesticides application or by discharge as final disposal in landfills of industrial products. These lipophilic substances move in soil rather slowly by leaching, runoff, and volatilization. The main factor that influences bioaccumulation process in the soil is the biodisponibility. This property is conditioned by the adsorption/desorption capacity of the different soils and by the chemical nature of the pollutant. This process is driven by the stronger or weaker binding forces involved, which influence the amount that is bioavailable for plant uptake of these lipophilic pollutants.

The main flow of POP generally occurs toward organic matter from soil particles and not to biota, a process called preferential partition. A negative correlation was observed between the adsorbtion coefficient related to soil organic carbon (*Koc*) and the bioaccumulation factor by plants. This means that in organic rich soils, bioaccumulation in plants is rather small.[2] A similar situation occurs with microorganisms in which previous bioaccumulated organochorines can move out from the cell to the soil.[11] The preferential partition toward soil can be the reason why a lack of toxicity on soil microorganisms by pesticide applications was frequently observed, even in high concentrations of pesticides. Little information about bioaccumulation in soil could be observed, but nevertheless cotransport of some organochlorine accumulated in microorganisms in sand aquifers with low organic content was reported.[12]

Soil invertebrates such as earthworms, beetles, slugs, and others can bioaccumulate lipophilic pesticides. The bioaccumulation process could be seen as a soil to soil–water equilibrium followed by a soil–water to worm equilibrium.[2] Consumers of this biota in animals of higher trophic levels such as birds can biomagnificate these chemicals. Plants can adsorb and bioaccumulate products from the soil with incorporation of residues mainly in the root. The translocation from root to foliage depends on plant species and on the chemical properties of the pollutant. Several evidences show that lipophilic compounds are sorbed onto the outer surface of roots of several plants, and in this case translocation is very low.

Bioaccumulation in plants can also occur with heavy metals. As safety rules, domestic waste and sludge from wastewater treatment stations with heavy metal contamination can be applied on soil for agriculture up to limited amounts to avoid pollution with hazardous toxicological effects. Contaminated grass, grain, and fruits can be accumulated by biomagnification when consumed by mammals, insects, and birds. Terrestrial animals have a plant mediated relationship with soil contamination.

In the Air

The main sources of atmospheric pollution are pesticide spraying with the reverse process of evaporation from soil to air, poliaromatics produced by burning of fuels, and plastic incineration. The rate of entry to the atmosphere and the distance of movement are principally dependent on the vapor pressure of the pollutants and metereological conditions. In some cases movement occurs on a global scale. The dynamic nature of the atmosphere can dilute pollutants in the air to exceptionally low concentrations and in these cases no significant bioaccumultion occurs.[13] Nevertheless urban and industrial areas, as well as the margins of roads with intense traffic, can have high concentrations of pollutants. Plants exposed to xenobiotics in the form of vapor, particles, aerosol or larger droplets, can undergo a passive process of foliage adsorption with an uptake mainly in the wax cuticle.[2] Bioaccumulation in plants can result in damage and can also affect higher trophic levels that consume these vegetables.

Direct contact of animals with these chemicals can enter by the respiratory organs, in mammals, or the outer body surface, mainly in insects.

Bioaccumulation Mechanism in Biota

In terrestrial animals, pollutants can enter by dermal contact, respiration, and food consumption. Atmospheric pollutants move to the lungs, where an equilibrium is difficult to be established, while generally atmosphere dilutes pollutant concentrations, unless there is an exposure to constant pollution sources. Chemicals move from lungs to circulatory fluid (plasma) and can be metabolized with further excretion. Another route is the storage mainly in rich lipid bodies such as brain and eggs in bird's.[2] If the entrance is by food consumption, the gastrointestinal tract can eliminate[10] these substances or degrade than to more polar compounds with further excretion, or can promote adsorption by plasma following the same route described earlier.

The uptake of heavy metals in microorganisms can occur by bioadsorbtion in capsular polysaccharides and cell-wall polymers, or cross these layers and cell membranes by an active enzymatic process involving phosphatases, reaching to the cell interior.[14] Some authors define bioaccumulation as only the process in which molecules reach cell interior. Many cells from animals, plants, fungi, yeast, and bacteria have metal-binding proteins with low molecular weight called metallothioneins. These proteins bind mainly to Cd and Zn and constitute a protection mechanisms to the toxic effects of these substances. Some other cell protective mechanisms exist such as enhancement of efflux from cell to the outside. Metals bind on different macromolecules and change enzymatic activities with inhibition or stimulation effects.

Lipophilic hydrocarbons in microorganisms cross polysaccharides from capsule and cell wall polymers with adsorption mainly by the lipids of the membranes.[15] Compounds that are inserted in cell lipids are more difficult to be degraded by chemical or enzymatic processes getting higher persistence.[16] The probable mechanism of action seems to be nonspecific, this means not related to a specific target.

Applications and Future Perspectives

From the scientific point of view a better understanding of the integration between the different environment compartments and biota including modeling systems is an important approach that needs more development. Another possibility is to use bioaccumulation for environmental monitoring, based on the accumulation capacity of many pollutants in specific plants or animals, allowing chemical measurements that otherwise in water or air are below the analytic detection capacity.[17] This method has the possibility to integrate all pollutant exposure of plants or animals in a specific environment and can be in the future an important parameter for ecotoxicological evaluations.

After the disaster of the mercury pollution in the Minamata Bay in Japan in which more than 630 people died and many became physically and mentally disabled, the magnitude of poisoning effects due to bioaccumulation was recognized for the first time.[18] This was the beginning of a scientific research that produced a large amount of information. With this knowledge it became clear that bioaccumulation is a natural process that cannot be stopped by man but can be avoided with a more efficient control of pollutant release. To overcome the economic, social, and political difficulties for better pollution control together with the development of more ecological technologies are our challenge for today and for the future.

References

1. Connell, D.W. General Characteristics of Organic Compounds which Exhibit Bioaccumulation. In *Bioaccumulation of Xenobiotic Compounds*; Connell, D.W., Ed.; CRC Press: Boca Raton, Florida, 1990; 47–57.
2. Connell, D.W. Bioamagnification of Lipophilic Compounds in Terrestrial and Aquatic Systems. In *Bioaccumulation of Xenobiotics Compounds;* Connell, D.W., Ed.; CRC Press: Boca Raton, Florida, 1990; 145–185.
3. Brooke, D.N.; Dobbs, A.J.; Williams, N. Octanol: Water partition coefficients (P): measurement estimation and interpretation, particularly for chemicals with $P > 105$. Eco-toxicol. Eviron. Saf. **1986**, *11*, 251.
4. Goodyear, K.L.; McNeill, S. Bioaccumulation of heavy metals by aquatic macro-invertebrates of different feeding guilds: a review. Sci. Total Environ. **1999**, *229*, 1–19.
5. Bourg, A.C.M. Speciation of Heavy Metals in Soils and Groundwater and Implications for Their Natural and Provoked Mobility. In *Heavy Metals: Problems and Solutions*; Salomons, N., Förstner, U., Mader, P., Eds.; Springer Verlag: Berlin, 1995; 17–31.
6. Tsezos, M.; Bell, J.P. Comparison of the biosorption and desorption of hazardous organic pollutants by life and dead bioamass. Wat. Res. **1989**, *23* (5), 561–568.

7. Connell, D.W.; Hawker, D.W. Predicting the distribution of persistent organic chemicals in the environment. Chem. Aust. **1986**, *53,* 428.

8. Carbonell, G.; Ramos, C.; Pablos, M.V.; Ortiz, J.A.; Tarazona, J.V. A system dynamic model for the assessment of different exposure routes in aquatic ecosystems. Sci. Total Environ. **2000**, *247,* 107–118.

9. Froescheis, O.; Looser, R.; Cailliet, G.M.; Jarman, W.M.; Ballschmiter, K. The deep-sea as a final global sink of semivolatile persistent organic pollutants part I: PCBs in surface and deep-sea dwelling fish of the north and south atlantic and the Monterey bay canion (California). Chemo- sphere **2000**, *40,* 651–660.

10. Lin, K.H.; Yen, J.H.; Wang, Y.S. Accumulation and elimination kinetics of herbicides butachlor, thiobencarb and Chlomethoxyfen by *Aristichthys nobilis.* Pestic. Sci. **1997**, *49,* 178–184.

11. Brunninger, B.M.; Mano, D.M.S.; Scheunert, I.; Langenbach, T. Mobility of the organochlorine compound dicofol in soil promoted by *Pseudomonas fluorescens.* Ecotoxic. Environ. Safety **1999**, *44,* 154–159.

12. Jenkins, M.B.; Lion, L.W. Mobile bacteria and transport of polynuclear aromatic hydrocarbons in porous media. Appl. Environ. Microbiol. **1993**, *59* (10), 3306–3313.

13. Connel, D.W. Environmental Routes Leading to the Bioaccumulation of Lipophilic Chemicals. In *Bioaccumulation of Xenobiotic Compounds;* Connell, D.W., Ed.; CRC Press: Boca Raton, FL, 1990; 59–73.

14. Gomes, N.C.M.; Mendonca-Hagler, L.C.S.; Savvaidis, I. Metal bioremediation by microorganisms. Rev. Microbiol. **1998**, *29,* 85–92.

15. Mano, D.M.S.; Langenbach, T. [14C]Dicofol association to cellular components of azospirillum. Pestic. Sci. **1998**, *53,* 91–95.

16. Mano, D.M.S.; Buff, K.; Clausen, E.; Langenbach, T. Bioaccumulation and enhanced persistence of the acaricide dicofol by *Azospirillum lipoferum.* Chemosphere **1996**, *33* (8), 1609–1619.

17. Maagd, P.G.J. Bioaccumulation test applied in whole effluent assessment: a review. Env. Toxic. Chem. **2000**, *19* (1), 25–35.

18. Takashi, H.; Tsubaki, T. Epidemiology: Mortality in Minamata Disease. In *Recent Advance in Minamata Disease Studies; Methylmercury Poisoning in Minamata and Niigata Japan;* Kodansha Ltd.: Tokyo, 1986; 1–23.

34

Biodegradation

Sven Erik Jørgensen

Introduction: Overview of the Properties of Toxic Chemical Compounds of Particular Importance for Environmental Management

Slightly more than 100,000 chemicals are produced in such an amount that they threaten or may threaten the environment. They cover a wide range of applications: household chemicals, detergents, cosmetics, medicines, dye stuffs, pesticides, intermediate chemicals, auxiliary chemicals in other industries, additives to a wide range of products, chemicals for water treatment, and so on. They are (almost) indispensable in modern society and cover many more or less essential needs in the industrialized world, which has increased the production of chemicals about 40-fold during the last five decades. A minor or major proportion of these chemicals are inevitably reaching the environment through their production, during their transportation from the industries to the end user, or by their application. In addition, the production or use of chemicals may cause more or less unforeseen waste or by-products, for instance, chloro-compounds from the use of chlorine for disinfection. As we would like to have the benefits of using the chemicals but cannot accept the harm they may cause, this conflict raises several urgent questions, which we already have discussed in other entries.

We cannot answer these crucial questions without knowing the properties of the chemicals. Organization for Economic Cooperation and Development (OECD) has made a review of the properties that we should know for all chemicals. We need to know the boiling point and melting point to know in which form (solid, liquid, or gas) the chemical will be found in the environment. We must know the dispersion of the chemicals in the five spheres: the hydrosphere, the atmosphere, the lithosphere, the biosphere, and the technosphere (the part of the earth that is controlled and under the influence of human technology). This will require knowledge about their solubility in water; the water/lipid partition coefficient; Henry's constant (the constant in Henry's Law, which indicates the distribution of the chemical between air and water); the vapor pressure; the rate of degradation by hydrolysis, photolysis, chemical oxidation, and microbiological processes; and the adsorption equilibrium between water and soil—all as functions of the temperature. We need to discover the interactions between living organisms and the chemicals, which implies that we should know the biological concentration factor (BCF), the magnification through the food chain, the uptake rate and the excretion rate by the organisms, and where in the organisms the chemicals will be concentrated, not only for one organism but for a wide range of organisms. We must also know the effects on a wide range of different organisms. It means that we should be able to find the Lethal Concentration causing 50% mortality oif the test animals (LC50) and Lethal Dose causing 50% mortality of the test animals (LD50) values; the Maximum Allowable Concentration (MAC) and Non-effect Concentration (NEC) values (for the abbreviations and the definitions used, see Appendix 5); the relationship between the various possible sublethal effects and concentrations; the influence of the chemical on fecundity; and the carcinogenic and teratogenic properties. We should also know the effect on the ecosystem level. How do the chemicals affect populations and their development and interactions, i.e., the entire network of the ecosystem? A reduction of one population may for instance influence the entire ecosystem because all populations are bound together in an ecological network.

Table 1 gives an overview of the most relevant physical-chemical properties of organic compounds and their interpretation with respect to the behavior of the environment. It is clear from the table that a high water solubility is not desirable as it implies that the chemical compound is very mobile. On the other hand, a low water solubility means a high solubility in fat tissue and, therefore, a high bioaccumulation and a high biomagnification. The biodegradability may, however, frequently be considered an even more important property than water solubility or K_{ow}. If a compound is biodegraded fast, it will be decomposed before it harms the environment. It means that a fastbiodegradation will, so to speak, neutralize the (harmful) effect of a high water solubility and a high solubility in fat tissue. On the other hand, if a compound is biodegraded slowly, it will stay in the environment for a very long time, which implies that a high mobility and a high risk of bioaccumulation and biomagnifications will be harmful. Therefore, it is almost possible to conclude that a compound with high biodegradability will clearly be much less harmful than a compound with a low biodegradability. Biodegradability is therefore a very crucial property for the estimation of a chemical compound's environmental effects.

The list of properties needed to give an adequate answer to the six questions mentioned above could easily be made longer (see, for instance, the list recommended by OECD). To provide all the properties corresponding to the list given here is already a huge task. More than 10 basic properties should be known for all 100,000 chemicals and organisms, which would require 1,000,000 pieces of information. In addition, we need to know at least 10 properties to describe the interactions between 100,000 chemicals and organisms. Let us say, modestly, that we use 10,000 organisms to represent the approximately 10 million species on earth. This gives a total of $1,000,000 + 100,000*10,000*10 =$ in the order of 10^{10} properties to be quantified! Today, we have determined less than 1% of these properties by measurements,

TABLE 1 Overview of the Most Relevant Environmental Properties of Organic Compounds and Their Interpretation

Property	Interpretation
Water solubility	High water solubility corresponds to high mobility.
K_{ow}	High K_{ow} means that the compound is lipophilic. It implies that it has a high tendency to bioaccumulate and be sorbed to soil sludge and sediment. BCF and Koc are correlated with K_{ow}.
Biodegradability	This is a measure of how fast the compound is decomposed to simpler molecules. A high biodegradation rate implies that the compound will not accumulate in the environment, while a lowbiodegradation rate may create environmental problems related to the increasing concentration in the environment and the possibilities of a synergistic effect with other compounds.
Volatilization, vapor	A high rate of volatilization (high vapor pressure) implies that the pressure compound will cause an air pollution problem.
Henry's constant, H	H determines the distribution between the atmosphere and the hydrosphere.
pK	If the compound is an acid or a base, pH determines whether the acid or the corresponding base is present. As the two forms have different properties, pH becomes important for the properties of the compounds.

Note: K_{ow} = Ratio solubility in octanol (represent fat tissue) divided by the solubility in water; Koc = express the adsorption ability to soil consisting of 100% organic carbon, can also with good approximation be considered as the concentration in soil with 1100% orgniac carbon divided with the concentration in water at equilibrium; H = Henry's constant in Henry's Law; pK = − log (equilibrium constant for the dissociation process of acids: HA = A⁻ + H⁺).

and with the present rate of generating new data, we can be certain that during the 21st century, we shall not be able to reach 10% even with an accelerated rate of ecotoxicological measurements.

Environmental risk assessments require, among other inputs, information about the properties of the chemicals and their interactions with living organisms. It is maybe not necessary to know the properties with the very high accuracy that can be provided by measurements in a laboratory, but it would be beneficial to know the properties with sufficient accuracy to make it possible to utilize the models for management and risk assessment. Therefore, estimation methods have been developed as an urgently needed alternative to measurements. They are to a great extent based on the structure of the chemical compounds, the so-called QSAR and SAR methods (Quantitative Structure-Activity Relationship, it means estimation methods of chemical properties based on the chemical structure), but it may also be possible to use allometric principles to transfer rates of interaction processes and concentration factors between a chemical and one or a few organisms to other organisms.[1]

It may be interesting in this context to discuss the obvious question: why is it sufficient to estimate a property of a chemical in an ecotoxicological context with 20%, or sometimes 50% or higher, uncertainty? Ecotoxicological assessment usually gives an uncertainty of the same order of magnitude, which means that the indicated uncertainty may be sufficient from, for instance, the viewpoint of ecological modeling or ecological indicators, but can results with such an uncertainty be used at all? The answer in most (many) cases is "yes," because we want in most cases to assure that we are (very) far from a harmful or very harmful level. We use a safety factor of 100–1000 in many cases. When we are concerned with very harmful effects, such as, for instance, complete collapse of an ecosystem or a health risk for a large human population, we will inevitably select a safety factor that is very high. In addition, our lack of knowledge about synergistic effects and the presence of many compounds in the environment at the same time force us to apply a very high safety factor. In such a context, we will usually go for a concentration in the environment that is magnitudes lower than that corresponding to a slightly harmful effect or considerably lower than the NEC. It is analogous to civil engineers constructing bridges. They make very sophisticated calculations (develop models) that account for wind, snow, temperature changes, and so on, and afterward, they multiply the results by a safety factor of 2–3 to ensure that the bridge will not collapse. They use safety factors because the consequences of a bridge collapse are unacceptable.

The collapse of an ecosystem or a health risk to a large human population is also completely unacceptable. Thus, we should use safety factors in ecotoxicological modeling to account for the uncertainty. Due to the complexity of the system, the simultaneous presence of many compounds, and our present knowledge, or rather, lack of knowledge, we should, as indicated above, use 10–100 or sometimes even 1000 as safety factor. If we use safety factors that are too high, the risk is only that the environment will be less contaminated at maybe a higher cost. Besides, there are no alternatives to the use of safety factors. We can, step by step, increase our ecotoxicological knowledge, but it will take decades before it may be reflected in considerably lower safety factors. A measuring program of all processes and components is impossible due to the high complexity of the ecosystems. This does not, of course, imply that we should not use the information of measured properties available today. Measured data will almost always be more accurate than the estimated data. Furthermore, the use of measured data within the network of estimation methods will improve the accuracy of estimation methods. Several handbooks on ecotoxicological parameters are, fortunately, available. References to the most important are given below. Estimation methods for the physical-chemical properties of chemical compounds were already applied 40–60 years ago, as they were urgently needed in chemical engineering. They are, to a great extent, based on contributions to a focal property by molecular groups and the molecular weight: the boiling point, the melting point, and the vapor pressure as a function of the temperature. In addition, a number of auxiliary properties result from these estimation methods, such as the critical data and the molecular volume. These properties may not have a direct application as ecotoxicological parameters in environmental risk assessment but are used as intermediate parameters, which may be used as a basis for estimation of other parameters.

The water solubility, the octanol/water partition coefficient, K_{ow}, and Henry's constant are crucial parameters in our network of estimation methods, because many other parameters are well correlated with these three parameters. The three properties can fortunately be found for a number of compounds or be estimated with reasonably high accuracy by use of knowledge of the chemical structure, i.e., the number of various elements, rings, and functional groups. In addition, there is a good relationship between water solubility and K_{ow}.[2] Particularly in the last decade, many good estimation methods for these three core properties have been developed.

During the last couple of decades, several correlation equations have been developed based upon a relationship between the water solubility, K_{ow}, or Henry's constant on the one hand and physical, chemical, biological, and ecotoxicological parameters for chemical compounds on the other. The most important of these parameters are the following: the soil/water adsorption isotherms; the rate of the chemical degradation processes (hydrolysis, photolysis, and chemical oxidation); the BCF; the ecological magnification factor (EMF); the uptake rate; the excretion rate; and a number of ecotoxicological parameters. The ratio of concentrations both in the sorbed phase and in water at equilibrium Ka and BCF may often be estimated with a relatively good accuracy from expressions like Ka or BCF = a log K_{ow} + b. Numerous expressions with different a and b values have been published.[3–5]

Biodegradation

It was concluded in the previous section that biodegradation is probably the most important property for the estimation of a chemical compound's environmental effect. This section is therefore devoted to the presentation of this important property.

Biodegradation rates may be expressed in several ways. Microbiological biodegradation may, with good approximation, be described as a Monod equation:[5]

$$dc/dt = -dB/Ydt = -\mu max^* Bc/Y(Km+c) \tag{1}$$

where c is the concentration of the compound considered, Y is the yield of biomass B per unit of c, B is the biomass concentration, μmax is the maximum specific growth rate, and Km is the half saturation constant. If $c \ll$ Km, the expression is reduced to a first-order reaction scheme:

$$dc/dt = -K'Bc \tag{2}$$

where $K' = \mu max/(Y\, Km)$. B is in nature determined by the environmental conditions. In aquatic ecosystems, B is for instance highly dependent on the presence of suspended matter. B may therefore under certain conditions[5] be considered a constant, which reduces the rate expression to

$$dc/dt = -kc \tag{3}$$

An indication of k in the unit 1/hr, 1/24hr, 1/week, 1/mo, or 1/yr can therefore be used to describe the rate of biodegradation. If the biological half-life time is denoted t, we get the following relation:

$$\ln 2 = 0.7 = k\,t \tag{4}$$

This implies that the biological half-life time also can be used to indicate the biodegradation rate.

The biodegradation in waste treatment plants is often of particular interest, in which case the % of the Theoretical Oxygen Demand (ThOD) or the theoretical biological oxygen demand (BOD) may be used as a suitable reference. Most often, however, the 5-day BOD as percentage of the theoretical BOD is used. It may also be indicated as the BOD5fraction. For instance, a BOD5fraction of 0.7 will mean that BOD5 corresponds to 70% of the theoretical BOD. It is, however, also possible to find an indication of percentage removal in an activated sludge plant. The biodegradation is, however, in some cases very dependent on the concentration of microorganisms as expressed in the above-shown equations. Therefore K' indicated in the unit mg/(g dry wt 24 hr) will in many cases be more informative and correct.

In the microbiological decomposition of xenobiotic compounds, an acclimatization period from a few days to 1–2 mo should be foreseen before the optimum biodegradation rate can be achieved. We distinguish between primary and ultimate biodegradation. Primary biodegradation is any biologically induced transformation that changes the molecular integrity. Ultimate biodegradation is the biologically mediated conversion of organic compounds to inorganic compounds and products associated with complete and normal metabolic decomposition.

To conclude, the biodegradation rate is expressed by application of a wide range of units:

1. As a first-order rate constant (1/24 hr).
2. As half-life time (days or hours).
3. mg per g sludge per 24 hr[mg/(g 24 hr)].
4. mg per g bacteria per 24 hr[mg/(g 24 hr)].
5. mL of substrate per bacterial cell per 24 hr[mL/(24 hr cells)].
6. mg COD per g biomass per 24 hr[mg/(g 24 hr)].
7. mL of substrate per gram of volatile solids inclusive microorganisms [mL/(g 24 hr)].
8. BODx/BOD∞, i.e., the biological oxygen demand in x days compared with complete degradation (-), named the BODxcoefficient.
9. BODx/COD, i.e., the biological oxygen demand in x days compared with complete degradation, expressed by means of COD (-).

Estimation of Biodegradation

The biodegradation rate in water or soil is difficult to estimate because the number of microorganisms varies several orders of magnitudes from one type of aquatic ecosystem to the next and from one type of soil to the next. Artificial intelligence has been used as a promising tool to estimate this important parameter. However, a (very) rough, first estimation can be made on the basis of the molecular structure and the biodegradability. The following rules can be used to set up these estimations:

1. Polymer compounds are generally less biodegradable than monomer compounds; 1 point for a molecular weight >500 and ≤1000, 2 points for a molecular weight >1000.
2. Aliphatic compounds are more biodegradable than aromatic compounds; 1 point for each aromatic ring.
3. Substitutions, especially with halogens and nitro groups, will decrease the biodegradability; 0.5 point for each substitution, although 1 point if it is a halogen or a nitro group.
4. The introduction of a double or triple bond will generally mean an increase in the biodegradability (double bonds in aromatic rings are of course not included in this rule); -1 point for each double or triple bond.
5. Oxygen and nitrogen bridges [– O – and – N – (or=)] in a molecule will decrease the biodegradability; 1 point for each oxygen or nitrogen bridge.
6. Branches (secondary or tertiary compounds) are generally less biodegradable than the corresponding primary compounds; 0.5 point for each branch.

Find the number of points and use the following classification:

≤1.5 points: the compound is readily biodegraded. More than 90% will be biodegraded in a biological treatment plant.

2.0–3.0 points: the compound is biodegradable. Probably about 10–90% will be removed in a biological treatment plant. BOD5 is 0.1–0.9 of the theoretical oxygen demand.

3.5–4.5 points: the compound is slowly biodegradable. Less than 10% will be removed in a biological treatment plant. BOD10≤0.1 of the theoretical oxygen demand.

5.0–5.5 points: the compound is very slowly biodegradable. It will hardly be removed in a biological treatment plant, and a 90% biodegradation in water or soil will take ≥6 mo.

≥6.0 points: the compound is refractory. The half-life time in soil or water is counted in years. The structure of dichlorodiphenyltrichloroethane (DDT) corresponds, for instance, to about 7 points, and the biological half-life of DDT in soil is about 14 years.

Several useful methods for estimation of biological properties are based upon the similarity of chemical structures. The idea is that if we know the properties of one compound, it may be used to find the properties of similar compounds. If for instance we know the properties of phenol, which is named the parent compound, it may be used to give more accurate estimation of the properties of monochloro- phenol, dichloro-phenol, trichloro-phenol, and so on and for the corresponding cresol compounds. Estimation approaches based on chemical similarity give generally more accurate estimation but of course are also more cumbersome to apply, as they cannot be used generally in the sense that each estimation has a different starting point, namely, the compound, named the parent compound, with known properties.

Allometric estimation methods presume[6] that there is a relationship between the value of a biological parameter and the size of a considered organism.

The various estimation methods, including estimation methods applicable for biodegradation, may be classified into two groups:

1. General estimation methods based on an equation of general validity for all types of compounds, although some of the constants may be dependent on the type of chemical compound, or they may be calculated by adding contributions (increments) based on chemical groups and bonds.
2. Estimation methods valid for a specific type of chemical compound, for instance, aromatic amines, phenols, aliphatic hydrocarbons, and so on. The property of at least one key compound is known. Based upon the structural differences between the key compound and all other com- pounds of the considered type(for instance, two chlorine atoms have substituted hydrogen in phenol to get 2,3-dichloro-phenol) and the correlation between the structural differences and the differences in the considered property, the properties for all compounds of the considered type can be found. These methods are based on chemical similarity.

Methods of class 2 are generally more accurate than methods of class 1, but they are more cumber- some to use as it is necessary for each type of chemical to find for each property the right correlation. Furthermore, the requested properties should be known for at least one key component, which some- times may be difficult when a series of properties are needed. If estimation of the properties for a series of compounds belonging to the same chemical class is required, it is tempting to use a suitable collection of class 2 methods.

Methods of class 1 form a network that facilitates possibilities of linking the estimation methods together in a computer software system, for instance,WINTOX. [1] An updated version named Estimation of Ecotoxicological Properties (EEP) is now available. EEP can estimate the biodegradability, in contrast to WINTOX. The software is easy to use and can rapidly provide estimations. Each relationship between two properties is based on the average result obtained from a number of different equations found in the literature. There is, however, a price for using such "easy-to-go" software. The accuracy of the estima- tions is not as good as with more sophisticated methods based upon similarity in chemical structure, but in many, particularly modeling, contexts, the results found by WINTOX and EEP can offer sufficient accuracy. In addition, it is always useful to come up with a first intermediate guess. It could, for instance, be used to estimate whether a chemical compound would be decomposed by biological treatment.

The software also makes it possible to start the estimations from the properties of the chemical compound already known. The accuracy of the estimation from use of the software can be improved considerably by having knowledge about a few key parameters, for instance, the boiling point and

Henry's constant. WINTOX and EEP are based on average values of results obtained by simultaneous use of several estimation methods for most of the parameters. It implies increased accuracy of the estimation, mainly because it gives a reasonable accuracy for a wider range of compounds. If several methods are used in parallel, a simple average of the parallel results has been used in some cases, while a weighted average is used in other cases where it has been found beneficial for the overall accuracy of the program. When parallel estimation methods are giving the highest accuracy for different classes of compounds, use of weighting factors seems to offer a clear advantage. It is generally recommended to apply as many estimation methods as possible for a given case study to increase the overall accuracy. If the estimation by WINTOX and EEP can be supported by other recommended estimation methods, it is strongly recommended to do so.

References

1. Jørgensen, S.E.; Halling-Sørensen, B.; Mahler, H. *Handbook of Estimation Methods in Ecotoxicology and Environmental Chemistry;* Taylor and Francis Publ.: Boca Raton, FL, 1998; 230 pp.
2. Jørgensen, S.E. *Principles of Pollution Abatement;* Elsevier: Amsterdam, 2000; 520 pp.
3. Jørgensen, S.E.; Jørgensen, L.A.; Nors Nielsen, S. *Handbook of Ecological and Ecotoxicological Parameters*; Elsevier: Amsterdam, 1991; 1380 pp.
4. Jørgensen, L.A.; Jørgensen, S.E.; Nors Nielsen, S. *Ecotox,* CD; Elsevier: Amsterdam, 2000; corresponding to 4000 pp.
5. Jørgensen, S.E.; Bendoricchio, G. *Fundamentals of Ecological Modelling,* 3rd Ed.; Elsevier: Amsterdam, 2001; 530 pp.
6. Peters, R.H. *The Ecological Implications of Body Size*; Cambridge University Press: Cambridge, 1983; 329 pp.

35

Biological Control of Vertebrates

Peter Kerr and
Tanja Strive

Introduction

Biological control is the deliberate use of one organism to control another, normally a pest species, by reducing its population size, rate of increase or geographic spread and so diminishing its environmental and economic impacts. In classical biological control, usually applied to insect and plant pest species, predators, disease agents or parasites are used to diminish the impact of the pest sometimes with considerable success. Biological control of vertebrate pests has a much more limited history and very few success stories. The advantage of biological control is that while it may be expensive to search for and evaluate biological control agents, a successful biological control provides ongoing control at little cost. Against this must be balanced the risk of unforseen ecological consequences due to the introduced agent or control of the pest and the likely diminution of the impact of the control over time. These principles, successes and limitations are illustrated by examining the natural history of biological control of vertebrate pest species together with a discussion of more recent attempts to use biotechnology to develop novel biological controls.

Vertebrate Pests

Vertebrate pest species may be invasive, feral, deliberately or accidentally introduced, or native to the area; the definition of what constitutes a pest species will vary from location to location, across time, social groups and perspectives. Pest species range from jawless fish to mammals and exist on local and continental scales. Even iconic species such as Australian koalas (*Phascolarctos cinereus*) or African elephants (*Loxodonta africana*) can have local pest status and require some form of management. Similarly, options for control of vertebrate pests will inevitably be influenced by a wide range of social, ecological and economic pressures. A particular consideration, not always addressed, is that unlike plants or insect pests, vertebrate pests are capable of feeling pain and distress, so welfare must be taken into account when control measures are proposed.

Conventional control measures for vertebrate pests can basically be summarized as exclusion by some form of barrier or deterrent (physical, chemical, habitat destruction, guardian dogs, predators) or removal, which could include capture and relocation for iconic species but will more commonly be lethal, including hunting, (commercial, recreational, directed culling), trapping and poisoning. Fertility control, using surgical, hormonal, chemical or immunological means, has been applied, at least experimentally, for some species (Pickard and Holt 2007). The inability to deliver long-acting targeted reproductive controls cheaply and on large scales currently limits the application of fertility control to localized regions or high-value iconic species.

The common problem with all conventional measures is that unless the pest can be permanently eliminated, ongoing control is necessary with the recurrent costs competing with other social and political priorities. These shifting priorities frequently result in poorly sustained control.

Biological Control

Predators as Biological Control Agents for Vertebrate Pests

Biological control of vertebrate pests using introduced predators has a long if not particularly glorious history. The classic example is the use of cats (*Felis catus*) to control rodents around human settlements, which extends back into prehistory (Driscoll et al. 2007). If pest populations are low enough then predators can prevent an increase in population size (Pech et al. 1992). However, introduction of exotic predators frequently leads to unintended consequences. For example, cats have been deliberately introduced onto many islands to control rodents or European rabbits (*Oryctolagus cuniculus*) often with unforseen and devastating consequences for bird and reptile populations (Courchamp et al. 2003, Nogales et al. 2004).

Mongooses (*Herpestes auropunctatus*) were introduced into the West Indies and subsequently Hawaiian Islands to control rodents and became major pests endangering ground-dwelling birds and many other species while in Hawaii, the rats moved into the trees and threatened arboreal birds (Courchamp et al 2003).

Similarly, in an attempt to control introduced European rabbits, weasels (*Mustela nivalis*), ferrets (*Mustela furo*), stoats (*Mustela erminea*) and cats were widely released in New Zealand during the later part of the 19th century and ferret release continued into the 1920s (Gibb and Williams 1994). Ferrets, cats and stoats became well established with serious and ongoing impacts on native birds and reptiles. In addition, ferrets have provided a reservoir host for bovine tuberculosis with trade and public health consequences. It appears unlikely that these predators had a significant effect on rabbit populations at the time of release, but following intensive rabbit control measures from the 1950s onwards and the release of rabbit hemorrhagic disease virus (RHDV) in 1997, predators may be important for maintaining populations at low levels (Gibb and Williams 1994, Reddiex et al. 2002, Henning et al. 2008).

In a further example, foxes (*Dusicyon griseus*) were introduced into Tierra del Fuego in 1951 in an attempt to control an estimated 30 million European rabbits, the endemic fox species *Dusicyon culpaeus* having been driven almost to extinction. It is improbable that the 24 introduced foxes and their progeny had any impact on this rabbit population. However, the introduction of myxomatosis in 1954 dramatically reduced rabbit numbers, and it has been suggested that the introduced foxes together with the preexisting *D. culpaeus* may have helped to prevent buildup of the rabbit population, although rabbits formed less than 2% of the diet of *D. griseus* (Jaksic and Yanez 1983).

There are, however, two small-scale success stories with predators. Gray wolves (*Canis lupus*) reintroduced to Yellowstone National Park in the United States have provided a high-order predator to control elk (*Cervus elaphus*) populations. Elk browsing had prevented regeneration of aspen (*Populus tremuloides*) and other trees since the 1920s when wolves were eliminated, leading to erosion and environmental degradation (Ripple and Beschta 2007). The presence of wolves has reduced grazing pressure on lowland aspen as the elk have changed their behavior or had their numbers reduced. However, this introduction could be seen more as a restoration of the previous ecology of the system rather than as biological control.

In a more direct example of biological control, European red foxes (*Vulpes vulpes*) were deliberately introduced onto two small islands in the Aleutians and appear to have eliminated populations of previously introduced arctic foxes (*Alopex lagopus*) although whether this was by competition for limited resources, direct predation or both is not clear (Bailey 1992). To prevent breeding, only male red foxes were released on one island and on the second, five vasectomised males and five females were released.

Reintroduction of high-order predators, such as dingoes (*Canis familiaris*) in Australia and leopards (*Panthera pardus*) in South Africa, has been proposed as a means of control for multiple vertebrate pest species. This could be potentially coupled with deployment of guardian dogs to protect livestock (Minnie et al. 2015, Newsome et al. 2016, Allen et al. 2019). Introduction of predator fish has also been investigated to control European carp (*Cyprinus carpio*) in Australia and the United States (Poole and Bajer 2019).

Biological Control of Vertebrate Pests Using Parasites

Parasites in the broadest sense are essentially another form of predator, albeit often nonlethal. Parasites are traditionally divided into microparasites (bacteria, viruses, protozoa, fungi) and macroparasites (helminths, arthropods). The earliest documented experiment with a microparasite for biological control is the use of chicken cholera bacteria (*Pasteurella multocida*) by Louis Pasteur in 1887 to exterminate European rabbits on an 8 Ha walled estate in France (Pasteur 1888). However, there are only three examples of vertebrate biocontrol by a microparasite on any large scale: myxomatosis in European rabbits, rabbit hemorrhagic disease in European rabbits and feline panleukopenia in cats.

There is some evidence that macroparasites may regulate vertebrate populations in the wild (Hudson et al. 1998, Tompkins and Begon 1999, Albon et al. 2002, Redpath et al. 2006). However, the interactions of parasites including macroparasites with other biotic and abiotic factors are complex (Tompkins et al. 2011). The only successful example of biological control with macroparasites is the release of two species of flea, *Spilopsyllus cuniculi* and *Xenopsylla cunicularis*, in Australia, which, by enhancing the transmission of the virus causing myxomatosis, act as indirect biological controls for European rabbits (see Chapter 13 Myxoma virus and Rabbit hemorrhagic disease virus as biological controls for rabbits).

Other attempts to use macroparasites as biological controls for vertebrates have been less successful. In one example, field, laboratory and modeling studies on the nematode *Capillaria hepatica* suggested it could act as a biological control to prevent irruptions of house mice (*Mus musculus*) in the cereal growing areas of south-eastern Australia (Singleton and McCallum 1990). However, intensive field releases of *C. hepatica* failed to show any impact on wild mouse populations (Singleton et al. 1995, Singleton and Chambers 1996).

Lungworms (*Rhabdias* spp.) have been proposed as possible biological controls for two invasive amphibians, cane toads (*Bufo marinus*) in Australia and coquis (*Eleutherodactylus coqui*) in Hawaii (Kelehear et al. 2009, Marr et al. 2010). However, preliminary studies with coquis were unpromising, and it is difficult to see how lungworms could effectively be introduced into the invading cane toad population on a landscape scale in the Australian outback.

The Natural History of Successful Biological Controls

Biological Control of Rabbits

The European rabbit originated in the Iberian Peninsula but has been spread deliberately or accidentally into the wild throughout much of Europe, Britain, North Africa, parts of Chile and Argentina, New Zealand, over 800 islands and, perhaps most famously, Australia (Flux 1994). All domestic breeds of rabbits are derived from these European rabbits. Wild rabbits from Britain were introduced into Australia in 1859 and within 50 years had spread across most of the nontropical parts of the continent constituting the most dramatic and rapid biological invasion ever documented for a mammalian pest (Flux 1994).

Rabbits are highly adaptive generalist herbivores that substantially modify their habitat (Thompson 1994, Myers et al. 1994). Their burrows and warrens enable survival under many different climatic extremes and high reproductive rates mean that rabbit populations can rapidly expand when rainfall triggers food availability (Myers et al. 1994). In dry areas of Australia, even quite low densities of rabbits (less than 1 per hectare) can prevent regeneration of shrubs and trees leading to local species extinctions, soil erosion, weed invasion and loss of other vertebrate species which are unable to survive in the modified habitat (Williams et al. 1995, Bird et al. 2012, Cooke 2012). In Australia, rabbits are a food source for two recently introduced predators, the European red fox and the feral cat (Woinarski et al. 2015). By supporting populations of foxes and cats, rabbits contribute to the endangerment and extinction of small marsupials by these predators (Dickman 1996, Kearney et al. 2018). In addition to the major ecological damage caused by rabbits, there is also a significant cost due to lost agricultural production from grazing pressure, weed invasion, soil erosion and crop destruction (Myers et al. 1994, Williams et al. 1995) as well as the economical and ecological costs of conventional controls such as habitat destruction and widespread poisoning (Williams et al. 1995, Cooke 2012).

Biological control of rabbits was successfully undertaken with myxoma virus (MYXV), the causative agent of myxomatosis, following its release in 1950. While initially highly successful, subsequent rabbit-virus coevolution led to a diminution of the effect of myxomatosis and a rebound of rabbit numbers, particularly in the low rainfall rangelands. In 1995, the escape of RHDV from a study site, introduced a second biological control for rabbits, which again substantially reduced rabbit numbers. These viruses and their impacts are described in detail in Chapter 13 Myxoma virus and Rabbit hemorrhagic disease virus as biological controls for rabbits.

Biological Control of Cats

The domestic cat (*F. catus*) adapts quickly to a feral lifestyle and has become a serious problem following its deliberate or accidental introduction onto islands (Courchamp et al. 2003, Nogales et al. 2004). Feral cats have also been associated with species loss or endangerment in Australia and New Zealand (Dickman 1996, Fitzgerald and Gibb 2001, Woinarski et al. 2015, Murphy et al. 2019).

Marion Island, the larger of the Prince Edward Islands, is a South African possession in the southern Indian Ocean (46°54′S; 37°45′E). This sub-Antarctic island, about 1900 km south east of Cape Town, has an area of 290 km² and hosts major breeding populations of penguins and seabirds (Bester et al. 2002). Five domestic cats were deliberately released in 1949 to control introduced house mice around a meteorological station (van Aarde and Skinner 1981, Bester et al. 2002).

In 1975, the cat population was estimated at 2139±290 (van Aarde and Skinner 1981, van Aarde 1984). These cats would have needed to consume over 450,000 burrowing petrels a year just to satisfy their minimum energy requirements (van Aarde and Skinner 1981). At least one species of petrel was believed to have been driven to local extinction by 1965 (Bester et al. 2002). The isolation, rugged landscape and the climate of Marion Island meant that conventional control of the cat population would be difficult and so biological control was an attractive option (Bester et al. 2002). Feline panleukopenia virus (FePV) was chosen as a biological control after a limited serological survey indicated that it was not present in the cat population (van Aarde 1984).

Feline Panleukopenia Virus

FePV (family *Parvoviridae*; sub-family *Parvovirinae*; genus *Protoparvovirus*; species *Carnivore protoparvovirus 1*) is a small (25 nm diameter) icosahedrally symmetrical, non-enveloped virus with a single-stranded DNA genome of 5200 nucleotides (Murphy et al. 1999, Stuetzer and Hartmann 2014). The virus is a natural pathogen of cats; it infects most if not all Felidae; closely related viruses infect dogs, mink and other carnivores (Steinel et al. 2001). Because it needs components of the host DNA polymerase to replicate its genome, FePV only replicates in the nucleus of actively dividing cells. This means it has a tropism for lymphoid tissue, bone marrow and the dividing epithelial cells in the intestinal crypts. Destruction of these cells leads to the clinical signs of gastroenteritis and panleukopenia (massive loss of white blood cells and platelets) seen in cats infected with FePV, (in neonates and fetal kittens the virus infects cells of the developing brain and heart rather than the gut) (Truyen et al. 2009).

Infected cats shed virus in feces, urine, saliva and vomit. Estimates of the duration of shedding range from 7 days to several months (Murphy et al. 1999, Steinel et al. 2001). Very high titers of virus (10^9 ID_{50} per g of feces) may be excreted and the virus can persist in the environment in an infectious state for many months. There are thus two modes of transmission: direct from an infected to a susceptible cat and indirect from the environment (Berthier et al. 2000).

Infection is mostly via the oropharynx with virus initially replicating in the pharyngeal lymphoid tissue, followed by a viremia that distributes virus to all tissues of the infected animal. Case fatality rates have been estimated at 20% in adult cats and 80% in kittens (Berthier et al. 2000). Cats that recover from infection are likely to be immune for life with immunity primarily due to neutralizing antibody. Kittens born to immune queens are protected for some weeks after birth by maternal antibody, delivered in the colostrum (Truyen et al. 2009). Kittens infected while still partially protected by maternal antibody may have an enhanced survival rate.

Feline Panleukopenia Virus on Marion Island

In March 1977, FePV was inoculated into 96 previously trapped cats which were released back into a population estimated at 3400 (Howell 1984, van Aarde 1984). This was at the end of the breeding season when kitten numbers and sub-adult numbers were likely to be high. The virus had a case fatality rate of approximately 50% in a small number of cat passages; both cat-passaged and cell culture-passaged virus were released (Howell 1984).

Following the release, cat sightings dropped from 102 to 18.9 per sampling period, suggesting that the virus had established and was having a significant impact on the cat population. In 1978, 45 of 57 cats sampled had antibodies to FePV (Van Rensburg et al. 1987), confirming that the virus had established and spread in the population. Subsequent serological testing indicated that it was persisting in the cat population (Van Rensburg et al. 1987).

Population density was estimated to have decreased by 65% 3 years after the release from 3400 to 600 (van Aarde 1984). As would be expected, the disease had its main impact on the young and sub-adult populations causing a marked shift in age distribution, with recovered and immune cats forming a long-lived breeding population and epizootics of the disease occurring in susceptible kittens each breeding season as protection from maternal antibodies waned. By 1982, it was considered that the rate of population decrease had slowed or stabilized.

Sea bird breeding had improved in areas where cats had been eliminated, but there was still sufficient predation to impede re-establishment of some species (Bester et al. 2002). Therefore, it was decided to proceed with the complete eradication of cats from Marion Island using conventional means. Cats were finally eliminated in 1991 following intensive hunting, trapping and poisoning campaigns (Bester et al. 2002). Ironically, cats were introduced to Marion Island as a form of biological control for the house mouse. Subsequent to the eradication of cats, the house mouse is now regarded as a threatening species (van Aarde et al. 1996).

Prospects for Biological Control of Cats

Because it does not have a prolonged carrier state, FePV is unlikely to be introduced onto islands by small founder populations. However, the success of FePV on Marion Island has not seen more use of this virus in cat control. The only documented example being on Jarvis Island in the Pacific (Rauzon 1985). On this 414 Ha atoll, 31 cats were inoculated with FePV and released into a population estimated to be <200 cats. No attempt was made to determine whether the virus was already present in the population. At least 10 of 19 marked cats survived the infection, and there was no good evidence that the virus established. Cats were subsequently eradicated from Jarvis Island using conventional means (Rauzon 1985).

Modeling studies (Courchamp and Sugihara 1999) suggest that two feline retroviruses which cause persistent infections, *feline leukaemia virus* and *feline immunodeficiency virus*, could be effective biological controls on island populations with feline leukemia virus capable of eradicating cats under some conditions. FePV was regarded as unsuitable, despite the apparent success on Marion Island, because of its short transmission time. However, this is at least partially compensated by the prolonged infectivity of FePV in the environment (Berthier et al. 2000).

Potential Biological Control Agents

Koi (Carp) Herpesvirus as a Potential Biological Control for the European Carp

The introduced European carp (*C. carpio*) has become the dominant fish species in inland waterways of south-eastern Australia. Their high fecundity, ecological adaptability and omnivorous diet have made carp a highly visible pest species, with impacts due to increased water turbidity, decreased vegetation and possibly increased frequency of algal blooms due to higher nutrient levels in waterways (Koehn 2004, McColl et al. 2007). European carp are also considered a pest species in North America and New Zealand but in other parts of the world are highly valued (Saunders et al. 2010).

Cyprinid herpesvirus 3 (CyHV-3; order: *Herpesvirale*; family: *Alloherpesviridae*; genus: *Cyprinivirus*) is a widely distributed emerging pathogen of European carp that is absent from Australia. The virus has been associated with massive die-offs of koi and common carp with estimates of mortality ranging from 70–100%. It appears to be species-specific for common and koi carp although it has been suggested that goldfish (*Carassius auratus*) may be asymptomatically infected (El-Matbouli and Soliman 2011). High virulence, species specificity, and apparently ready transmission in water, together with absence from Australia suggested that this virus may have potential as a biological control agent (McColl et al. 2007). CyHV-3 was imported into the high-security Australian Animal Health laboratory for evaluation (McColl et al. 2007, Saunders et al. 2010, McColl et al. 2016). Larval carp appear highly susceptible with mortality rates of 96–100% following infection. Adults had a lower mortality rate (Saunders et al. 2010). Plans have been developed for a virus release in 2019–2020; however, there are still questions being raised about the strategy (Kopf et al. 2017, Lighten and van Oosterhout 2017, Marshall et al. 2018), and it is likely that any release will be delayed. It is considered necessary that biological control would be combined with conventional control measures and possibly future biotechnological controls (Thresher et al. 2014, McColl et al. 2018).

Future Biocontrols and Biotechnology for Control of Vertebrate Pests

Biotechnology has the potential to engineer existing parasites to enhance virulence, modify antigenicity or to deliver other antigens or regulatory molecules such as interfering RNA. Any such modification for vertebrate pest control is likely to be controversial particularly when a vertebrate pest in one country, such as the rabbit in Australia, is a valued keystone species in its natural range and because of the risk of altering the host range of the parasite. More recently, genetic biocontrols such as gene drives, based around DNA editing CRISPR/Cas9-type and similar systems, which would directly modify pest species have been investigated as potential controls for vertebrate pest species.

Virally Vectored Immunocontraception

Biological controls that inhibit reproduction or development are an attractive target for pest control. Virally vectored immunocontraception (VVIC) aimed to use a genetically engineered transmissible agent to deliver an immunocontraceptive antigen to a pest species. Infection with the recombinant organism would stimulate an immune response to the immunocontraceptive antigen which would block fertility. Potential antigens include components of sperm, the zona pellucida surrounding the egg or peptide hormones such as gonadotrophin-releasing hormone. Direct injection of such antigens is already used in some wildlife population control programs (Naz and Saver 2016). However, direct injection is not feasible for control of widespread invasive species such as rabbits, mice and foxes.

In Australia, extensive studies were undertaken to develop VVIC for three key invasive species: house mice, European rabbits and European red foxes. However, none of the technologies developed were sufficiently effective for field use (Hardy et al. 2006). While mice could be rendered sterile by inoculation with recombinant murine cytomegalovirus, the virus did not transmit effectively from mouse to mouse (Redwood et al. 2007). In rabbits, >90% of females could be rendered infertile following inoculation with recombinant MYXV, but in around 50% of cases, infertility was short-lived (Mackenzie et al. 2006, van Leeuwen and Kerr 2007). Field experiments using surgical sterilization of female rabbits indicated that levels of immunocontraception efficacy >80% would be required to have an impact on the population and reduce ecological damage (Twigg et al. 2000, Williams et al. 2007). Despite considerable effort, no effective product was developed for foxes (Reubel et al. 2005, Strive et al. 2006, Strive et al. 2007). All the recombinant viruses tested expressed zona pellucida proteins as antigens. The use of possibly more effective antigens such as gonadotropin-releasing hormone was not explored.

Prospects for Virally Vectored Immunocontraception

Although potentially a means of biological control, immunocontraception vectored by a recombinant organism has efficacy and safety hurdles to overcome before it could be a reality. These include: (1) an immunocontraceptive antigen must stimulate a long-lasting immune response that cross-reacts with a self-antigen to cause sustained infertility at least in females. (2) The recombinant organism must be able to establish in a population, transmit and be maintained in the population in competition with field strains of the organism or be able to be regularly reintroduced, possibly in a bait. As demonstrated with murine cytomegalovirus, recombinant viruses may not transmit effectively. (3) The immune response should not need boosting. (4) The vector organism and preferably the antigen should be species-specific. (5) The risk of the organism jumping species by mutation or recombination should be minimal.

These scientific and technical conditions have not yet been satisfied in any species. Beyond the technical issues, there are national and international social, environmental and political issues that would need to be addressed before vector-delivered fertility control could be a reality.

Biological Controls for Cane Toads

The cane toad (*Rhinella marinus*) is native to South America and was introduced to Australia in 1935 to control beetles in sugarcane crops (Shine 2010). While the toads failed as beetle-biocontrol agents, they proved to be excellent invaders and have spread from their original release in coastal northern Queensland south to NSW and across the tropical north of Australia as far as Western Australia and are continuing to expand towards the Indian Ocean (Shine 2010).

All life stages of the cane toad are toxic to predators due to compounds collectively referred to as bufadienolides. As a result, cane toad invasions have led to declines in populations of native predator species, such as snakes, freshwater crocodiles, goannas and quolls at the invasion front although the long-term impact is more difficult to quantify (Shine 2010).

Between 1990 and 1993, an extensive search for natural pathogens was undertaken in the toad's native range in Venezuela. A number of virus isolates from cane toads and a frog were made, all belonging to the family *Iridoviridae*, genus *Ranavirus* and considered to represent a single species. While these viruses killed cane toad tadpoles, they were also lethal to at least one Australian native frog species and considered unsuitable as biocontrol agents (Shanmuganathan et al. 2010). More recently, metagenomic studies have identified a novel rhinovirus and an endogenous retrovirus in cane toads from Australia (Russo et al. 2018).

In a biotechnological strategy, it was proposed to use a recombinant virus to deliver cane toad adult-specific proteins into tadpoles. This approach was based on the hypothesis that expression of adult-specific proteins in tadpoles would create an immune response, since tadpoles are not immunologically tolerant to adult proteins. By selecting an antigen that was critical for metamorphosis, it was proposed that maturation would be inhibited. Although proof of concept had been shown many years earlier in tadpoles of the American bullfrog immunized with adult hemoglobin (Maniatis et al. 1969), this was unsuccessful in cane toads and the work has been discontinued (Shanmuganathan et al. 2010).

To date, no biocontrol is available to effectively reduce cane toad populations or halt their spread across the Australian continent.

Genetic Biocontrol of Invasive Species

The use of genetic biocontrols such as gene drives to spread an allele through a population is receiving considerable attention (Esvelt et al. 2014). The basic concept is to use a genetic construct that copies itself into homologous chromosomes, thus creating super-Mendelian inheritance whereby more than 50% of the offspring inherit the drive element. An element that disabled a key gene involved in reproduction or caused sex bias, for example, by turning all individuals into phenotypic males, could be introduced into the population causing population crashes. The advent of CRISPR/Cas9 and other bacterial nuclease/guide systems as gene editing tools has turned this from a largely theoretical approach to a potentially practical means of controlling populations. Genetic biocontrol of pest populations would have the advantages of being self-disseminating and strictly species-specific (since only transmitted by sexual reproduction). In addition, genetic biocontrol should be much more humane than conventional lethal controls or pathogens.

Proof of concept has been demonstrated in mosquito populations under containment conditions (Kyrou et al. 2018). In vertebrates, there have been proposals to use gene drives to control invasive rodent species on islands and for invasive fish (Thresher et al. 2014, Leitschuh et al. 2018). Modeling studies have examined the opportunities and constraints, particularly the selection of resistant alleles, of gene drives for rodent control (Prowse et al. 2017). A germline-expressed gene drive has been demonstrated in laboratory mice; however, super-Mendelian inheritance occurred only in female mice (Grunwald et al. 2019). Although currently in its infancy, it is likely that this technology will rapidly improve. In the meantime, there will be considerable debate on the risks associated with driving a sterilizing gene into a population and whether these can be satisfactorily managed (Esvelt et al. 2014, National Academy of Sciences 2016).

Conclusions

A number of conclusions can be drawn from the history of biological control of vertebrate pests that should inform the search for and use of biological control agents.

1. It is challenging to find effective biological controls for vertebrates and those that have been most successful, myxomatosis and rabbit hemorrhagic disease, were emerging diseases with extreme case fatality rates not pre-existing diseases of the European rabbit in its natural range.

2. Searching for biological control agents in the original home range of the vertebrate pest, as has been so successful for plant and insect pests, may be unsuccessful for vertebrate pests because it could well be that predation or resource limitation was more important in population control than parasitism particularly by a single organism (Tompkins et al. 2011). This was exemplified by the search for cane toad pathogens in Venezuela. FePV is a natural pathogen of cats but has been successfully used on only one island and may not be species-specific.

3. Laboratory trials may be poor predictors of success in the field.

4. As the early experience with myxomatosis in Australia indicates, it may be difficult to establish a novel pathogen in a population, particularly if the epidemiology of the pathogen is not well understood.

5. The initial impact of a biological control is unlikely to be sustained.

   At the very least, the pathogen may undergo local extinction when host populations fall to low numbers. For rapidly reproducing invasive pest species, this can allow local recovery of populations. Unlike insects and plants, vertebrates surviving infection develop adaptive immunity that can protect from subsequent reinfection. This has two impacts: it provides an immune pool of breeding animals, and passive antibody transfer may protect their offspring for some weeks after birth and hence alter the impact and timing of epizootics. At some level, there will likely be selection for enhanced resistance to the pathogen and, depending on the life history of the pathogen, there may be selection for more attenuated strains of the pathogen.

6. Biological controls should not be seen as "silver bullets" that remove the need for ongoing integrated pest management strategies. However, successful biological controls such as myxomatosis and rabbit hemorrhagic disease (RHD) can significantly suppress populations on continental scales.

7. The introduction of novel high-order predators to control populations has generally not been successful due to the low species specificity of the predator and the need to reduce pest numbers to very low levels before predators can keep them in check.

8. Biotechnological approaches to vertebrate biocontrol have so far been unsuccessful. These have a lot of hurdles to overcome, not least scientific and technical, but also regulatory and social. However, given the difficulty in finding natural biocontrol agents, biotechnology should not be ignored and gene biocontrol technologies offer potentially powerful new means of controlling populations if efficacy and safety can be achieved.

Bibliography

Albon, S. D., A. Stein, R. J. Irvine, R. Langvatn, E. Ropstad et al. 2002. The role of parasites in the dynamics of a reindeer population. *Proc. Roy. Soc. Lond. B.* 269:1625–1632.

Allen, B. L., L. R. Allen, G. Ballard, M. Drouilly, P. J. S. Fleming et al. Animal welfare considerations for using large carnivores and guardian dogs as vertebrate biocontrol tools against other animals. *Conserv. Biol.* 232:258–270.

Bailey, E. P. 1992. Red foxes, *Vulpes vulpes*, as biological control agents for introduced Arctic foxes, *Alopex lagopus* on Alaskan islands. *Can. Field. Nat.* 106:200–205.

Berthier, K., M. Langlais, P. Auger, and D. Pontier. 2000. Dynamics of a feline virus with two transmission modes within exponentially growing host populations. *Proc. Roy. Soc. Lond. B* 267:2049–2056.

Bester, M. N., J. P. Bloomer, R. J. van Aarde, B. H. Erasmus, P. J. J. van Rensburg et al. 2002. A review of the successful eradication of feral cats from sub-Antarctic Marion Island, Southern Indian Ocean. *S. Af. J. Wildl. Res.* 32:65–73.

Bird, P., Mutze, G., Peacock, D., Jennings, S. 2012. Damage caused by low-density exotic herbivore populations: the impact of introduced European rabbits on marsupial herbivores and *Allocasuarina* and *Bursaria* seedling survival in Australian coastal shrubland. *Biol Invasions* 14: 743–755.

Cooke, B. D. 2012. Rabbits: manageable environmental pests or participants in new Australian ecosystems? *Wildl. Res.* 39:279–289.

Courchamp, F., J. L. Chapuis, and M. Pascal. 2003. Mammal invaders on islands: impact, control and control impact. *Biol. Rev.* 78:347–383.

Courchamp, F., and G. Sugihara. 1999. Modeling the biological control of an alien predator to protect island species from extinction. *Ecol. Appl.* 9:112–123.

Dickman, C. R. 1996. Impact of exotic generalist predators on the native fauna of Australia. *Wildl. Biol.* 2:185–195.

Driscoll, C. A., M. Menotti-Raymond, A. L. Roca, K. Hupe, W. E. Johnson et al. 2007. The near eastern origin of cat domestication. *Science* 317:519–523.

El-Matbouli, M., and H. Soliman. 2011. Transmission of Cyprinid herpesvirus-3 (CyHV-3) from goldfish to naive common carp by cohabitation. *Res. Vet. Sci.* 90:536–539.

Esvelt, K. M., A. L. Smidler, F. Catteruccia, and G. M. Church. 2014. Concerning RNA-guided gene drives for the alteration of wild populations. *eLife* 3:e03401.

Fitzgerald, B. M., and J. A. Gibb. 2001. Introduced mammals in a New Zealand forest: long-term research in the Orongorongo Valley. *Biol. Cons.* 90:97–108.

Flux, J. E. C. 1994. World distribution. In *The European Rabbit. The History and Biology of a Successful Colonizer*, edited by H. V. Thompson and C. M. King, 8–21. Oxford: Oxford University Press.

Gibb, J. A., and J. M. Williams. 1994. The rabbit in New Zealand. In *The European Rabbit. The History and Biology of a Successful Colonizer*, edited by H. V. Thompson and C. M. King, 158–204. Oxford: Oxford University Press.

Grunwald, H. A., V. M. Gantz, G. Poplawski, X-R. S. Xu, E. Bier et al. 2019. Super-Mendelian inheritance mediated by CRISPR-Cas9 in the female mouse germline. *Nature* 566:105–109.

Hardy, C. M., L. A. Hinds, P. J. Kerr, A. J. Redwood, G. R. Shellam et al. 2006. Biological control of pest animals using virally-vectored immunocontraceptive vaccines. *J. Reprod. Immunol.* 71:102–111.

Henning, J., D. U. Pfeiffer, P. R. Davies, M. A. Stevenson, and J. Meers. 2008. Mortality patterns over 3 years in a sparse population of wild rabbits (*Oryctolagus cuniculus*) in New Zealand, with an emphasis on rabbit haemorrhagic disease (RHD). *Eur. J. Wild. Res.* 54:619–626.

Howell, P. G. 1984. An evaluation of the biological control of the feral cat *Felis catus* (Linnaeus, 1758). *Acta Zool. Fennica* 172:111–113.

Hudson, P. J., A. P. Dobson, and D. Newborn. 1998. Prevention of population cycles by parasite removal. *Science* 282:2256–2258.

Jaksic, F. M., and J. L. Yanez. 1983. Rabbit and fox introductions in Tierra del Fuego: history and assessment of the attempts at biological control of the rabbit infestation. *Biol. Conserv.* 26:367–374.

Kearney, S. G., J. Cawardine, A. E. Reside, D. O. Fisher, M. Maron et al. 2018. The threats to Australia's imperilled species and implications for a national conservation response. *Pac. Conserv. Biol.* https://doi.org/10.1071/PC18024.

Kelehear, C., J. K. Webb, and R. Shine. 2009. *Rhabdias pseudosphaerocephala* infection in *Bufo marinus*: lung nematodes reduce viability of metamorph cane toads. *Parasitology* 136:919–927.

Koehn, J. D. 2004. Carp (*Cyprinus carpio*) as a powerful invader in Australian waterways. *Freshwater Biol.* 49:882–894.

Kopf, R. K., D. G. Nimmo, P. Humphries, L. J. Baumgartner, M. Bode et al. 2017. Confronting the risks of large-scale invasive species control. *Nature Ecol. Evol.* 1:0172.

Kyrou, K., A. M. Hammond, R. Galizi, N. Kranjc, A. Burt et al. 2018. A CRISPR-Cas9 gene drive targeting *doublesex* causes complete population suppression in caged *Anopheles gambiae* mosquitoes. *Nat. Biotech.* 36:1062–1066.

Leitschuh, C. M., D. Kanavy, G. A. Backus, R. X. Valdez, M. Serr et al. 2018. Developing gene drive technologies to eradicate invasive rodents from islands. *Journal of Responsible Innovation* 5 (Sup 1):S121–S138.

Lighten, J., and C. van Oosterhout. 2017. Biocontrol of common carp in Australia poses risks to biosecurity. *Nat. Ecol. Evol.* 1:0087.

Mackenzie, S M., E. A. McLaughlin, H. D. Perkins, N. French, T. Sutherland, et al. 2006. The immunocontraceptive effects on female rabbits (*Oryctolagus cuniculus*) infected with recombinant myxoma virus expressing rabbit ZP2 and ZP3. *Biol Reprod* 74:511–521.

Maniatis, G. M., L. A. Steiner, and V. M. Ingram. 1969. Tadpole antibodies against frog hemoglobin and their effect on development. *Science* 165:67–69.

Marr, S. R., S. A. Johnson, A. H. Hara, and M. E. McGarrity. 2010. Preliminary evaluation of the potential of the helminth parasite *Rhabdias elegans* as a biological control agent for invasive Puerto Rican coquis (*Eleutherodactylus coqui*) in Hawaii. *Biol. Control.* 54:69–74.

Marshall, J., A. J. Davison, R. K. Kopf, M. Boutier, P. Stevenson et al. 2018. Biocontrol of invasive carp: risks abound. *Science* 359:877.

McColl, K. A., A. Sunarto, and E. C. Holmes. 2016. Cyprinid herpesvirus 3 and its evolutionary future as a biological control for carp in Australia. *Virol. J.* 13:206.

McColl, K. A., A. Sunarto, and M. J. Neave. 2018. Biocontrol of carp: more than just a herpesvirus. *Front. Microbiol.* 9:2288.

McColl, K. A., A. Sunarto, L. M. Williams, and St-J. Crane M. 2007. Koi Herpes virus: dreaded pathogen or white knight? *Aquaculture Health* 2007 (9):4–6.

Minnie, L., A. F. Boshoff, and G. I. H. Kerley. 2015. Vegetation type influences livestock predation by leopards: implications for conservation in agro-ecosystems. *Afr J Wildlife Res* 45:204–214.

Murphy, F. A., E. P. J. Gibbs, M. C. Horzinek, and M. J. Studdert. 1999. *Veterinary Virology*. 3 ed. San Diego: Academic Press.

Murphy, B. P., L-A. Woolley, H. M. Geyle, S. M. Legge, R. Palme et al. 2019. Introduced cats (*Felis catus*) eating a continental fauna: the number of mammals killed in Australia. *Biol Conservation* 237:28–40.

Myers, K., I. Parer, D. Wood, and B. D. Cooke. 1994. The rabbit in Australia. In The European *Rabbit. The History* and *Biology* of a *Successful Colonizer*, edited by H. V. Thompson and C. M. King, 108–157. Oxford: Oxford University Press.

National Academy of Sciences, Engineering & Medicine. 2016. Gene drives on the horizon: advancing science, navigating uncertainty and aligning research with public values. Washington, DC: National Academies Press.

Naz, R. J., and A. E. Saver. 2016. Immunocontraception for animals; current status and future perspective. *Am. J. Reprod. Immunol.* 75:426–439.

Newsome, T. M., G-A. Ballard, M. S. Crowther, J. A. Dellinger, P. J. S. Fleming et al. 2016. Resolving the value of the dingo in ecological restoration. *Restoration Ecology* 23:201–208.

Nogales, M., A. Martin, B. R. Tershy, C. J. Donlan, D. Veitch et al. 2004. A review of feral cat eradication on islands. *Conserv. Biol.* 18:310–319.

Pasteur, L. 1888. Sur la destruction des lapins en Australie et dans la Nouvelle-Zelande. *Annales de l'Institut de Pasteur (Paris)* 2:1–8.

Pech, R. P., A. R. E. Sinclair, A. E. Newsome, and P. C. Catling. 1992. Limits to predator regulation of rabbits in Australia: evidence from predator removal experiments. *Oecologia* 89:102–112.

Pickard, A. R., and W. V. Holt. 2007. Contraception in wildlife. *J. Fam. Plann. Reprod. Health. Care.* 33:48–52.

Poole, J. R., and P. G. Bajer. 2019. A small native predator reduces reproductive success of a large invasive fish as revealed by whole-lake experiments. *PLoS ONE* 14:e0214009.

Prowse, T. A. A., P. Cassey, J. V. Ross, C. Pfitzner, T. A. Wittmann et al. 2017. Dodging silver bullets: good CRISPR gene-drive design is critical for eradicating exotic vertebrates. *Proc. R. Soc. Lond. B* 284:20170799.

Rauzon, M. J. 1985. Feral cats on Jarvis Island: their effects and their eradication. *Atoll Research Bulletin* No. 282:1–30.

Reddiex, B., G. J. Hickling, G. L. Norbury, and C. M. Frampton. 2002. Effects of predation and rabbit haemorrhagic disease on population dynamics of rabbits (*Oryctolagus cuniculus*) in North Canterbury, New Zealand. *Wildl. Res.* 29:627–633.

Redpath, S. M., F. Mougeot, F. M. Leckie, D. A. Elston, and P. J. Hudson. 2006. Testing the role of parasites in driving the cyclic population dynamics of a gamebird. *Ecol. Lett.* 9:410–418.

Redwood, A. J., L. M. Smith, M. L. Lloyd, L. A. Hinds, C. M. Hardy et al. 2007. Prospects for virally-vectored immunocontraception in the control of wild house mice (*Mus domesticus*). *Wildl. Res.* 34:530–539.

Reubel, G. H., S. Beaton, D. Venables, J. Pekin, J. Wright et al. 2005. Experimental inoculation of European red foxes with recombinant vaccinia viruses expressing zona pellucida C proteins. *Vaccine* 23:4417–4426.

Ripple, W. J., and R. L. Beschta. 2007. Restoring Yellowstone's aspen with wolves. *Biol. Conserv.* 138:514–519.

Russo, A. G., J-S. Eden, D. Enosi Tuipulotu, M. Shi, D. Selechnik et al. 2018. Viral discovery in the invasive Australian cane toad (*Rhinella marina*) using metatranscriptomic and genomic approaches. *J. Virol.* 92:e00768-18.

Saunders, G., B. Cooke, K. McColl, R. Shine, and T. Peacock. 2010. Modern approaches for the biological control of vertebrate pests: an Australian perspective. *Biol. Control* 52:288–295.

Shanmuganathan, T., J. Pallister, S. Doody, H. McCallum, T. Robinson et al. 2010. Biological control of the cane toad in Australia: a review. *Animal Conserv.* 13:16–23.

Shine, R. 2010. The ecological impact of invasive cane toads (*Bufo marinus*) in Australia. *Q. Rev. Biol.* 85:253–291.

Singleton, G.R., and L. K. Chambers. 1996. A manipulative field experiment to examine the effect of *Capillaria hepatica* (Nematoda) on wild mouse populations in southern Australia. *Int. J. Parasitol.* 26:383–398.

Singleton, G.R., L. K. Chambers, and D. M. Spratt. 1995. An experimental field study to examine whether *Capillaria hepatica* (Nematoda) can limit house mouse populations in eastern Australia. *Wildl. Res.* 22:31–53.

Singleton, G.R., and H. I. McCallum. 1990. The potential of *Capillaria hepatica* to control mouse plagues. *Parasitol. Today* 6:190–193.

Steinel, A., C. R. Parrish, M. E. BLoom, and U. Truyen. 2001. Parvovirus infections in wild carnivores. *J. Wildl. Dis.* 37: 594–607.

Strive, T., C. M. Hardy, N. French, J. D. Wright, N. Nagaraja et al. 2006. Development of canine herpesvirus based antifertility vaccines for foxes using bacterial artificial chromosomes. *Vaccine* 24:980–988.

Strive, T., C. M. Hardy, and G. H. Reubel. 2007. Prospects for immunocontraception in the European red fox (*Vulpes vulpes*). *Wildl. Res.* 34:523–539.

Stuetzer, B., and K. Hartmann. 2014. Feline parvovirus infection and associated diseases. *Vet J.* 201:150–155.

Thompson, H. V. 1994. The rabbit in Britain. In The European *Rabbit*. The *History* and *Biology* of a *Successful Colonizer*, edited by H. V. Thompson and C. M. King, 64–107. Oxford: Oxford University Press.

Thresher, R., J. van de Kamp, G. Campbell, P. Grewe, M. Canning et al. 2014. Sex-ratio-biasing constructs for the control of invasive lower vertebrates. *Nat. Biotech.* 32:424–427.

Tompkins, D. M., and M. Begon. 1999. Parasites can regulate wildlife populations. *Parasitol. Today* 15:311–313.

Tompkins, D. M., A. M. Dunn, M. J. Smith, and S. Telfer. 2011. Wildlife diseases: from individuals to ecosystems. *J. Anim. Ecol.* 80:19–38.

Truyen, U., D. Addie, S. Belak, C. Boucraut-Baralon, H. Egberink et al. 2009. Feline Panleukopenia. ABCD guidelines on prevention and management. *J. Fel. Med. Surg.* 11:538–546.

Twigg, L. E., T. J. Lowe, G. R. Martin, A. G. Wheeler, G. S. Gray et al. 2000. Effects of surgically imposed sterility on free-ranging rabbit populations. *J. App. Ecol.* 37:16–39.

van Aarde, R. J. 1984. Population biology and the control of feral cats on Marion Island. *Acta Zool. Fennica* 172:107–110.

van Aarde, R. J., S. Ferreira, T. Wassenaar, and D. G. Erasmus. 1996. With the cats away the mice may play. *S. Af. J. Sci.* 92:357–358.

van Aarde, R. J., and J. D. Skinner. 1981. The feral cat population at Marion Island: characteristics, colonization and control. Colloque sur les Ecosystemes Subantarctiques. C.N.F.R.A. no. 51.

van Leeuwen, B. H., and P. J. Kerr. 2007. Prospects for fertility control in the European rabbit (*Oryctolagus cuniculus*) using myxoma virus-vectored immunocontraception. *Wildl. Res.* 34:511–522.

Van Rensburg, P. J. J., J. D. Skinner, and R. J. Van Aarde. 1987. Effects of feline panleucopaenia on the population characteristics of feral cats on Marion Island. *J. App. Ecol.* 24:63–73.

Williams, C. K., C. C. Davey, R. J. Moore, L. A. Hinds, L. E. Silvers et al. 2007. Populations responses to sterility imposed on female European rabbits. *J. Appl. Ecol.* 44:291–301.

Williams, K., I. Parer, B. Coman, J. Burley, and M. Braysher. 1995. *Managing Vertebrate Pests: Rabbits.* Canberra: Bureau of Resource Sciences. Australian Government Publishing Services.

Woinarski, J. C. Z., A. A. Burbidge, and P. L. Harrison. 2015. Ongoing unraveling of a continental fauna: decline and extinction of Australian mammals since European settlement. *Proc. Natl. Acad. Sci. USA.* 112:4531–4540.

36

Biological Controls

Heikki Hokkanen

Introduction

Biological control of pests has been actively practiced for the control of pests, weeds, and plant diseases for more than 100 years, and it has had some 150 spectacular successes,[1] which in economic terms have been just as impressive as in ecological terms: the calculated return for investment is 32:1, while for other control methods the ratio is around 2.5:1.[2,3] However, the obtained successes are only the tip of the iceberg of all the work carried out in the field. To date, more than 6000 introductions of alien natural enemies have been carried out, worldwide.[4] It is estimated that only about 35% of all introduced biocontrol agents have become ecologically established in the target ecosystem, and only 60% of these have provided any economic or biocontrol success.[3,5] Of all the individual biocontrol projects, only 16% have resulted in complete control of the target pest.[6] A major ecological and economic challenge is to improve the ratio of successes in biological control, while retaining the excellent safety record of this approach to pest control.

General Principles

While it has been shown that biological control can be effective in any climate, ecosystem, and crop, the factors determining success or failure remain largely unknown, and often are economic rather than ecological in nature.[1] Very few general principles to improve the efficacy and predictability of biological control have emerged; these include better ecological background knowledge, genetic improvement (in particular, genetic engineering) of biocontrol agents, and the utilization of new ecological associations in selecting the biocontrol agents.[7] The genetic engineering of biocontrol agents, especially insects, is still in its infancy and cannot be expected to improve the success ratios in the foreseeable future. In contrast, the new association principle has—usually unknowingly—been used for a long time, and is increasingly employed to find more effective natural enemies for current biological control programs.

New Associations

The standard biological control principle is to reestablish the ecological balance between an exotic pest and its natural enemies occurring in their country of origin (the "old association approach").[2,3] It has been argued, however, that this is an inefficient way of practicing biological control, because due to an evolved long-term equilibrium between the pest and the natural enemy, the control agent only seldom is very efficient.[6] To find more effective enemies one should search among agents that do not share an evolutionary history with the target pest (the "new association" approach). Such natural enemies can be found, for example, for the target pest in areas where the pest has been introduced only recently, or among enemies attacking related species in other geographical areas.[6]

Evidence for Improved Efficacy

Analysis of past biocontrol successes and failures have indicated that when employing the new association principle it is possible to increase the success ratio by at least 75%.[6] More detailed studies showed that some natural enemy groups may be particularly attractive as new association agents (Table 1). Such analyses are, however, often confounded by the fact that new association agents seldom have been considered as the primary choice in biological control, and consequently, usually five to seven old association agents are introduced before a new association agent is tried. In addition, on average much greater numbers (two-to fourfold) of old association agents are normally introduced (Table 1), which further increases the probability of biocontrol success, and biases the analyses against new association agents. Therefore, the estimate for improving the success ratio appears to be conservative.

Several spectacular, well-documented biocontrol successes that have employed new association control agents are known, and these include the complete control of serious pests such as the sugarcane borer *Diatraea saccharalis* in the Caribbean, coconut spike moth *Levuana iridescens* in Fiji, southern green stink bug *Nezara viridula* in Hawaii, the moth *Oxydia trychiata* in Colombia, and several scale insect species in California, Greece, and Australia.[7] Further, more detailed examples will be given below on new research with good prospects of success utilizing this approach.

TABLE 1 Comparisons of Biological Control Introductions with Old and New Association Control Agents Utilizing Tachinidae, Braconidae and Eulopidae

	Proportion (%) of Successes of All Cases (Introductions)		Total Number of Cases		Bias in the Release Numbers[a]
	Old	New	Old	New	
Tachinidae	10.9	17.1	92	41	3.8-fold
Braconidae	17.2	14.4	169	97	1.6-fold
Eulopidae	28.6	35.7	56	28	1.7-fold
Overall	17.4	18.7	317	166	

Source: Hokkanen, H. M. T., unpublished data.

[a] Indicates how many more individuals on average of old association agents were released in the introduction projects, compared with new association agents. In the case of Tachinidae and Braconidae the mean number of released new association agents was below 5000 individuals, which is considered to be the necessary number to ensure a fair chance for the natural enemies to establish themselves.

Recent Cases Employing New Associations

Eurasian Watermilfoil

The Eurasian watermilfoil (*Myriophyllum spicatum*) was introduced into North America several decades, possibly 100 years, ago. It grows rapidly, forms a dense canopy on the water surface, and often interferes with recreation, inhibits water flow, and impedes navigation. Herbicides and mechanical harvesting have been used to control infestations, costing $150–$2000 per acres annually in Minnesota.[8]

Sometimes naturally occurring declines of the watermilfoil have been observed. The main causal agent proved to be a native beetle *Euhrychiopsis lecontei*, the milfoil weevil, which subsequently has shown control potential in controlled field experiments. The weevil is a specialist herbivore of watermilfoils, but prefers the Eurasian to its native host, the northern watermilfoil (*M. sibiricum*). Research is in progress to use the milfoil weevil effectively as a biocontrol agent against the Eurasian watermilfoil in North America.[8]

Lantana

Lantana camara is a serious weed of Mexican or Caribbean origin, affecting cropping lands and forest areas in 47 countries. Lantana was the focus of the first weed biocontrol effort in history (1902), and there is an enormous literature on Lantana biocontrol. Several complexes of herbivores have been credited for exerting some degree of biocontrol of the weed (e.g., in Hawaii), many employing new association agents jointly with old association agents. Latest research gives data on the good efficacy and release in Australia of the moth *Ectaga garcia* originating from South America, where it feeds on the related weed *Lantana montevidensis*.[9]

Triffid (Siam) Weed

The triffid weed (*Chromolaena odorata*) is a perennial shrub native of tropical America. In recent decades it has become a serious pest of humid tropics around the world.[10] It spreads rapidly in lands used for forestry, pasture, and plantation crops and can reach a height of three meters in open situations and up to eight meters in forests. For more than two decades the triffid weed has been the subject of intensive research as a target for biological control. However, so far all attempts at biocontrol of *C. odorata* have failed. Recently the new association biological control agent, arctiid moth *Pareuchaetes aurata aurata* collected from *C. jujuensis* in South America, was considered as more promising than the related moth *P. pseudoinsulata*, an old association control agent previously thought of as one of the best biocontrol candidates.[10]

Southern Green Stink Bug

The biological control of the southern green stink bug (the green vegetable bug) (*Nezara viridula*) in Australia, New Zealand, and Hawaii has been heralded as a landmark example of classical biological control.[11] An egg parasitoid—old association agent—*Trissolcus basalis* and a tachinid fly—new association agent—*Trichopoda pennipes* have jointly provided these successes. Control by the fly has been considered as relatively more important, and indeed, in Australia where the fly has failed to establish, the control is poor and the bug remains a serious pest. Currently in Australia another new association tachinid fly, *Trichopoda giacomellii,* is being released after research showed it has excellent potential for control.[12]

Citrus Leafminer

Citrus agroecosystems have numerous potentially damaging pests often maintained under substantial to complete biological control by both old and new association agents. The citrus leafminer *Phyllocnistis citrella*, native to Asia, has spread rapidly throughout the citrus growing areas of the world in recent years.[13] It arrived in Florida in 1993 and in less than one year invaded and colonized the entire state. An old association parasitic wasp *Ageniaspis citricola* was introduced in 1994, and after establishment it has held the pest under control with significant help from native parasitoids such as *Pnigalio minio* (new association agent).[13] In some other areas native parasitoids similarly have shown significant control effect on the citrus leafminer (e.g., in Italy). This example illustrates well the fact that invading species often do not become pests, because effective local natural enemies keep them in check.

Tarnished Plant Bug

An ongoing study in the United States has identified as the most important parasitoid of the native pest *Lygus lineolaris*, the tarnished plant bug, the exotic species *Peristenus digoneutis*, originally introduced for the control of related introduced mirid plant bugs.[14] This example serves well to point out the importance of native pests, which in most if not all areas form the majority of all pest species. As old association biological control agents seldom can be utilized for the control of native pests, their biocontrol by introduced natural enemies has attracted relatively little attention and, indeed, only three decades ago was considered an impossible task. Several recent examples, usually utilizing new association control agents, show that biological control can work against native pests just as well as against exotic ones.

Future Prospects

Compared with chemical control, the success rates of biological control are outstanding. While only about one out of 15,000 tested chemicals ends up as a chemical pesticide meeting the requirements of efficacy and safety, approximately one out of seven introductions of natural enemies has been successful using old associations.[15] Using new association control agents this rate could still be increased to about one out of four, while the array of potential natural enemies is also substantially larger providing a wider choice. In addition, the potential uses for natural enemy introductions are broadened to include the control of native pests.

A major concern with respect to all biological control introductions is the question of nontarget safety. Biological control has an excellent record of safety[3,16] and it covers the new association agents as well: there have been some 1500–2000 introductions already (out of 6000) that have involved new association agents.[7] Those extremely few cases where a negative nontarget effect has been suspected as a result of biological control, all involve old association agents; therefore it is clear that new associations can safely be used to help obtain biological control successes at an increasing rate.

References

1. Hokkanen, H.M.T. Success in classical biological control. CRC Crit. Rev. Plant Sci. **1985**, *3*, 35–72.
2. Cullen, J.M.; Whitten, M.J. Economics of Classical Biological Control: A Research Perspective. In *Biological Control: Benefits and Risks*; Hokkanen, H.M.T., Lynch, J.M., Eds.; Cambridge University Press: Cambridge, U.K., 1995; 270–276.
3. *Biological Control: Benefits and Risks*; Lynch, J.M., Hokkanen, H.M.T., Eds.; Cambridge University Press: Cambridge, U.K., 1995.
4. Waage, J. In *Agendas, Aliens and Agriculture*; Global Biocontrol in the Post UNCED Era, Cornell Community Conference on Biological Control, http://www.nysaes.cornell.edu/ent/bcconf/talks/waage.html (accessed Jan 5, 1999).

5. Hokkanen, H.M.T. Pest Management, Biological Control. In *Encyclopedia of Agricultural Science*; Academic Press, Inc.: San Diego, 1994; 3, 155–167.

6. Hokkanen, H.M.T.; Pimentel, D. New associations in biological control: theory and practice. Can. Entomol. **1989**, *121,* 829–840.

7. Hokkanen, H.M.T. New Approaches in Biological Control. In *CRC Handbook of Pest Management in Agriculture*, 2nd Ed., Pimentel, D., Ed.; CRC Press: Boca Raton, FL, 1991; II, 185–198.

8. Newman, R.M. Biological Control of Eurasian Watermilfoil. http://www.fw.umn.edu/research/milfoil/milfoilbc.html (accessed Feb 5, 1999).

9. Day, M.D.; Wilson, B.W.; Latimer, K.J. The life history and host range of *Ectaga garcia,* a biological control agent for *Lantana camara* and *L. montevidensis* in Australia. BioControl **1998**, *43,* 325–338.

10. Kluge, R.L.; Caldwell, P.M. The biology and host specificity of *Pareuchaetes aurata aurata* (Lepidoptera: Arctiidae), a "New Association" biological control agent for *Chromolaena odorata* (Compositae). Bull. Entomol. Res. **1993**, *83,* 87–94.

11. Caltagirone, L.E. Landmark examples in classical biological control. Annu. Rev. Entomol. **1981**, *26,* 213–232.

12. Coombs, M. *Biological Control of Green Vegetable Bug in Australia and PNG*; Pest Management Current Programs and Projects. http://www.ento.csiro.au/research/pestmgmt/pmp16.htm (accessed Oct 5, 1999).

13. Timmer, L.W. *Citrus Leafminer Proves to be an IPM Success*; IPM Florida, 1996 *Winter.* http://www.ias.ufl.edu/~FAIRSWEB/IPM/IPMFL/v2n4/leafminer.htm (accessed Feb 5, 1999).

14. Day, W.H. Host preferences of introduced and native parasites (Hymenoptera: Braconidae) of phytophagous plant bugs (Hemiptera: Miridae) in Alfalfa-Grass Fields in the Northeastern USA. BioControl **1999**, *44,* 249–261.

15. Hokkanen, H.M.T. Role of Biological Control and Transgenic Crops in Reducing Use of Chemical Pesticides for Crop Protection. In *Techniques for Reducing Pesticide Use*; Pimentel, D., Ed.; John Wiley and Sons: New York, 1997; 103–127.

16. *Evaluating Indirect Ecological Effects of Biological Control*; Scott, J.K., Quimby, P.C., Wajnberg, E., Eds.; CABI Publishing: Wallingford, U.K., 2001; 261.

37

Biological Factors Impeding Recovery of Predatory Fish Populations

Catalina Chaparro
Pedraza

Introduction

Historically, global fisheries have relied significantly on large predatory fish species (Myers and Worm 2003). The previous seemingly inexhaustible fish abundance contributed to the public's perception that fish stocks such as tuna, billfish, cod, salmon, and pike are almost extinction-proof. However, this perception proved to be wrong. Currently, 33% of global fish populations are exploited at biologically unsustainable levels (FAO 2018). In the second half of the 20th century, the fraction of large predatory fishes in global catches, that have increased by fivefold, have declined (Worm et al. 2009). By the turn of the millennia, large predatory fish biomasses were estimated to be only about one tenth of pre-industrial abundance (Myers and Worm 2003).

There has been much public concern over declining fish stocks on a global scale, spurring multiple United Nations resolutions on the conservation and management of fish stocks (UN 2019). Most management measurements have focused on reducing exploitation rates, which generally result in an improvement for fully exploited and overexploited stocks (Worm et al. 2009). However, reductions in fishing pressure, although necessary, are not always sufficient for recovery of depleted populations, which are estimated to be 13% of global exploited fish populations (Garcia et al. 2018). Indeed, a long-term dataset from more than 230 fish populations shows that most collapsed populations exhibit little or no recovery 15 years after collapse (Hutchings and Reynolds 2004). Only 12% of collapsed marine populations had fully recovered in this period, all of them clupeids (Hutchings and Reynolds 2004), which typically have a low trophic level in marine food webs. In contrast, even when fishing mortality has been reduced, collapsed fish populations of species in higher trophic levels exhibit little or no recovery as much as 15 years later (Hutchings and Reynolds 2004). An emblematic example is the Canadian cod collapse, which despite a prolonged moratorium in effect, exhibits no recovery after two decades (Maroto and Moran 2014; Pedersen et al. 2017).

Factors impeding recovery of depleted fish populations can be numerous including biological, environmental and socioeconomic factors (Garcia et al. 2018). Some biological factors preventing a successful recovery of depleted fish stocks are altered ecological interactions and intrinsic characteristics of the populations, i.e., life history traits. Environmental factors include loss of habitat, in particular, coastal and coral reef degradation, as well as natural oscillations in climatic and productivity variables that condition the recruitment into the depleted stock (Friedland et al. 2009). Several socioeconomic factors have been identified to cause the failure of recovery of depleted fish stocks, for instance, the inability to find alternative livelihoods for fishermen after the depletion preventing fishing mortality to be reduced, social organization and cohesion that may facilitate compliance or illegal behavior (Wakeford et al. 2009). In this chapter, I will focus on the biological factors that impair recovery of depleted predatory fish populations in more detail.

There is general consensus that life history traits affect the vulnerability and capacity to recover of fish populations. In particular, traits associated with maximum population growth include fecundity, body growth rate, age and size at maturity (Hutchings and Reynolds 2004). Although low fecundity is characteristic of fish species with low population growth rate (Musick 1999), high fecundity does not ensure recovery (Hutchings and Reynolds 2004; Dulvy et al. 2003; Hutchings and Kuperinen 2014). In fact, highly fecund species, such as Atlantic cod (DFO 2003), exhibit lack of recovery following a decrease in fishing pressure. Rapid body growth rate is associated with high potential to recover (Hutchings and Reynolds 2004). Age and size at maturity are negatively associated with maximum population growth and, thus, with recovery potential (Roff 2002; Denney et al. 2002; Dulvy and Reynolds 2002; Dulvy et al. 2003).

Biological Characteristics of Predatory Populations

Although some life history traits are specific of species or even populations, most predatory fish species are characterized by a significant increase in body size through ontogeny that can span four orders of magnitude (Werner and Gilliam 1984). This increase in body size results in changes in ecological interactions throughout an individual's life. In fact, the majority of predatory fishes do not have a predatory diet as small juveniles because its body size at birth is often similar to that of their prey (de Roos and Persson 2013). Therefore, individuals of predatory fish species prey on smaller fish species only when they reach a large body size, while small individuals necessarily feed upon other type of food that might be the same of their future prey. As a result, individuals of a predator and prey species may be engaged in a competitive or a predatory interaction depending on the life stage of the predator individual interacting. Furthermore, throughout life history, predatory interactions can be reversed, for instance, while individuals of the predator species in the adult stage feed upon large juveniles or adults of the prey species, the latter may prey upon small juveniles of the predator species (Gårdmark et al. 2015).

Depensation

As a consequence of the life history specifics, large predatory fish species described above may exhibit a reduced population growth capacity at low densities (i.e., below a threshold), a phenomenon known as depensation or Allee effect (Figure 1) (de Roos and Persson 2002).

Strong depensation effects are caused by the existence of alternative stable states. In this view, a high and a low population density constitute alternative states stabilized by internal feedbacks, and therefore, a population depleted to its low density by overexploitation cannot return to its high-density state when the exploitation rate decreases to the point at which the population has collapsed. Instead, the exploitation rate should be reduced to lower levels for the population to recover (Figure 2). The lag between the exploitation rate at which collapse and recovery occurs is known as hysteresis.

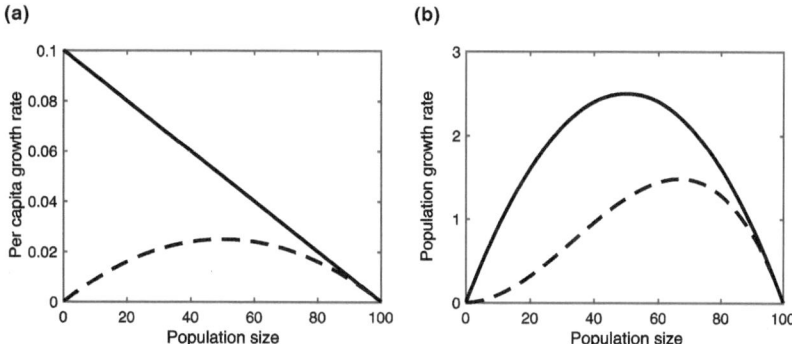

FIGURE 1 Depensation or Allee effects occur when population growth rate is reduced at small population size. (a) The per capita growth rate decreases linearly with population size when there are no Allee effects (solid line), while it is reduced at low and at high population size when there are Allee effects (dashed line). (b) As a consequence, in the absence of Allee effects (solid line), the population grows at its maximal rate at intermediate population size, while in the presence of Allee effects (dashed line), the population grows at its maximal rate when the population is large. Both populations have a carrying capacity of 100.

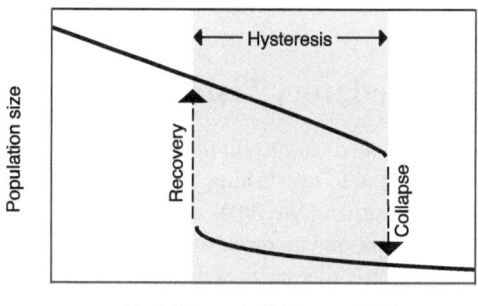

Exploitation rate (fishing mortality)

FIGURE 2 Alternative stable population states cause depensation. A population that is unexploited or exploited at low rate has a large size. As exploitation rate gradually increases, the population size gradually decreases until the exploitation rate is near the value at which the population collapses. Owing to hysteresis caused by alternative stable states, after the collapse, it is not enough for the population to recover to reduce the exploitation rate to the value where the collapse occurred. A further reduction in exploitation rate is required for the population to recover.

Several mechanisms have been hypothesized to cause depensation in fish populations that can be classified into three categories: reproduction difficulties, impaired social relations and altered ecological interactions, which are addressed in turn below.

Reproduction difficulties may emerge from: (1) sex ratio abnormalities, for instance, when fishing removes faster one sex than the other (Garcia et al. 2018); (2) insufficient spawner densities for effective egg fertilization and successful mating encounters and (3) removal of larger individuals because of their increased value and fishing tactics (i.e., minimum mesh and fish landing size (Hutchings and Reynolds 2004)). It is important to bear in mind that fecundity increases exponentially with female fish size (McIntyre and Hutchings 2003). Hence, larger fish can contribute disproportionately to population birth rate through greater weight-specific fertility and larval quality (Sogard 1997; Venturelli et al. 2012; Trippel 2004). As a consequence, populations composed of small individuals will have reduced reproductive potential (Scott et al. 2011) (migratory fish populations may be an exception when facing high costs of breeding migration (Chaparro-Pedraza and de Roos 2019)).

Depensation may be caused by impaired social relations at low population densities. For instance, by altering social behaviors such as schooling or migration, fishing may affect the reproductive process or survival of the remaining population as well as transmission of knowledge about suitable feeding and breeding areas and migration pathways from old-year classes to young ones (Petitgas et al. 2010).

Ecological interactions can cause depensation through various mechanisms. One proposed mechanism arises from the interaction between a size-selective predator and preys with density-dependent growth (de Roos and Persson 2002). Another mechanism emerges from competition between a predator and its prey in early life stages. As mentioned above, adult predatory fish individuals feed on prey fish, whereas small predatory juveniles and prey fish may compete for the same food resources (de Roos and Persson 2013). At high population density, adult predatory individuals, by controlling the prey population at low or intermediate densities through predation, reduce competition pressure on the predatory juveniles (Gårdmark et al. 2015). A release of the top-down control results in high density of prey population that controls the predatory population recruitment through competition. Alternatively, a third mechanism emerges when adult prey individuals feed upon early life stages of the predatory fish population, the recruitment in the latter is controlled by predation of abundant adult prey individuals when the predatory population is at low density.

Although theoretical work indicates that depensation through these mechanisms is likely (de Roos and Persson 2002; Maroto and Moran 2014), it has been difficult to demonstrate empirically (Garcia et al. 2018). However, recent empirical studies have revealed evidence supporting these depensatory mechanisms.

Depensation in Wild Predatory Fish

The northwest population of Atlantic cod (*Gadus morhua*) is an emblematic case of collapse and little recovery in fisheries. This collapse, due to overfishing, had devastating socioeconomic impacts. After dramatic declines of the population during the 1980s and 1990s, Canada instituted a moratorium on fishing in 1992 that left 40,000 workers unemployed, thousands left fishery to work in other trades or professions, in many cases, displaced to other parts of the country (Garcia et al. 2018). Despite this long moratorium, cod population have shown only slight increases (Pedersen et al. 2017). An analysis of the population dynamics of depleted fish stocks concluded that the lack of recovery in the Canadian cod is consistent with a depensatory phenomenon at low population densities (Maroto and Moran 2014). Likewise, the depensatory phenomenon was confirmed by a population model, which concluded that the lack of recovery of the cod population was mainly caused by predation by gray seals driving an Allee effect (Neuenhoff et al. 2018).

At the beginning of 1990s, the cod population in the Baltic Sea also collapsed as a result of overfishing concurrent with poor environmental conditions (Gårdmark et al. 2015). Following this collapse, the population did not recover during the late 1990s and early 2000s. The lack of recovery has been postulated to result from increased competition between cod larvae and sprat (*Sprattus sprattus*), a main prey of adult cod (Möllmann et al. 2009). As a consequence of the collapse of the cod population, the sprat population increased due to release from predation pressure. This caused stronger competition pressure on cod larvae for zooplankton prey. A shift in the regulation of zooplankton dynamics, from a control by limiting resources (bottom up) before the collapse of the cod population to a control by predation of sprat population (top down) after the collapse, supports this hypothesis (Casini et al. 2008). Gårdmark et al. (2015) demonstrate that the lack of top-down regulation in the sprat population after the cod collapse caused a change in the distribution of size classes in this population. As a consequence of this change, the food available for the cod in larval and adult stages was diminished and, therefore, the viability of this population reduced.

In the mid-1990s, pike (*Esox lucius*) population, a large top predator, declined substantially in the Baltic Sea (Bergström et al. 2015; Eriksson et al. 2011; Ljunggren et al. 2010; Nilsson et al. 2004). Pike is a major predator on adult threespine stickleback fish (*Gasterosteus aculeatus*) (Donadi et al. 2017).

Therefore, the decline of pike has favored the stickleback population by predator release, causing the stickleback population to grow exponentially about a decade after the collapse of the pike population (Bergström et al. 2015; Bergström et al. 2016; Donadi et al. 2017; Sieben et al. 2011). Following the collapse of the pike population in the Baltic Sea, various conservation measurements have been adopted with the aim of enabling the recovery of the pike population, including partial closure of the pike fishery and habitat restoration (Engstedt et al. 2017; Nilsson et al. 2014). In spite of this, the recovery of the pike population has been very slow, if at all (Bergström et al. 2016). A recent experimental study suggests that the cause of the impaired capacity of the pike population to recover is a depensatory effect. This study demonstrates that adult stickleback predation on pike larvae heavily reduces the recruitment of the pike population, suggesting this as the mechanism causing depensation and impeding recovery of the pike population (Nilsson et al. 2019).

The possibility of depensatory effects in severely depleted fish predatory populations may be high. However, the evidence remains scarce beyond few examples, perhaps because few populations have been depleted to densities where the population growth rate is heavily reduced (the depensatory effect or Allee effect is conspicuous) (Garcia et al. 2018). In fact, pike population in the Baltic Sea and the Canadian Cod population were depleted below 10% (Nilsson et al. 2019) and 1% (Myers and Worm 2005), respectively, of their original abundance. The collapse of these populations is, in fact, beyond the precautionary biomass limit usually imposed by fisheries that is 20%; at this limit, populations are considered heavily depleted (Garcia et al. 2018).

Evolutionary Responses to Exploitation

Most fisheries target specific age or size classes within a population. Typically, larger and older, individuals are more likely to be caught. There is compelling evidence supporting claims that large fish were caught more frequently in the early years of industrial fisheries (Allendorf and Hard 2009). For instance, cod individuals weighting more than 30 kg were frequently caught in New England waters before 1900 and even a gigantic individual of nearly 100 kg was taken off Massachusetts in 1895 (Jordan and Evermann 1902). Such selective mortality imposed on the targeted phenotypes brings about evolutionary change if the characteristics that make these phenotypes more vulnerable are heritable. Accordingly, rapid changes in phenotypic traits of fish populations have been documented as a result of fishing mortality (Haugen and Vøllestad 2001; Law 2001; Conover et al. 2007; Allendorf and Hard 2009).

Besides the immediate negative impact in the population birth rate due to loss of high fecundity potential, increased mortality of large individuals results in selection against fast body growth rate (Biro and Post 2008; Swain et al. 2007); as opposed to natural selection that usually favors this trait in unexploited populations. As a consequence of this evolutionary change in response to selective pressures induced by fishing, depleted populations are composed mainly of phenotypes with slow body growth rate, a trait associated with low potential to recover (Hutchings and Reynolds 2004). In addition to changes in growth rate, fishing pressure can alter size and age at maturity (Heino and Dieckmann 2008; Heino 1998). These and other evolutionary responses tend to improve resilience of the exploited populations to fishing mortality as slow-growth and long-lived species accelerate their turnover (Pauly 1995). However, simultaneously, these evolutionary responses may reduce trait variation and, thus, resilience of these populations to the environmental oscillations to which they were adapted through natural selection.

Since mortality caused by fishing is usually considerably higher than natural mortality, selection pressures are higher in the former. As a consequence, evolutionary changes induced by fishing pressure may emerge in few decades; however, once fishing pressure is released, it may take centuries to be reversed (Enberg et al. 2009). There is growing concern that undermined genetic variation caused by fisheries-induced evolutionary change might increase the range of hysteresis between collapse and recovery trajectories and thus enhance the effect of depensation through the mechanisms mentioned in the previous section. Therefore, evolutionary processes that may improve resilience of exploited populations, may lessen the capacity to recover once depleted.

Population Size, Environmental and Demographic Stochasticity

In addition to the negative effects on genetic variation, reduced population size makes depleted populations more vulnerable to extinction as a consequence of demographic and environmental stochasticity. Small populations are largely governed by the misfortunes of each individual. Hence, the persistence of small populations is largely dependent on realizations of random events that determine individual survival and reproduction. These random events are the result of unpredictable factors that influence the survival and fecundity of some (demographic stochasticity) or all (environmental stochasticity) individuals in a population (Lande 1993).

Along with the decline of a population, its geographical range tends to contract as individuals remain in the most favorable habitats (Swain and Wade 2010). Such geographical range contraction can further exacerbate the risk of extinction in small populations as a consequence of stochastic events. Northern cod provides an example of how such range contractions increase the risk of exposing collapsed populations to environmentally stochastic events. In April 2003, about 1500 tones of mature cod were frozen to death in a very small area of Trinity Bay along the northeast coast of Newfoundland (Hoag 2003; Hutchings and Reynolds 2004). An unpredictable oceanographic event that trapped cod in supercooled water provides the most likely explanation for this cull mass (Hutchings and Reynolds 2004). Relative to the small geographical area where the event took place, it had a disproportionally large negative effect on the population recovery because the spawning population of northern cod in 2003 was estimated to be only 14,000 tones (DFO 2003). Hence, this environmentally stochastic event in a very small geographical area probably resulted in the death of more than 10% of the breeding population.

Rebuilding Predatory Fish Populations after Depletion

Most plans for rebuilding fisheries focused on populations that have been driven below acceptable levels of productivity. These plans advocate measures that are adequate to avoid depletion, and it is often assumed that if the population has been driven to depletion, the same measures, more effectively applied, are also those that will enable the rebuilding of the population (Garcia et al. 2018). However, this assumption may be misleading. When a population has been depleted to very low level, a set of additional measures are required in order to avoid extinction and, if possible, increase the chances of recovery (Garcia et al. 2018).

Measures to reduce fishing mortality are necessary for the rebuilding of a depleted population. Rebuilding plans relying on a substantial reduction at the onset rather than on incremental minor reductions over time are more successful (Murawski 2017). However, socioeconomic and governance factors can prevent fishing mortality to be substantially reduced in the onset. For instance, the inability to find alternative livelihoods for fishermen following the depletion causes exploitation to continue, and the lack of social control facilitates illegal catch (Wakeford et al. 2009). Therefore, a successful rebuilding plan should ensure environmental and social sustainability pursuing the rebuilding of the natural resource and the people's livelihoods in concert (FAO 1995).

In addition to reductions in fishing mortality, protection of critical habitats and enhancement of stock resilience are common measures in rebuilding plans for populations that have been only lightly overfished. These measures are certainly necessary also for rebuilding populations that have been deeply overfished. However, if depensation effects govern the dynamics of the depleted population, the recovery may be slower than expected (Murawski 2017), which increases socioeconomic costs associated to the rebuilding process. Hence, implementing additional measures aim at destabilizing the present poor (small population size) stable state may speed up the rebuilding process and, thus, increase its chances of success.

Depensatory, evolutionary and stochastic effects have hampered the recovery of predatory fish populations in spite of efforts to reduce the fishing pressure and restore the habitat of these species. This indicates that, perhaps, a small population size (i.e., low density) is a stable state for these populations, and thus, additional actions that destabilize this state may be required. For instance, Persson et al. (2007)

demonstrated experimentally in a freshwater lake that the interaction between a size-selective predator and preys with density-dependent growth that has been proposed as one of the mechanisms causing depensation in predatory fish populations (de Roos and Persson 2002) can be destabilized by culling of the prey population. In this experiment, the removal of prey individuals proved to be an effective measure to trigger the recovery of the predator population.

Similarly, counterintuitive measures to reverse the negative evolutionary effects caused by size-selective fishing mortality may improve the capacity to recover of the predatory population. As mentioned before, natural selection is usually weaker than selection imposed by fishing mortality, therefore increasing the strength of selection in the same direction than natural selection may speed up the process to reverse the evolutionary effects of selection induced by fishing. For instance, while increased mortality of large size classes due to fishing results in selection against fast growing phenotypes, increased mortality of small size classes will favor these phenotypes. Indeed, experimental evidence shows that juvenile growth rates increases with high mortality of small size classes as a consequence of genetic change in somatic growth rate (Conover and Munch 2002). Furthermore, this study found that populations in which small individuals are harvested have higher yield in the long term than populations in which large individuals are harvested. Therefore, selection in favor of fast growing phenotypes through removal of small individuals in a depleted population may rapidly raise the frequency of fast growing phenotypes, increasing the potential to recover.

In this chapter, I have presented biological factors that affect the recovery of predatory fish populations. It is important to bear in mind that fisheries are social-ecological systems in which the natural and human components tightly interact and coevolve (Berkes et al. 2000). The biological factors described in this chapter interact with economic, social and governance factors in a complex manner. Therefore, the measures targeting biological factors that condition the rebuilding of predatory fish populations must account for the interactions of these with socioeconomic factors.

References

Allendorf, F. W., and J. J. Hard. 2009. "Human-Induced Evolution Caused by Unnatural Selection through Harvest of Wild Animals." *Proceedings of the National Academy of Sciences* 106 (Supplement_1): 9987–9994. doi:10.1073/pnas.0901069106.

Bergström, L., O. Heikinheimo, R. Svirgsden, E. Kruze, L. Ložys, A. Lappalainen, L. Saks, et al. 2016. "Long Term Changes in the Status of Coastal Fish in the Baltic Sea." *Estuarine, Coastal and Shelf Science* 169: 74–84. doi:10.1016/j.ecss.2015.12.013.

Bergström, Ulf, Jens Olsson, Michele Casini, Britas Klemens Eriksson, Ronny Fredriksson, Håkan Wennhage, and Magnus Appelberg. 2015. "Stickleback Increase in the Baltic Sea – A Thorny Issue for Coastal Predatory Fish." *Estuarine, Coastal and Shelf Science* 163 (PB): 134–142. doi:10.1016/j.ecss.2015.06.017.

Berkes, Fikret, Carl Folke, and Johan Colding. 2000. Linking Social and Ecological Systems: Management Practices and Social Mechanisms for Building Resilience. Cambridge: Cambridge University Press, 2000.

Biro, P. A., and J. R. Post. 2008. "Rapid Depletion of Genotypes with Fast Growth and Bold Personality Traits from Harvested Fish Populations." *Proceedings of the National Academy of Sciences* 105 (8): 2919–2922. doi:10.1073/pnas.0708159105.

Casini, M., J. Hjelm, J.-C. Molinero, J. Lovgren, M. Cardinale, V. Bartolino, A. Belgrano, and G. Kornilovs. 2008. "Trophic Cascades Promote Threshold-like Shifts in Pelagic Marine Ecosystems." *Proceedings of the National Academy of Sciences* 106 (1): 197–202. doi:10.1073/pnas.0806649105.

Chaparro-Pedraza, P. Catalina, and André M. de Roos. 2019. "Low Marine Food Levels Mitigate High Migration Costs in Anadromous Populations." *BioRxiv*, January, 577353. doi:10.1101/577353.

Conover, David O., and Stephan B. Munch. 2002. "Sustaining Fisheries Yields Over Evolutionary Time Scales." *Science* 297 (5578): 94–96. doi:10.1126/science.1074085.

Conover, David O., Stephan B. Munch, and David Conover. 2007. "Sustaining Fisheries Yields Over Evolutionary Time Scales." *Time* 297 (5578): 94–96. doi:10.1126/science.1074085.

de Roos, A. M., and L. Persson. 2002. "Size-Dependent Life-History Traits Promote Catastrophic Collapses of Top Predators." *Proceedings of the National Academy of Sciences* 99 (20): 12907–12912. doi:10.1073/pnas.192174199.

de Roos, André M., and Lennart Persson. 2013. Population and Community Ecology of Ontogenetic Development. Princeton: Princeton University Press.

Denney, Nicola H., Simon Jennings, and John D. Reynolds. 2002. "Life-History Correlates of Maximum Population Growth Rates in Marine Fishes." *Proceedings of the Royal Society B: Biological Sciences* 269 (1506): 2229–2237. doi:10.1098/rspb.2002.2138.

DFO, Deparment of Fisheries and Oceans. 2003. "Northern (2J+3KL) Cod." Ottawa (Canada).

Donadi, S., B. K. Eriksson, P. Jacobson, J. P. Hansen, Å. N. Austin, M. van Regteren, U. Bergström, J. S. Eklöf, and G. Sundblad. 2017. "A Cross-Scale Trophic Cascade from Large Predatory Fish to Algae in Coastal Ecosystems." *Proceedings of the Royal Society B: Biological Sciences* 284 (1859): 20170045. doi:10.1098/rspb.2017.0045.

Dulvy, Nicholas K., and John D. Reynolds. 2002. "Predicting Extinction Vulnerability in Skates." *Conservation Biology* 16 (2): 440–450. doi:10.1046/j.1523-1739.2002.00416.x.

Dulvy, Nicholas K., Yvonne Sadovy, and John D. Reynolds. 2003. "Extinction Vulnerability in Marine Populations." *Fish and Fisheries* 4 (1): 25–64. doi:10.1046/j.1467-2979.2003.00105.x.

Enberg, Katja, Christian Jørgensen, Erin S. Dunlop, Mikko Heino, and Ulf Dieckmann. 2009. "Implications of Fisheries-Induced Evolution for Stock Rebuilding and Recovery." *Evolutionary Applications* 2 (3): 394–414. doi:10.1111/j.1752-4571.2009.00077.x.

Engstedt, O., J. Nilsson, and P. Larsson. 2017. "Habitat Restoration – A Sustainable Key to Management." In Biology and Ecology of Pike, edited by C. Skov and A. Nilsson, 248–268. Boca Ratón: CRC Press, 2018, 1. p. 248-268.

Eriksson, Britas Klemens, Katrin Sieben, Johan Eklf, Lars Ljunggren, Jens Olsson, Michele Casini, and Ulf Bergstrm. 2011. "Effects of Altered Offshore Food Webs on Coastal Ecosystems Emphasize the Need for Cross-Ecosystem Management." *Ambio* 40 (7): 786–797. doi:10.1007/s13280-011-0158-0.

FAO. 1995. "Code of Conduct for Responsible Fisheries." Food & Agriculture Organization.

FAO. 2018. The State of World Fisheries and Aquaculture. doi:10.1126/science.aaw5824.

Friedland, Kevin D., Julian C. MacLean, Lars P. Hansen, Arnaud J. Peyronnet, Lars Karlsson, David G. Reddin, Niall Ó. Maoiléidigh, and Jennifer L. McCarthy. 2009. "The Recruitment of Atlantic Salmon in Europe." *ICES Journal of Marine Science* 66 (2): 289–304. doi:10.1093/icesjms/fsn210.

Garcia, S. M., Yimin Ye, Rice Jake, and A. Charles. 2018. Rebuilding of Marine Fisheries. Part 1: Global Review. FAO Fisheries and Aquaculture Technical Paper 630/1. FAO.

Gårdmark, Anna, Michele Casini, Magnus Huss, Anieke Van Leeuwen, Joakim Hjelm, Lennart Persson, and André M. de Roos. 2015. "Regime Shifts in Exploited Marine Food Webs: Detecting Mechanisms Underlying Alternative Stable States Using Sizestructured Community Dynamics Theory." *Philosophical Transactions of the Royal Society B: Biological Sciences* 370 (1659): 1–10. doi:10.1098/rstb.2013.0262.

Haugen, Thrond O., and Leif Asbjørn Vøllestad. 2001. "A Century of Life-History Evolution in Grayling." *Genetica* 112–113: 475–491. doi:10.1023/A:1013315116795.

Heino, Mikko. 1998. "Management of Evolving Fish Stocks." *Canadian Journal of Fisheries and Aquatic Sciences* 55 (8): 1971–1982. doi:10.1139/f98-089.

Heino, Mikko, and Ulf Dieckmann. 2008. "Detecting Fisheries-Induced Life-History Evolution: An Overview of the Reaction-Norm Approach." *Bulletin of Marine Science* 83 (1): 69–93.

Hoag, H. 2003. "Atlantic Cod Meet Icy Death." *Nature* 422 (6934): 792. doi:10.1038/422792a.

Hutchings, J. A., and A. Kuperinen. 2014. "Ghosts of Fisheries-Induced Depletions: Do They Haunt Us Still?" *ICES Journal of Marine Science* 71 (6): 1467–1473. doi:10.1093/icesjms/fst048.

Hutchings, J. A., and J. D. Reynolds. 2004. "Marine Fish Population Collapses: Consequences for Recovery and Extinction Risk." *BioScience* 54 (4): 297–309. doi:10.1641/0006-3568(2004)054[0297:mfpccf] 2.0.co;2.

Jordan, D. S., and B. W. Evermann. 1902. "American Food and Game Fishes: A Popular Account of All the Species Found in America North of the Equator, with Keys for Ready Identification, Life Histories and Methods of Capture." Doubleday, New York.

Lande, Russell. 1993. "Risks of Population Extinction from Demographic and Environmental Stochasticity and Random Catastrophes." *The American Naturalist* 142 (6): 911–927.

Law, R. 2001. "No Phenotypic and Genetic Changes Due to Exploitation." In Conservation of Exploited Species, edited by J. D. Reynolds, G. M. Mace, K. H. Redford, and J. G. Robinson, 323–342. Cambridge: Cambridge University Press.

Ljunggren, Lars, Lars Ljunggren, Alfred Sandstrom, Ulf Bergstrom, Johanna Mattila, Antti Lappalainen, Gustav Johansson, et al. 2010. "Recruitment Failure of Coastal Predatory Fish in the Baltic Sea Coincident with an Offshore Ecosystem Regime Shift." *ICES Journal of Marine Science* 67 (8): 1587–1595. doi:10.1093/icesjms/fsql09.

Maroto, Jose M., and Manuel Moran. 2014. "Detecting the Presence of Depensation in Collapsed Fisheries: The Case of the Northern Cod Stock." *Ecological Economics* 97: 101–109. doi:10.1016/j.ecolecon.2013.11.006.

McIntyre, Tara M., and Jeffrey A. Hutchings. 2003. "Small-Scale Temporal and Spatial Variation in Atlantic Cod (Gadus Morhua) Life History." *Canadian Journal of Fisheries and Aquatic Sciences* 60 (9): 1111–1121. doi:10.1139/f03-090.

Möllmann, Christian, Rabea Diekmann, Bürbel Müller-karulis, Georgs Kornilovs, Maris Plikshs, and Philip Axe. 2009. "Reorganization of a Large Marine Ecosystem Due to Atmospheric and Anthropogenic Pressure: A Discontinuous Regime Shift in the Central Baltic Sea." Global *Change Biology* 15 (6): 1377–1393. doi:10.1111/j.1365-2486.2008.01814.x.

Murawski, Steven A. 2017. "Keynote Rebuilding Depleted Fish Stocks : The Good, the Bad, and, Mostly, the Ugly." *ICES Journal of Marine Science*, 67(9), 1830–1840.

Musick, J A. 1999. "Life in the Slow Lane: Ecology Andconservation of Long-Lived Marine Animals." In American Fisheries Society Symposium, Vol. 23, American Fisheries Society, Bethesda, Maryland, USA.

Myers, R. A., and Boris Worm. 2003. "Rapid Worldwide Depletion of Predatory Fish Communities." *Nature* 423 (May): 280–283.

Myers, Ransom A., and Boris Worm. 2005. "Extinction, Survival or Recovery of Large Predatory Fishes." *Philosophical Transactions of the Royal Society B: Biological Sciences* 360 (1453): 13–20. doi:10.1098/rstb.2004.1573.

Neuenhoff, Rachel D., Douglas P. Swain, Sean P. Cox, Murdoch K. McAllister, Andrew W. Trites, Carl J. Walters, and Mike O. Hammill. 2018. "Continued Decline of a Collapsed Population of Atlantic Cod (Gadus Morhua) Due to Predation-Driven Allee Effects." *Canadian Journal of Fisheries and Aquatic Sciences* 76 (1): 168–184. doi:10.1139/cjfas-2017-0190.

Nilsson, Jonas, Jan Andersson, Peter Karås, and Olof Sandström. 2004. "Recruitment Failure and Decreasing Catches of Perch (Perca Fluviatilis L.) and Pike (Esox Lucius L.) in the Coastal Waters of Southeast Sweden." *Boreal Environment Research* 9 (4): 295–306.

Nilsson, Jonas, Olof Engstedt, and Per Larsson. 2014. "Wetlands for Northern Pike (Esox Lucius L.) Recruitment in the Baltic Sea." *Hydrobiologia* 721 (1): 145–154. doi:10.1007/s10750-013-1656-9.

Nilsson, Jonas, Henrik Flink, and Petter Tibblin. 2019. "Predator-prey Role Reversal May Impair the Recovery of Declining Pike Populations." *Journal of Animal Ecology* 88: 927–939. doi:10.1111/1365-2656.12981.

Pauly, Daniel. 1995. "Anecdotes and the Shifting Baseline Syndrome of Fisheries." *Trends in Ecology & Evolution* 10 (10): 430. doi:10.1016/S0169-5347(00)89171-5.

Pedersen, Eric J., Heike Link, Patrick L. Thompson, Zofia E. Taranu, Andrew Gonzalez, R. Aaron Ball, Charlotte Moritz, et al. 2017. "Signatures of the Collapse and Incipient Recovery of an Overexploited Marine Ecosystem." *Royal Society Open Science* 4 (7): 170215. doi:10.1098/rsos.170215.

Persson, L., P.-A. Amundsen, A. M. De Roos, A. Klemetsen, R. Knudsen, and R. Primicerio. 2007. "Culling Prey Promotes Predator Recovery--Alternative States in a Whole-Lake Experiment." *Science* 316 (5832): 1743–1746. doi:10.1126/science.1141412.

Petitgas, P., D. H. Secor, I. McQuinn, G. Huse, and N. Lo. 2010. "Stock Collapses and Their Recovery: Mechanisms That Establish and Maintain Life-Cycle Closure in Space and Time." *ICES Journal of Marine Science* 67: 1841–1848. doi:10.1093/icesjms/fsq082.

Roff, Derek A. 2002. "Life History Evolution". Sunderland, Massachusetts: Sinauer Associattes.

Scott, Beth, Gudrun Marteinsdottir, and Peter Wright. 2011. "Potential Effects of Maternal Factors on Spawning Stock–Recruitment Relationships under Varying Fishing Pressure." *Canadian Journal of Fisheries and Aquatic Sciences* 56 (10): 1882–1890. doi:10.1139/f99-125.

Sieben, Katrin, Lars Ljunggren, Ulf Bergström, and Britas Klemens Eriksson. 2011. "A Meso-Predator Release of Stickleback Promotes Recruitment of Macroalgae in the Baltic Sea." *Journal of Experimental Marine Biology and Ecology* 397 (2): 79–84. doi:10.1016/j.jembe.2010.11.020.

Sogard, Susan M. 1997. "Size Selective Mortality in the Juvenile Stages of Teleost Fishes: A Review." *Bulletin of Marine Science* 60 (3): 1129–1157.

Swain, Douglas P., Alan F. Sinclair, and J. Mark Hanson. 2007. "Evolutionary Response to Size-Selective Mortality in an Exploited Fish Population." *Proceedings of the Royal Society B: Biological Sciences* 274 (1613): 1015–1022. doi:10.1098/rspb.2006.0275.

Swain, D. P., and E. J. Wade. 2010. "Density-Dependent Geographic Distribution of Atlantic Cod (Gadus Morhua) in the Southern Gulf of St. Lawrence." *Canadian Journal of Fisheries and Aquatic Sciences* 50 (4): 725–733. doi:10.1139/f93-083.

Trippel, Edward A. 2004. "Egg Size and Viability and Seasonal Offspring Production of Young Atlantic Cod." *Transactions of the American Fisheries Society* 127 (3): 339–359. doi:10.1577/1548-8659(1998)127<0339:esavas>2.0.co;2.

UN. 2019. "General Assembly Resolutions and Decisions." *Oceans and the Law of the Sea in the General Assembly of the United Nations.* https://www.un.org/Depts/los/general_assembly/general_assembly_resolutions.htm.

Venturelli, Paul A., Brian J. Shuter, and Cheryl A. Murphy. 2012. "Evidence for Harvest-Induced Maternal Influences on the Reproductive Rates of Fish Populations." *Proceedings of the Royal Society B: Biological Sciences* 276 (1658): 919–924. doi:10.1098/rspb.2008.1507.

Wakeford, Robert C., David J. Agnew, and Christopher C. Mees. 2009. "Review of Institutional Arrangements and Evaluation of Factors Associated with Successful Stock Recovery Plans." *Reviews in Fisheries Science* 17 (2): 190–222. doi:10.1080/10641260802667075.

Werner, E. E., and J. F. Gilliam. 1984. "The Ontogenetic Niche and Species Interactions in Size-Structured Populations." *Ecology* 15: 393–425.

Worm, Boris, Ray Hilborn, Julia K. Baum, Trevor A. Branch, Jeremy S. Collie, Christopher Costello, Michael J. Fogarty, et al. 2009. "Rebuilding Global Fisheries." *Science* 325 (5940): 578–585. doi:10.1126/science.1173146.

38

Bioremediation

Ragini Gothalwal

Introduction

After the industrial revolution, a significant number of industries were set up, which released organic compounds in the environment. These chemical compounds (pesticides, fertilizers, plastic, dyes, and heavy metals) are present in the form of metal, non-metal, metalloid, inorganic compounds, and organic compounds. Organic compounds are divided into aliphatic, alicyclic, aromatic, polyaromatic, and mixed types. These compounds have a tendency to transform into another compound and reach the ecosystem. Owing to the inadequacy and increased cost of physical and chemical methods available, the need for ecofriendly and cheaper methods has been strongly realized. Bioremediation has the potential to lower the cost of remediation by orders of magnitude over alternative technologies, and governmental agencies and the public as taxpayers are interested in cost-effective cleanup technologies.

Biodegradation and bioremediation are matching processes, to the extent that they are both based on the conversion or metabolism of pollutants by microorganisms. The difference between these two is that biodegradation is a natural process whereas bioremediation is a technology. Bioremediation requires an efficient bacterial strain that can degrade the largest pollutant to a minimum level. Microbial diversity offers an immense field of environmental friendly options for mineralization of contaminants or their transformations into less harmful or non-hazardous compounds. Molecular biology methods are now being employed to study bioremediation. Conjugative gene transfer occurs, through plasmid-borne catabolic genes, between bacteria in oil and the competitive, indigenous bacterial population. Therefore, it is important to understand the role of catabolic genes, by molecular cloning and characterization, in the degradation of a particular organic compound, so that it can be applied for bioremediation. This review emphasizes the distribution and extent of environmental contaminants from the ecosystem. It also accentuates the current status of research in the area of biodegradation and bioremediation and presents the new approaches available for microbial and phyto tracking in the environment.

Principle of Bioremediation

The understanding of the bioremediation process is derived from the combination of biochemical and microbiological processes,[1,2] summarized as follows:

1. The relationship of comparative biochemistry applies equally to the axenic culture of microorganisms and those in field soil.
2. Microbial growth requirements are the same whether in laboratory culture or in the field.
3. Limitations resulting from ecological interactions include the need to accommodate the presence of other microbes as well as adaptation to the physical and chemical properties of the microsite wherein the microbes live.
4. A microbe amended in a soil ecosystem not only should pass the requisite genetic information and be capable of expressing that capability *in situ*, but it must also have the capacity to become a part of the overall soil microbial community.[1] Such microbes are referred to as competent rhizospheric bacteria with intrinsic bioremediation potential.

Targets of bioremediation must be based not only in terms of structure but also in terms of the matrix containing the target. Thus, the applicability of bioremediation can be considered for each of the environmental status of matter: 1) solid (soil, sediment, sludge); 2) liquid (ground water, industrial wastewater); 3) gas (industrial air emission); and 4) subsurface environment (saturated and vadose zones). The general approaches to bioremediation (Figure 1) are as follows:

1. Bioaugmentation;
2. Biostimulation (environmental modification, through the nutrient application process);
3. Phytoremediation (addition of microbes).

The biological community for bioremediation generally consists of the natural soil microflora. However, higher plants can also be manipulated to enhance toxicant removal (phytoremediation), especially for remediation of metal-contaminated soils.[3,4]

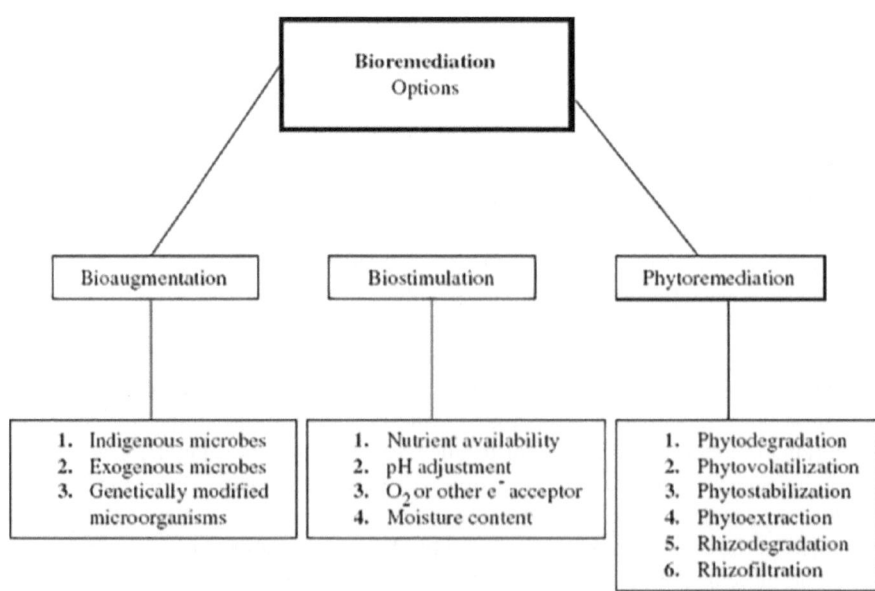

FIGURE 1   Bioremediation approaches for environmental cleanup.
Source: Musarrat and Zaidi.[1]

Microbial Degradation of Organic Pollutants

The soil microbial diversity is regarded as a major factor responsible for many biochemical transformations, including degradation of diverse organic compounds.[5-7] The chemical structures of organic pollutants exert profound influence on the metabolic abilities of microbes. Thus, harvesting the microbiological activities for biodegradation of recalcitrant hazardous chemicals and its implementation forms the basis for the increased use of biotreatment systems (Table 1). Predominantly, noxious chemicals, such as polychlorinated biphenyls (PCBs), trichloroethylene (TCE), polycyclic aromatic hydrocarbons (PAHs), and pesticide residue in soils and water are the targets. Removal of unwanted residues, as well as pesticide efficacy, is ultimately dependent on the presence, number, and enzymatic capability

TABLE 1 Examples of Bacteria, Actinomycetes, Fungi, and Cyanobacteria Engaged in Pesticide Degradation

Organisms	Organic Compounds or Pesticides
Bacteria and actinomycetes	
Alcaligenes denitrificans	Fluoranthene (PAH)
Alcaligenes faecalis	Arylacetonitriles
Archromobacter	Carbofuran
Arthrobacter	EPTC, glyphosate, pentachlorophenol (PCP)
Bacillus sphaericus	Urea herbicides
Brevibacterium oxydans IH35A	Cyclohexylamine
Burkholderia sp. P514	1,2,4,5-TeCB
Clostridium	Quinoline, glyphosate
Comamonas testosteroni	Arylacetonitriles
Corynebacterium nitrophilous	Acetonitrile, carboxylic acid, ketones
Dehalococcoides ethenogenes 195	Trichloroethylene (TCE)
Desulfitobacterium dehalogenans	Hydroxylated PCBs
Desulfovibrio sp.	Nitroaromatic compound
Flavobacterium	PCP
Geobacter sp.	Aromatic compound
Klebsiella pneumoniae	3 and 4 Hydroxybenzoate
Methylococcus capsulatus (Bath)	Trichloroethylene
Methylosinus trichosporium OB 3b	1,1,1-Trichloroethane (TCA)
Moraxella	Quinoline, glyphosate
Nitrosomonas europaea	TCA
Nocardia	Quinoline, glyphosate
Pseudomonas aeruginosa	Nitriles, biphenyl, parathion
Pseudomonas sp.	Quinoline, glyphosate
Pseudomonas stutzeri	Parathion
Pseudomonas cepacia	2,4,5-T
Pseudomonas paucimobilis	PCP
Pseudomonas putida 6786	Propane
Pseudomonas striata	Propham, chloropham
Rhodococcus chlorophenolicus	PCP
Rhodococcus corallinus	S-triazines
Rhodococcus rhodochrous	Propane
Rhodococcus sp.	Propane, TCA
Rhodococcus UM1	Pyrene

(Continued)

TABLE 1 (*Continued*)   Examples of Bacteria, Actinomycetes, Fungi, and Cyanobacteria
Engaged in Pesticide Degradation

Organisms	Organic Compounds or Pesticides
Fungi	
Aspergillus flavus	DDT
Aspergillus parasiticus	DDT
Aspergillus niger	2,4-D
Candida tropicalis	Phenol
Chrysosporium lignorum	3,4-Dichloroaniline
Fusarium solani	Acylamilide
Fusarium oxysporum	DDT
Hendersonula toruloidea	2,4-D
Hydrogenomonas + Fusarium sp.	DDM, nitrile
Mucor alternans	DDT
Penicillium	Acylamilide
Penicillium megasporum	2,4-D
Phallinus weirii	DDT
Phanerochaete chrysosporium	PAH, 2,4,6-trinitrotoluene, PCP, DDT, 2,4,5-T, lindane
Pleurotus ostreatus	DDT
Polyporus versicolor	DDT
Pullularia	Acylamilide
Rhodotorula	Benzaldehyde
Stereum hirsutum	Phenanthrene
Trametes versicolor	Dieldrin
Trichoderma sp.	Nitrile
Trichoderma viride	DDT
Trichosporon cutaneum	Phenol
Yeast	Paraquat
Ectomycorrhizal fungi	
Tylospora fibrillosa	Mefluidide
Thelephora terrestris	
Suillus variegatus	
Suillus granulatus	
Suillus luteus	
Hymenoscyphus ericae	
Paxillus involutus	
Cyanobacteria	
Cylindrospermum sp.	BHC
Aulosira fertilissima	lindane
Plectonema boryanum	diazinon, endrin
Nostoc muscorum	Carbofuran
Wollea bhardlvajae	hexachlorocyclohexane (HCH)
Nostoc musorum	HCH
Mastigocladus laminosus	Tolkan
Tolypothrix tenuis	fluchloralin
Anabaena doliolum,	Butachlor

(Continued)

TABLE 1 (*Continued*) Examples of Bacteria, Actinomycetes, Fungi, and Cyanobacteria Engaged in Pesticide Degradation

Organisms	Organic Compounds or Pesticides
Nostoc muscorum	
Anabaena ARM 286	BHC
Anabaena ARM 310	Ekalux
Anabaena variabilis	Bavistin
Aulosira fertilissima	eenlate
Scytonema chiastum	captan
Scytonema stuposum	dithane
	cyathion
Anabaena khannae	Butachlor
Calothrix marchica	benthiocarb
Nostoc calcicola	pandimethalin
Tolypothrix limbata	oxadiazon
Aulosira fertilissima ARM 68	Monocrotophos
Nostoc muscorum ARM 221	malathion, dichlorovos, phosphomidon, quinolphos

Source: Data from Adhikary[9] and Gothalwal and Bisen.[10]

of soil microbes. Mineralization of these compounds takes place only when environmental condition, water activity, presence of O_2, temperature, and pH are suitable for the growth and survival of the organism (Table 2). Degradation fails when the target compound is either very concentrated or much diluted. Successful biological cleanup of soil and water contaminants takes place in the presence of aerobic microorganisms. A dichlorinating anaerobic microbial population can also be helpful in cleanup processes. Genes for complete mineralization of some of the haloaromatic compounds are also reported and effectively utilized for recruitment of microbes and environmental applications. Degradation of naphthalene is more difficult than that of bionuclear compounds such as biphenyl, dibenzofuran, and dibenzo-*p*-dioxin, and of mononuclear aromatic compounds such as aniline, benzene, salicylate, phenoxyacetate, and toluene. Peripheral or funneling, central degradative and oxoadipate pathway sequences are necessary for the complete degradation of the compounds.[8] Pentachlorophenol (PCP) and PCB are major recalcitrant compounds. The major problems of PCP degradation are the formation of toxic end products during the metabolism and substitutions of chlorine.

Organochlorine pesticides (DDT, endosulfan, hexachlorobenzene, and hexachlorocyclohexane) are degraded by microbes by means of reductive dechlorination, dehydrochlorination, oxidation, and isomerization of the parent molecule. The principal reactions involved in the breakdown of phosphotriesters (organophosphate insecticides) are hydrolysis, oxidation, alkylation, and dealkylation.[11] Microbial degradation through hydrolysis of p-O-alkyl and p-O-aryl bonds is considered to be the most significant step in the degradation of parathions, methyl parathion, and p-nitrophenol. Reductive dechlorination of organochlorine is an important microbial reaction. A classic example of this is the conversion of DDT to DDD; this reaction occurs in several species of bacteria such as *Pseudomonas, Bacillus, Arthrobacter, Clostridium;* in soil actinomycetes; in yeasts; and in fungi such as *Trichoderma viridae, Mucor alterans,* white rot fungi, *Pleurotus australis, Phellinus weirii,* and *Polyporus versicolor* (Table 3). The formation of DDE from DDT through dehydrochlorination is commonly observed in algae, diatoms, and phytoplankton (*Cylindrotrea dentorium, Cyclotella nana, Isochryeii gabana, Nitzschia* spp.). Microbial isomerization reactions involve the conversion of γ-BHC to α-BHC, diedrin to photodieldrin, and D-keto andrin to endrin by *Pseudomonas putida*.[12]

Lindane is very persistent in the environment and resistant to microbial degradation.[13] It is degraded aerobically as well as anaerobically. *Pseudomonas paucimobilis* UT26 is a unique microbe

TABLE 2 Major Factors Affecting Bioremediation

Microbial
- Growth until critical biomass is reached
- Mutation and horizontal gene transfer
- Enrichment of the capable microbial populations
- Production of toxic metabolites

Environmental
- Depletion of preferential substrates
- Lack of nutrients
- Inhibitory environmental conditions

Substrate
- Too low concentration of contaminants
- Chemical structure of contaminants
- Toxicity of contaminants
- Solubility of contaminants

Biological aerobic vs. anaerobic process
- Oxidation/reduction potential
- Availability of electron acceptors
- Microbial population present in the site

Growth substrate vs. co-metabolism
- Type of contaminants
- Concentration
- Alternate carbon source present
- Microbial interaction (competition, succession, and predation)

Physico-chemical bioavailability of pollutants
- Equilibrium sorption
- Irreversible sorption
- Incorporation into humic matter

Mass transfer limitations
- Oxygen diffusion and solubility
- Diffusion of nutrients
- Solubility/miscibility in/with water

Source: Data from Boopathy.[15]

TABLE 3 Mechanisms of Radionuclide Bioremediation

S.N.	Mechanism	Microorganisms	Remediated Radio
1.	Biosorption	*Rhizopus arrhizus, S. cerevisae, Penicillium americium, Aspergillus, Aeromonas hydrophila, Candida utilis*	Uptake
2.	Engineering biosorption	Eukaryotic metallothien, *E. coli* Lan B	
3.	Bioaccumulation	*Citrobacter* sp.	UO_2^{2+}
		Rhodococcus erythropali CS98	Cesium
		Rhodococcus sp. strain CS 402	Cesium
		Micrococcus luteus	Neptunium
		Radioresistant bacteria	
		Deinococcus radiodurans	U (Vi), Tc (Vii)
		D. geothermalis	Radioactive waste

that utilizes hexachlorocyclohexane (HCH) as its sole source of carbon and energy under aerobic conditions.[14] Five structural genes (*lin A, lin B, lin C, lin D*, and *lin E*) and one regulatory gene (*lin R*) are involved in degradation of α-HCH in UT 26. *lin A, lin B, lin C*, and *lin D* codes for dehydrochlorinase, halidohydralone, dehydrogenase, and reductive dehalogenase, respectively, and *lin E* encodes ring cleavage oxygenase. Microbial degradation of γ-HCH has also been reported in *Anabaena* sp. PCC 7120 and *Nostoc ellipsosorum*.[16] The enzyme responsible for catalyzing the hydrolysis step in parathion is organophosphate hydrolase (OPH), which is encoded by the *opd* gene. This gene has been isolated from several bacteria. A naturally occurring variant of OPH enzyme, designated as opdA, capable of degrading a broad range of organophosphates was isolated from an *Agrobacterium radiobacter* strain.[17] The slow growth and low culture yields of native OPH-producing strains make them uneconomical for practical use. Therefore, efforts have been made by several researches to improve the applicability of OPH for pesticide bioremediation. A consortium comprising two bacteria that were genetically engineered (*E. coli* and *P. putida* KT 2440) efficiently worked together to break down parathion and prevent accumulation of p-nitrophenol [PNP].[18]

Catechol is a terminal metabolite formed during the degradation pathways of various compounds, and a variety of potential degraders of catechol have been reported by Kim et al.[19] Biodegradation of phenol and toluene by *Pseudomonas* sp., *Bacillus* sp., and *Staphylococcus* sp. was studied by Prasanna et al.[20] These strains were isolated from pharmaceutical industrial effluents. Mixed cultures showed more efficient degradation than pure strains within 5–7 hr at lower concentrations. Polycyclic aromatic hydrocarbon-degrading bacteria and ligninolytic and non-ligninolytic fungi are ubiquitously distributed in the natural environment such as soils and woody materials. The principal mechanism for aerobic bacterial metabolism of PAH is the initial oxidation of the benzene ring by the action of dioxygenase enzyme to form *cis*-dihydrodiol intermediates, which can then be further metabolized via catechols to CO_2 and H_2O.[21] Many bacterial, fungal, and algal strains have been shown to degrade a wide variety of PAHs (Table 1). There are limited reports on degradation of high molecular weight PAHs with more than four benzene rings. In general, high molecular weight PAHs are degraded slowly by indigenous or augmented microorganisms, as the persistence of PAHs increases with their molecular size.

The recalcitrant nature of chloroaromatic haloalkanes is due to the low electron density at the aromatic ring which makes the enzyme oxygenase unable to attack this compound. Many soil microorganisms (*Pseudomonas* and *Alcaligenes)* which synthesize the halogenase can utilize Halogenated alkonic acids [HAA].[22] Anaerobic methane-oxidizing bacteria can degrade TCE in pure *Pseudomonas* culture through a co-metabolic process.[23] The PCP-degrading *Pseudomonas* sp. strain IST 103 has been isolated, which was found to be capable of utilizing PCP as a carbon source. The enzyme PCP-4 monooxygenase was found to be responsible for the dechlorination of PCP.[24] The gene for the degradation may be plasmid encoded or present on the chromo-some.[25,26] Attempts have been made to enhance PCB biodegradation by modifying oxygenase.[27] One of the most efficient methods of biodegradation consists of sequential anaerobic and aerobic treatments for highly chlorinated compounds. Biochemical and genetic engineering approaches for dehalogenase and oxygenase could lead to "super bugs" that could be used for the bioremediation of chlorinated compounds.[28] Modified degradative genes could be introduced into the original strain and/or major indigenous strains isolated from contaminated sites, and it is hoped that these super bugs could have application in bioremediation in the near future, confirming their usefulness and safety. Raji et al.[29] have isolated a bacterial culture able to grow on benzoate and useful for remediation of PCB-contaminated sites. *Arthrobacter* sp. IFL YN 10 demonstrated mineralization of C^{14} ringlabeled atrazine. This isolates can be used to develop a consortium for bioremediation of pesticides.[30]

To improve the biodegradation efficiency and implementation, integrating various components such a microbial strain in consortium, solid O_2 source, and appropriate role of nutrients with controlled release pattern into a granule formulation with an oleophilic matrix, may provide an ideal approach to improve remediation of crude oil pollutants.[31] Abed et al.[32] reported that salinity and temperature are important environmental parameters that influence the degradation process of petroleum compounds. The inhibitory effect of salinity was shown to be more pronounced for aromatic than for aliphatic compounds.[33] Higher temperatures also reduce the viscosity of crude oil, which increases its diffusion through sediments, a process that render oil components accessible to bacteria. The possibility of the involvement of catabolic plasmid in the degradation of anthracene by *Pseudomonas* sp. isolated from an oil filling station was investigated by Kumar et al.[34] Many γ and β proteobacterial groups (*Halophaga, Geo-thrix, Acidobacterium*) and green non-sulfur bacteria with a strong potential to degrade hydrocarbons were present in benthic cyanobacterial mats.[35] The aliphatic fraction of petroleum hydrocarbon[36] is degraded by *Arthrobacter, Alcaligenes, Flavobacterium,* and *Bacillus.* Kniemeyer et al.[37] came across a green *Methanospirillium* that is able to degrade aromatic hydrocarbons. Singh and Lin[38] isolated 10 indigenous microorganisms from oil-containing soil; five isolates achieved 86.94% diesel degradation in 2 weeks. The results strongly indicate that the environmental condition of the contaminated site plays a crucial role in the degradation. Cohen[39] reported the development of cyanobacterial mats in oil-contaminated courts. Cyanobacterial polysaccharides play a major role in the emulsification of oil, actually breaking the oil into small droplets, which are subsequently attacked by the heterotrophs. Bioremediations of high fat and oil wastewater by selected lipase-producing bacteria such as *Bacillus subtilis, B. lichenformis, B. amyloliquifaciens, Serratia marcescens, P. aeruginosa,* and *Staphylococcus aureus* were carried out in wastewater from palm oil mill, dairy, slaughterhouse, soap industry, and domestic wastewater. After 12 days of bioremediation, the least biological oxygen demand and lipid content was observed in consortia, and the lipid degradation capacity of *P. aeruginosa* was higher than that of other bacteria.[40] Verma et al.[41] proved the biotechnological importance and advantage of using *P. aeruginosa* SL72 and *Acinetobacter* sp. SL-3 individually or as a consortium for waste treatment, resulting in substantial removal of the crude oil within a week, using a low-cost, efficient, and environment-friendly technique.

Spent wash is dark brown due to the recalcitrant melanoidin pigment. *Pseudomonas* sp. was selected for degradation of the pigment by Chavan et al.[42] Chuphal and Thakur[43] have characterized an alkalophilic bacterial consortium (*Micrococcus luteus, Deinococcus radiothilus, Micrococcus diversue, P. syringae P. myricure*) for *ex situ* bioremediation of color and adsorbable organic halogens in pulp and paper mill effluent. Nanda et al.[44] employed *Nostoc* sp. for bioremediation of tannery effluents; the main economic advantage of this system is the lack of a serious sludge disposal problem, consequently resulting in a much cheaper operating cost. No microorganism has been found to degrade polythene without an additive such as starch. The discovery of new enzymes and the cloning of genes for synthetic polymer-degrading enzyme from *Pseudomonas* sp. were reviewed by Premraj and Doble.[45]

Chemotaxis has been postulated to play an important role in enhancing biodegradation as it increases the bioavailability of pollutants to bacteria. Some toxic organic compounds are chemoattractants for different bacterial species, which can lead to improved biodegradation of these compounds. A *Ralstonia* sp. was chemotactic toward different Nitro aeromatic compounds (NACs), i.e., p-nitrophenol (PNP), p-nitrobenzoate (PNB), and o-nitrobenzene (ONB).[46]

Mycoremediation

The key to mycoremediation is determining the right fungal species to target specific pollutants. Certain strains have been reported to successfully degrade the nerve gases VX and sarin. Battelle in a plot of soil contaminated with diesel oil was inoculated with mycelia of oyster mushrooms; within weeks, more than 95% of PAH had been reduced to non-toxic components. Mycofiltration is a similar process using fungal mycelia to filter toxic waste and microorganisms from waste in soil. Breakdown (70%–100%) of anthracene oil found in PAH was reported in 27 days in an N_2-limited culture of *Phaenerochaete chrysosporium*. Pentachlorophenol is an important constituent of paper mill effluents, and *Phaenochaste chrysorperiucm* immobilized on rotating biological contactor disk efficiently degrades 2,4-dichlorophenol, 2,4,6-trichlorophenol (TCP), polychlorinated quiacole, and several chlorinated vanillins.[47]

Bioremediation of Inorganic Contaminants

Microbes encounter metals such as Cr, Mn, Fe, Co, Ni, Cu, Zn, Ni, Ag, Cd, Pb, and Au, and metalloids such as As, Se, and Sb, having a diverse nature, in the environment. Microbes can detoxify metals by valence transformation, extracellular chemical precipitation, or volatilization. Such microbes combat high concentrations of heavy metals by the following processes:

1. Inactivation of metals;
2. Alteration of the site of inhibition;
3. Enhancement in impermeability of metals; and
4. Other by-pass mechanisms.[48]

Bacterial biomass can also bioaccumulate heavy metals both in live and dead states through intracellular accumulation and extracellular absorption, respectively, giving an effective alternative for small-scale remediation purposes.[49] Bacteria can remediate heavy metals by a variety of mechanisms, including bioaccumulation, biosorption, and bioremediation (Figure 2). Bioaccumulation is the retention and concentration of a substance by an organism through the cell membrane into the cytoplasm, where the metal is separated and immobilized. However, in biosorption, the negatively charged metal ions are separated through adsorption to the negative ionic groups on the cell surface (carboxyl residue, phosphate residue, SH groups, or hydroxyl group) such as capsule or slim layers. The charged functional groups[50] serve as nucleation sites for the deposition of various metalbearing precipitatesi[51] *Bacillus* SJ-101 exhibits a much higher capacity of intracellular Ni accumulation, which is attributed to the anionic nature of its cell surface. In bioremediation, biologically catalyzed redox reactions lead to immobilization of metals. Microorganisms are known to reduce a wide variety of multivalent metals that pose environmental problems.[52,53] The reduced species are highly insoluble and precipitate out from solution.

Bioremediation is one such promising option that harnesses the impressive capabilities of microbes associated with roots to degrade organic pollutants and transform toxic metals. Since it is a plant-based *in situ* phytorestoration technique, it is proven to be economicaly efficient and easy to implement under field conditions.[54] All plant growth-promoting rhizobacterial strains (*Azotobacter chroococcum, Bacillus megatorium, B. mucilaginosus, B. subtilis* SJ-101, *Pseudomonas* sp., *P. fluorescens, Rhizobium leguminosarum, Kluyvera ascorbata* SUD165)[55] can be used for bioremediation of metals. Rhizobacteria associated with hyperaccumulators (*B. subtilis, B. pumilus, P. pseudoalcaligens, and Brevibacterium halotolerans*) are also widely used in bio and rhizoremediation of multimetal-contaminated sites.[56] The rhizobacteria are used or manipulated with three main objectives for bioremediation of metal-contaminated soils: 1) hyperaccumulation of metals in plants; 2) reduction of the uptake of metals; and 3) *in situ* stabilization of

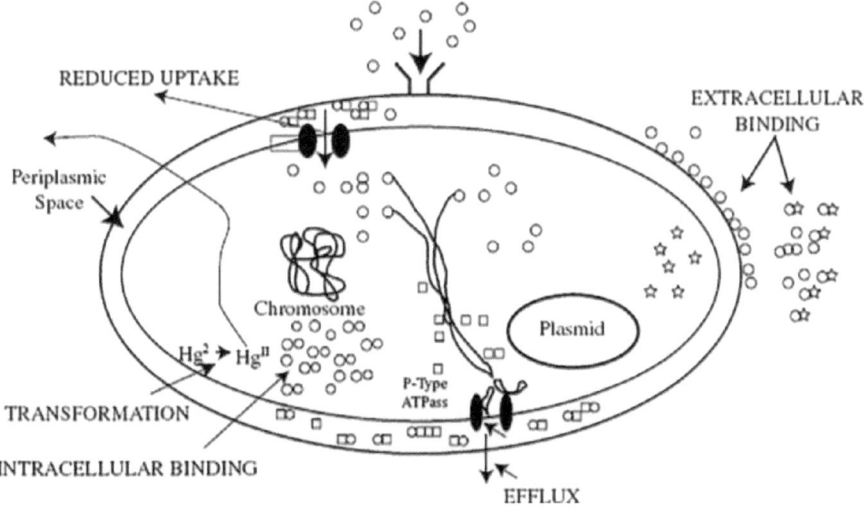

FIGURE 2 Different metal resistance mechanisms in bacteria.
Source: Singh and Srivastava.[66]

the metals as organocomplexes (Figure 2). The chemical conditions of the rhizosphere differ from bulk soil as a consequence of various processes induced by plant roots as well as by rhizobacteria,[57] such as secretion of organic acids followed by reduction in pH and production of siderophores, phytochelains, amino acid, and ACC deaminase. *Pseudomonas maltophilia* was shown to reduce the mobile and toxic Cr (VI) to non-toxic and immobile Cr (III) and also minimize the environmental mobility of other toxic ions (Hg, Pb, Cd).[58] Al Ayely et al[59] studied the effect of increasing levels of As and P on fern injected with mycorrhiza. The greater diversity of plant species may be responsible in part for the greater bacterial diversity in the bulk soils.

Redox Reaction Leading to Immobilization

The direct enzymatic reduction of soluble uranium (Vi), Cr (Vi), and Tc (Vii) to insoluble species is well documented.[60,61] A number of Cr (Vi)-reducing microbial strains, including *Oscillatoria* sp., *Arthrobacter* sp., *Agrobacter* sp., *Pseudomonas ambigua*, *Chlamydomonas* sp., *Chlorella vulgaris*, *Zoogloea ramigera*, *P. aeruginosa*,[62] and anaerobic sulfate-reducing bacteria have been isolated from chromate contaminated soil water and sediment.[63]

Redox Reaction Leading to Solubilization

Solubilization of adsorbed and co-precipitated metals may occur by direct or indirect microbial processes. The solubilization of toxic heavy metals and radionuclide from coprecipitates requires at least partial solubilization of oxide minerals.[64] Presumably, the organic acids formed by the metabolic activity of microbes lowers the pH of the system so that it interferes with the electrostatic forces that hold heavy metals and radionuclides on the surface of iron or Mn oxide minerals. Extracellular polymeric substances serve as biosorbing agents by accumulating nutrients from the surrounding environment, and also play a crucial role in biosorption of heavy metals. Being polyanionic in nature, exopolysaccharide forms complexes with metal cations, resulting in metal immobilization with the exopolymeric matrix.[65] Biofilm formation is a strategy that microorganisms might use to survive a toxic flux in these inorganic compounds; biofilm populations are protected from toxic metals by the combined action of chemical, physical, and physiological phenomena, which are in some instances linked to phenotypic variations among the constituent biofilm cells. Harrison et al.[67] have prepared a multifunctional model by which a biofilm population can withstand metal toxicity by the process of cellular diversification.

An effective ecofriendly approach for the removal of Mn from e-waste by the fungus *Helminthosporium solani* was shown by Savitha et al.[68] through the process of nonmetabolism-dependent biosorption under different environmental conditions of metal concentration, pH, and dry biomass concentration. "Metallothionein" has become a generic term applied to low molecular weight proteins or polypeptides that bind metal ions in metal thiolate clusters, and whose synthesis increases in response to elevated concentrations of certain metals.[69] The maximum cobalt removal efficacy (1 µg of $^{60}Co/g$, dry wt) of bacterial mass (*B. megaterium*, *P. putida*, *Flavobacteriuam devorans*, *Salmonella typhimurium*, *Streptomyces gresius*, *Rhizopus* sp., *Rhodococcus* sp., *E. coli*) could be achieved within 6 hr[70] compared with the 8–500 ng/g attained after 24–48 hr. The gum kondagogu, a natural carbohydrate polymer, was investigated[71] for its adsorptive removal of the toxic metal ions Cd^{2+}, Cu^{2+}, Fe^{2+}, Pb^{2+}, Ni^{2+}, Zn^{2+}, and Hg^{2+} present in industrial effluents. Kondogogu has a potential to be used as an effective, non-toxic, economical, and efficient biosorbent cleanup matrix for the removal of toxic ions with re-adsorption capacity at 90% level even after three cycles of desorption. *Klebsiella oxytoca* was able to biodegrade cyanide to a non-toxic end product (ammonia) using cyanide as the sole N2 source, which might preceed using ammonia as an assimilatory substrate.[72]

Microorganisms interact with radionuclides via several mechanisms (Table 3), some of which are used as the basis of potential bioremediation strategies. Based on the different types of effluents generated, a single technology cannot be suitable to address the problems. Cyanobacteria have the capacity to utilize nitrogenous compounds as well as phosphates; in addition, they pick up metal ions such as Cr, Co, Cu, and Zn very effectively. It has been observed that immobilized cyanobacteria have greater potential than their free cells. Various natural polymers such as alginate agar and carrageenan, and synthetic polymers such polyacrylamide and polyurethane have already been tried.[73] Novel methods of immobilization, including coimmobilizations of various species, are required for the symbiotic interaction among themselves, which will result in synergistic enhancement or removal capabilities. A gene cluster composed of nine open reading frames involved in Ni^{2+}, Co^{2+}, and Zn^{2+} sensing and tolerance in *Synechocystis* sp. PCC 6803 has been identified by Mario Garcia et al.[74] The biosorption of Hg^{2+} by *Spirulina platensis* and *Aphanothece flocculosa* was studied under a batch stirred reaction system,[75] and more than 90% from *A. flocculosa* and 100% mercury recovery from *S. platen-sis* can be achieved for 4 and 1 cycle, respectively. Protein and total non-protein thiols were measured as stressresponsive metabolites in response to Ni in *Anabaena doliolum*.[76] As *A. doliolum* is a high-biomass-producing strain, it can be conveniently separated from the solution by filtration and can be used at pilot-scale removal of Ni from wastewater.

Phytoremediation: A Beneficial Alternative

Phytoremediation is a promising green technology for accelerated decontamination of soil and water. It is a natural process in which plants are used to remove pollutants (pesticides, solvents, explosives, crude oil, PAHs) from soil and water. Plants can degrade or transform both the organic and metal (landfill leachates and radionuclide) contaminants by acting as filters or traps. Plants and associated microbes can degrade the pollutants or at least limit their spread in the environment. There are several ways in which plants can be used for the phytoremediation of organic contaminants viz. phytoextraction, phytodegradation, rhizodegradation, and phytovolatization (Figure 3). They can be compared to solar driven pumps for extraction and concentration of elements from the environment.[4] The plants can also be used for "phytomining." Rugh et al.[77] inserted an altered mercuric ion reductive gene (*mer A*)

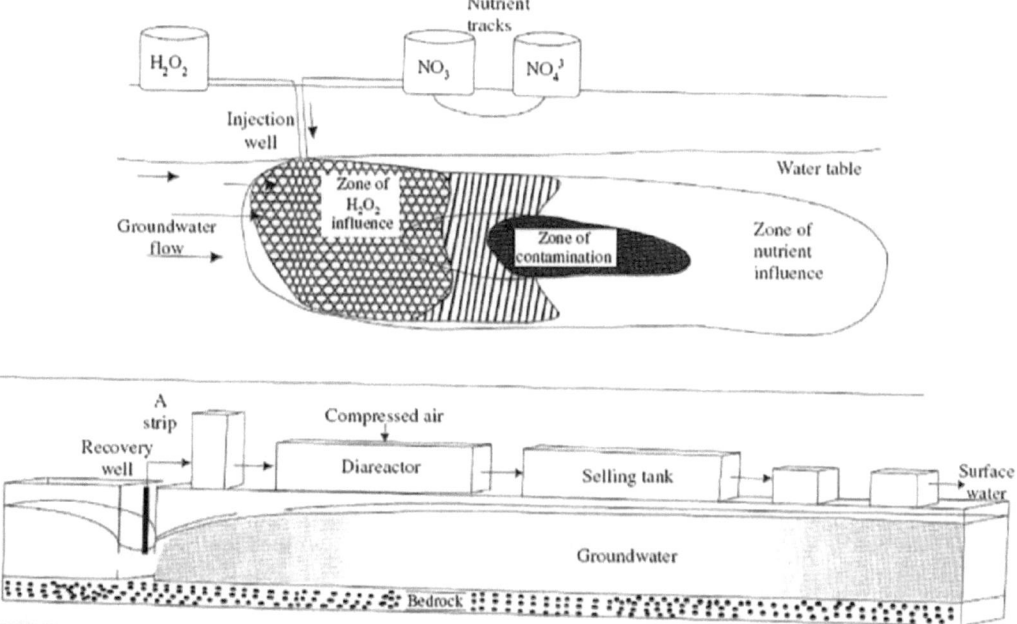

FIGURE 3 *In situ* bioremediation processes of bioaugmentation and biostimulation.
Source: Thakur.[81]

into *Arabidopsis thaliana,* creating a transgenic plant that volatilizes mercury to the atmosphere. Phytoremediation depends on a variety of factors, including physical conditions, chemical properties of the contaminant, and relative tolerance of the plant to the contaminant. Morphological characteristics of plant uptake are controlled by soil factors such as clay content, organic matter, soil moisture, and pH. The plant root can alter the pH in the rhizosphere by secreting organic acids, thus effecting the bioavailability of certain compounds, i.e., ionic pollutants whose solubility and desorption from soil colloids are pH dependent. Certain plants also produce biosurfactants that may increase the solubility of more lipophilic compounds.

Rugh[78] reported laboratory model plants such as *Thale cress* and tobacco to enhance phytoremediation of organomercurials, TCE, and nitroaromatic explosives that have been engineered with a non-plant transgene. Hyperaccumulator plants are also grown along with non-hyperaccumulators to enhance the heavy metal uptake by intermingling the roots and induce the colonization of efficient rhizobacteria. Recombinant *Mesorhizobium huakii,* by incorporating the phytochelatinase gene from *Arabidopsis thaliana* into *M. huakii* subsp. *rengei* 133, increased Cd accumulation by 1.5-fold in *Astragalus sinicus.*[79] The phytoremediation potential of water hyacinth (*Eichhornia crassipes*) in the treatment of tannery effluents was evaluated by Athaullah et al.[80] who found a promising risk reduction, cost-effective technology for water sanitation and conservation.

Bioremediation Technology

Bioremediation technology using microorganisms was invented by George M. Robinson in 1960. Bioremediation technologies can be generally classified as *in situ* or (microbial ecological approach) or *ex situ* (microbial approach) (Figure 3); some examples of these technologies include bioventing, land farming, bioreactor, composting, bioaugmentation, rhizofiltration, biostimulation, and solid phase bioremediation. In the development of the technology, the following points should be considered:

1. Heterogeneity of the contaminant
2. Concentration of the contaminant and its effect on the biodegradative microbe
3. Persistence and toxicity of the contaminant
4. Behaviour of the contaminant in soil
5. Conditions favorable for the biodegradative microbe or microbial population

Bioremediation is one of the most promising technologies for treating military sites, industrial wastes, municipal/urban wastes, mining waste, chemical spills and hazardous wastes, etc. *In situ* bioremediation can be implemented in many treatment modes, including aerobic, anoxic, anaerobic, and co-metabolic. The aerobic mode has proved to be most effective in reducing contaminant levels of aliphatic and aromatic compounds. Biofiltration is best suited for airstream containing volatile organic compounds.

Pollution remediation of tannery effluents is very complex. A multiprong treatment is thus required for a combination of nanotechnology and microbial technology, with a prior proceeding such as cycloning, flotation, microflotation, or electroflotation.

Bioremediation can be done on site, is often cheap, with minimal site disruption, eliminates waste permanently, eliminates long-term liability, has greater public acceptance, and can be coupled with other physical or chemical methods.[15] Table 2 shows the major factors affecting bioremediation. Another technique employed for bioremediation involves bioaugmentation, i.e., pumping genetically engineered microorganisms or microorganisms with enhanced degradation ability into the subsurface. Timely bioremediation of petroleum-contaminated soils is possible with innovative engineering and environmental manipulation to enhance microbial activity beyond the natural effective season.[82]

The first field slurry process used to remediate the nitroaromatic herbicide dinoterb used SABRE (sequential anaerobic bioremediation process), which was renamed as the FAST (fermentative anaerobic soil treatment) process.[83] Hence, the optimized remediation strategies exploring microbial diversity are being executed *in situ* successfully.

The aim is to translate research findings from the laboratory into viable technologies for remediation in the field. In the injection method, bacteria and nutrients are injected directly into the contaminated aquifer, or nutrient or enzymes are often referred fertilizer, that stimulate the activity of the bacteria are added. Bioreactors using immobilized cells have several advantages over conventional effluent-treatment technologies. Degradation of 4-chlorophenol by anaerobes attached to granular activated carbon in a biofilm reactor was evaluated during both open and closed modes of operation. Continuous flow fluidized bed reactor and bench scale continuous flow activated sludge reactor were used to study the removal of TCP, TeCP and PCP[84] which are used as sole source of carbon and energy. The ability of *Arthrobacter* cells to degrade PCP was evaluated for immobilized, non-immobilized, and co-immobilized cells. Fixed film bioreactor has been used with mixed bacterial communities for the treatment of pulp and paper mill effluent. The effluent after treatment showed a removal of color (80%), Chemical oxygen demand (71%), and chlorinated organic compound (68%).[85] The fungal slurry was successfully applied for compost preparation and biomediation of the Cr-contaminated tannery soil. The treated effluent was used for seed germination of crops without any phytotropic influence.[86]

Biocolloid formation methods have been adapted for bioremediation of metals by bacteria and fungi without the need *ex situ* treatment. Electrokinetically enhanced *in situ* soil decontamination and dispersing by chemical reaction together with microbes. Chlorophenol-contaminated saw mill soil used composting without bioaugmentation in a cheap and feasible method.[89] The use of biphenyl as an *in situ* co-substrate is expensive and environmentally problematic,[90] thus there was a need to investigate the ability of alternative cosubstrates to support the co-metabolic degradation to PCBs.

Biocapsules have been tested for various applications and can be produced for site-specific applications (Bioremediation Applied Bioscience http://www.bioprocess.com). *In situ* groundwater biodegradation in the United States has been carried out through various processes by numerous companies, as shown in Table 4. The removal effectiveness can reach 100% (Figure 4).

TABLE 4 In Situ Ground Water Biodegradation Examples of Process and Companies

S.N.	Name of Company	Technology Used	Types of Environmental Pollutant Remediated
1.	Bio-genesis Technology	Custom-blended microbial culture (GT-1000 series)	Oil, BTEX, diesel fuel
2.	Biopim	Biological sand filter	BTEX, TPH, phenol monochlorobenzene, metals
3.	ENSR Consulting and Engineering Technology (remediation beneath building foundation)	Steam injection, soil vapor extraction, ground water extraction, and air stripping	TCA, TCE, DWAPLS
4.	Petro Clean Bioremediation System	Indigenous microorganisms	Gasoline, diesel fuel, aviation fuel, solvents, PNAs, VOCs, and other organic compounds
5.	Kemron Environmental Services	Engineered, site-specific, groundwater recirculation system	Petroleum products solvents, halogenated volatiles and semivolatiles, BTEX, polynuclear aromatics, and organic acids
6.	Remediation Technologies Inc.	Treating groundwater in saturated zone injection, recovery well, monitoring wells	Dissolved contaminants
7.	SBP Technologies	Encapsulated cell inserted into a well	PAH (high mol. wt.), chlorinated aromatics (PCP), and pesticides
8.	OHM Remediation Services Corporation	Aquifer	Petroleum hydrocarbons, BTEX, chlorinated and non-chlorinated solvents

(Continued)

TABLE 4 (*Continued*)   In Situ Ground Water Biodegradation Examples of Process and Companies

S.N.	Name of Company	Technology Used	Types of Environmental Pollutant Remediated
9.	Electrokinetics Inc. (no need to add microorganism)	Electro osmosis or electro chemical migration	TCE, BTEX, PAH
10.	Geo Microbial Technologies Inc.	Removal of H_2S anaerobic process	—
11.	EODT Services Inc.	Use of biodispersant (also same as high-energy nutrients for microorganisms)	—
12.	Ecology Technologies International Inc.	Use of FyreZyme (multifactorial liquid agents)	Organic contaminants
13.	Gaia Resource Inc.	—	Hazardous and radioactive waste
14.	IT Corporation	Pump and treat system	Industrial effluents (hazardous organic compounds
15.	Ground Water Tech Inc.	Destructive technology	Hazardous compounds
16.	Yellow Stone Environmental Science Inc.	Pump and treat, aerobic processes, denitrification, sulfate reduction	Aromatic hydrocarbon, halogenated hydrocarbon, VOC, BTEX, phenol, cresol, CCl_4, PCE, vinyl chloride
17.	Waste Stream Tech. Inc.	Bioaugmentation	Organic compounds
18.	Micro Bac International Inc.	Batch and continuous feed treatment using M-1000 microbial consortium	Host specific
19.	Kuzanci Environmental Techniques (Figure 4)[87]	Microlife DCB series	Petroleum derivatives, stops foul odor

Source: Data from Bioremediation 1999.[88]

Before   After

Before   After

FIGURE 4   Treatment of soil hydrocarbon-polluted soils by Microlife DCB series bioremediation products. The progress is clearly seen in the photos.
Source: http://www.microbial-products.com/microbial-bacterial products bioremedation.asp/geli (accessed December 2011).[87]

Molecular Probes in Bioremediation

The traditional method of bacterial enumeration is often insufficient for monitoring the specific microbe's biochemical reaction in mixed microbial communities. It has became apparent that a significant number of microbes in these systems are viable but non-culturable. The catabolic enzymes, genes, and proteins expressed in the microbes can be exploited for the detection of the fate and effect of microbes in the bioremediation process.[91] Antibody- and fatty acid-based probes, nucleic acid sequences, and DNA probes can be used to detect genes in the bacterial genome or on plasmid, or to detect mRNA or tRNA. Other relevant techniques employed are PCR, repetitive sequence-based PCR, 16S rDNA, random amplified polymorphic DNA, and fluorescence *in situ* hybridization.[81] In a previous study, a specific synthetic PCR-amplifiable DNA fragment was introduced into a *Pseudomonas* chromosome to allow genetically engineered microbes to be identified easily.

The development of a new field of metabolic engineering involves the improvement of cellular activities by manipulation of enzymatic, transport, and regulatory function of the cell by using recombinant DNA technology. Advances in the field of genetic engineering, sequencing of the whole genome of several organisms, and developments in bioinformatics have speed up the process of gene cloning and transformation. Furthermore, many powerful analytical techniques have been developed for metabolic pathway analysis and analysis of cellular functions, such as gas chromatography (GC), gas chromatography-mass spectrometry (GC-MS), nuclear magnetic resonance, 2D gel electrophoresis, matrix-assisted laser desorption/ionization time of flight (MALDI-TOF), liquid chromatography-mass spectrometry (LC-MS), and DNA chips. Metabolic engineering is, therefore, an effort to improve the ability of microorganisms. Bioremediation require the integration of huge amount of data from different sources. Pazos et al.[92] developed "Meta Router," a system for maintaining heterogeneous information related to bioremediation in a framework that allows its query, administration, and mining. The system can be accessed and administered through a web interface for studying and representing the global properties of the bioremediation network. Bioinformatics require the study of microbial genomics, proteomics, systems biology, computational biology, phylogenetic trees, and data mining, and the application of major bioinformatics tools for determining the structure and biodegradation pathway of xenobiotic compounds. Bioinformatics has taken on a new glittering by entering the field of bioremediation.[93] The limitations of bioremediation has paved the way for the development of Genetically engineered microorganisms (GEMs), or designer biocatalysts harboring artificially designed catabolic pathways.[94] Database such as the University of Minnesota Biocatalysts/Biodegradation database provide a scope for *in silico* designing of biocatalysts for in vivo construction followed by *in situ* application. In the era of functional genomics, it is easy to construct GEMs by reshuffling the gene(s), promoter, etc., to enhance their performance in situ.

Conclusion

The popularity of bioremediation is further enhanced because it is perceived as being more "green" than other remediation technologies. As a result, bioremediation companies have a viable future regardless of the long-term effectiveness of the process. Special emphasis is required on the exploitation of biotechnological innovations to improve presently available biocatalysts, and for the evaluation of future effects of microorganisms and their proper application in the optimization of *in situ* bioremediation. The use of enzymes for degradation of pesticides can be developed as a technology for bioremediation. A super strain can be created to achieve the required result in a short time frame. One important characteristic of this technology is that it is carried out in a non-sterile open environment, which contains a host of microbes. Therefore, a strategy should be tailored in such a manner that due consideration be given to the various environmental constraints (type and amount of pollutant, climatic condition, hydrogeodynamics) that affect a particular location. Feasibility studies are essential and can have an enormous impact on the cost of full-scale remediation. Rhizoremediation can contribute to the restoration of polluted sites. Phytoremediation will require an integration of activities by plant scientists, microbiologists, chemists, and engineers, so that these systems that can be used to prevent and remediate pollution can become a reality. Environmental friendly processed need to be developed to clean up the environment without creating harmful waste products. For the development of economically usable technologies, scientists and technologists would have to offer creative solutions for either introducing new capabilities or enhancing current efficiencies.

References

1. Musarrat, J.; Zaidi, S. Bioremediation of agrochemicals and heavy metals in soil. In *Biotechnological Applications of Microorganisms. A Techno-Commercial Approach;* Mahesh-wari, D.K., Dubey, R.C., Eds.; I.K. International Publishing House, Pvt. Ltd.: New Delhi, India, 2006; 311–331.
2. Tate, R.L. *Soil Microbiology;* John Wiley & Sons: New York, 2000; 464–494.
3. Salt, D.E.; Smith, R.D.; Raskin, I. Phytoremediation. Annu. Rev. Plant Physiol. Plant Mol. Biol. **1998**, *49,* 643–668.
4. Alkorta, I.; Garbisu, C. Biores. Technol. **2001**, *77,* 229–236.
5. Spain, J.C. *Biodegradation of Nitro Aromatics Compounds;* Plenum Press: New York, 1995.
6. Stoner, D.L. *Biotechnology for the Treatment of Hazardous Waste;* Lewis Publisher: Boca Raton, FL, 1994.
7. Unterman, R. A history of PCB biodegradation. In *Bioremediation Principle and Application;* Crawford, R.L., Crawford, D.L., Eds.; University Press, Cambridge, U.K., 1996; 209–253.
8. Timmis, K.N.; Steffen, R.J.; Unterman, R. Degrading microorganisms for the treatment of toxic waste. Annu. Rev. Microbiol. **1994**, *48,* 527–557.
9. Adhikary, S.P. General introduction. In *Blue Green Algae Survival Strategies in Extreme Environment;* Printer Publication: Jaipur, India, 2006; 7–17.
10. Gothalwal, R.; Bisen, P.S. Bioremediation. In *Encyclopedia of Pest Management;* Pimentel, D., Ed.; Marcel Dekker Pub. Ltd.: New York, 2002; 89–93.
11. Singh, N.; Asthana, R.K.; Kayastha, A.M.; Pandey, S.; Chaudhary, A.K.; Singh, S.P. Thiol: An exo-polysaccharide production in a cyano bacterium under heavy metal stress. Process. Biochem. **1999**, 35, 63–68.
12. Benezet, H.J.; Matsumura, F. Isomerization of γ-BHC to α-BHC in the environment. Nature, **1973**, *243,* 480–481.
13. Agnihotri, N.P.; Gajbhiye, V.T.; Kumar. M.; Mahapatra, S.P. Environ. Manage. Assess. **1994**, *30,* 105–112.
14. Imai, R.; Nagata, Y.; Fukuda, M.; Takagi, M.; Yano, K. Molecular cloning of *Pseudomonas paucimobilis* gene encoding a 17-kilodalton polypeptide that eliminates HCL molecules from γ-hexachlorodihexane. J. Bacteriol. **1989**, *173,* 6811–6819.
15. Boopathy, R. Factor limiting bioremediation technologies. Bioresour. Technol. **2000**, *74,* 63–67.
16. Kuritz, T.; Wolk, P. Use of filamentous cyanobacteria for biodegradation of organic pollutants. Appl. Environ. Microbiol. **1995**, *61,* 234–238.
17. Horne, I.; Sutherland, T.D.; Harcourt, R.L.; Rusell, R.J.; Oakes Hott, J.G. Identification of an *opd* (organophosphate degradation) gene in an *Agrobacterium* isolate. Appl. Environ. Microbiol. **2002**, *68,* 3371–3376.
18. Gilbert, E.S.; Walker, A.W.; Keasling, J.D. A constructed microbial consortium for biodegradation of the organophosphorus insecticide parathion. Appl. Microbiol. Biotechnol. **2003**, *61,* 77–81.
19. Kim, K.P.; Lee, J.S.; Park, S.I.; Rhee, M.S.; Kim, C.K. Isolation and identification of *Klebsiella oxytoca* C302 and its degradation of aromatic hydrocarbons. Korean J Microbiol. **2000**, *36,* 58–63.
20. Prasanna, N.; Sarvanan, N.; Geetha, P.; Shanmugaprakash, M.; Rajasekardan, P. Biodegradation of phenol and toluene by *Bacillus* sp., *Pseudomonas* sp. and *Staphylococcus* sp. isolated from pharmaceutical industrial effluent. Adv. Biotech. **2008**, *7,* 20–24.

21. Mueller, J.G.; Cerniglia, C.E.; Pritchard, P.H. Bioremediation of environments contaminated by polycyclic aromatic hydrocarbons. In *Bioremediation: Principle and Applications;* Crawford, R.L., Crawford, D.L., Eds.; Cambridge Univ. Press: Idaho, 1996; 125–194.

22. Hardman, D.J.; Gowland, P.C.; Slater, J.H. Large plasmids from soil bacteria enriched on halogenated alkonic acids. Appl. Environ. Microbiol. **1986**, *51*, 44–51.

23. Little, C.D.; Palumbo, A.U.; Herbes, S.E.; Lidstrom, M.E.; Tyndall, R.L.; Gilmer, P.J. Trichloroethylene biodegradation by a methane-oxidizing bacterium. Appl. Environ. Microbiol. **1988**, *54*, 951–956.

24. Thakur, I.S.; Verma, P.K.; Upadhyaya. K.C. Molecular cloning and characterization of pentachlorophenol-degrading monooxygenase genes of *Pseudomonas* sp. from the chemostat. Biochem. Biophys. Res. Commun. **2002**, *290*, 770–774.

25. Khan, A.; Tewari, R.; Walia, A. Molecular cloning of 3- phenyl catechol dioxygenase involved in the catabolic pathway of chlorinated biphenyl from *Pseudomonas putida* and its expression in **E. coli.** Appl. Environ. Microbiol. **1988**, *54*, 2664–2671.

26. Pritchard, P.H. Fate of pollutants. J. Water Pollut. Control Fed. **1986**, *58*, 635–645.

27. Furukawa, K. Engineering dioxygenase for efficient degradation of environmental pollutants. Curr. Opin. Biotechnol. **2000**, *11*, 244–249.

28. Furakawa, K. 'Superbugs' for bioremediation. Trends Biotechnol. **2003**, *21*, 187–190.

29. Raji, S.; Mitra, S.; Sumathi, S. Dechlorination of chlorobenzoates by an isolated bacterial culture. Curr. Sci. **2007**, *93*, 1126–1129.

30. Sagarkar, S.; Nouriainen, A.; Bijorklot, K.; Purohit, H.J.; Jargensen, R.S.; Kapley, A. Bioremediation of atrazine in agricultural soil. In *52nd Annual Conference of AMI.* International Conference of Microbial Biotechnology for Sustainable Development, Nov 3–6, 2011, Chandigarh, 140.

31. Wang, Q.; Zhang, S.; Li, Y.; Klassen, W. Potential approaches to improving biodegradation of hydrocarbon for bioremediation of crude oil pollution. J. Environ. Protect. **2011**, *2*, 47–55.

32. Abed, R.M.M.; Thukain, A.A.; de Beer, D. Bacterial diversity of a cyanobacterial mat degrading petroleum compounds at elevated salinities and temperatures. FEMS Microbiol. Ecol. **2006**, *57*, 290–301.

33. Milli, G.; Almallah, M.; Bianchi, M.; Wambeke, F.V.; Bertrand, J.C. Effect of salinity on petroleum biodegradation. Fresenius J. Anal. Chem. **1991**, *339*, 788–791.

34. Kumar, G.; Singla, R.; Kumar, R. Plasmid associated anthracene degradation by *Pseudomonas* sp. isolated from filling station site. Nat. Sci. **2010**, *8*, 89–94.

35. Margesin, R.; Labbe, D.; Schinner, F.; Green, C.W.; Whyte, L.G. Characterization of hydrocarbon degrading microbial population in contaminated and pristine alpine soils. Appl. Environ. Microbiol. **2003**, *69*, 3985–3092.

36. Mishra, S.; Jyot, J.; Kuhad, R.C.; Lal, B. Evaluation of inoculum addition to stimulate in situ bioremediation of oily sludge contaminated soil. Appl. Environ. Microbiol. **2001**, *67*, 1675–1681.

37. Kniemeyer, O.; Fischer, T.; Wilkes, H.; Glockner, F.O.; Widdel, F. Anaerobic degradation of ethylbenzene by a new type of marine sulfate—Reducing bacterium. Appl. Environ. Microbiol. **2003**, *69*, 760–768.

38. Singh, C.; Lin, J. Isolation and characterization of diesel oil degrading indigenous microorganisms in Kwazulu-Natal, South Africa. Afr. J. Biotechnol. **2008**, *7*, 1927–1932.

39. Cohen, Y. Bioremediation of oil by marine microbial mats. Int. Microbiol. **2002**, *5*, 189–193.

40. Prasad, M.P.; Manjunath, K. Comparative study on biodegradation of lipid rich waste water using lipase producing bacterial species. Indian J. Biotechnol. **2011**, *10*, 121–124.

41. Verma, S.; Lata; Saxena, J.; Sharma, V. Bioremediation of crude oil contaminated waste using mixed consortium of biosurfactant and lipase producing strains. In *52 Annual Conference of AMI;* Int. Conf. on Microbial Biotechnology for Sustainable Development, Nov 3–6, 2011, Chandigarh, 48.

42. Chavan, M.N.; Kulkarni, M.V.; Zope, V.P.; Mahulikar, P.P. Microbial degradation of melanoidine in distillery spent wash by an indigenous isolate. Indian J. Biotechnol. **2006**, *5*, 416–421.

43. Chuphal, Y.; Thakur, I.S. Characterization of alkalophilic bacterial consortium for *ex-situ* bioremediation of color and adsorbable organic halogen in pulp and paper mill effluent. In *Microbial Diversity Current Perspective and Potential Application*; Satyanarayan, T., Johri, B.N., Eds.; I.K. International Pvt. Ltd.: New Delhi, India, 2005; 573–584.

44. Nanda, S.; Sarangi, P.K.; Abraham, J. Cyanobacterial remediation of industrial effluents. N. Y. Sci. J. **2010**, *3*, 32–36.

45. Premraj, R.; Doble, M. Biodegradation of polymers. Indian J. Biotechnol. **2005**, *4*, 186–193.

46. Pandey, G.; Chauhan, A.; Samanta, S.K.; Jain; R.K. Che-motoxin of a *Ralstonia* sp. SJ98 toward co-metabolizable nitroaromatic compounds. Biochem. Biophys. Res. Commun. **2002**, *299*, 404–409.

47. Arora, D.S.; Chander, M. Biotechnological application of white rot fungi in biodegradation of various pollutants. In *The Environment in Biotechnological Applications of Microbes*; Verma, A., Podila, K.I.K., Eds.; International Pvt. Ltd.: New Delhi, India, 2007; 262–280.

48. Belliveau, B.H.; Trevors, J.T. Mercury resistance and detoxification in bacteria. Appl. Organometal. Chem. **1989**, *3*, 283–294.

49. Chang, J.S.; Hang, J. Biotechnol. Bioeng. **1944**, *44*, 9991006.

50. Beveridge, T.J. Biotechnol. Bioeng. Sym. N. Y. **1986**, *52*, 127–140.

51. Volseky, B.; Chang, K.H. Biotechnol. Bioeng. **1995**, *42*, 451–460.

52. Wildung, R.E.; Gorby, Y.A.; Krupka, K.M.; Hess, N.J.; Li, S.W.; Plymale, A.E.; McKinloj, J.P.; Fredrickson, J.K. Effect of electron donor and solution chemistry on products of dissimilarity reduction of technetium by **Shewanella putrefaciens**. Appl. Environ. Microbiol. **2000**, *66*, 2451–2460.

53. Weilingo, B.; Mizuba, M.M.; Hansel, C.M.; Fendrof, S. Environ Sci. Technol. **2006**, *34*, 522–527.

54. Kamaludeen, S.P.B.; Ramasamy, K. Rhizoremediation of metals: Harnessing microbial communities. Indian J. Microbiol. **2008**, *40*, 80–88.

55. Zhuang, X.; Chen, J.; Shim, H.; Bai, Z. New advances in plant growth promoting rhizobacteria for bioremediation. Environ. Int. **2007**, *33*, 406–413.

56. Abou-Shanab, R.A.; Ghanem, K.; Ghanem, N.; Al-Kalaibe, A. The role of bacteria on heavy metal extraction and uptake by plants growing on multi-metal-contaminated soil. World J. Microbiol. Biotechnol. **2008**, *24*, 253–262.

57. Marschner, H. *Mineral Nutrition of Higher Plants*; Academic Press: London, 1995; 889.

58. Blake, R.C.; Choate, D.M.; Bardhan, S.; Revis. N.; Barton, L.L; Zocco, T.G. Chemical transformation of toxic metals by a *Pseudomonas* strain from a toxic waste site. Environ. Toxicol. Chem. **1993**, *12*, 1365–1376.

59. Al Ayely, A.; Sylvia, D.M.; Ma, L. Mycorrhiza increase arsenic uptake by the hyper accumulator Chinese brake fern (**Pleris vittata** L.). J. Environ. Qual. **2005**, *34*, 2181–2186.

60. Henrot, J. Health Phys. **1989**, *57*, 239–245.

61. Sabaty, M.; Avazeri, C.; Pignol, D.; Vermeiglio, A. Characterisation of the reduction of selenate and tellurite by nitrate reductases. Appl. Eviron. Microbiol. **2001**, *67*, 5122–5126.

62. Chatterjee, S.; Ghosh, I.; Mukherjea, K.K. Uptake and removal of toxic Cr (VI) by *Pseudomonas aeruginosa*: Physico-chemical and biological evaluation. Curr. Sci. **2011**, *101*, 645–652.

63. Mabbett, A.N.; Lloyd, J.R.; Macaskie, L.E. Effect of complexing agents on reduction of Cr(VI) by **Desulfovibrio vulgaris** ATCC 29579. Biotechnol. Bioengg. **2002**, *79*, 389–397.

64. Lovely, D.R.; Phillips, E.J.P. Bioremediation of uranium contamination with enzymatic uranium reduction. Environ. Sci. Technol. **1992**, *26*, 2228–2234.

65. Pal, A.; Paul, A.K. Microbial extracellular polymeric substances; central elements in heavy metal bioremediation. Indian J. Microbiol. **2008**, *48*, 49–64.

66. Singh, S; Srivastava, S. Rhizobia and its role in rhizoremediation. In *Microbial Diversity: Current Perspective and Potential Applications*. Satynarayana, T., Johri, B.N., Eds.; I.K. International Pvt. Ltd., New Delhi, India, 2005; 655–676.

67. Harrison, J.J.; Ceri, H.; Turner, R.J. Multimetal resistance and tolerance in microbial biolfilms. Nature **2007**, *5*, 928–938.
68. Savitha, J.; Sahana, N.; Prakash, V.K. Metal biosorption by *Helminthosporium solani*—A simple microbiological technique to remove metal from e-waste. Curr. Sci. **2010**, *98*, 903–904.
69. Kojima, Y. Definitions and nomenclature of metallothioneins. Methods Enzymol. **1991**, *205*, 8–10.
70. Rashmi, K.; Haritha, A.; Balaji, V.; Tripathi, V.S.; Venkateswaran, G.; Maruthi Mohan, P. Bioremediation of ^{60}Co from simulated spent decontamination solution of nuclear power reactors by bacteria. Curr. Sci. **2007**, *92*, 1407–1409.
71. Vinod, V.T.P.; Sashidhar, R.B. Bioremediation of industrial toxic metals with gum kondagogu (*Cochlospermum gassypium*): A natural carbohydrate biopolymer. Indian J. Bio-technol. **2011**, *10*, 113–120.
72. Kao, C.M.; Liu, J.K.; Lou, H.R.; Lin, C.S.; Chen, S.C. Biotransformation of cyanide to methane and ammonia by *Klebsiella oxytoca*. Chemosphere **2003**, *50*, 1055–1061.
73. Prakasham, R.S.; Ramakrishna, S.V. The role of cyanobacterium in effluent treatment. J. Sci. Ind. Res. **1998**, *57*, 258–265.
74. Mario Garcia, D.; Luis, L.M.; Francisco, J.F.; Jose, C.R. A gene cluster involved in metal homeostasis in the cyanobacterium *Synechocystis* sp. strain PCC 6803. J. Bacteriol. **2000**, *182*, 1507–1514.
75. Cain, A.; Vannela, R.; Keith Woo, L. Cyanobacteria as a biosorbent for mercuric ion. Bioresour. Technol. **2008**, *99*, 6578–6586.
76. Shukla, M.K.; Tripathi, R.D.; Sharma, N.; Dwivedi, S.; Mishra, S.; Singh, R.; Shukla, O.P.; Rai, U.N. Responses of cyanobacteria *Anabaena doliolum* during nickel stress. J. Environ. Biol. **2009**, *30*, 871–876.
77. Rugh, C.L.; Senecoff, J.F.; Meagher, R.B.; Merkle, S.A. Development of transgenetic yellow poplar for mercury phytoremediation. Nat. Biotechnol. **1998**, *16*, 925–928.
78. Rugh, C.L. Genetically engineered phytoremediation: One man's trash in another man's transgene. Trends Biotechnol. **2004**, *22*, 496–498.
79. Sriprang, R.; Hayashi, M.; Ono, H.; Takayi, M.; Hirata, K.; Murooka, Y. Enhanced accumulation of Cd (2) by a ***Me-sorhizobium*** sp. transformed with a gene from *Arabidopsis thaliana* coding for phytochelation synthase. Appl. Environ. Microbiol. **2003**, *69*, 1791–1796.
80. Athaullah, A.; Asrarsheriff, M.; Sultan Mohideen, A.K. Phytoremediation of tannery effluent using *Eichhornia crassipes* (Mart.) Solms. Adv. Bio Tech. **2011**, *3*, 10–12.
81. Thakur, I.S. Microbial bioremediation of pollutant chlorinated in the environment. In *Biotechnological Application of Microbes*; Verma, A., Podila, K.I.K., Eds.; International Pvt. Ltd.: New Delhi, India, 2007; 239–261.
82. Filler, D.M.; Lindstrom, J.E.; Braddock, J.F.; Johnson, R.A.; Nickalashi, R. Integral biopile components for successful bioremediation in the Arctic. Cold Reg. Sci. Tech-nol. **2001**, *32*, 143–156.
83. Crawford, R.L. The microbiology and treatment of Nitro aromatic compounds. Curr. Opin. Biotechnol. **1995**, *6*, 329–336.
84. Thakur, I.S.; Verma, P.; Upadhyaya, K. Involvement of plasmid in degradation of pentachlorophenol by *Pseudomonas* sp. from a chemostat. Biochem. Biophys. Res. Commun **2001**, *286*, 109–113.
85. Thakur, I.S. Screening and identification of microbial strains for removal of colour and adsorbable organic halogen in pulp and paper mill effluent. Process Biochem. **2004**, *39*, 1693–1699.
86. Shah, S.; Thakur, I.S. Enrichment and characterization of a microbial community from tannery effluent for degradation of pentachlorophenol. World J. Microbiol. Biotechnol. **2002**, *18*, 693–698.
87. Available at http://www.microbial-products.com/microbial-bacterial-products-bioremedation.asp (accessed December 2011).
88. Bioremediation 1999, available at http://www.inweh.unu.edu/447/lectures/bioremediation.htm (accessed December 2011).

89. Hamada, M.F; Haddad, A.I.; Abd-E-L-Bury, M. Treatment of phenolic wastes in an aerated sub-merged fixed-film (ASFF) bioreactor. J. Biotechnol. **1987**, *5*, 279–292.
90. Lajoie, C.A.; Zylstra, G.J.; Deflaun, M.F.; Strom, P.F. Development of field application vector for bioremediation of soil contaminated with polychlorinated biphenyls. Appl. Environ. Microbiol. **1993**, *59*, 1735–1741.
91. Thakur, I.S. Structural and functional characterization of a stable, 4-chlorosalicyclic acid degrading bacterial community in a chemostat. World J. Microbiol. Biotechnol. **1995**, *119*, 643–645.
92. Pazos, F.; Gurjas, D.; Valencia, A.; Lorenzo, V.D. Meta Router: Bioinformatics for bioremediation. Nucleic Acid Res. **2005**, *33*, D.588–D592.
93. Fluekar, M.H.; Sharma, J. Bioinformatics applied in bioremediation. Innov. Rom. Food Biotcchnol. **2008**, *2*, 28–36.
94. Urbance, J.W.; Coli, J.; Saxena, P.; Tiedje, J.M. Nucleic Acid Res. **2003**, *31*, 152–155.

Nídia Sá Caetano

Introduction: Biological Waste (Biowaste)

Solid waste composition has varied since ancient times depending on the activity that originates it. One of the most important fractions is municipal solid waste (MSW), which represents a heterogeneous collection of wastes produced in urban areas, the nature of which varies from region to region.

Biowastes (biodegradable wastes) arise from living or once-living sources from several human, agricultural, horticultural, and industrial sources, in three groups—waste of directly animal origin (manures), plant materials (grass clippings and vegetable peelings), and processed material (food industry and slaughterhouse wastes and paper and paperboard wastes)—and include any waste that is capable of undergoing anaerobic or aerobic decomposition. Different terminologies have been applied to this kind of waste, such as *putrescible, green, food, yard, biosolids, garden,* or simply *organic wastes,* but chemically speaking, biowaste is characterized by high carbon content in the form of cellulose, hemicelluloses, and lignin, or even proteins and fat, that can be biologically degraded into carbon dioxide, methane, and water.[1]

Disposal Problems Associated with Biowaste

Disposal of biowaste either through uncontrolled landfill or if abandoned, presents some health and pollution issues that should be addressed. In fact, as biowaste is biologically degradable, it is a free food source for every kind of microorganism, including pathogens that could endanger an entire population.

Leachate

As water percolates through biowaste, it leaches out inorganic and organic compounds, with the risk of soil and groundwater contamination. Also, some persistent pathogens that can be found in long-term deposits of biowaste could endanger population. Landfill leachate is an organicrich liquor that is an excellent food source for heterotrophic microorganisms, but being so concentrated, its treatment is hardly achieved. Also, the existence of heavy metal contamination is toxic to microorganisms that could otherwise be successful in performing the leachate treatment.

Methane

The second pollution issue from disposal of biowaste is that methane (CH_4) is produced under anaerobic conditions that naturally occur in landfills. The problems with methane are that it is a greenhouse gas, with more than 20 times the damaging effect of carbon dioxide (CO_2), and that it remains in the atmosphere for approximately 9–15 years.

Regulatory Issues of Waste Management

Europe is committed to recycling as one of the main objectives of the waste management policy. Through recycling, materials contained in solid waste are reintroduced in the production cycle, leading to raw materials and energy savings and reducing the cost of landfill disposal.[2] Current European Directive on solid waste management[3] demands that countries adopt appropriate waste treatment methods, aiming to reduce the amount of waste sent to its final destination—landfill. Taking this into consideration, waste valorization through reusing and/or recycling, or by using other processes (energetic, organic), is also intended. The organic fraction of solid waste can be valorized by composting or anaerobic digestion and, according to the established in the Council Directive on the landfill of waste,[4] should be diverted from the flux of wastes to landfill. Taking the year 1995 as baseline, member states should reduce landfilling of biodegradable MSW to 75% by 2010, to 50% by 2013, and to 35% by 2020.

The organic matter (OM) in solid waste (currently constitutes about 40% of the MSW in Portugal, 25% in the United States,[5] almost 50% in Abu Dhabi City,[6] 50% in France,[7] and almost 77% in Brazil[8]) can be recycled to useful products (compost, methane gas, etc.,) through biological treatment processes. MSW valorization is mainly achieved by recycling constituents such as glass, metals, plastics, paper and cardboard, and OM, which is only possible when these residues are mostly collected selectively, although construction of mechanical biological treatment (MBT) facilities, comprising screening and other physical separation units, can contribute to the achievement of the established targets.

Composting is seen as a valuable recycling process for the organic fraction of MSW (OFMSW) and, thus, is of particular importance given the already existing systems and the potential to grow. Nevertheless, in the European Union (EU), it was not yet possible to come to an agreement on biological treatment of biowaste, in spite of the long work that has been done and that resulted in the publication of a Working Document on Biological Treatment of Biowaste in 2001.[9] This is not only due to the enormous differences in the degree of development of waste management of the various member states but also, in part, due to the existing lobbies.

Biological Aerobic Waste Treatment: Composting

Composting is the biological process used most often for the controlled aerobic conversion of OFMSW and any kind of solid and semisolid organic waste to a humus-like material, known as compost. Overall, the composting process can be represented by the reaction in Figure 1.

This exothermic process is realized in the presence of oxygen by a biological consortia of microorganisms and takes place in two distinct phases: a first phase, in which predominantly thermophilic biochemical degrading reactions occur (temperature rises as a result of the heat produced biologically), and a second phase, in which the humification/stabilization processes occur.[10,11] (Hogan and collaborators[12] suggested that the temperature rise results from the low thermal conductivity of waste.) Compost resulting from this process is a stable product, free of pathogens and plant seeds that can be applied, with benefits to the soil. This definition is intended to distinguish the composting process from the ordinary decomposition that occurs in nature.[13] Moreover, Bertoldi[14] clarified that the stabilization phase corresponds to a humification process that can be prevented under conditions of oxygen scarcity and substrate inadequacy.

Depending on the feedstock nature, the nitrogen, phosphorus, and potassium content of the compost may be insufficient for its classification as an organic fertilizer, allowing instead for its usage as a soil improver. This means that compost properties allow for soil pH amendment, acting as a source of OM that can enhance the water retention and cation exchange capacity of the soil and improve soil aeration.[15]

These are the reasons why composting is currently known as a process of recycling the OM in the solid waste and why using compost in soil represents the reintroduction of OM in soils, reducing erosion and thus desertification that is increased due to intensive land use.

Composting can be successfully applied to garden waste, separated MSW, mixed MSW, co-composting with sludge from urban and industrial wastewater treatment plants (WWTPs), and agricultural and livestock residues.

Composting Backgrounds

Organic soil correction with agricultural and livestock waste dates back to the utilization of soil for crop production, having been the principal means of restoring the nutrient balance in soil.[16] Composting is known, for a long time, by farmers as a method that allows obtaining an organic fertilizer from domestic waste. There are records of composting in piles in China for more than 2000 years, and there are even biblical references on the practice of soil correction. About 1000 years ago, Abu Zacharia described these procedures that have been practiced 3000 years earlier in the manuscript of *El Doctor Excellente Abu Zacharia Iahia de Sevilla,* translated from Arabic into Spanish by the order of King Carlos V, and published in 1802 as *El Libro de Agricultura.* In this book, Abu Zacharia insisted that animal manure should not be applied too fresh or directly to the soil, but only after mixing with 5 to 10 times its weight of vegetable and animal bedding waste.[17]

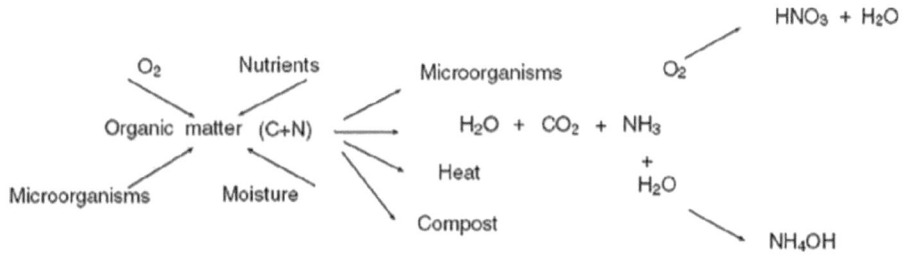

FIGURE 1 Schematics representation of the composting process.

In the growing cities in Europe, during the 18th and 19th centuries, farmers exchanged their products by MSW, using them as soil improver. Until the mid-20th century, MSWs were almost completely recycled through agriculture and did not pose an accumulation problem.[17] Composting of organic wastes and residues was envisaged as more of an art than a science until about the 1930s. By then, several developments of mechanical or intensive systems were achieved in Europe (Itano process in 1928, Beccari in 1931, and VAM in 1932). The Europeans continued to develop and install composting systems in Europe, South America, and Asia, and it was only in 1974 that the U.S. Department of Agriculture at Beltsville, Maryland, developed the "static pile" method that was currently used until the 1990s in the United States.[18]

By the end of 1960, composting was considered an attractive process for stabilization of the OFMSW with the final product being sold profitably as soil improver.[19] However, by the end of the same decade, composting was no longer that interesting for MSW management, not only due to the lower quality of solid waste but also due to the lack of market for compost. Recent stress on usage of less environmental impacting methods has redirected interest into the composting process, particularly concerning the recycling of MSW and urban and industrial WWTPs.

Despite the fact that implementing this process to agricultural wastes is ancient, scientific support was only established in the early 20th century, mainly due to the work done by an English agronomist (Sir A. Howard, 1924–1931) in Indore, India. This agronomist established the fundamental principles for the maintenance of a microbial population in optimal conditions of activity: the need to mix vegetable and animal wastes, the need for neutralization of the fermenting biomass, and the need for provision of adequate amounts of air and water. Thus, materials were stacked in piles (windrows) that have a dimension of 9×4.2 m and a height of 60 cm, which allows for maintaining adequate levels of heat and humidity.[20] Construction of these windrows was made by successive layers of waste (manure, soil, and straw) and moistening them conveniently. Windrows were revolved 16, 30, and 60 days after the start of the procedure that needed 90 days before completion and incorporation of the product in soil could be done.

Although windrow composting was the most common practice, during 1950–1960, there was a huge amount of publicity for projects involving composting in reactors.[21,22] However, these projects had almost universally poor results and had an abrupt end due to bad performance and high economic costs. Bad performance was generally the result of an inadequate project or operation and not of the process or technology itself. In the 1950s, there were already more than 20 patented composting systems.[23] In this period, mechanical separation for removal of any non-compostable from the waste stream was initiated.[14]

Co-composting of MSW and biosolids (composting simultaneously MSW and sludge) has attracted increasing attention.[24] MSWs are used as bulking agent and the biosolids act as readily available nitrogen and humidity source. This technique was investigated and applied fully and systematically in the 1950s,[25,26] but moisture content (96%) of the digested solids was a limiting factor as biosolids dehydration was not frequent by then.[27]

In the EU, production of agricultural and food wastes exceeds 1 billion tons/yr, which is 3 times larger than the production of sludge and 6 times higher than the production of MSW.[28] Intensive livestock farming incrementation worsened the problem through production of large amounts of animal waste, often in specific locations. Despite this, the relatively reduced application of composting in the management of this waste should be noticed, especially when compared with the application of this process in the management of MSW.[29] This is probably due to the fact that the agricultural wastes are often applied directly to soil, without any previous treatment, and that composting is generally considered a process for pollution control rather than a beneficial and efficient process for nutrient recycling.

Taking into account the requirements of EU legislation regarding waste management and environmental protection, and because there is nowadays a greater awareness of the importance of controlling the loss of nutrients in the waste treatment processes, recently, there has been a greater research effort to develop strategies that can control gaseous emissions, stabilize OM, and ensure nutrient retention and the absence of toxic products or pathogens in the compost. There are currently several groups of European researchers working on specific issues of composting of livestock and agricultural waste (such as kinetics of composting of MSW, effect of the composition of agricultural waste on compost quality and composting kinetics, effect of contaminants on composting process and on compost quality, etc.), aiming to produce good quality compost.[24,29-34] It is expected and highly desirable an increase in the world-wide application of composting for the treatment of these wastes and the use of its compost for agricultural purposes.

Composting has been successfully applied, including for the waste treatment of animal slaughter and carcasses.[35] Mesquita[36] enumerated some composting projects, where different raw materials were successfully used, and Williams[37] identified several large-scale separate composting schemes implemented in Europe.

Advantages and Drawbacks of Composting

If correctly used, the composting process for MSW treatment presents several advantages. When it is a part of a MSW integrated management system, these advantages are even more important; thus, valorization of the OFMSW by the integrated management can be achieved, and consequently, the final product quality can be improved.

A relevant issue is that besides being a process of effective organic waste treatment, composting is also an excellent recycling process, retaining in the final product (compost) macro- and micronutrients that may return to the soil and supplement other energy cycles in nature.[38]

The main advantages of the composting process are as follows:

- Economy and natural resource preservation and reuse of OM and of macronutrients (N, P, K, Ca, and Mg) and micronutrients (Fe, Mn, Cu, Zn, etc.,) owing to composting being a recycling process.
- Environmental benefits as a result of elimination of air, soil, and water pollution due to proper disposal of organic wastes or their appropriate confinement in controlled landfills. Also, there are additional benefits from using organic compost (instead of artificial fertilizers and soil improvers): in the recovery of degraded, contaminated sites and salty soil, in reforestation, in soil erosion control, etc.
- Public health benefits, due to the elimination of pathogens. In developing countries, this issue contributes to child mortality control.
- As an offshoot of health benefits, e.g., prevention and elimination of diseases, economic benefits arise, contributing to the reduction of costs of treatment and increased productivity. Another benefit is the elimination of costs for land remediation of uncontrolled dump sites.
- Social benefits arise from the fact that in addition to eliminating the practice of using organic waste as food for animals, composting promotes employment either in the treatment units (selective collection, sorting, processing) or in compost utilization and application.

The main disadvantages arising from composting are as follows:

- Being a labor-intensive process, composting gives rise to operational costs (compensated with positive social impact).
- Producing compost from selective organic collection entails high costs.
- The possibility of compost contamination (if the adopted process is not technologically appropriate and poorly conducted).

Methods of Composting

Composting is a simple technology that requires a fairly low intervention and has modest initial, operational, and support costs. It can be very attractive to authorities charged with biowaste management but with reduced budgets.

There are many factors that can influence the decision on the specific details of composting methods that should be adopted, but composting of OM in MSW or similar materials will most naturally be based in one of two options, either source-separated biowaste or biowaste recovered from MBT plants. Either of these options will ultimately be put in practice in home composting systems or in centralized composting facilities.

While home composting takes advantage of many distributed and local production units, centralized or large-scale facilities comprise two large groups of systems, both open and closed.[39] Among open systems of composting, three types can be distinguished: windrow (or revolved piles), the static aerated piles, and vermicomposting.[30] Closed systems of composting include reactors (either horizontal or vertical) of different sizes and shapes equipped with different feeding and aeration systems, aimed at accelerating the startup of the oxidation process, especially in cold climate countries, thus allowing for a more efficient control of odor emissions and minimizing the land area requirements.[37]

The choice of a specific type of system relies essentially on socioeconomic factors, on the amount of biowaste to compost, and on its final destination. Open systems are generally more suitable for developing countries[40,41] and for agro/livestock facilities[42] due to their easier operation and lower mechanization, building, maintenance, and operational costs.

Processing of agricultural and livestock waste using any of these technologies not only presents substantial benefits from the health, economic, and environmental points of view but also allows for a safe and potentially useful final product that can be used as a soil conditioner, valorized as a fertilizer, or for the floriculture and horticulture industries. Its use for agricultural purposes is important, both in countries where soil is extremely poor in OM (such as those of southern Europe) and in countries where extensive use of inorganic soil fertilizers has endangered its structure.[40]

The most common technologies, particularly for composting the OFMSW and similar biowaste, are presented below.

Home Composting

Home composting is one of the most interesting ways of managing biowaste. In fact, while people compost at home, they are guaranteeing the removal of OM from MSW flux, thus facilitating the achievement of the goals established in the EU Landfill and EU Waste Management Directives, as well as contributing to the reduction of waste management costs by reducing the total amount of waste that needs treatment. Also, by using home-produced compost, households do not need to use other fertilizers or soil conditioners; hence, they also have lower garden maintenance costs. Home composting demands the direct involvement of households, leading to higher separation rates of other recyclables as well.

One of the biggest issues in home composting is the choice of the compost bin. Although there are several configurations available in the market (Figure 2), some of them are far more efficient than others

FIGURE 2 Home composters: (a) Wood, homemade, with two composting chambers, 1 m³ each. (b) Commercial plastic, with openings for air circulation (0.4 m³), offered by the Terra-à-Terra project.

and their cost is not always according to quality. In some cities, there are special programs that try to address this problem, reducing costs to the consumer. In Porto, Portugal, Lipor has been developing a project—Terra-à-Terra—that aims to promote organic waste reduction at the households of Lipor's municipalities. Householders with a garden or that work in companies with a garden in the project area can receive a free compost bin after attending a free composting training session. Through this project that started in 2007, more than 4500 composters have been delivered in 3 years, and it is estimated that there is a potential biowaste reduction of about 3000 tons/yr that will prevent the emission of 528 tons of CO_2 per year if this waste is treated at the energy recovery plant.[43]

Centralized Composting

Seasonal variation in the composition of the MSW requires a flexible and competent management of the composting system.[1,44] For this reason, careful choice of the system and a thorough control of the facilities are of primary importance.

Windrows

This technology is based on the Indore composting method. Rows of parallel piles of material with a height of 1.5–2 m and a width of 3–4 m are constructed until they reach 100 m or more. The shape of the piles is such that sufficient heat is generated to maintain temperatures while allowing for oxygen diffusion into the center of the pile, through natural convection. Batteries of piles must be constructed on an impermeable surface (e.g., cement) so as to facilitate the collection of leachate that eventually will form and so that piles can be easily turned. Windrows must be turned periodically (initially at the end of 5 days and then every 2 or 3 days), as shown in Figure 3. The convex shape of the windrows allows rainwater to drip along the surface and not infiltrate the pile of material. This system has some drawbacks such as being prone to anaerobiosis due to layer compaction, reduction of free space for aeration, and production of leachate that fills the remaining voids between particles, which further contributes to reducing aeration.

Aerated Static Piles

Piles are similar to those in windrow composting. However, the material is not mechanically aerated. Temperature control is achieved by natural air convection or forced aeration, either by introduction of compressed air or by vacuum induction (Figure 4).

The material for composting remains stacked during a very long period, which might result to collapsing and thus pore clogging. To prevent this, it is a common practice to mix biowaste with some bulking material (such as wood chips or sawdust) that should be more stable and have a higher particle size, thus providing a structure that will help prevent the material in the pile from collapsing. Aerated piles should be covered with a layer of compost or wood chips that confer some insulation, act as a biofilter, and reduce odor release.[42]

FIGURE 3 Schema of windrow composting and how to turn composting piles; the upper material will stay in the bottom after turning.

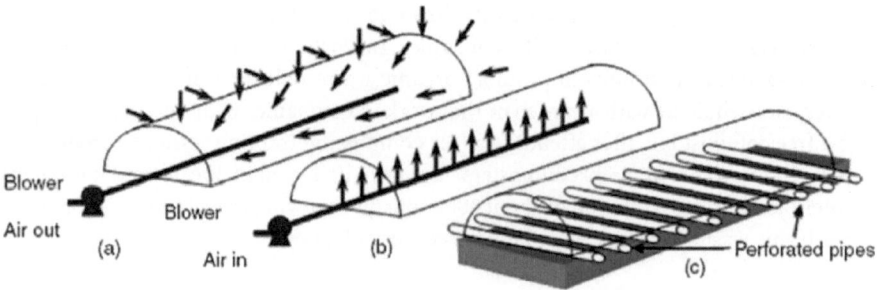

FIGURE 4   Static pile composting: (a) with suction of air from the outer layers to the base of the piles; (b) with compressed air injection through the base of the pile; and (c) natural convection.
Sources: (a) from Risse and Faucette;[45] (b) and (c) adapted from Graves and Hattemer.[46]

FIGURE 5   Indoor aerated static pile system with aeration through vacuum induction.

These composting technologies were traditionally implemented in open courtyards. Currently, and in regions where environmental conditions are particularly unsuited to this type of processing, these technologies are implemented in covered spaces or indoors (Figure 5), in which case, it is possible to perform air treatment in biofilters, achieving a dramatic reduction in odorous emissions.

In-vessel

Due to being enclosed, in-vessel systems allow for tighter control of temperature, moisture, aeration, and biowaste mixing. In-vessel systems include different configurations such as tunnels, rotating drums, reactor tanks, silos or towers, agitated bays and beds, enclosed halls, or even containers.[37] Although the operating systems are essentially the same as for windrows or aerated piles, these reactors are much more efficient, thus needing much less area.[1]

Silos or Towers

This technology is based on the use of silos or other types of structures with a height exceeding 4 m. These reactors are fed at the top, via a distribution mechanism, and the material progresses towards the bottom by way of gravity, usually after 14 days of composting. After this period, a curing time of ca. 2 months is needed to stabilize the compost.[42] Process control is commonly performed by air injection at the base of the reactor, moving upwards countercurrent to the biowaste (Figure 6).[47] Due to the great height of the reactor, it is necessary to use large airflow per unit surface area of distribution, which

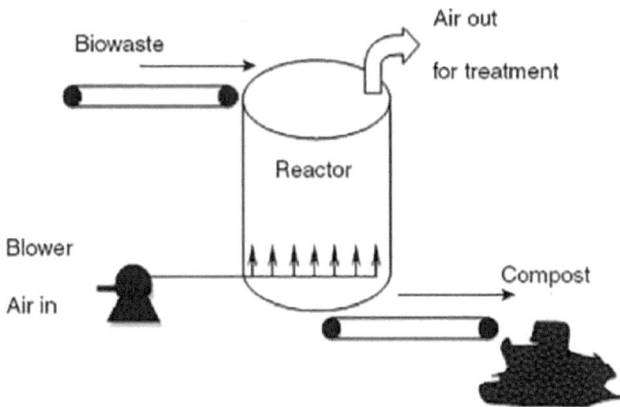

FIGURE 6   Schematic drawing of vertical reactors system for composting.

makes it difficult to control the process.[48] It is not easy to keep the temperature and oxygen at optimal levels, due to material compaction and to inexistence of mixing in the silo.[42] To minimize these problems, material should be thoroughly mixed prior to feeding the silo and air distribution and collecting systems must be improved, by changing the direction of airflow from vertical to horizontal between alternating sets of inlet and exhaust pipes.[44]

This type of reactor has been successfully used for composting of sludge from WWTPs (where the addition of bulking materials for porosity enhancement and a uniform feed facilitates process control) but is rarely used for processing heterogeneous materials, such as the OFMSW.[44]

Horizontal Reactors (Tunnels and Agitated Beds)

These reactors allow controlling the occurrence of high temperature, moisture, and oxygen gradients, which are frequent in vertical reactors, because air needs to travel lower distances through the material. There are a variety of configurations, comprising static or agitated reactors, with aeration by air injection or vacuum induction. Agitated systems usually take advantage of the revolving process to displace the material along the reactor continuously, while the static systems usually need loading and unloading mechanisms. Handling equipment can also grind the material, exposing new areas for decomposition, but excessive grinding can also reduce porosity. Typically, aeration systems are installed at the base of the reactor and can use temperature and/or oxygen concentration as control variables. The agitated systems with less than 2 to 3 m thick beds have been successfully used in heterogeneous MSW treatment (Figure 7, left).

FIGURE 7   Horizontal composting systems: (left) tunnel composting system composed of horizontal parallel beds; (right) turning machine for a tunnel composting system.

Tunnels are formed in long parallel channels (Figure 7, left), made of concrete walls with a movable porous floor, like a grid to allow for greater airflow (usually forced aeration) under the compost, and covered by a roof. Biosolids are revolved by a turning machine (Figure 7, right) running on tracks along the concrete walls between channels. The turner movement along the bed displaces the compost until it is ejected at the end of the bed.[49] The duration of the composting process is determined by the length of the bed and the turning frequency, ranging from 6 to 20 days, after which the compost must be further processed in windrows or aerated static piles for 1 to 2 months. Agitated bed systems operation ranges from 2 to 4 weeks; also, an extended curing period is needed to stabilize compost.[18]

Rotary Drums

In these reactors, biomaterial stays for only a few hours or days. Their effect is essentially homogenization and grinding of materials, only allowing start of composting by temperature control. As the reactors (diameter, 2 to 4 m; length, up to 45 m) have a slight inclination and have a rotating movement (0.5 to 2 rpm), the flow of material is continuous, countercurrent to the air supply; thus, the material leaving the reactor is cooled by fresh air and the material entering the drum makes contact with warm air, which favors bacterial growth (Figure 8).[42] These reactors are particularly interesting in processing the OFMSW for composting[46] as they allow for a faster startup of the process while reducing the malodorous emissions to the atmosphere. The DANO drum is a commercial rotating cylinder that has retention times of about 3 days with eventual interest in modern MBT plants. Most of the biological process has to be realized after leaving the reactor.

All of the systems described above can be used for composting different types of materials. Obviously, capital and operational costs of reactor processes are significantly higher than those of a windrow or aerated static pile process; thus, the residence time in the reactors is rarely enough to obtain mature compost. For this reason, a reactor is used in the early stages of composting to allow an easier control of the process, which facilitates odor emission reduction. After this processing, the material that comes out of the reactor must be stabilized in windrows or aerated static piles.[44] As the OFMSW treatment is difficult due to high cellulosic carbon content and sometimes low moisture and low porous structure, it may take as long as 6 months for compost maturation and stabilization, unlike compost from WWTPs that usually only takes 2–3 mo to be stabilized.

Vermicomposting

Vermicomposting is an alternative technology to conventional composting in which selected species of earthworms (*Lumbricus terrestris* and *Dendrobaena veneta*) and red worms (*Eisenia foetida,* Figure 9) eat organic material, absorb the nutrients they need, and excrete the rest, producing a humus-like material, known as vermicompost.[1,50] Vermicomposting should be applied preferentially to residues of fruit and vegetables, tea leaves, tea bags, coffee grounds, paper, and shredded green garden waste and has large application in agricultural wastes, including manure wastes.[51] Vermicompost produced from

FIGURE 8   Composting rotary drum: (left) side view and (right) view from the waste admission side.

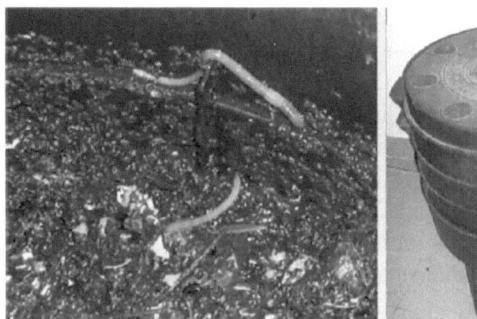

FIGURE 9   *Eisenia foetida* in a home/commercial vermicomposter.

OFMSW is a "nutritive biofertilizer" 4–5 times more powerful than conventional composts and even better than chemical fertilizers for better crop growth and safer food production.[52]

Vermicompost systems can range from inexpensive wood or plastic boxes to sophisticated modular units (Figure 9) that are self-contained and fully automated to keep controlled environmental conditions while processing waste into earthworm castings. Open systems range from windrows or beds of variable scales to open field operations. Usually, windrows are built no more than 30 cm deep for easy aeration of the beds.[51] Vermicompost has the advantage of being a material that growers can produce "on-farm" and use as a biofertilizer, using own feedstock and implementing either a midscale vermicomposting unit technology or a modular unit.[1]

Worm beds can be as long as 50 m; feedstock can be the resulting product of a MBT plant, whereas OFMSW is processed in a rotary drum. Worms should be kept at temperatures of 10–35°C. Moisture is fundamental to worms; they breathe oxygen through their moist skin.[51] Resulting worm casts can be used as a biofertilizer as they contain 5 times more N, 7 times more P, 1.5 times more Ca, 11 times more K, and 3 times more exchangeable Mg than the soil. These casts are also rich in humic acids (which condition the soil), have a perfect pH balance, and have plant growth factors similar to those found in seaweed.[51]

Under the optimum temperature (20–30°C) and moisture (60%–70%) conditions for worm breeding, about 5 kg of worms (ca. 10,000) can process 1 ton of waste into vermicompost in 30 days.[53]

Worms can be easily recovered for further processing because they do not like light (a few hours of direct exposure to sunlight can cause paralysis or death), dryness, or even some specific odors (onions), migrating naturally for places where fresh wastes have been added. Worms can also be harvested by using a trommel screen.[54]

Vermicompost produced from organic wastes, such as food and yard wastes, have enormous economic potential for increasing crop yields, suppressing attacks from pests, and controlling the spread of diseases.[55–57]

Evaluation of a Composting System

Among the various criteria used for assessing the system efficiency, the most important for MSW composting are product quality, rejected ratio, and recycling rate. Compost quality can be evaluated by its appearance and by functional characteristics and the nature of the contaminants, characteristics that are critical for its commercialization. Rejected ratio and recycling rate affect the quality of the product, as they affect the contaminant's concentration and the amount of final waste that must be disposed off in landfill.

Compost quality should constitute the main objective of the composting facility. Some of the aspects of quality, such as the degree of maturation and size of particles, can be corrected at the end of the process by increasing the curing time and by using physical processes for particle size reduction. However, other issues may be far more difficult to remedy, as when there is chemical contamination.

A composting factor should be not only thoroughly thought about when conceiving the biowaste management system but also observed during the initial phases of processing. Chemical contamination can be caused by heavy metals or other chemicals, often from domestic hazardous waste. Contamination can also be of physical type; that is, the product may be contaminated with inert materials (glass, plastics and metals, brick and concrete, etc.,) that arise from domestic or commercial waste either voluntarily or by mistake. Removal of physical contaminants may not be very difficult from a technological point of view but represents a high fraction of the economic costs of a composting facility, which cannot be completely eliminated because even source-separated waste can be contaminated. However, the use of a source-separated raw material is of primary importance for high-quality assurance of the compost.

Phases of Composting

Throughout the composting process, the microbial population will vary and different types of microorganisms will play distinct roles during the various stages of the process.[38] Thus, in the first step, there will be predominantly mesophilic bacteria that hydrolyze the easily fermentable organic material. The reactions are exothermic, leading to heat release and temperature rise. Thermophilic microorganisms (bacteria, fungi, and actinomycetes) begin to develop on a large scale—from 40°C.[58] The thermophilic microorganisms multiply, and as soon as the temperature reaches 55–60°C, the attack on complex molecules (carbohydrates, proteins, etc.,) starts, which will result in their transformation into simpler products (simple sugars, amino acids, etc.,) that are used by other microorganisms (Figure 10). If there is no external control, the temperature of the composting material can reach 80°C, which represents the limit for the thermophilic population, resulting in microbiological activity inhibition and survival only of spores of bacteria. As degradation is reduced, and temperature lowers, mesophilic bacteria, actinomycetes, and fungi (mostly those that were in the outer areas of the pile) gain further activity, attacking the most resistant compounds (such as lignin and cellulose, the less readily biodegradable components of biowaste).[1] Complex enzymatic reactions are responsible for humus production, mainly by lignin and protein combination. At this stage, protozoa and some higher organisms (worms, nematodes, and millipedes) can be found in compost.

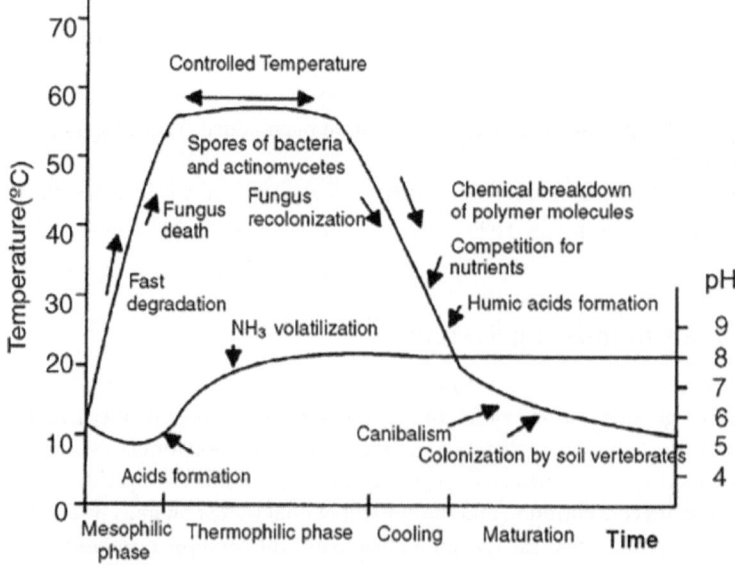

FIGURE 10　Microbiological transformations and temperature and pH profiles during a controlled composting process. (*Source:* Adapted from Neto and Mesquita.[38])

Process Parameters in Composting

Preparation of a composting process is not a simple task, especially for achieving optimal results. For this reason, in commercial applications, mechanization is a key factor, allowing for efficiently controlling the most important project factors (temperature, pH, moisture, C/N ratio, aeration rate, particle size, and mixing/turning).[58]

As any biological process, composting is conditioned by factors that affect the activity of the microorganisms involved in the process.

Temperature

Temperature is the process parameter that best describes the biological equilibrium and shows the efficiency of the process. Aerobic composting systems operate under mesophilic (20–40°C) and thermophilic (40–60°C) conditions that change along the process as a result of the respiration and metabolic activities of the organisms involved in the composting process.[1]

Temperature control of the composting material under static pile or in-vessel processes can be achieved through temperature monitoring and control of the air flow rate. In these processes, temperature can be controlled in the thermophilic range, around 60°C, which allows for the development of a more complex microbiological consortia, responsible for increasing the rate of decomposition of the OM (higher process efficiency), better compost sanitation or pathogen elimination, and elimination of weed seeds, insect larvae, and parasite eggs, among other advantages of the process.[38]

In windrow composting systems, temperature can only be controlled indirectly, by changing the turning frequency of the composting material. After turning, temperature lowers 5–10°C, rising again after only a few hours, as a result of the increase in oxygen availability. On the other hand, if temperature rises above 70°C, most of the microorganisms will die or enter the dormant phase, slowing the composting process and resulting to lower-quality compost. Also, high temperature can lead to humidity and nitrogen loss (through ammonia volatilization at a pH of 7.5).

In a system adequately controlled, temperature in the composting pile will rise up to 40–60°C from the second to the fourth day and will lower down to 35–38°C after 10 to 15 days because easily biodegradable organic material has already been converted. The increase in temperature also depends on factors such as nutrient availability, moisture content, particle size, aeration, turning of the pile, and the thermal insulation of the system (either exposed systems or confined systems).

Typical temperature profiles in non-controlled composting systems (with and without turning) are shown in Figure 11.

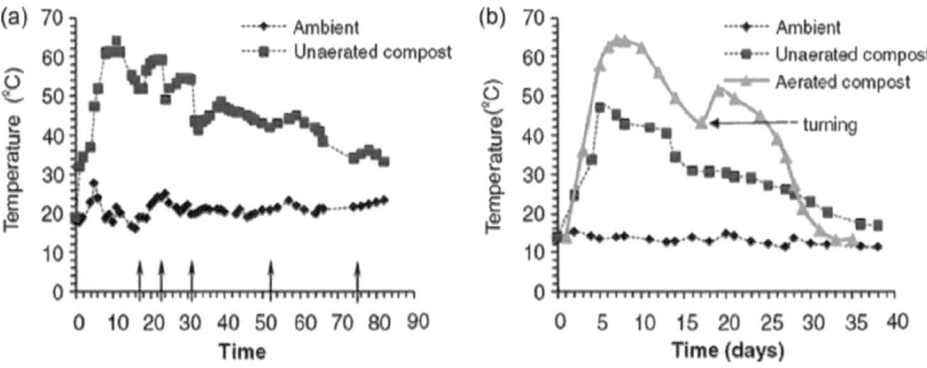

FIGURE 11 Typical temperature history observed in (a) windrow composting systems with turning[59] and (b) static piles with and without forced aeration.[59]

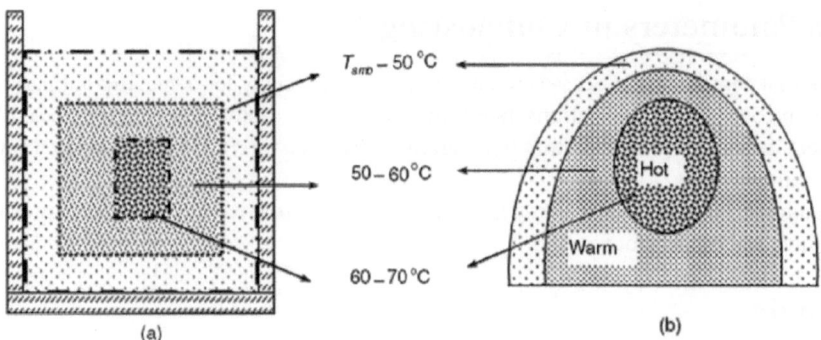

FIGURE 12 Temperature profile in a composting pile: (a) in-vessel/tunnel; (b) windrow.

It should also be noted that temperature is usually not uniform along the composting pile, the inner material having a higher temperature than the outer layers, as shown in Figure 12.

Moisture

Decomposition of the OM mainly depends on moisture content to enable microbiological activity. Microorganisms have, in their structure, about 90% water that is needed not only for new cell production but also and specially for the dissolution of the nutrients needed for cellular metabolism.

Moisture content can be adjusted either through mixing different materials or through water addition to an optimum between 40% and 70%. Below 30% humidity, composting rate dramatically decreases, but more than 70% moisture content may lead to pore filling with water, thus preventing oxygen from reaching the OM and inducing anaerobic conditions. This is why the most recommended value is 60%.[1,42]

Also turning of the composting material can help control moisture content. If the organic material has 55%–60% humidity and the composting period is 15 days, the first turning should be done on the third day and then every other day.[58]

Aeration

Aeration is very important in composting and holds two main functions: the first, and most important one, is to supply aerobic microorganisms with the oxygen they need for their respiration and metabolic activity, and the second function is to control temperature and, consequently, the rate of oxidation of the OM as well as odor emission.

Theoretically, the optimal rate of aeration would be the one that would allow for biological oxygen demand (BOD) supply during all the different composting steps. Nevertheless, parameters such as the nature of the composting material, particle size, aeration technology, moisture content, and porosity of the material in the pile may prevent full satisfaction of this requirement.[23] On the other hand, excessive aeration may cause excessive loss of humidity and heat leading to the misidentification of the end of the composting process.

In static pile systems, initial oxygen concentration in the pores is similar to that in the surrounding air (ca. 17%), because this is the air that was trapped in these spaces when the pile was built, whereas CO_2 concentration is significantly lower (ca. 0.5%–5%). As aerobic degradation occurs, oxygen concentration will decrease and CO_2 concentration will increase. Anaerobic conditions will develop at O_2 concentration lower than 5%, which can be prevented through introduction of air, vacuum induction, or turning of the material in the composting pile (which will also allow for a more efficient distribution of nutrients and microorganisms).

In systems with forced aeration, air flowrate and total air needed in the process are fundamental project parameters that help maintain the level of available O_2 very close to the required level throughout the whole process. The total amount of oxygen required and the total amount of CO_2 produced in the process can be estimated through the mass balance in Equation 1, where the composition of the OM for degradation and of the compost can be represented as $C_aH_bO_cN_d$ and $C_wH_xO_yN_z$, respectively.[58]

$$C_aH_bO_cN_d + 0.5(ny + 2s + e - c)O_2 \rightarrow$$

$$nC_wH_xO_yN_z + sCO_2 + rH_2O + (d - nz)NH_3 \qquad (1)$$

with $r = 0.5\,[b - nx - 3\,(d - nz)]$ and $s = a - nw$.

If ammonia produced in the process is converted into nitrate (nitrification), some more oxygen will be needed, as estimated using Equation 2.

$$NH_3 + 2O_2 \rightarrow HNO_3 + H_2O \qquad (2)$$

Particle Size

It could be expected that the lower the particle size of the organic material, the higher would be the rate of biochemical reactions, due to the increase in surface area exposed to microorganisms and to oxygen, which translates to the composting period being reduced. However, if particle size is too small, porosity will be reduced along with oxygen and CO_2 diffusion, which is fundamental during the thermophilic phase of the process, when oxygen consumption is higher.

An optimum particle size of 25–50 mm has been accepted for most of the materials, but this can vary between 13 and 75 mm depending on the nature of the composting material. A shredding operation may be needed prior to composting, which will entail an increased operational and financial cost.

Porosity of the composting material can be enhanced through introduction of bulking material that will allow for easier aeration while keeping structural characteristics that are essential for the construction of composting piles. Also, the height of the composting piles is based on particle size, in order to avoid excessive compaction during composting.

Feedstock Composition

The biomaterials used in composting have such a diversified composition that can usually supply all the elements that are fundamental for microbial metabolism (carbon, nitrogen, phosphorus, potassium, calcium, iron, copper, etc.).

Of these, the most important is undoubtedly the C/N ratio; an optimal ratio of 30/1 is recommended. Carbon is the source of energy and an essential factor for multiplication of cell material, and nitrogen is present (and available) in such an amount that it can be used for protein, amino acid, and nucleic acid formation.

Mixing of wastes with different compositions is usual since hardly ever a substrate has all the characteristics required for efficient and effective composting, leading to a compost with the recommended values of carbon/nitrogen ratio (C/N), moisture content, particle size, density, etc. Usually, these mixtures are prepared using carbon-rich vegetable waste and separate solid material (newsprint paper, manure, slurry or sludge, yard wastes) characterized by containing high levels of nitrogen, in order to obtain a C/N ratio of about 25/1 to 30/1[30] (Table 1). If the C/N ratio is lower than 30/1, there will be a loss of nitrogen as ammonia, causing malodorous emissions, but if C/N is higher than 80/1, nitrogen concentration will be so low that it will become a limiting factor. It must also be taken into consideration that in the OM, not all carbon will be in the readily available form, i.e., biodegradable, in contrast to nitrogen

TABLE 1   Nitrogen Content and C/N Ratio of Common Materials Used in Composting (Dry Basis)

Material	N (%, dry weight)	C/N Ratio[a]
Food processing wastes		
Fruit wastes	1.52	34.8
Mixed slaughterhouse wastes	7.0–10.0	2.0
Potato tops	1.5	25.0
Coffee grounds		20.0
Vegetable wastes	2.5–4.0	11–13
Manure		
Cow manure	1.7	18.0
Horse manure	2.3	25.0
Pig manure	3.75	20.0
Poultry manure	6.3	15.0
Sheep manure	3.75	22.0
Sludge		
Digested activated sludge	1.88	15.7
Raw activated sludge	5.6	6.3
Wood and straw		
Lumber mill wastes	0.13	170.0
Oat straw	1.05	48.0
Sawdust	0.10	200.0–500.0
Wheat straw	0.3	128.0
Wood (pine)	0.07	723.0
Paper		
Mixed paper	0.25	173
Newsprint	0.05	983
Brown paper	0.01	4490
Trade magazines	0.07	470
Junk mail	0.17	223
Yard wastes		
Grass clippings	2.15	20.1
Leaves (fleshly fallen)	0.5–1.0	40.0–80.0
Tree trimmings	3.1	16
Biomass		
Water hyacinth	1.96	20.0
Bermuda grass	1.96	24

Source: Rynk et al.[42] and Tchobanoglous et al.[58]

[a] C/N ratio based on total dry weight.

that is readily available. Thus, lignocellulosic material that is rich in carbon is very hardly biodegradable, so when it is present, the most adequate value for the C/N ratio would be 35/1 to 40/1.[60]

During the composting period, C/N ratio slowly decreases because when OM is degraded, about 65% of the carbon is released as CO_2 and the remainder is used with nitrogen in cell, which is released only when there is cellular death. The final compost should have a C/N ratio of 15/1.

pH Control

Although authors agree that pH affects biological processes, composting studies using MSW and wastewater sludge where the initial pH of the composting material was varied have shown that a self-regulation of pH occurs;[10] thus, pH is not a critical factor in the composting process. Most of the times, there is no need for pH correction, although in few particular situations, lime can be added for pH correction.

Most of the bacteria involved in the process operate better in the pH range of 6 to 7.5, but fungi prefer a pH of 7.5 to 9. In the composting process, there is a pH variation across time. Thus, in the first phase of the process, simple organic acids are produced, lowering pH to 5 or even less. After only 3 days, pH starts rising up to 8–8.5, as organic acids are being further transformed. In the cooling phase, the pH of the compost will be slightly reduced until 7–8 (Figure 13). Ideally, pH must be less than 8.5 so that nitrogen loss (as ammonia) is minimized.[58]

Odor Control

Odor problems arise from development of anaerobic conditions that allow for formation of organic acids, or ammonia release, most of which are extremely malodorous. Nevertheless, material in the pile can act as a biofilter that retains and degrades part of these compounds.

Pathogen Control

Pathogen destruction is of primary importance in composting systems as it is affected by temperature profile and aeration process. Pathogens can be eliminated under different conditions (*Salmonella* is destroyed in 15–20 min at 60°C, but at 55°C, it will take 1 hr). As a recommendation, if biowaste is kept at 70°C for 1–2 hr, all pathogens will be eliminated. It should be noted that in the second stage of the composting process (humification), the *Penicillium* fungus that destroys pathogens, acting as an antibiotic, appears.

Most European countries recommend a temperature of 55–65°C for 4–15 days to guarantee compost sanitization.[1,58]

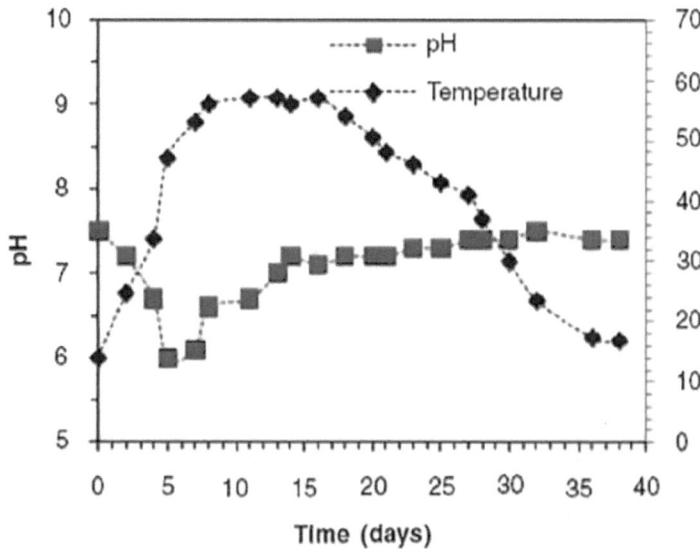

FIGURE 13   Typical temperature and pH range in windrow composting. (*Source:* Fonseca and Amorim.[59])

Evaluation of Maturation and Compost Maturity

As stated before, compost cannot be produced in record time; the length of the composting process depends on several factors, such as the feedstock material, particle size, nutrient balance, moisture content, and the composting technique. Although some authors and technologies claim to compost wastes in 7 or 20 days, this is not possible because compost produced in this time is not stabilized as total humification of the OM can only be achieved after about 110 days.[23] Thus, composting of garden or food waste materials can be successfully achieved in 3 mo under aerated, in-vessel, or even turned windrow systems but will take as long as 1 year in an unaerated static pile.

Compost maturation is fundamental in order to prevent ammonia liberation into the roots of cultures (at very low C/N ratios), biochemical reduction of soil nitrogen (when residual carbon in high C/N ratio compost is used by microorganisms), production of toxins that inhibit plant metabolism and seed germination, and microbiological activity that may lead to oxygen demand as well as other toxic effects to plants.[10]

The degree of maturation and stabilization of compost can be assessed using different methodologies: 1) final drop in temperature and self-heating capacity (the absence of temperature increase after turning and humidifying the biowaste up to 50% moisture content means that the process has ended); 2) amount of decomposable and resistant OM in the compost (lignin content higher than 30% means that compost is stable);[58] 3) chemical oxygen demand (COD) (a COD value below 350 mg O_2/g compost shows that the compost is stable); 4) oxygen uptake rate [it should be less than 40 mg of O_2 per kilogram of dry matter per hour for compost to be considered stable; a specific oxygen uptake rate (SOUR) test[61] can be performed]; 5) CO_2 production (it should be small, meaning that there is no significant microorganism activity); 6) C/N ratio (it usually decreases along the process, but it should be interpreted taking into account the initial characteristics of the OM and composting conditions. If OM has an initial C/N ratio of 35–40, the stabilized compost should have a C/N ratio of 18–20; however, if the initial C/N ratio is 10 or even lower as in manure wastes, an increase of the C/N ratio should be expected during the composting process);[17] 7) growth of the fungus *Chaetomium gracilis*;[58] and 8) the absence of ammonia in the compost and the presence of nitrates.[38]

Commercialization Issues

To be marketed, compost must have a uniform size; must be free of contaminants such as glass, plastics, and metals; and must be free from unpleasant odors. Typically, compost is a dark color product, has a wet soil smell, has a nice texture, and does not show the original raw material (Figure 14). The most common treatments before compost packaging are crushing and screening, although sometimes compost is commercialized as pellets.

FIGURE 14 How compost looks like: (left) from OFMSW; (right) from co-composting of slaughterhouse with sawdust and agricultural waste.

Conclusion

Methods that can be used to perform composting over biowastes of different nature have been presented, from the simpler and less expensive ones, to some more sophisticated and expensive technologies. Most of the composting processes are based on reactions involving microorganisms, but worms can also be used to perform the same task of degrading OM and producing a stabilized and mature product (that is sterilized and of good quality) that can be used as a soil conditioner or fertilizer. With CO_2 and H_2O being the only by-products, composting can be considered environmentally benign and can contribute to OM recycling. The influence of the main operational parameters has been discussed, and it was shown that under appropriate conditions, almost any OM can be composted.

References

1. Evans, G. *Biowaste and Biological Waste Treatment*; James and James (Science Publishers) Ltd: London, 2001.
2. Gama, P.S. *Recolha Selectiva e Reciclagem de Resíduos Sólidos Urbanos*. Curso sobre Valorização e Tratamento de Resíduos. Prevenção, Recolha Selectiva, Compostagem e Confinamento em Aterro. LNEC/APESB, Lisbon, 10th–12th December, 1996.
3. Directive 2008/98/EC of the European Parliament and of the Council, of 19 November 2008 on waste and repealing certain Directives, Official Journal of the European Union, Brussels, 2008; L312/3-L312/30.
4. Directive 99/31/EC of the Council, of 26 April 1999 on the landfill of waste, Official Journal of the European Union, 1999; L182/1-L182/19.
5. Davis, M.L.; Masten, S.J. *Principles of Environmental Engineering and Science*, 1st Ed.; McGraw-Hill Companies, Inc.: New York, 2004.
6. Abu Qdais, H.A.; Hamoda, M.F.; Newham, J. Analysis of residential solid waste at generation sites. Waste Manage Res. **1997**, *15* (4), 395–405.
7. Francou, C.; Lineres, M.; Derenne, S.; Le Villio-Poitrenaud, M.; Houot, S. Influence of green waste, biowaste and paper–cardboard initial ratios on organic matter transformations during composting. Bioresour. Technol. **2008**, *99* (18), 8926–8934.
8. Lino, F.A.M.; Ismail, K.A.R. Energy and environmental potential of solid waste in Brazil. Energy Policy **2011**, *39* (6), 3496–3502.
9. Directorate-General Environment A.2, CEC. Working document on biological treatment of biowaste—2nd draft. European Commission, Brussels, 2001; 1–22, available at http://www.ymparisto.fi/download.asp?contentid=5765 (accessed June 8, 2011).
10. Neto, J.T.P. On the Treatment of Municipal Refuse and Sewage Sludge Using Aerated Static Pile Composting—A Low Technology Approach. Ph.D. Thesis: University of Leeds, 1987.
11. Haug, R.T. *The Practical Handbook of Compost Engineering*; Lewis Publishers: Boca Raton, FL, 1993.
12. Hogan, J.A.; Miller, F.C.; Finstein, M.S. Physical modelling of composting ecosystem. Appl. Environ. Microbiol. **1989**, *55* (5), 1082–1092.
13. Golueke, C.G. *Biological Reclamation of Solid Wastes;* Rodale Press: Emmaus, PA, 1977; 249 pp.
14. Bertoldi, M. Compost quality and standard specifications: European perspective. In *Science and Engineering of Composting: Design, Environmental, Microbiological and Utilization Aspects*; Hoitink, H.A.J., Keener, H.M., Eds.; The Ohio State University, Renaissance Publications: Ohio, 1993; 523–535.
15. Epstein, E.; Taylor, J.M.; Chaney, R.L. Effects of sewage sludge and sludge compost applied to soil on some physical and chemical properties. J. Environ. Qual. **1976**, *5*, 422–426.
16. Avnimelech, Y. Organic residues in modern agriculture. In *The Role of Organic Matter in Modern Agriculture*; Chen, Y., Avnimelech, Y., Eds.; Martinus Nijhoff Publishers: Netherlands, 1986; 1–10.

17. Brito, L.M. Taxas de Mineralização da Matéria Orgânica nos Resíduos Sólidos Urbanos: Efeitos Agronómicos e Ambientais. In Seminário sobre Produção de Correctivos Orgânicos a Partir de Resíduos Sólidos Urbanos—Sua Importância para a Agricultura Nacional, Exponor (Matosinhos), April 8, 1997.

18. Corbitt, R.A. *Standard Handbook of Environmental Engineering,* 2nd Ed.; McGraw-Hill Book Company: New York, 1998; 1248 pp.

19. Mays, D.A.; Giordano, P.M. Landspreading municipal waste compost. BioCycle **1989,** *30* (3), 37–39.

20. Howard, A.; Wad, Y.D. *The Waste Products of Agriculture: Their Utilisation as Humus*; Oxford University Press: London, 1931.

21. Golueke, C.G. Composting refuse at Sacramento, California. Compost Sci. **1960,** *1* (3), 12–15.

22. McGauhey, P.H. Refuse composting plant at Norman, Oklahoma. Compost Sci. **1960,** *1* (3), 5–8.

23. Neto, J.T.P. Aspectos Tecnológicos da produção e Quali-dade do Composto Orgânico a partir da Compostagem de Lixo Urbano. In Seminário sobre Produção de Correctivos Orgânicos a Partir de Resíduos Sólidos Urbanos—Sua Importância para a Agricultura Nacional, Exponor (Matosinhos), April 8, 1997.

24. Lu, Y.; Wu, X; Guo, J. Characteristics of municipal solid waste and sewage sludge co-composting. Waste Manage. **2009,** *29* (3), 1152–1157.

25. Black, R.J. The solid waste problem in metropolitan areas. California Vector Views **1964,** *11* (9), 51.

26. Diaz, L.F. Combining experience with common sense. BioCycle **1982,** *30* (10), 48–49.

27. Golueke, C.G.; Diaz, L.F. Historical review of composting and its role in municipal waste management. In Proceedings of the Science of Composting; Bertoldi, M., Sequi, P., Lemmes, B., Papi, T., Eds.; Chapman & Hall: Glasgow, 1996; 3–14.

28. Ferrero, G., L'Hermite, P. Composting—Progress to date in the European Economic Community and prospects for the future. In *Composting of Agriculture and other Wastes;* Gasser, J.K., Ed.; Elsevier Applied Science Pub.: Amsterdam, 1985; 3–10.

29. Lopez-Real, J.M. Composting of agricultural wastes. In Proceedings of the Science of Composting; Bertoldi, M., Sequi, P., Lemmes, B., Papi, T., Eds.; Chapman & Hall: Glasgow, 1996; 542–550.

30. Mesquita, M.M.F. Compostagem de material sólido—Principais vantagens e aspectos mais problemáticos. In Seminário para apresentação e discussão do Plano de adaptação à legislação ambiental pelo sector da suinicultura. LNEC e Federação Portuguesa das Associações de Suinicultores: LNEC, Lisbon, Jan. 22–23, 1996.

31. Hamoda, M.F.; Abu Qdais, H.A.; Newham, J. Evaluation of municipal solid waste composting kinetics. Resour., Conserv. Recycl. **1998,** *23* (4), 209–223.

32. John Paul, J.A.; Karmegam, N.; Daniel, T. Municipal solid waste (MSW) vermicomposting with an epigeic earthworm, *Perionyx ceylanensis* Mich. Bioresour. Technol. **2008,** *102* (12), 6769–6773.

33. Elango, D.; Thinakaran, N.; Panneerselvam, P.; Sivanesan, S. Thermophilic composting of municipal solid waste. Appl. Energy **2009,** *86* (5), 663–668.

34. Tintner, J.; Smidt, E.; Böhm, K.; Binner, E. Investigations of biological processes in Austrian MBT plants. Waste Manage. **2010,** *30* (10), 1903–1907.

35. Sims, J.T.; Murphy, D.W.; Handwerker, T.S. Composting of poultry wastes—Implications for dead poultry disposal and manure management. J. Sustainable Agric. **1992,** *2* (4), 67–82.

36. Mesquita, M.M.F. Aplicações Inovadoras da Compostagem e Especificações Técnicas para Avaliação da Qualidade do Composto. In Curso sobre Valorização e Tratamento de Resíduos. Prevenção, Recolha Selectiva, Compostagem e Confinamento em Aterro. LNEC/APESB, Lisbon, Dec 10–12, 1996.

37. Williams, P.T. *Waste Treatment and Disposal,* 2nd Ed.; John Wiley & Sons, Ltd.: Chichester, West Sussex, 2005.

38. Neto, J.T.P.; Mesquita, M.M.F. *Compostagem de Resíduos Sólidos Urbanos. Aspectos Teóricos, Operacionais e Epidemiológicos*; LNEC: Lisbon, 1992.

39. de Bertoldi, M.; Zucconi, F. Composting of organic residues. In *Bioenvironmental Systems*; Wise, D.L., Ed.; CRC Press: Boca Raton, Florida, 1987; Vol. III.

40. Loehr, R.C. *Pollution Control for Agriculture,* 2nd Ed.; Academic Press: Orlando, Florida, 1984.

41. Neto, J.P.; Stentiford, E.I. The main process constraints in composting. In Proceedings of the First Italian-Brazilian Symposium of Sanitary Engineering, Rio de Janeiro, Brazil, Mar 29–Apr 3, 1992.

42. Rynk, R.; van de Kamp, M.; Willson, G.B; Singley, M.E.; Richard, T.L.; Kolega, J.J.; Gouin, F.R.; Laliberty, Jr., L.; Kay, D.; Murphy, D.W.; Hoitink, H.A.J.; Brinton, W.F. *On Farm Composting.* Northeast Regional Agricultural Engineering Service. Cooperative extension. NRAES-54, Rynk, R., Ed., NRAES: Ithaca, New York, 1992, available at http://infohouse.p2ric.org/ref/24/23702.pdf (accessed May 20, 2012).

43. Lipor, Terra-à-Terra Project, 2007–2011, available at http://www.hortadaformiga.com/gb/conteudos.cfm?ss=8 (accessed August 2011).

44. Richard, T.L. MSW Composting: Biological Processing. Cornell Department of Agricultural and Biological Engineering, New York State College of Agriculture and Life Sciences: Cornell University, Ithaca, 1998, available at http://www.cals.cornell.edu/dept/compost/MSW.FactSheets/msw.fs2.html (accessed August 2011).

45. Risse, M.; Faucette, B. Food Waste Composting. Institutional and Industrial Applications. University of Georgia College of Agricultural and Environmental Sciences, 2000, available at http://www.caes.uga.edu/applications/publications/files/pdf/B%201189_2.PDF (accessed August 2011).

46. Graves, R.E.; Hattemer, G.M. Chapter 2—Composting, In *National Engineering Handbook,* Part 637—Environmental Engineering, United States Department of Agriculture, Natural Resources Conservation Service, 2010.

47. Misra, R.V.; Roy, R.N.; Hiraoka, H. *On-Farm Composting Methods*; Food and Agriculture Organization of the United Nations: Rome, 2003.

48. Diaz, L.F., Savage, G.M., Eggerth, L.L.; Golueke, C.G. *Composting and Recycling Municipal Solid Waste*; Lewis Publishers, CRC Press: Boca Raton, Florida, 1993.

49. Diaz, L.F.; Savage, G.M.; Eggerth, L.L. Solid Waste Management, United Nations Environment Programme, UNEP IETC and CalRecovery Inc., 2005, available at http://www.unep.or.jp/ietc/Publications/spc/Solid_Waste_Management/ (accessed August 2011).

50. Arancon, N.Q.; Edwards, C.A. The use of earthworms in the breakdown of organic wastes to produce vermicomposts and animal feed protein. In *Earthworm Ecology*; Edwards, C.A. Ed.; CRC Press: Boca Raton, Florida, 2004; 345–379.

51. Munroe, G. *Manual of On-Farm Vermicomposting and Vermiculture;* Organic Agriculture Centre of Canada, 2007.

52. Sinha, R.K.; Agarwal, S.; Chauhan, K.; Soni, B.K. Vermiculture technology: Reviving the dreams of Sir Charles Darwin for scientific use of earthworms in sustainable development programs. J. Technol. Investment **2010,** *1* (3), 155–172.

53. Sinha, R.K, Herat, S., Valani, D.; Chauhan, K. Earth-worms—The environmental engineers: Review of vermiculture technologies for environmental management and resource development. Int. J. Global Environ. Issues **2010,** *10* (3/4), 265–292.

54. Bogdanov, P. Commercial Vermiculture: How to Build a Thriving Business in Redworms; VermiCoPress: Oregon, 1996.

55. Atiyeh, R.M.; Subler, S.; Edwards, C.A.; Bachman, G.; Metzger, J.D.; Shuster, W. Effects of vermicomposts and composts on plant growth in horticulture container media and soil. Pedobiologia **2000,** *44,* 579–590.

56. Arancon, N.Q.; Edwards, C.A. Effects of vermicomposts on plant growth. In Proceedings of the Vermi-Technologies Symposium for Developing Countries, Department of Science and Technology—Philippine Council for Aquatic and Marine Research and Development, Los Banos, Philippines, 2006, http://www.slocountyworms.com/wp-content/uploads/2010/12/EFFECTS-OF-ERMICOMPOSTS-ON-PLANT-GROWTH.pdf (accessed August 2011).

57. Arancon, N.A.; Edwards, C.A.; Babenko, A.; Cannon, J.; Galvis, P.; Metzger, J.D. Influences of vermicomposts, produced by earthworms and microorganisms from cattle manure, food waste and paper waste, on the germination, growth and flowering of petunias in the greenhouse. Appl. Soil Ecol. **2008,** *39* (1), 91–99.

58. Tchobanoglous, G.; Theisen, H.; Vigil, S. *Integrated Solid Waste Management. Engineering Principles and Management Issues*; McGraw-Hill Book Company: Singapore, 1993.

59. Fonseca, E.; Amorim, J. Quantificação dos Resíduos Sólidos Gerados no ISEP: Projecto e Operação de Uma Unidade de Compostagem Para Tratamento da Fracção Biodegradável, *CESE em Engenharia Química—Tecnologias de Protecção Ambiental/Projecto*, ISEP/IPP: Porto, 1999.

60. Cross, F.; O'Leary, P.; Walsh, O. Operation and maintenance considerations for waste-to-energy systems. Waste Age **1987,** *18* (8).

61. Lasaradi, K.E.; Stentiford, E.I. A simple respirometric technique for assessing compost stability. Water Res. **1998,** *32* (12), 3717–3723.

40

Insects and Mites: Biological Control

Ann E. Hajek

Introduction

Biological control is defined as the use of natural enemies to suppress a pest population, making the pests and their associated damage less abundant. Natural enemies were first used to control insect pests when farmers in ancient China and Yemen moved colonies of predaceous ants to control pests of tree crops. Today, the natural enemies used to control insect and mite pests include a diversity of predators, parasitoids, and pathogens. Specific strategies have been developed for release of natural enemies or enhancement of their persistence and activity. Biological control has been used very successfully for permanent suppression of introduced pests. Among natural enemies applied for shorter-term control in 1990, even the most widely used biological control agent, *Bacillus thuringiensis*, accounted for <1% of the insecticide market. However, biological control agents are widely used for control in environmentally sensitive areas or controlled environments and constitute important components of integrated pest management programs.

Strategies for Using Biological Control

Natural enemies can be used in a variety of very different ways. The first major uses of natural enemies for pest control were directed at control of introduced insect pests. Natural enemies from the land of origin of introduced pests were released in areas of pest introduction. This strategy, called classical biological control, now also includes introduction of exotic natural enemies to control native pests. In all cases, a high degree of host specificity is required in the natural enemies to be introduced. After the exotic natural enemy is established in the new location, its effectiveness is based on population increases in response to increasing densities of pest populations. Classical biological control can be dramatically effective, with 34% of insect natural enemies that are released becoming established and 17% completely controlling devastating pests. Classical biological control is known to be extremely cost effective with cost benefit estimates of up to 200:1, if a program is successful at establishing an effective natural enemy.

A second strategy, augmentation, involves releasing natural enemies for pest control, usually in instances where natural enemies can be effective but are not sustained in the environment at high enough densities to provide control. Inundative augmentation is used when only the natural enemies that are released in high numbers are expected to exert control. Under inoculative augmentation, control effects are more delayed and are predominantly exerted by the progeny of the released organisms. Natural enemies used for augmentation are often mass reared, so understanding requirements for mass production of high quality natural enemies that are healthy and active after shipment and release is critical for the use of this strategy.

The third major strategy, conservation, involves manipulations to enhance the persistence and activity of natural enemies already occurring in the environment. This strategy takes on a diversity of forms based on requirements of the individual natural enemies. To cause less mortality of natural enemies, use of synthetic chemical pesticides that kill natural enemies can be altered in different ways ranging from eliminating their use to selecting pesticides with less impact on natural enemies to timing pesticide applications to minimize the effect on natural enemies. Alternatively, natural enemy populations can be increased by maintaining or improving the environment to provide ideal conditions. For example, irrigating, strip-harvesting, intercropping, retaining vegetation adjacent to crops, and planting cover crops all have been shown to provide favorable habitats and food to maintain or increase populations of natural enemies. In a program to control the brown plant- hopper on rice in south and southeast Asia, the activity of a suite of native natural enemies, aided by host plant resistance and application of insecticides only when absolutely necessary, provided better control than pesticides alone.

Types of Natural Enemies

Predators

Predators are generally larger than their prey and each usually consumes several prey individuals either for growth of immatures or for subsistence and reproduction of adults. The predatory life style is very common among insects and mites but predators with the most importance for biological control belong to four insect orders (Hemiptera, Coleoptera, Diptera, and Hymenoptera) and eight mite families (Figure 1). Predators feed on a diversity of prey life stages, from eggs to adults, and display a range of host specificity, from feeding only on one prey species to generalized feeding on many prey species. One of the most famous examples of classical biological control is the introduction of the highly host-specific Vedalia beetle that was imported from Australia and released against outbreak populations of the introduced cottony cushion scale threatening the southern California citrus industry. After the 1888–1889 releases of this predator, cottony cushion scale populations decreased dramatically and, by 1890, scale populations had been decimated. The immense success of this early program was instrumental in building interest in use of natural enemies for biological control. As a second example, in more recent years, phytoseiid mites attacking tetranychid spider mites have been developed for mass release in greenhouses or on some outdoor crops. Pesticide-resistant mite strains have also been developed for use against spider mites attacking tree crops.

Parasitoids

Parasitoids develop at the expense of a single host and usually kill their hosts. Parasitoids have been used extensively for biological control because, in contrast to predators, the impressive degree of host specificity often characteristic of parasitoids leads to sensitive responses to changes in host density. Parasitoids used for biological control are predominantly in the Order Hymenoptera (Figure 2) with less common use of Diptera. The immature parasitoid is usually a featureless larva associated with the host while winged adults disperse to mate and find new hosts. To enable their close association with hosts, parasitoids have adopted amazing and diverse life cycles. Different species of parasitoids attack different life stages of hosts (egg through adult) and can develop either externally on hosts or internally within

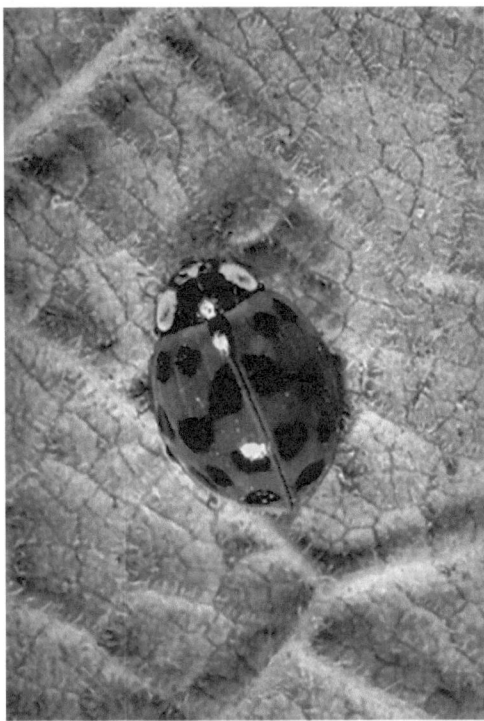

FIGURE 1   An adult of the multicolored Asian lady beetle (Harmonia axyridis), which was introduced from Asia for control of aphids and scales (length ca. 1 cm). Both larvae and adults are predatory. (Photo by J. Ogrodnick.)

FIGURE 2   An adult of the parasitoid Muscidifurax raptor (length ca. 2 mm) parasitizing house flies. Females lay eggs in fly puparia, larvae grow while consuming the fly pupae and winged adults then emerge to mate and find more hosts. These flies can be purchased and released to augment naturally occurring populations. (Photo by S. Long.)

hosts. One to many parasitoid individuals of one or more species can develop within a host. Parasitoids have been more widely used for classical biological control than either predators or pathogens. In recent years, the tiny wasp Epidinicarsis lopezi was released by land and air in 34 countries in Africa to control the introduced cassava mealybug. Due to the activity of this wasp, cassava mealybug is no longer considered a problem, saving African farmers hundreds of millions of dollars in reduced crop losses. Some parasitoids widely used for augmentative biological control are tiny species of Trichogramma attacking eggs of Lepidoptera and Encarsia formosa, a member of the Aphelinidae that attacks whiteflies. Both of these tiny wasps are mass produced in insectaries and shipped to users for release against pest populations threatening crops.

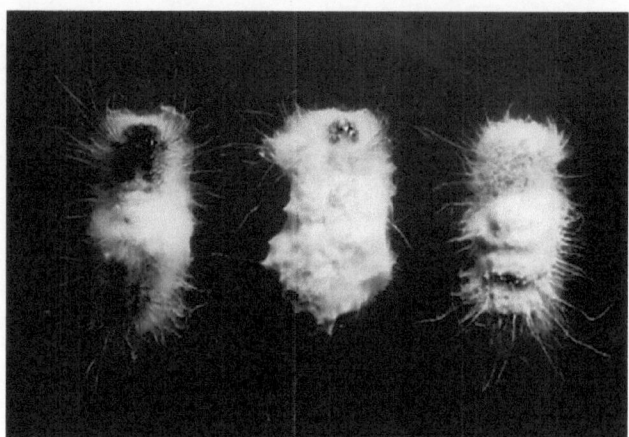

FIGURE 3   Gypsy moth (Lymantria dispar) larvae killed by the entomopathogenic fungus Paecilomyces farinosus (each larva ca. 3 cm in length). When infecting, this fungus penetrates externally through the larval cuticle, then increases within the host and, after host death, grows out through the integument to produce spores that will infect healthy hosts. (Photo by T. Ebaugh.)

Pathogens

Microorganisms that are parasitic, referred to as pathogens, are masters at exploiting insect and mite hosts (Figure 3). Pathogens important for biological control include a diversity of viruses, bacteria, fungi, protozoa, and nematodes. This range of types of pathogens exhibits a comparable medley of diverse interactions with their hosts. Of primary importance, viruses, bacteria, and most protozoa must be ingested by hosts in order to infect, while fungi and some protozoa can penetrate directly through the host cuticle. The nematodes of greatest importance to biological control, Steinernema and Heterorhabditis, can enter hosts through body openings although some possess the ability to penetrate directly through the cuticle. While some pathogens have mechanisms for active dispersal to find new hosts, host finding is generally not directed and these pathogens rely principally on their production of huge numbers of progeny in order to be assured of locating healthy hosts. Associations between pathogens and hosts range from facultative to obligate but pathogens important for biological control are all specialized for infecting only insects and mites. Pathogens have been used for classical biological control relatively infrequently although some programs have provided complete control. Much of the development of pathogens has been directed toward inundative augmentation of mass- produced microbes. The bacterium *Bacillus thuringiensis* is applied more than any other biological control agent. Strains of this bacterium predominantly kill Lepidoptera, Diptera, and Coleoptera through the activity of a toxin destroying the integrity of the gut. For many years, this bacterium was applied principally as a spray but recently several crop plants have been engineered to express genes encoding the toxin.

Biological Control in Practice

There is great demand for use of biological control programs to eliminate insect and mite pests, especially in environmentally sensitive areas and areas where humans live. Use of natural enemies to control pests can be highly effective due to the diversity of types of natural enemies and approaches. However, because biological control involves management of living organisms, it can be somewhat unpredictable. Therefore, biological control programs generally are tailored to specific pest systems and to optimize control, often require knowledge of the biology and ecology of the insect or mite host. Biological control has proven to be most effective under certain conditions (Table 1), although these generalities should not prevent investigations of use of biological control for alternative situations.

TABLE 1 Characteristics of Systems and Conditions More Commonly Associated with Successful Biological Control

Highly efficient natural enemy
Less mobile pest living in an exposed location
Perennial crop, natural habitat, or controlled environment, for example, a greenhouse
Crop or environment where some pest damage is tolerated
Controls for other pests do not interfere with the activity of natural enemies

Introduced pests are not insignificant, comprising 39% of the 600 major arthropods pests in the United States. By 1990, more than 4300 introductions of exotic parasitoids and predators had been made to control insect pests, many of which were introduced. Due to its low cost and permanent effectiveness when successful, classical biological control continues as one of the first control strategies to be investigated after a new pest has been introduced.

Augmentation and conservation are now often employed as important parts of integrated pest management programs. Although since the late 1940s (the start of the DDT era), for most pests synthetic chemical pesticides have been the first control strategy considered, use of natural enemies for control is increasing, especially for specific applications and systems.

Bibliography

1. Barbosa, P., Ed.; *Conservation Biological Control*; Academic Press: San Diego, 1998; 396.
2. Bellows, T.S.; Fisher, T.W.; Caltagirone, L.E.; Dahlsten, D.L.; Gordh, G.; Huffaker, C.B., Eds.; *Handbook of Biological Control: Principles and Applications of Biological Control*; Academic Press: San Diego, 1999; 1046.
3. Evans, H.F., Ed.; *Microbial Insecticides: Novelty or Necessity?*; British Crop Protection Council: Surrey, U.K., 1997; 301.
4. Flint, M.L.; Dreistadt, S.H. *Natural Enemies Handbook: An Illustrated Guide to Biological Pest Control*; University of California Press: Berkeley, CA, 1998; 154.
5. Greathead, D.; Waage, J., Eds.; *Insect Parasitoids*; Academic Press: London, 1986; 389.
6. Lacey, L.A.; Goettel, M.S. Current developments in microbial control of insect pests and prospects for the early 21st century. Entomophaga **1995**, 40 (1), 3–27.
7. Lynch, J.M.; Hokkanen, H.M.T., Eds.; *Biological Control. Benefits and Risks*, Cambridge University Press: Cambridge, U.K., 1995; 304.
8. New, T.R. *Insects As Predators*; New South Wales University Press: Kensington, Australia, 1991; 178.
9. Van Driesche, R.G.; Bellows, T.S., Jr. *Biological Control*; Chapman and Hall: New York, 1996; 539.

41

Invasion Biology

Jennifer Ruesink

Introduction

Biological invasions occur when species or distinct populations breach biogeographical barriers and extend their ranges to areas where they were not historically present.[1–3] Invasion biology concerns the causes and consequences of these new species, which are also referred to as invasive, introduced, alien, exotic, nonnative, or nonindigenous species.

Biological invasions occur in five steps: arrival, establishment, population growth, population spread, and impact (Table 1). Only a small proportion of species that arrive actually establish, and so forth; thus, each step acts as a filter for the invasion process. Successful invasion depends on characteristics of the invading species, recipient environment, and process by which the two are brought together.[4] However, no widely accepted method currently exists for identifying the characteristics that promote invasion a priori.[5]

TABLE 1 Reasons for and Responses to Five Steps of an Invasion

Step	Why?	Control
Arrival	Intentional releases, intentional imports, unintended hitchhiking (by product)	Risk assessment—choice of species, treatment, or quarantine of vectors
Establishment	Suitable abiotic conditions, available resources, propagule pressure	Reduce pathways, containment
Population growth	Intrinsic rate of reproduction, apomixis/vegetative reproduction	Early eradication
Population spread	Mobility—dispersal, home range size, transport by humans	Eradicate new populations, reduce human transport
Impact	Density (abundant resources), few enemies, per capita effect (new role), alteration of resource base	Effective screening prior to introduction; mechanical, chemical, or biological control; make environment less suitable

Why Do New Species Invade?

Species have always expanded their ranges, but the pace of invasion has accelerated recently due to increased human travel and trade.[5–7] Humans transport species in three ways: (1) on purpose, with the intention that they will grow in outdoor environments (fish and game, plantation trees, and biological control agents), (2) on purpose but with no intention that they will establish (pets, horticultural and agricultural plants, aquaculture, and sterile releases), and (3) accidentally (hitchhiking on packages, live imports, and people). The contribution of each of these main pathways varies among taxa. Of South Africa's weeds, 89% were intentionally introduced[8] but only 11% of insect invaders in North America were intentional.[9] Ducks, pheasants, pigeons, finches, and parrots have more introduced species worldwide than would be expected by chance because these bird families include many pet or game species.[10]

Which Species Invade?

Propagule Pressure

Species can invade a new area if abiotic (especially climatic) conditions are suitable and an exploitable resource exists. Those species that can invade, will invade if given sufficient opportunity.[11–13] High rates of arrival have been termed "propagule pressure" and can occur either through numerous releases or releases of many individuals. Some of the best evidence that propagule pressure affects invasion come from compilations of biocontrol introductions: the successful establishment of insect predators rose seven times when the number of introductions doubled, and releases >31,200 individuals were eight times more successful than those of <5000 individuals.[14]

Species Traits

For a given propagule pressure, some species may be more likely to invade than others. Traits promoting invasion could include the ability to increase rapidly from low density, a generalist diet, and broad climatic tolerance. Although statistical relationships between species traits and invasibility are often weak,[15] analyses of certain taxa introduced to particular environments have been successful.[16,17] For instance, invasive species of pines in South Africa tend to have small seeds, short intervals between reproductive bouts, and short times to maturity, whereas noninvasive species show the opposite traits.[18] For woody plants, those that have become invasive in North America often reproduce vegetatively, germinate easily, and have a long fruiting period.[19]

Where Do Species Invade?

In addition to species traits, characteristics of the recipient community may also influence the ease with which new species invade. Indeed, particular conditions in native environments (e.g., disturbance and species interactions) may select for species that are effective invaders of other areas, thus resulting in asymmetric patterns of invasion. For instance, European insects have invaded forests in North America but not vice versa,[20] and there has been a unidirectional appearance of tillering grasses in bunchgrass habitats worldwide.[21] Assemblages may be more resistant to invasion if they are undisturbed[22,23] or contain many natural enemies.[24,25] For many years, it was a rule of thumb that species-poor islands were invasible and species-rich tropics were not.[1] However, surveys of plant assemblages indicate that areas of naturally high species richness tend also to have numerous invaders, perhaps because soil and climate conditions are generally conducive to plant growth.[26] Although species richness per se may not influence invasion, loss of species could make systems easier to invade.

Habitat alteration (flooding, drought, fire, wind, eutrophication, and channeling) creates new conditions that are suitable for a new suite of species. These species are often introduced. For instance, western North America hosts many conspicuous invasive fishes in part because once fast-flowing rivers are now lakes separated by dams.[27] Roadsides in North America contain a high proportion of European plant species, probably because European plants have had millenia to evolve to take advantage of disturbance, whereas disturbance in North American habitats has risen recently.[28] Based on an Australian study of two invaders, both physical disturbance and nutrient addition improve the performance of invasive plants.[29]

Which Species Have Effects and Where?

Impacts of invaders relevant to pest management include the ecological and/or economic damage from pests and the effectiveness of biocontrol agents.[30] Impacts are expected to be particularly pronounced when species reach high abundances or have high per capita effects.[31,32] High abundances can occur when species escape limits to population growth (abundant resources or few enemies). High per capita effects may arise when a species plays a new role, especially by altering the resource base. For instance, many of the worst plant invaders of natural areas are nitrogen fixers (which alter nutrients) or climbing vines (which alter light).[33]

A disproportionate number of invasive species have harmful effects, more so if introduced accidentally rather than intentionally. In Japan, for instance, 8% of native insects are considered pests, but 72% of introduced insects are pests.[34] Of agricultural weeds in North America, 50–75% are nonindigenous.[6] On the other hand, only 1–6% of nonindigenous plant species in Great Britain have become weedy or widespread.[8] Some introduced species may be problematic due to an absence of natural enemies, but this escape from control does not appear to be entirely general. Based on a compilation of life tables for 124 holometabolous insects, mortality rates due to parasitoids, predators, and diseases do not differ for native and introduced species.[35]

Species introduced accidentally often have harmful effects, and, conversely, species introduced intentionally have beneficial effects less often than desired. Of 463 grasses and legumes introduced to Australia to improve pasture, only 5% have raised productivity.[36] About a third of established biocontrol insects actually reduce the target organism.[37] Although impact in the native environment is not necessarily a useful indicator of what a species will do once introduced, one indicator of potential impact is the fate of prior introductions.[19]

Pest status tends to be based on economic considerations, but introduced species also cause ecological damage. Invasive species contribute to endangerment of nearly 50% of species listed under the United States Endangered Species Act,[38,39] and they have dramatically altered the structure and function of ecosystems.[40,41] Ecological and economic effects have the same root causes—abundance and high per capita effects of invaders—but the affected habitats can be quite distinct. Only 25% of plants that cause problems in natural areas are also agricultural weeds.[33] Thus, screening procedures that keep out economic pests would fail to restrict many species that cause ecological harm.

Time Lags and Surprises

Introduced species have occasionally surprised researchers by expanding to previously intolerable places or irrupting after remaining localized and rare for many years. "Boom and bust" patterns have also been observed, in which an invader initially reaches high abundance and then declines, sometimes even going extinct. Tolerance of new conditions (e.g., temperature or host plants) may require genetic adaptation, which could result in time lags before invasion.[42] Of 184 woody species currently considered invasive near Brandenburg, Germany, 51% did not appear to be invasive until >200 years after their initial introduction.[43] However, only 7% of 627 cases involving biocontrol introductions showed time lags before population increase, whereas 28% increased and 27% went extinct immediately.[44]

Regardless of frequency, cases in which invaders have unexpected impacts have become well known, especially for biocontrol agents with nontarget effects such as feeding on endangered species (e.g., the weevil *Rhinocyllus conicus* on thistles and the moth *Cactoblastis* on *Opuntia* cacti)[45-47] or competing with natives (e.g., the ladybird beetle *Coccinella septempunctata*)[5]

Can Invasions Be Controlled?

The first steps of an invasion can be controlled by limiting entry or vigilantly eliminating newly established populations of invaders. For instance, an assessment of species associated with raw logs indicated a potential loss of billions of dollars due to forest pests if Siberian larch was not treated prior to import into the United States.[48] For many years, medflies (*Ceratitis capitata*) have epitomized the notion that "an ounce of prevention is worth a pound of cure." California spent $100 million to eradicate an incipient invasion in 1981, thereby preventing nine times that amount of crop damage. By 1996, however, medflies were apparently established and spreading. The extent of the invasion makes eradication unlikely.[49] Humans simply have to learn to live with these naturalized species.[50]

Control during the last steps of an invasion usually involves chemical, biological, or mechanical reduction of unwanted species in areas where effects are most serious. Control efforts at this stage would benefit from considerations of demography and behavior of invaders. For instance, seedling competition among annual plants is often fierce, so efforts to reduce seed production will not reduce plant numbers. Instead, control efforts should be directed at reducing seedling growth and survival.[51] Knowing how insects move among microhabitats could aid in trap placement or crafting habitats that promote desirable species and discourage undesirable ones.[52] Knowing encounter and feeding rates could aid in calculating the number of consumers necessary for effective biological control.

Species invasions are a form of ecological gambling in which the consequences of any particular introduction are uncertain, despite an emerging framework of factors contributing to high risk invasions. The influx of new species can be slowed by reducing pathways for introduction and intentionally introducing species only when beneficial effects will be large and native alternatives do not exist.[53] Distinct biotas are valuable and intriguing but increasingly difficult to maintain under pressures of globalization.

References

1. Elton, C.S. *The Ecology of Invasions by Animals and Plants*; Chapman and Hall: London, 1985.
2. Williamson, M. *Biological Invasions*; Chapman and Hall: London, 1996.
3. Mack, R.N.; Simberloff, D.; Lonsdale, W.M.; Evans, H.; Clout, M.; Bazzaz, F.A. Biotic invasions: causes, epidemiology, global consequences, and control. *Ecol. Appl.* **2000**, *10* (3), 689–710.
4. Lodge, D.M. Biological invasions: lessons for ecology. *Tr. Ecol. Evol.* **1993**, *8*, 133–137.
5. Ruesink, J.L.; Parker, I.M.; Groom, M.J.; Kareiva, P.M. Reducing the risks of nonindigenous species introductions: guilty until proven innocent. *Bioscience* **1995**, *45* (7), 465–477.
6. Office of Technology Assessment. *Harmful Non-Indigenous Species in the United States*; U.S. Government Printing Office: Washington, DC, 1993.
7. Cohen, A.N.; Carlton, J.T. Accelerating invasion rate in a highly invaded estuary. *Science* **1998**, *279*, 555–558.
8. Crawley, M.J.; Harvey, P.H.; Purvis, A. Comparative ecology of the native and alien floras of the British isles. *Phil. Trans. R. Soc. Lond. B.* **1996**, *351*, 1251–1259.
9. Sailer, R.I. History of insect introductions. In *Exotic Plant Pests and North American Agriculture*; Wilson, C.L., Graham, C.L., Eds.; Academic Press: New York, 1983; 15–38.
10. Lockwood, J.L. Using taxonomy to predict success among introduced avifauna: relative importance of transport and establishment. *Conserv. Biol.* **1999**, *13* (3), 560–567.

11. Veltman, C.J.; Nee, S.; Crawley, M.J. Correlates of introduction success in exotic New Zealand birds. *Amer. Nat.* **1996**, *147* (3), 542–557.

12. Green, R.E. The influence of numbers released on the outcome of attempts to introduce exotic bird species to New Zealand. *J. Anim. Ecol.* **1997**, *66*, 25–35.

13. Duncan, D.P. The role of competition and introduction effort in the success of passeriform birds introduced to New Zealand. *Amer. Nat.* **1997**, *149* (5), 903–915.

14. Beirne, B. Biological control attempts by introductions against pest insects in the field in Canada. *Can. Entomol.* **1975**, *107*, 225–236.

15. Goodwin, B.J.; McAllister, A.J.; Fahrig, L. Predicting invasiveness of plant species based on biological information. *Conserv. Biol.* **1999**, *13* (2), 422–426.

16. Panetta, F.D. A system of assessing proposed plant introductions for weed potential. *Plant Prot. Quarterly* **1993**, *8*, 10–14.

17. White, P.S.; Schwarz, A.E. Where do we go from here? The challenges of risk assessment for invasive plants. *Weed Technol.* **1998**, *12* (4), 744–751.

18. Rejmanek, M.; Richardson, D.M. What attributes make some plant species more invasive. *Ecology* **1996**, *77* (6), 1655–1661.

19. Reichard, S.H.; Hamilton, C.W. Predicting invasions of woody plants introduced to North America. *Conserv. Biol.* **1997**, *11* (1), 193–203.

20. Niemela, P.; Mattson, W.J. Invasions of North American forests by European phytophagous insects. *Bioscience* **1996**, *46* (10), 741–753.

21. Mack, R.N. Alien plant invasion into the intermountain west. A case history. In *Ecology of Biological Invasions of North America and Hawaii*; Mooney, H.A., Drake, J.A., Eds.; Springer-Verlag: New York, 1986; 191–213.

22. Orians, G.H. Site characteristics favoring invasions. In *Ecology of Biological Invasions of North America and Hawaii*; Mooney, H.A., Drake, J.A., Eds.; Springer-Verlag: New York, 1986; 133–148.

23. Hobbs, R.J.; Huenneke, L.F. Disturbance, diversity, and invasion: implications for conservation. *Conserv. Biol.* **1992**, *6*, 324–337.

24. Goeden, R.D.; Louda, S.M. Biotic interference with insects imported for weed control. *Annu. Rev. Entomol.* **1976**, *21*, 325–342.

25. Mack, R.N. Predicting the identity and fate of plant invaders: emergent and emerging approaches. *Biol. Conserv.* **1996**, *78*, 107–121.

26. Stohlgren, T.J.; Binkley, D.; Chong, G.W.; Kalkhan, M.A.; Schell, L.D.; Bull, K.A.; Otsuki, Y.; Newman, G.; Bashkin, M.; Son, Y. Exotic plant species invade hot spots of native plant diversity. *Ecol. Monogr.* **1999**, *69* (1), 25–46.

27. Moyle, P.B.; Light, T. Fish invasions in California: do abiotic factors determine success? *Ecology* **1996**, *77* (6), 1666–1670.

28. Pysek, P. Is there a taxonomic pattern to plant invasions? *Oikos* **1998**, *82*, 282–294.

29. Hobbs, R.J. The nature and effects of disturbance relative to invasions. In *Biological Invasions: A Global Perspective*; Drake, J.A., Mooney, H.A., di Castri, R., Kruger, F., Groves, R., Rejmanek, M., Williamson, M., Eds.; John Wiley and Sons: Chichester, 1986; 389–405.

30. Pimentel, D.; Lach, L.; Zuniga, R.; Morrison, D. Environmental and economic costs of nonindigenous species in the United States. *Bioscience* **2000**, *50*, 53–65.

31. Parker, I.M.; Simberloff, D.; Lonsdale, W.M.; Goodell, K.; Wonham, M.; Kareiva, P.M.; Williamson, M.H.; Von Holle, B.; Moyle, P.B.; Byers, J.E.; Goldwasser, L. Impact: toward a framework for understanding the ecological effects of invaders. *Biological Invasions* **1999**, *1* (1), 3–19.

32. Ricciardi, A.; Hoopes, M.F.; Marchetti, M.P.; Lockwood, J.L. Progress toward understanding the ecological impacts of nonnative species. *Ecol. Monogr.* **2013**, *83*, 263–282.

33. Daehler, C.C. The taxonomic distribution of invasive angiosperm plants: ecological insights and comparisons to agricultural weeds. *Biol. Conserv.* **1998**, *84*, 167–180.

34. Morimoto, N.; Kiritani, K. Fauna of exotic insects in Japan. *Bull. Nat. Inst. AgroEnvironmental Sci.* **1995**, *12*, 87–120.

35. Hawkins, B.A.; Cornell, H.V.; Hochberg, M.E. Predators, parasitoids, and pathogens as mortality agents in phytophagous insect populations. *Ecology* **1997**, *78* (7), 2145–2152.

36. Lonsdale, W.M. Inviting trouble: introduced pasture species in Northern Australia. *Australian J. Ecol.* **1994**, *19* (3), 345–354.

37. Williamson, M.; Fitter, A. The varying success of invaders. *Ecology* **1996**, *77* (6), 1661–1666.

38. Foin, T.C.; Riley, S.P.D.; Pawley, A.L.; Ayres, D.R.; Carlsen, T.M.; Hodum, P.J.; Switzer, P.V. Improving recovery planning for threatened and endangered species. *Bioscience* **1998**, *48*, 177–184.

39. Wilcove, D.S.; Rothstein, D.; Dubow, J.; Phillips, A.; Losos, E. Quantifying threats to imperiled species in the United States. *Bioscience* **1998**, *48*, 607–615.

40. Vitousek, P.M.; D'Antonio, C.M.; Loope, L.L.; Westbrooks, R. Biological invasions as global environmental change. *Am. Scientist* **1996**, *84* (5), 468–478.

41. Mack, M.C.; D'Antonio, C.M. Impacts of biological invasions on disturbance regimes. *Tr. Ecol. Evol.* **1998**, *13* (5), 195–198.

42. Secord, D.; Kareiva, P. Perils and pitfalls in the host specificity paradigm. *Bioscience* **1996**, *46* (5), 448–453.

43. Kowarik, I. Time Lags in Biological Invasions with Regard to the Success and Failure of Alien Species. In *Plant Invasions—General Aspects and Special Problems*; Pysek, P., Prach, K., Rejmanek, M., Wade, M., Eds.; SPB Academic Publishing: Amsterdam, 1995; 15–38.

44. Crawley, M.J. The population biology of invaders. *Phil. Trans. R. Soc. Lond. B.* **1986**, *314*, 711–731.

45. Louda, S.M.; Kendall, D.; Connor, J.; Simberloff, D. Ecological effects of an insect introduced for the biological control of weeds. *Science* **1997**, *277* (5329), 1088–1090.

46. Johnson, D.M.; Stiling, P.D. Distribution and dispersal of Cactoblastis cactorum (Lepidoptera: Pyralidae), an exotic opuntia-feeding moth, in Florida. *Florida Entomol.* **1998**, *81*, 12–22.

47. Cory, J.S.; Myers, J.H. Direct and indirect ecological effects of biological control. *Tr. Ecol. Evol.* **2000**, *15* (4), 137–139.

48. U.S. Department of Agriculture. Pest Risk Assessment of the Importation of Larch from Siberia and the Soviet Far East; USDA Forest Service, Misc. Publ. No. 1495. U.S. Government Printing Office: Washington, DC, 1991.

49. Carey, J.R. The future of the Mediterranean fruit fly *Ceratitus capitata* invasion of California: a predictive framework. *Biol. Conserv.* **1996**, *78* (1–2), 35–50.

50. Myers, J.H.; Savoie, A.; Van Randen, E. The irradication and pest management. *Annu. Rev. Entomol.* **1998**, *43*, 471–491.

51. McEvoy, P.B.; Rudd, N.T. Effects of vegetation disturbance on insect biological control of tansy ragwort, *Senecio jacobea*. *Ecol. Appl.* **1993**, *3* (4), 682–698.

52. Holway, D.A.; Suarez, A.V. Animal behavior: an essential component of invasion biology. *Tr. Ecol. Evol.* **1999**, *14* (8), 328–330.

53. Ewel, J.J.; O'Dowd, D.J.; Bergelson, J.; Daehler, C.C.; D'Antonio, C.M.; Gomez, L.D.; Gordon, D.R.; Hobbs, R.J.; Holt, A.; Hopper, K.R.; Hughes, C.E.; LaHart, M.; Leakey, R.B.; Lee, W.G.; Loope, L.L.; Lorence, D.H.; Louda, S.M.; Lugo, A.E.; McEvoy, P.B.; Richardson, D.M.; Vitousek, P.M. Deliberate introductions of species: research needs. *Bioscience* **1999**, *49* (8), 619–630.

Index